Matrices

$$\mathbf{A} \times \mathbf{B} = \begin{vmatrix} \hat{\mathbf{x}} & \hat{\mathbf{y}} & \hat{\mathbf{z}} \\ A_x & A_y & A_z \\ B_x & B_y & B_z \end{vmatrix} \qquad \mathbf{A} \cdot (\mathbf{B} \times \mathbf{C}) = \begin{vmatrix} A_x & A_y & A_z \\ B_x & B_y & B_z \\ C_x & C_y & C_z \end{vmatrix} \qquad \nabla \times \mathbf{A} = \begin{vmatrix} \hat{\mathbf{x}} & \hat{\mathbf{y}} & \hat{\mathbf{z}} \\ \dfrac{\partial}{\partial x} & \dfrac{\partial}{\partial y} & \dfrac{\partial}{\partial z} \\ A_x & A_y & A_z \end{vmatrix}$$

Triple Products

$(\mathbf{A} \times \mathbf{B}) \cdot \mathbf{C} = \mathbf{A} \cdot (\mathbf{B} \times \mathbf{C})$

$\mathbf{A} \times (\mathbf{B} \times \mathbf{C}) = \mathbf{B}(\mathbf{A} \cdot \mathbf{C}) - \mathbf{C}(\mathbf{A} \cdot \mathbf{B})$

Mero AA

Nera glacc

First Derivatives

$\nabla(fg) = f\nabla g + g\nabla f$

$\nabla(f/g) = -\dfrac{f}{g^2}\nabla g + \dfrac{1}{g}\nabla f$

$\nabla(\mathbf{A} \cdot \mathbf{B}) = (\mathbf{B} \cdot \nabla)\mathbf{A} + (\mathbf{A} \cdot \nabla)\mathbf{B} + \mathbf{B} \times (\nabla \times \mathbf{A}) + \mathbf{A} \times (\nabla \times \mathbf{B})$

$\nabla \cdot (f\mathbf{A}) = \mathbf{A} \cdot \nabla f + f\nabla \cdot \mathbf{A}$

$\nabla \cdot (\mathbf{A} \times \mathbf{B}) = \mathbf{B} \cdot (\nabla \times \mathbf{A}) - \mathbf{A} \cdot (\nabla \times \mathbf{B})$

$\nabla \times (f\mathbf{A}) = f\nabla \times \mathbf{A} - \mathbf{A} \times \nabla f$

$\nabla \times (\mathbf{A} \times \mathbf{B}) = (\mathbf{B} \cdot \nabla)\mathbf{A} - (\mathbf{A} \cdot \nabla)\mathbf{B} + \mathbf{A}(\nabla \cdot \mathbf{B}) - \mathbf{B}(\nabla \cdot \mathbf{A})$

Second Derivatives

$\nabla \cdot \nabla f = \nabla^2 f$

$\nabla \times \nabla f = 0$

$\nabla \cdot (\nabla \times \mathbf{A}) = 0$

$(\nabla \times \nabla) \cdot \mathbf{A} = 0$

$(\nabla \times \nabla) \times \mathbf{A} = 0$

$\nabla \times (\nabla \times \mathbf{A}) = \nabla(\nabla \cdot \mathbf{A}) - \nabla^2 \mathbf{A}$

Integrals

$\displaystyle\int \nabla f \, dv = \oint f \, d\mathbf{a}$

$\displaystyle\int \nabla f \times d\mathbf{a} = -\oint f \, d\mathbf{l}$

$\displaystyle\int \nabla \cdot \mathbf{A} \, dv = \oint \mathbf{A} \cdot d\mathbf{a}$ (Divergence theorem)

$\displaystyle\int \nabla \times \mathbf{A} \, dv = -\oint \mathbf{A} \times d\mathbf{a}$

$\displaystyle\int (\nabla \times \mathbf{A}) \cdot d\mathbf{a} = \oint \mathbf{A} \cdot d\mathbf{l}$ (Stokes' theorem)

$\displaystyle\int (d\mathbf{a} \times \nabla) \times \mathbf{A} = -\oint \mathbf{A} \times d\mathbf{l}$

The reviews are in for Good's
Classical Electromagnetism

"Good's Classical Electromagnetism *is an outstanding alternative to existing texts: clear, easy to understand, and very readable."*

Chris Vuille, *Embry-Riddle Aeronautical University*

"This book is by far the best E & M text that I have seen in terms of examples and problems it presents. The examples are excellent, and I especially like the constant emphasis on numerical values to illustrate the theoretical results. I believe that the large number of examples is very helpful."

Ronald E. Jodoin, *Rochester Institute of Technology*

"The pacing is nice, and I particularly liked the emphasis on giving real numbers for sizes of quantities."

Leonard M. Sander, *University of Michigan*

"I like the arrangement of the material: an early introduction of the Maxwell's Equations, the successive consideration of the electric and magnetic dipoles, and also the successive consideration of dielectrics and magnetics (this helps the students to develop the unified approach to description of these media)."

Mike Gershenson, *Rutgers the State University of New Jersey-New Brunswick*

Classical
Electromagnetism

Robert H. Good
California State University, Hayward

SAUNDERS GOLDEN SUNBURST SERIES

Saunders College Publishing
Harcourt Brace College Publishers

Fort Worth   Philadelphia   San Diego   New York   Orlando   Austin
San Antonio   Toronto   Montreal   London   Sydney   Tokyo

Vice-President Publisher: Emily Barrosse

Vice-President Publisher Editor: John Vondeling

Product Manager: Pauline Mula

Developmental Editor: Ed Dodd

Project Editor: Sarah Fitz-Hugh

Production Manager: Charlene Catlett Squibb

Art Director and Text Designer: Kathleen Flanagan

Cover Credit: *Slac bubble chamber, particle tracks produced when high-speed proton strikes atom. (Photo Researchers, Inc.)*

CLASSICAL ELECTROMAGNETISM
ISBN: 0–03-022353-9
Library of Congress Catalog Card Number: 98-88791

Address for domestic orders
Saunders College Publishing, 6277 Sea Harbor Drive, Orlando, FL 32887-6777
1-800-782-4479

Address for international orders
International Customer Service, Harcourt Brace & Company
6277 Sea Harbor Drive, Orlando, FL 32887-6777
(407) 345-3800
Fax (407) 345-4060
email hbintl@harcourtbrace.com

Address for editorial correspondence
Saunders College Publishing, Public Ledger Building, Suite 1250, 150 S. Independence Mall West, Philadelphia, PA 19106-3412

Web Site Address
http://www.hbcollege.com

Printed in the United States of America

9012345678   039   10 987654321

Introduction

A text in classical electromagnetism consists largely of the development and exploitation of Maxwell's equations. "Maxwell's theory is Maxwell's equations," said Hertz. This undergraduate text follows the usual development, in general, but it introduces Maxwell a little sooner than other texts. We begin with the basic experimental results which involve the three laws we need: Gauss, Chapter 2; Ampère, Chapter 3; and Faraday, Chapter 4. These, plus Maxwell's insight, lead directly into Maxwell's equations, Chapter 5.

After Chapter 5, with Maxwell in hand, we can proceed to derive the important phenomena of electromagnetism with logic and consistency. Starting with Maxwell's equations means never having to say you're sorry. All the derivations and developments can be correct, general, and economical; things flow smoothly. Energies and materials introduce themselves naturally, and radiation follows rather easily.

Basic applications of relativistic concepts in electromagnetism are introduced all along. Thus, after Coulomb's law, we show a picture of the field of a relativistic charge which doesn't follow Coulomb's law (Section 2.4). And after finding the magnetic force on a moving charge, we point out the relativistic implications (Sections 3.1, 3.3). Later on, we devote a chapter to relativity in electromagnetism.

This text is intended for students who are familiar with vectors and with differential and integral calculus. They should have studied physics for a year or so, but they needn't know much about electromagnetism; we'll be starting at the beginning. We employ the usual and practical SI units. A class can hardly cover the whole book in only a year, so the instructor should consider which parts to skim over lightly: every chapter has some items that should be called essential, but here and there we find sections that even the author might regard as being of lesser interest. Our pace is more or less "uniformly accelerated": in Chapter 2, we work out examples in great detail, but by the time we are halfway through the book we are moving fairly briskly, with many references to preceding explanations.

We solve our differential equations as we go along, and we develop the requisite formalism of vector calculus as we need it. This author and most of his students find it easier to learn mathematics in connection with physical models, rather than in the abstract. Consequently the divergence is introduced in Section 2.4 in connection with the electric field, the curl in Section 3.6 with the magnetic field, and the gradient in Section 6.2 with the scalar potential. (In this approach we follow Purcell.) For those who prefer to get the math out of the way first, we devote Chaper 1 to vector analysis. A cross-referenced computer appendix (Appendix C) discusses numerical and graphical methods, with examples in C++.

In the abstract field of electromagnetism, students need plenty of practical problem solving, especially quantitative. It has been said that you shouldn't start to solve a problem formally until you already know what kind of answer to expect. What order-of-magnitude voltage will be found for a single coil flipped in a magnetic field? For friction in electrostatics? For the Hall effect? (Answers: .01V, 1000 V, and 10^{-4} V). It is by solving plenty of problems with reasonable numerical values that students learn to recognize intuitively whether a particular device is likely to be putting out kilovolts or millivolts. Correspondingly we provide many such examples in the text itself, as well as a wide range of exercises from simple confidence-builders to fairly challenging problems (marked with an asterisk*). Conscientious students should be able to finish most of them in a course of reasonable duration. The odd-numbered ones have answers in the back of the book; the even ones often have matching odd ones, or worked examples in the text.

A text must be more than a picture book, but students should understand that with a good picture the problem is already half-solved. The text often implicitly suggests lecture demonstrations; in the words of Purcell, "One must be alert for every opportunity to bring the students into the world where an electric field is not a symbol merely, but something that crackles."

Instructor's Manual

The Instructor's Manual imports all of the end-of-chapter questions and relevant figures from the text and provides complete, worked-out solutions.

Acknowledgments

I am grateful for the thoughtful comments given to me by the following reviewers:

Michael Bershadsky, *Harvard University*
Jeffrey S. Dunham, *Middlebury College*
Stefan K. Estreicher, *Texas Tech University*
Lawrence Evans, *Duke University*
Richard Fitzpatrick, *University of Texas at Austin*
Mike Gershenson, *Rutgers University*
William H. Ingham, *James Madison University*
Ronald E. Jodoin, *Rochester Institute of Technology*

Michael Moloney, *Rose-Hulman Institute of Technology*
James D. Patterson, *Florida Institute of Technology*
Richard W. Robinett, *Pennsylvania State University*
Leonard M. Sander, *The University of Michigan*
Daniel Sperber, *Rensselaer Polytechnic Institute*
Chris Vuille, *Embry-Riddle Aeronautical University*
Richard Wolfson, *Middlebury College*

I also wish to express my appreciation for the support of California State University, Hayward; for the encouragement of Dr. Harper; and for Saunders College Publishing's competent and attractive realization of the book

ROBERT H. GOOD
HAYWARD, CA
November, 1998

Contents

CHAPTER 1

VECTOR ANALYSIS

The study of electromagnetic theory will require considerable facility with vector operations. This chapter is no substitute for a good mathematical course in vector analysis, but it will provide you with the tools you will need for this book.

Much of the material presented here will be repeated in later chapters as it becomes applicable to electromagnetic phenomena. For example, when we encounter the divergence of the electric field in Section 2.4, we will once again develop the divergence theorem of Section 1.5, with a somewhat different point of view. Consequently, you have your choice, depending on your mathematical talents and tendencies: Learn the math here, and use the later development of the same material as a kind of review, or postpone it until there is appropriate physical motivation, and use this chapter as a repository and summary to which you can return for review from anywhere in the book.

1.1. Vector Algebra

A **scalar** is completely characterized by its magnitude and has no associated direction; examples are mass, time, and so on. A **vector,** on the other hand, has both magnitude and direction: for example, velocity, force, and so on. (These represent descriptions rather than proper definitions; the definitions involve transformation properties that we needn't introduce here.) We employ **bold face** letters for vectors and *italics* for scalars; thus the vector $\mathbf{A}$ has a magnitude ("length") of $|\mathbf{A}| = A$, which is a scalar.

We will discuss the algebra of vectors in terms of Cartesian (rectangular) coordinates, Figure 1.1a, to make the operations clearer; however, the results given are not necessarily limited to those coordinates. Our coordinate system will of course be **right-handed,** Figure 1.1b: If we curl our fingers so as to push the x axis into the y axis, our outstretched thumb points in the direction of the z axis.

In Cartesian coordinates, a vector $\mathbf{A}$ may be presented in terms of components A_x, A_y, A_z, Figure 1.1c, and three **unit vectors** $\hat{\mathbf{x}}, \hat{\mathbf{y}}, \hat{\mathbf{z}}$ that are parallel to the three

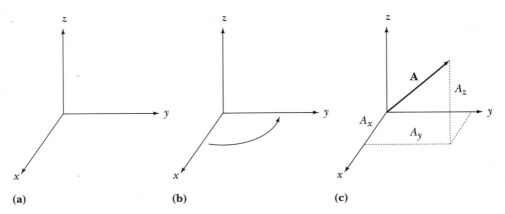

Figure 1.1   **(a)** Cartesian coordinates. **(b)** Right-handed system. **(c)** Components.

corresponding coordinate axes:

$$\mathbf{A} = A_x\hat{\mathbf{x}} + A_y\hat{\mathbf{y}} + A_z\hat{\mathbf{z}} \qquad (1.1)$$

A unit vector wears a hat (circumflex, "ˆ"); it is dimensionless and its magnitude is one. A unit vector may point in an arbitrary direction; that is, the unit vector corresponding to the arbitrary vector **A** points in the direction of **A**:

$$\hat{\mathbf{A}} = \frac{\mathbf{A}}{|\mathbf{A}|} = \frac{\mathbf{A}}{A}$$

In terms of Cartesian coordinates, the magnitude of a vector is given by the three-dimensional form of the Pythagorean theorem,

$$A = \sqrt{A_x^2 + A_y^2 + A_z^2} \qquad (1.2)$$

The vector operations are similar in some ways to ordinary scalar operations, as we shall now see. We will discuss, in order, addition, subtraction, multiplication, and division.

Addition

We cannot add or subtract a scalar and a vector—which component would the scalar go with? However, we can add two vectors **A** and **B**; the sum or "resultant" is **A** + **B**, which we can write in terms of components:

$$\mathbf{A} + \mathbf{B} = (A_x + B_x)\,\hat{\mathbf{x}} + (A_y + B_y)\,\hat{\mathbf{y}} + (A_z + B_z)\,\hat{\mathbf{z}}$$

Conceptually, we are adding the vectors "head-to-tail": We move **B** parallel to itself until its tail meets the head of **A**, and the sum is the vector from the tail of **A** to the head of **B**, Figure 1.2. It is clear from this result that

$$\mathbf{A} + \mathbf{B} = \mathbf{B} + \mathbf{A}$$

that is, addition is commutative. It is also associative; if we throw in a third vector

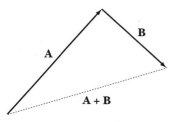

Figure 1.2   Head-to-tail.

C, we easily find that

$$(\mathbf{A} + \mathbf{B}) + \mathbf{C} = \mathbf{A} + (\mathbf{B} + \mathbf{C}) \tag{1.3}$$

Subtraction

To subtract **B** from **A** we just reverse the direction of **B** and add:

$$\mathbf{A} - \mathbf{B} = \mathbf{A} + (-\mathbf{B})$$

To reverse **B** we can change the sign of each component.

Multiplication

Multiplication by the scalar c presents no problem: We just make the vector longer (shorter, or whatever) by the factor of c. This can be accomplished by multiplying each component by c. The process is distributive:

$$c\,(\mathbf{A} + \mathbf{B}) = c\,\mathbf{A} + c\,\mathbf{B} \tag{1.4}$$

For the multiplication of two vectors, it turns out to be convenient to deal with two somewhat complementary cases: Roughly speaking, the first depends on the extent to which the vectors are mutually parallel, and the other involves to what extent they are perpendicular. We have, first, the **dot product** or scalar product,

$$\mathbf{A} \cdot \mathbf{B} = AB \cos \theta \tag{1.5}$$

where θ is the angle between the vectors. This is most easily visualized when one vector is moved, parallel to itself, until its tail touches the tail of the other, as shown in Figure 1.3. The dot product is commutative (since $\cos \theta = \cos(-\theta)$):

$$\mathbf{A} \cdot \mathbf{B} = \mathbf{B} \cdot \mathbf{A} \tag{1.6}$$

Physically, it represents the magnitude of the projection of one vector onto the other, times the magnitude of the other, and it has no direction in and of itself. It

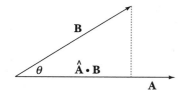

Figure 1.3   Dot product.

is zero when the vectors are mutually perpendicular. In Figure 1.3, $\hat{\mathbf{A}} \cdot \mathbf{B}$ is the scalar length of the projection of $\mathbf{B}$ onto $\mathbf{A}$.

When we take the dot product of a vector with itself, we find

$$\mathbf{A} \cdot \mathbf{A} = A^2 \cos(0) = A^2$$

since the angle between parallel vectors is zero. This gives us another expression for the magnitude:

$$A = \sqrt{\mathbf{A} \cdot \mathbf{A}}$$

The dot product may also be expressed in terms of the vector components:

$$\mathbf{A} \cdot \mathbf{B} = (A_x\hat{\mathbf{x}} + A_y\hat{\mathbf{y}} + A_z\hat{\mathbf{z}}) \cdot (B_x\hat{\mathbf{x}} + B_y\hat{\mathbf{y}} + B_z\hat{\mathbf{z}})$$
$$= A_x B_x \, \hat{\mathbf{x}} \cdot \hat{\mathbf{x}} + A_y B_y \, \hat{\mathbf{y}} \cdot \hat{\mathbf{y}} + A_z B_z \hat{\mathbf{z}} \cdot \hat{\mathbf{z}} + A_x B_y \, \hat{\mathbf{x}} \cdot \hat{\mathbf{y}} + \text{(other cross terms)}$$

Now the dot product $\hat{\mathbf{x}} \cdot \hat{\mathbf{x}} = 1$ because these unit vectors are parallel (so $\cos \theta = 1$), whereas $\hat{\mathbf{x}} \cdot \hat{\mathbf{y}} = 0$ because these are perpendicular. So the cross terms disappear and we are left with

$$\mathbf{A} \cdot \mathbf{B} = A_x B_x + A_y B_y + A_z B_z \tag{1.7}$$

 EXAMPLE 1-1

Find the dot product $\mathbf{A} \cdot \mathbf{B}$ of $\mathbf{A} = 2\hat{\mathbf{x}} + 3\hat{\mathbf{y}} + 4\hat{\mathbf{z}}$ and $\mathbf{B} = 5\hat{\mathbf{x}} + 6\hat{\mathbf{y}} + 7\hat{\mathbf{z}}$. Then find the angle θ between them.

ANSWER The dot product is

$$\mathbf{A} \cdot \mathbf{B} = 2 \times 5 + 3 \times 6 + 4 \times 7 = 56$$

so

$$AB \cos \theta = 56$$

Now the magnitudes of these vectors are

$$A = \sqrt{2^2 + 3^2 + 4^2} = 5.385 \quad \text{and} \quad B = \sqrt{5^2 + 6^2 + 7^2} = 10.488$$

so

$$\cos \theta = \mathbf{A} \cdot \mathbf{B}/AB$$
$$= 56/(5.385 \times 10.488) = 0.9915$$

which yields $\theta = \arccos(0.9915) = 7.5°$.

The other kind of product is the **cross product** or vector product,

$$\mathbf{A} \times \mathbf{B} = AB \sin \theta \, \hat{\mathbf{n}} \tag{1.8}$$

As before, θ is the angle between the vectors, Figure 1.4. Here $\hat{\mathbf{n}}$ is a unit vector

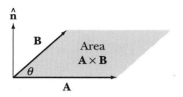

Figure 1.4 Cross product.

that is normal to the plane defined by **A** and **B**, and whose direction is provided by the right-hand rule: Curl your fingers in the direction of pushing **A** into **B**, and the extended thumb will point in the direction of n̂, Figure 1.4. The cross product is not commutative (since $\sin \theta = -\sin(-\theta)$):

$$\mathbf{A} \times \mathbf{B} = -\mathbf{B} \times \mathbf{A}$$

Physically, the cross product is the vector area of the parallelogram having **A** and **B** on adjacent sides (Fig. 1.4). Its magnitude is $|\mathbf{A} \times \mathbf{B}|$ and its direction is that of n̂.

When we take the cross product of a vector with itself, we find

$$|\mathbf{A} \times \mathbf{A}| = A^2 \sin(0) = 0$$

and in fact the cross product of any two parallel vectors is zero.

In terms of its vector components, the cross product may be written

$$\mathbf{A} \times \mathbf{B} = (A_x\hat{\mathbf{x}} + A_y\hat{\mathbf{y}} + A_z\hat{\mathbf{z}}) \times (B_x\hat{\mathbf{x}} + B_y\hat{\mathbf{y}} + B_z\hat{\mathbf{z}})$$
$$= (A_yB_z - A_zB_y)\,\hat{\mathbf{x}} + (A_zB_x - A_xB_z)\,\hat{\mathbf{y}} + (A_xB_y - A_yB_x)\,\hat{\mathbf{z}} \qquad (1.9)$$

Here we have employed $\hat{\mathbf{x}} \times \hat{\mathbf{y}} = \hat{\mathbf{z}}$ because these unit vectors are perpendicular (so $\sin \theta = 1$) and the direction is given by the right-hand rule; and $\hat{\mathbf{x}} \times \hat{\mathbf{x}} = 0$ because these are parallel.

In determinant form,

$$\mathbf{A} \times \mathbf{B} = \begin{vmatrix} \hat{\mathbf{x}} & \hat{\mathbf{y}} & \hat{\mathbf{z}} \\ A_x & A_y & A_z \\ B_x & B_y & B_z \end{vmatrix}$$

◢ EXAMPLE 1-2

Find the cross product $\mathbf{A} \times \mathbf{B}$ of $\mathbf{A} = 2\hat{\mathbf{x}} + 3\hat{\mathbf{y}} + 4\hat{\mathbf{z}}$ and $\mathbf{B} = 5\hat{\mathbf{x}} + 6\hat{\mathbf{y}} + 7\hat{\mathbf{z}}$. Then find the angle θ between them.

ANSWER $\mathbf{A} \times \mathbf{B} = (3 \times 7 - 4 \times 6)\hat{\mathbf{x}} + (4 \times 5 - 2 \times 7)\hat{\mathbf{y}} + (2 \times 6 - 3 \times 5)\hat{\mathbf{z}}$
$$= -3\hat{\mathbf{x}} + 6\hat{\mathbf{y}} - 3\hat{\mathbf{z}}$$

So

$$|\mathbf{A} \times \mathbf{B}| = \sqrt{3^2 + 6^2 + 3^2} = 7.348$$
$$= AB \sin \theta$$

From the preceding example, $A = 5.385$ and $B = 10.488$; so,

$$\sin \theta = |\mathbf{A} \times \mathbf{B}| / AB$$
$$= 7.348/(5.385 \times 10.488) = 0.1301$$

and

$$\theta = \arcsin(0.1301) = 7.48°$$

Division

We can of course divide a vector by a scalar: Just divide each component of the vector by the scalar. But there is no sort of division between vectors (as if to compensate for there being two kinds of multiplication).

Triple Products

We can multiply three vectors $\mathbf{A}$, $\mathbf{B}$, and $\mathbf{C}$ in several ways. Here is one:

$$\mathbf{A} \cdot (\mathbf{B} \times \mathbf{C}) = \begin{vmatrix} A_x & A_y & A_z \\ B_x & B_y & B_z \\ C_x & C_y & C_z \end{vmatrix} = (\mathbf{A} \times \mathbf{B}) \cdot \mathbf{C} \qquad (1.10)$$

It's the volume of a parallelepiped having vectors $\mathbf{A}$, $\mathbf{B}$, and $\mathbf{C}$ on adjacent sides.
And here is the "bac-cab" rule:

$$\mathbf{A} \times (\mathbf{B} \times \mathbf{C}) = \mathbf{B}(\mathbf{A} \cdot \mathbf{C}) - \mathbf{C}(\mathbf{A} \cdot \mathbf{B}) \qquad (1.11)$$

which is *not* the same as $(\mathbf{A} \times \mathbf{B}) \times \mathbf{C}$.

Derivations

To **derive** the relationships we have presented above, first resolve the quantities involved into Cartesian (x, y, z) components, and then just check the x components, assuming the relationships are symmetrical in x, y, z. (If they aren't symmetrical, they aren't general.) For "bac-cab," it goes like this:

$$[\mathbf{A} \times (\mathbf{B} \times \mathbf{C})]_x = [\mathbf{B}(\mathbf{A} \cdot \mathbf{C})]_x - [\mathbf{C}(\mathbf{A} \cdot \mathbf{B})]_x$$
$$A_y(B_xC_y - B_yC_x) - A_z(B_zC_x - B_xC_z) = B_x(A_xC_x + A_yC_y + A_zC_z)$$
$$- C_x(A_xB_x + A_yB_y + A_zB_z)$$

On the right, the two $A_xB_xC_x$ terms cancel. Then the two sides of the equation are the same, by inspection.

1.2. First Derivatives

Imagine that we have a **scalar field** defined as a scalar function of space and time; an example might be the temperature $T(x, y, z, t)$ in a room. It is specified at every point (x, y, z), and it may change with time. We can differentiate with respect to one of the four variables, keeping the other three constant. The result is a partial derivative, defined using a limiting procedure:

$$\frac{\partial T}{\partial x} = \lim \frac{T(x + \Delta x, y, z, t) - T(x, y, z, t)}{\Delta x}$$

$$\text{as } \Delta x \to 0$$

If we permit all variables to vary, we can get the total derivative form by a chain rule from elementary differential calculus; for example,

$$\frac{dT}{dt} = \frac{\partial T}{\partial x}\frac{dx}{dt} + \frac{\partial T}{\partial y}\frac{dy}{dt} + \frac{\partial T}{\partial z}\frac{dz}{dt} + \frac{\partial T}{\partial t}$$

An example of a **vector field** might be the wind velocity $\mathbf{v}(x, y, z, t)$ in a given region, having a magnitude and direction at every point in space and time. We can carry through the same operations as above, in a formal way, to arrive at derivatives of the vector function.

The above kinds of derivatives have their uses, but it will turn out that we have even greater interest in a kind of derivative operator that has directional characteristics in itself. That differential operator is called **"del"** and looks like this:

$$\boxed{\nabla \equiv \hat{\mathbf{x}}\frac{\partial}{\partial x} + \hat{\mathbf{y}}\frac{\partial}{\partial y} + \hat{\mathbf{z}}\frac{\partial}{\partial z}} \qquad (1.12)$$

(Although unit vectors typically go on the right of each term, we sometimes place them on the left when dealing with operators to remind ourselves that the operator is not operating on the unit vectors.) Del behaves in some ways like a vector; however, it is incomplete as it stands, it needs something to operate on. The three possible operations are similar to those of vector multiplication, described in the preceding section. We shall deal with these three cases in order.

Gradient ∇

The gradient of a scalar function f may be written ∇f or grad f. It is the vector function that results from the operation of del upon a scalar:

$$\text{grad } f = \nabla f = \frac{\partial f}{\partial x}\hat{\mathbf{x}} + \frac{\partial f}{\partial y}\hat{\mathbf{y}} + \frac{\partial f}{\partial z}\hat{\mathbf{z}} \qquad (1.13)$$

The gradient is analogous to multiplication of a vector by a scalar. For example, if we feed the scalar $f = 3y$ (Fig. 1.5) to the del operator, the result is simply

$$\nabla(3y) = \frac{\partial 3y}{\partial x}\hat{\mathbf{x}} + \frac{\partial 3y}{\partial y}\hat{\mathbf{y}} + \frac{\partial 3y}{\partial z}\hat{\mathbf{z}}$$

$$= 0 + 3\hat{\mathbf{y}} + 0 = 3\hat{\mathbf{y}}$$

which is a vector. In Figure 1.5, the shading represents a scalar field that has a magnitude, but no direction, at every point. It depicts something (temperature,

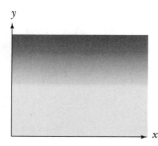

y

x   **Figure 1.5** $f = 3y$.

for example) that gets bigger as you go in the *y* direction, but not in the *x* direction, corresponding to the expression given, $f = 3y$.

Physical Significance The gradient points uphill, so to speak. That is, in the example above, the gradient, $3\hat{y}$, implies that the function, $3y$, increases in the *y* direction. In a more general approach, we start with the chain rule, above: The change in the scalar *f* (which could be temperature) as we move in an arbitrary direction is, at a fixed instant in time ($dt = 0$),

$$df = \frac{\partial f}{\partial x} dx + \frac{\partial f}{\partial y} dy + \frac{\partial f}{\partial z} dz$$

$$= \left(\frac{\partial f}{\partial x} \hat{x} + \frac{\partial f}{\partial y} \hat{y} + \frac{\partial f}{\partial z} \hat{z} \right) \cdot (\hat{x} \, dx + \hat{y} \, dy + \hat{z} \, dz)$$

so

$$df = (\nabla f) \cdot d\mathbf{l} \tag{1.14}$$

where *l* is the vector sum of arbitrary displacements in the *x*, *y*, and *z* directions. So in terms of magnitudes and the angle between the vectors,

$$df = |\nabla f| \, |d\mathbf{l}| \, \cos \theta$$

Now this says quite clearly that (for a given step size $d\mathbf{l}$) df is maximum when $\cos \theta = 1$, which means $\theta = 0$. We get the greatest change in *f* when the change in *l* is in the same direction as the gradient. The gradient ∇f tells you what direction to go in order to maximize the change in *f*. In other words, ∇f points uphill, in terms of *f*.

Note that we don't usually take a gradient of a vector (the result would be a tensor); when you apply del to a vector, you get a divergence or a curl, treated in subsequent paragraphs.

Divergence $\nabla \cdot$

The divergence of a vector **A** may be written $\nabla \cdot \mathbf{A}$ or div **A**. It is the scalar that results from operation of del upon a vector in a fashion analogous to the dot product of the preceding section:

$$\text{div } \mathbf{A} = \nabla \cdot \mathbf{A} = \frac{\partial A_x}{\partial x} + \frac{\partial A_y}{\partial y} + \frac{\partial A_z}{\partial z} \tag{1.15}$$

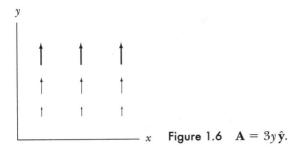

y

x **Figure 1.6** $\mathbf{A} = 3y\,\hat{\mathbf{y}}$.

For example, using the vector field $\mathbf{A} = 3y\hat{\mathbf{y}}$ (Fig. 1.6), whose x and z components are zero:

$$\nabla \cdot \mathbf{A} = \nabla \cdot (3y\hat{\mathbf{y}})$$

$$= \left(\hat{\mathbf{x}} \frac{\partial}{\partial x} + \hat{\mathbf{y}} \frac{\partial}{\partial y} + \hat{\mathbf{z}} \frac{\partial}{\partial z} \right) \cdot (0\hat{\mathbf{x}} + 3y\hat{\mathbf{y}} + 0\hat{\mathbf{z}})$$

$$= \frac{\partial 0}{\partial x} \hat{\mathbf{x}} \cdot \hat{\mathbf{x}} + \frac{\partial 3y}{\partial y} \hat{\mathbf{y}} \cdot \hat{\mathbf{y}} + \frac{\partial 0}{\partial z} \hat{\mathbf{z}} \cdot \hat{\mathbf{z}} = 3$$

which is a scalar, just as in the case of the dot product of two vectors. In Figure 1.6: Presence of a vector field implies that there is a vector having the magnitude and direction of the field at every point in space. We show only nine such vectors, at reasonable spacing, to assist the reader in visualization of the field. The vectors all point in the y direction, and they increase in size in the y direction, but not in the x direction, corresponding to the expression given, $\mathbf{A} = 3y\hat{\mathbf{y}}$.

Physical Significance Where there is a positive divergence, there is a source of a vector field. Looking at the example given, we see that the divergence is positive because the field both points in the y direction and increases in the y direction. If the field pointed in the y direction, but increased only in the x direction, the divergence would be zero (as it is in Fig. 1.7). So the divergence is related to how the field changes as you move in the direction of the field. This is also suggested by the mathematical forms involved: $\partial A_x / \partial x$ and so on. But this is only a qualitative and intuitive concept, and some cases may require further interpretation. For ex-

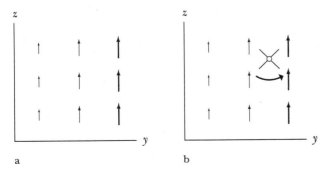

a

b

Figure 1.7   **(a)** $\mathbf{A} = 3y\,\hat{\mathbf{z}}$. **(b)** Paddle wheel test.

ample, the field described by $\hat{\mathbf{r}}/r^2$ points in the r (radial) direction, and decreases in the same direction; yet its divergence is zero (except at the origin). The true test for divergence lies in the mathematical analysis.

When $\nabla \cdot \mathbf{A} = 0$ everywhere, the field $\mathbf{A}$ is called **solenoidal.**

Curl $\nabla \times$

The curl of a vector $\mathbf{A}$ may be written $\nabla \times \mathbf{A}$ or curl $\mathbf{A}$. It is the vector that results from operation of the del operator upon a vector in a fashion analogous to the cross product of the preceding section:

$$\text{curl } \mathbf{A} = \nabla \times \mathbf{A} = \left(\frac{\partial A_z}{\partial y} - \frac{\partial A_y}{\partial z} \right) \hat{\mathbf{x}} + \left(\frac{\partial A_x}{\partial z} - \frac{\partial A_z}{\partial x} \right) \hat{\mathbf{y}} + \left(\frac{\partial A_y}{\partial x} - \frac{\partial A_x}{\partial y} \right) \hat{\mathbf{z}}$$

$$= \begin{vmatrix} \hat{\mathbf{x}} & \hat{\mathbf{y}} & \hat{\mathbf{z}} \\ \frac{\partial}{\partial x} & \frac{\partial}{\partial y} & \frac{\partial}{\partial z} \\ A_x & A_y & A_z \end{vmatrix} \tag{1.16}$$

For example, using the vector field $3y\hat{\mathbf{z}}$, Figure 1.7a, the curl is

$$\nabla \times (3y\hat{\mathbf{z}}) = \left(\frac{\partial 3y}{\partial y} - 0 \right) \hat{\mathbf{x}} + 0\,\hat{\mathbf{y}} + 0\,\hat{\mathbf{z}}$$

$$= 3\,\hat{\mathbf{x}}$$

The result is a vector, as would be expected for a cross product.

Physical Significance Where there is a curl, the corresponding vector field has a sort of rotational tendency. One intuitive test for the curl is the following: Imagine that the vector field represents flow of water, and place a small paddle wheel in it, Figure 1.7b; if the paddle wheel tends to go around, there is a curl. Looking at the example given, we see that the curl is nonzero because the field points in the z direction but increases in a different direction. If the field pointed in the same direction as that in which it increased, the curl would be zero, as it is in Figure 1.6. So the curl is related to how the field changes as you move across the field. This is also suggested by the mathematical forms involved: $\partial A_x/\partial y$ and so on. In this sense the divergence is complementary to the curl: The former provides information about change as you move along the field, and the other does the same across the field. However, these are only qualitative and intuitive concepts, and some cases may require further interpretation. For example, the field described by $\hat{\phi}/\rho$ (cylindrical coordinates, Section 1.7) points in the ϕ direction, and decreases across the field in the ρ direction; yet its curl is zero (except at the origin). The true test for curl lies in the mathematical analysis.

When $\nabla \times \mathbf{A} = 0$ everywhere, the field $\mathbf{A}$ is called **irrotational.**

Other Cases?

There is another plausible constellation of symbols,

$$\frac{\partial A_x}{\partial x} \hat{\mathbf{x}} + \frac{\partial A_y}{\partial y} \hat{\mathbf{y}} + \frac{\partial A_z}{\partial z} \hat{\mathbf{z}} \tag{1.17}$$

However, not only is it inaccessible using the ∇ operator, but also it turns out not to be physically meaningful (and it does not conform to the technical definition of a vector, Section 1.6). In fact, the grad, div, and curl presented before are the only such differential operations of physical significance.

Derivatives of Products

We shall employ symbols f and g for scalar functions, and $\mathbf{A}$ and $\mathbf{B}$ for vector functions.

It is easy to show, using elementary calculus, that

$$\nabla(fg) = f\nabla g + g\nabla f \tag{1.18}$$

and

$$\nabla\frac{f}{g} = \frac{1}{g}\nabla f - \frac{f}{g^2}\nabla g$$

just as in the case of ordinary derivatives.

To derive vector identities, resolve the vectors into their Cartesian components and check all x components (see Section 1.1). For scalar differential identities, check all x partial derivatives.

$$\nabla(\mathbf{A}\cdot\mathbf{B}) = (\mathbf{B}\cdot\nabla)\mathbf{A} + (\mathbf{A}\cdot\nabla)\mathbf{B} + \mathbf{B}\times(\nabla\times\mathbf{A}) + \mathbf{A}\times(\nabla\times\mathbf{B}) \tag{1.19}$$

$$\nabla\cdot(f\mathbf{A}) = \mathbf{A}\cdot\nabla f + f\nabla\cdot\mathbf{A}$$

$$\nabla\cdot(\mathbf{A}\times\mathbf{B}) = \mathbf{B}\cdot(\nabla\times\mathbf{A}) - \mathbf{A}\cdot(\nabla\times\mathbf{B})$$

$$\nabla\times(f\mathbf{A}) = f\nabla\times\mathbf{A} - \mathbf{A}\times\nabla f$$

$$\nabla\times(\mathbf{A}\times\mathbf{B}) = (\mathbf{B}\cdot\nabla)\mathbf{A} - (\mathbf{A}\cdot\nabla)\mathbf{B} + \mathbf{A}(\nabla\cdot\mathbf{B}) - \mathbf{B}(\nabla\cdot\mathbf{A})$$

The expression $(\mathbf{A}\cdot\nabla)\mathbf{B}$ means exactly what it seems to:

$$(\mathbf{A}\cdot\nabla)\mathbf{B} = \left(A_x\frac{\partial}{\partial x} + A_y\frac{\partial}{\partial y} + A_z\frac{\partial}{\partial z}\right)(B_x\hat{\mathbf{x}} + B_y\hat{\mathbf{y}} + B_z\hat{\mathbf{z}})$$

It is a vector with nine terms:

$$(\mathbf{A}\cdot\nabla)\mathbf{B} = \left(A_x\frac{\partial B_x}{\partial x} + A_y\frac{\partial B_x}{\partial y} + A_z\frac{\partial B_x}{\partial z}\right)\hat{\mathbf{x}} \tag{1.20}$$

$$+ \left(A_x\frac{\partial B_y}{\partial x} + A_y\frac{\partial B_y}{\partial y} + A_z\frac{\partial B_y}{\partial z}\right)\hat{\mathbf{y}}$$

$$+ \left(A_x\frac{\partial B_z}{\partial x} + A_y\frac{\partial B_z}{\partial y} + A_z\frac{\partial B_z}{\partial z}\right)\hat{\mathbf{z}}$$

1.3.   Second Derivatives

Here are the derivatives of scalar and vector fields obtained by applying ∇ twice.

We begin with ∇f: f is a scalar function, and the gradient ∇f is a vector function. Taking the divergence, we find

$$\nabla \cdot (\nabla f) = (\nabla \cdot \nabla) f = \nabla^2 f$$

$$= \frac{\partial^2 f}{\partial x^2} + \frac{\partial^2 f}{\partial y^2} + \frac{\partial^2 f}{\partial z^2} \tag{1.21}$$

which is a scalar called the **Laplacian** of f.

If we take the curl of the gradient ∇f, we find

$$\nabla \times (\nabla f) = (\nabla \times \nabla) f = 0 \qquad \text{always} \tag{1.22}$$

This identity is suggested by the fact that, for a vector $\mathbf{A}$, $\mathbf{A} \times \mathbf{A} = 0$ always. However, that is not a proof; it is necessary to expand the product in Cartesian coordinates to show that in fact it is zero. In the case of $\nabla \times \nabla$, we find the x component to be

$$\frac{\partial}{\partial y} \frac{\partial}{\partial z} - \frac{\partial}{\partial z} \frac{\partial}{\partial y}$$

which always yields zero because the order of differentiation doesn't matter. Correspondingly, we can obviously write

$$(\nabla \times \nabla) \cdot \mathbf{A} = 0 \tag{1.23}$$

and

$$(\nabla \times \nabla) \times \mathbf{A} = 0$$

One of the beauties of the "del" notation is that it tends to suggest relationships that turn out to be true, as above. However, it doesn't always work. For example: we know that the cross product $\mathbf{A} \times \mathbf{B}$ is perpendicular to both $\mathbf{A}$ and $\mathbf{B}$. So, is the curl $\nabla \times \mathbf{A}$ perpendicular to $\mathbf{A}$? Not always, and here is a counterexample: if

$$\mathbf{A} = \hat{\mathbf{x}} + z\hat{\mathbf{y}}$$

then

$$\nabla \times \mathbf{A} = -\frac{\partial A_y}{\partial z}\hat{\mathbf{x}}$$

$$= -\hat{\mathbf{x}}$$

which is *not* perpendicular to $\mathbf{A}$.

Another second derivative we need to list here is the divergence of the curl,

$$\nabla \cdot (\nabla \times \mathbf{A}) = 0 \tag{1.24}$$

Once again, this identity is suggested by the notation; after all, $\mathbf{A} \cdot (\mathbf{A} \times \mathbf{B}) = 0$ because $\mathbf{A} \times \mathbf{B}$ is perpendicular to $\mathbf{A}$. But once again that is not a proof; we would need to expand it into its components to see that it is so.

Finally we turn to an important relationship among several second derivative forms:

$$\nabla(\nabla \cdot \mathbf{A}) - \nabla^2 \mathbf{A} = \nabla \times (\nabla \times \mathbf{A}) \tag{1.25}$$

The expression $\nabla^2 \mathbf{A}$ implies, as you would guess, that the Laplacian provided previously is applied to each component of the vector, which results in a total of nine terms (but this works only in Cartesian coordinates; see Section 1.7). We shall carry through the demonstration of this identity by expanding it and keeping only the x components:

$$\frac{\partial}{\partial x}(\nabla \cdot \mathbf{A})\,\hat{\mathbf{x}} - \nabla^2 A_x\,\hat{\mathbf{x}} = \left(\frac{\partial (\nabla \times \mathbf{A})_z}{\partial y} - \frac{\partial (\nabla \times \mathbf{A})_y}{\partial z} \right)\hat{\mathbf{x}}$$

$$\frac{\partial^2 A_x}{\partial x^2} + \frac{\partial^2 A_y}{\partial x \partial y} + \frac{\partial^2 A_z}{\partial x \partial z} - \frac{\partial^2 A_x}{\partial x^2} - \frac{\partial^2 A_x}{\partial y^2} - \frac{\partial^2 A_x}{\partial z^2}$$

$$= \frac{\partial^2 A_y}{\partial x \partial y} - \frac{\partial^2 A_x}{\partial y^2} - \frac{\partial^2 A_x}{\partial z^2} + \frac{\partial^2 A_z}{\partial x \partial z}$$

which is true, by inspection.

 **EXAMPLE 1-3**

Show that Equation 1.25 applies to

$$\mathbf{A} = x^2\,\hat{\mathbf{x}} + x^2\,\hat{\mathbf{y}} + yz\,\hat{\mathbf{z}}$$

ANSWER We will evaluate each term of Equation 1.25 in turn.

1. $\nabla(\nabla \cdot \mathbf{A})$: $\quad \nabla \cdot \mathbf{A} = \left(\hat{\mathbf{x}}\frac{\partial}{\partial x} + \hat{\mathbf{y}}\frac{\partial}{\partial y} + \hat{\mathbf{z}}\frac{\partial}{\partial z} \right) \cdot (x^2\,\hat{\mathbf{x}} + x^2\,\hat{\mathbf{y}} + yz\,\hat{\mathbf{z}})$

$$= 2x + 0 + y$$

so $\quad \nabla(\nabla \cdot \mathbf{A}) = 2\,\hat{\mathbf{x}} + \hat{\mathbf{y}}$

2. $\nabla^2\mathbf{A}$: $\quad \nabla^2\mathbf{A} = \left(\frac{\partial^2}{\partial x^2} + \frac{\partial^2}{\partial y^2} + \frac{\partial^2}{\partial z^2} \right) x^2\,\hat{\mathbf{x}}$

$$+ \left(\frac{\partial^2}{\partial x^2} + \frac{\partial^2}{\partial y^2} + \frac{\partial^2}{\partial z^2} \right) x^2\,\hat{\mathbf{y}} + \left(\frac{\partial^2}{\partial x^2} + \frac{\partial^2}{\partial y^2} + \frac{\partial^2}{\partial z^2} \right) yz\,\hat{\mathbf{z}}$$

$$= 2\,\hat{\mathbf{x}} + 2\,\hat{\mathbf{y}}$$

3. $\nabla \times (\nabla \times \mathbf{A})$: $\quad \nabla \times \mathbf{A} = \left(\frac{\partial A_z}{\partial y} - \frac{\partial A_y}{\partial z} \right)\hat{\mathbf{x}}$

$$+ \left(\frac{\partial A_x}{\partial z} - \frac{\partial A_z}{\partial x} \right)\hat{\mathbf{y}} + \left(\frac{\partial A_y}{\partial x} - \frac{\partial A_x}{\partial y} \right)\hat{\mathbf{z}}$$

$$= z\,\hat{\mathbf{x}} + 2x\,\hat{\mathbf{z}}$$

so $\quad \nabla \times (\nabla \times \mathbf{A}) = (1 - 2)\,\hat{\mathbf{y}}$

$$= -\hat{\mathbf{y}}$$

And now we combine them:

$$\nabla(\nabla \cdot \mathbf{A}) - \nabla^2\mathbf{A} = \nabla \times (\nabla \times \mathbf{A})$$
$$2\,\hat{\mathbf{x}} + \hat{\mathbf{y}} - (2\,\hat{\mathbf{x}} + 2\hat{\mathbf{y}}) = -\hat{\mathbf{y}}$$
$$-\hat{\mathbf{y}} = -\hat{\mathbf{y}}$$

So it works.

1.4.   Need for Both $\nabla \times \mathbf{A}$ and $\nabla \cdot \mathbf{A}$

In Section 1.2 we mentioned the complementary nature of the curl and divergence; they provide different kinds of information concerning their fields. Now that we have Equation 1.25, we can expand on this notion: We can prove that, roughly speaking, the curl and divergence usually contain all you need to know about a vector field. More exactly: A vector field having given values on the boundary of some region, and having a given curl and divergence within that region, is unique; no different field can have the same values there, and in that sense no further description of the field is necessary. For real physical systems, the boundary condition can simply be that the field approaches zero at infinity, and then the curl and divergence determine the field completely (at a given instant in time).

To prove this "uniqueness theorem," let us begin by assuming that there *are* two fields having the same curl and divergence in the region, as well as the same values on the boundary. Call the fields $\mathbf{A}_1$ and $\mathbf{A}_2$. Their difference is $\mathbf{A}_3$, with

$$\mathbf{A}_3 = \mathbf{A}_1 - \mathbf{A}_2 \tag{1.26}$$

Then

$$\nabla \times \mathbf{A}_3 = \nabla \times \mathbf{A}_1 - \nabla \times \mathbf{A}_2 = 0$$

since

$$\nabla \times \mathbf{A}_1 = \nabla \times \mathbf{A}_2$$

Similarly,

$$\nabla \cdot \mathbf{A}_3 = \nabla \cdot \mathbf{A}_1 - \nabla \cdot \mathbf{A}_2 = 0$$

So $\mathbf{A}_3$ is a field with zero curl and zero divergence. Consequently, using Equation 1.25,

$$\nabla^2 \mathbf{A}_3 = \nabla(\nabla \cdot \mathbf{A}_3) - \nabla \times (\nabla \times \mathbf{A}_3)$$
$$= 0 - 0 = 0$$

which means that the Laplacian of each component is zero; that is,

$$\nabla^2 A_{3x} = \frac{\partial^2 A_{3x}}{\partial x^2} + \frac{\partial^2 A_{3x}}{\partial y^2} + \frac{\partial^2 A_{3x}}{\partial z^2} = 0 \tag{1.27}$$

and the same for the y and z components. Further, $A_{3x} = 0$ on the boundary, since $\mathbf{A}_1 = \mathbf{A}_2$ there. The question is this: Under the circumstances, can A_{3x} be different from zero anywhere? The answer is no, of course. Laplace's equation, Equation 1.27, admits of no local maxima or minima (Section 12.1). For example, suppose that somehow A_{3x} achieved a positive value inside the region. It has to get back down to zero at the boundary. So it must have a maximum value somewhere inside the region. It's downhill in every direction from there; at such a maximum, the second derivatives must all be negative, and so their sum cannot be zero as required by Equation 1.27. So there cannot be a maximum, and consequently A_{3x} must be zero everywhere. The same holds for the other components of $\mathbf{A}_3$. Now if $\mathbf{A}_3 = 0$ everywhere, it means that $\mathbf{A}_1$ and $\mathbf{A}_2$ represent the same vector field. If there is a field $\mathbf{A}_2$ having the same curl, divergence, and boundary values as $\mathbf{A}_1$, then that field must be the *same* as $\mathbf{A}_1$. Thus we have proven this uniqueness theorem.

Also, it is easy to show, by counterexample, that neither the divergence nor the curl is sufficient by itself to define a vector field (see problems 1-21, 1-22); we will need to know about both of them. (This uniqueness theorem is related to the "Helmholtz theorem," which shows how any vector field $\mathbf{A}$ may be uniquely specified in terms of its curl and divergence, provided that they vanish at infinity.)

Boundary conditions play a crucial role in conjunction with this uniqueness theorem. For example, here are two vectors that have the same curl and the same divergence (and are equal at the origin):

$$\mathbf{A}_1 = 0$$
$$\mathbf{A}_2 = x\hat{\mathbf{x}} - y\hat{\mathbf{y}}$$

yet they represent very different fields because they are not equal on the boundary of any region. (For more examples, see Figs. 2.24c and 2.25, and Figs. 3.24c and 3.25, with the accompanying analyses.)

The preceding results are of particular significance in electromagnetism for the following reason: Maxwell's equations, which lie at the heart of classical electromagnetic theory, consist of four statements specifying the curls, divergences, and time derivatives of the electric and magnetic fields, $\mathbf{E}$ and $\mathbf{B}$ respectively (Section 5.1). Furthermore, for charge and current distributions of finite extent, these fields approach zero at infinity. So our results imply that Maxwell's equations do indeed constitute a complete description of the fields, within the confines of classical physics.

1.5.   Integrals

We will be particularly interested in three of the possible integrals involving the del operator, one each for grad, div, and curl.

Gradient

We will need the integral of the gradient ∇f along a path (line, contour) l from point a to point b, Figure 1.8, where $d\mathbf{l}$ is the vector sum of arbitrary displacements in the x, y, and z directions. In particular, we are interested in the component of ∇f that points along the path increment vector $d\mathbf{l}$; so we will need the dot product. In Section 1.2 (Eq. 1.14) we noted that

$$df = (\nabla f) \cdot d\mathbf{l}$$

So, integrating,

$$f(\mathrm{b}) - f(\mathrm{a}) = \int_{\mathrm{a}}^{\mathrm{b}} (\nabla f) \cdot d\mathbf{l} \qquad (1.28)$$

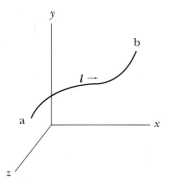

Figure 1.8 Line integral.

The result is just the difference between f values at the ends, regardless of the path of integration. **Physically,** this is something like saying that when you climb a hill, you gain the same total elevation regardless of what path you follow.

Some other ∇f integrals are shown inside the front cover, but we won't be needing them.

Divergence

In this case we will be interested in integrating the vector **A** over the entire surface **a** of an arbitrary volume v; here $d\mathbf{a}$ is a vector proportional to, and normal to, an area increment, and it points outward from the volume v. Mathematically, we want to calculate the **flux** Φ of **A**:

$$\Phi = \oint \mathbf{A} \cdot d\mathbf{a}$$

The symbol $\oint$ here implies integration over a closed surface, one that completely encloses a volume. Flux is like the flow of water through an imaginary surface, where the result depends on the dot product of the flow rate and the vector normal to the surface. Note that it is a scalar. Flux out of the volume is positive and flux inward is negative. So if, for example, the net flux over the whole surface is positive, it means that there is more flux out of the volume than in, for some reason.

We begin by breaking the volume into microscopic cubes of volume Δv, Figure 1.9, which are so small that the field derivatives change negligibly over the cube. After analyzing this, we will reconstitute the arbitrary volume out of little cubes.

First we look at the difference in flux $\Delta\Phi_x$ due to A_x, as shown. (Φ is no vector; the "x" subscript merely indicates that this is the contribution from A_x.) After we understand $\Delta\Phi_x$, the y and z directions will be easy. The net flux involving A_x is

$$\Delta\Phi_x = A_x(x + \Delta x, y, z)\, \Delta y\, \Delta z - A_x(x, y, z)\, \Delta y\, \Delta z \qquad (1.29)$$

To simplify the notation, we have placed the point (x, y, z) in the center of the left-hand face of the cube, and we assume that the component $A_x(x, y, z)$ represents a suitable average over that face. (We do this because y and z are not going to change in a significant way in this first calculation.)

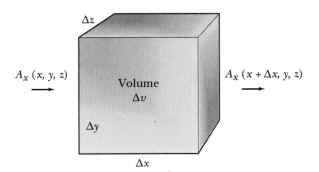

Figure 1.9   Small cube.

When we multiply and divide Equation 1.29 by Δx, a partial derivative form is obtained:

$$\Delta \Phi_x = \frac{A_x(x + \Delta x, y, z)}{\Delta x} \Delta x \, \Delta y \, \Delta z - \frac{A_x(x, y, z)}{\Delta x} \Delta x \, \Delta y \, \Delta z$$

$$= \left(\frac{A_x(x + \Delta x, y, z) - A_x(x, y, z)}{\Delta x} \right) \Delta x \, \Delta y \, \Delta z$$

$$\rightarrow \frac{\partial A_x}{\partial x} \Delta v, \text{ in the limit as } \Delta x \rightarrow 0$$

where the volume element is $\Delta v = \Delta x \, \Delta y \, \Delta z$.

For y and z the story is similar. We can just add them up; so the total difference in flux is

$$\Delta \Phi = \left(\frac{\partial A_x}{\partial x} + \frac{\partial A_y}{\partial y} + \frac{\partial A_z}{\partial z} \right) \Delta v \tag{1.30}$$

$$= \nabla \cdot \mathbf{A} \, \Delta v$$

Now on the left we have the net flux out of the cube, which is $\oint \mathbf{A} \cdot d\mathbf{a}$ over the entire surface, and on the right we find the total divergence throughout the enclosed volume, $\int \nabla \cdot \mathbf{A} \, dv$. So the integral relationship we want is

$$\boxed{\int \nabla \cdot \mathbf{A} \, dv = \oint \mathbf{A} \cdot d\mathbf{a}} \tag{1.31}$$

which is called the **divergence theorem** (sometimes called Gauss's theorem). **Physically,** in terms of water, the net flow through the surface is given by the flux Φ, which is the integral on the right. If it is not zero, then there is a source or sink of flux in the volume, represented by a divergence on the left. In the case of water and positive flux, it would suggest that there might be a source of water somewhere in the volume, perhaps a faucet or a spring, providing us with this excess flow, or the water might be expanding (boiling, perhaps), so that the mass of water inside the volume was becoming less.

To extend the preceding principle to an arbitrary volume: Let us stack a lot of our little cubes so as to construct an arbitrary volume. The divergence just adds up throughout the volume; but the flux out of one little cube goes right into the next one, canceling its contribution to the net flux, except on the surface of the arbitrary volume. It is only on the outer surface that we get any net contribution to the flux. So the divergence theorem holds for an arbitrary volume, with the integrals being carried out over the entire surface and volume.

Curl

Finally, we will want the line integral of **A** around a closed contour, $\oint \mathbf{A} \cdot d\mathbf{l}$, one that completely encloses an area. As before, we begin by examining the situation microscopically. Figure 1.10 shows a small square loop in the xy plane, in a region of field **A**. The area $\Delta x \, \Delta y$ is called Δa_z because the normal to the element of area points in the z direction, up out of the page, by the right-hand rule: Curl the fingers in the direction of the loop, and the thumb points in the direction of the area. After we deal with the xy plane, we can add the xz and yz planes. We will calculate the line integral of **A** around the loop.

We begin at the lower left corner and proceed counterclockwise around the square, multiplying the length of each side by the corresponding (parallel) field component. For the lines at the top and bottom, involving the x direction, we place the point (x, y, z) at the middle of the lower line, and we assume that the component $A_x(x, y, z)$ represents a suitable average over that side. The contribution from the bottom line is $A_x(x, y, z) \Delta x$. (Because of the dot product in the contour integral, only the A_x component is involved.) For the top, we get $-A_x(x, y + \Delta y, z)\Delta x$; it's negative because we're going in the negative x direction. For these two sides, then, the contribution to the line integral is

$$A_x(x, y, z)\Delta x - A_x(x, y + \Delta y, z)\Delta x$$

When we multiply and divide by Δy, it turns into a partial derivative form:

$$\frac{A_x(x, y, z)\Delta x \, \Delta y - A_x(x, y + \Delta y, z)\Delta x \, \Delta y}{\Delta y} \rightarrow -\frac{\partial A_x}{\partial y}\Delta a_z \qquad (1.32)$$

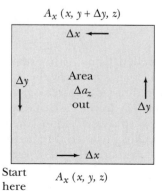

Start
here

$A_x(x, y, z)$

Figure 1.10   Small square.

in the limit of small Δy. The left and right sides of the loop behave similarly, so we have

$$\left(\oint \mathbf{A} \cdot d\mathbf{l}\right)_z = \left(\frac{\partial A_y}{\partial x} - \frac{\partial A_x}{\partial y}\right)\Delta a_z$$

$$= (\nabla \times \mathbf{A})_z \Delta a_z$$

Adding in the x and y components, we get a total of

$$\oint \mathbf{A} \cdot d\mathbf{l} = \int \nabla \times \mathbf{A} \cdot d\mathbf{a} \qquad (1.33)$$

which is **Stokes' theorem** for a simple loop. This represents, on the left, the line integral of $\mathbf{A}$ about the closed loop, and on the right, the integral of the curl of $\mathbf{A}$ over the area bounded by the loop. **Physically,** if we represent the field as a flow of water, the curl produces a kind of rotational tendency in the line integral which can be detected by a little paddle wheel (Fig. 1.7b).

We can construct a large and irregular loop out of such components. The curl just adds up over the area. The line elements of adjacent small loops cancel out, except at the very edge of the large loop. So Stokes' theorem holds for an arbitrary loop, the line integral being carried out over the perimeter of the loop, and the surface integral over the entire area of the loop.

1.6. **Tensors**

"Tensor" is a general sort of word, and you are already familiar with two kinds:

A **scalar** is a tensor of zero rank; it is just a simple quantity that is not burdened with any kind of direction. For example, in Newton's second law, $\mathbf{F} = m\mathbf{a}$, the mass m is a scalar that helps define the relationship between the vector force $\mathbf{F}$ and the vector acceleration $\mathbf{a}$.

A **vector** is a tensor of rank one. It has magnitude and direction. The vector equation $\mathbf{F} = m\mathbf{a}$ is a shorthand form comprising the three equations:

$$F_x = ma_x$$

$$F_y = ma_y$$

$$F_z = ma_z$$

The acceleration in the x direction depends on the mass and on the force in the x direction, and so on. The components don't get mixed up: An x component of acceleration doesn't affect a y component of force, and so on.

On the other hand, for the Lorentz force F_x on a charge Q moving at speed v_y across a magnetic field B_z we will find (Section 3.3):

$$F_x = Qv_yB_z$$

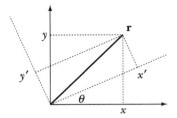

Figure 1.11   Coordinate rotation.

(This comes from the cross product $\mathbf{F} = Q\mathbf{v} \times \mathbf{B}$.) The force in the x direction depends on the speed in the y direction and the field in the z direction. To deal with such phenomena we are going to need some sort of formalism that interrelates components along various axes. In short, we will need a **tensor** of second rank, usually just called a tensor.

Previously, we have said that a vector is characterized by direction and magnitude. And that concept is useful, but now, in order to cope with tensors in general, we are going to introduce a more rigorous definition that depends on transformation properties.

To illustrate: Suppose we have a vector $\mathbf{r} = x\hat{\mathbf{x}} + y\hat{\mathbf{y}}$ in an unprimed two-dimensional frame, Figure 1.11. What will be its components in a primed frame that is rotated by an angle θ relative to the first? By projecting the unprimed components onto the primed axes, we find

$$x' = x\cos\theta + y\cos(90° - \theta) = x\cos\theta + y\sin\theta \tag{1.34}$$

$$y' = x\cos(90° + \theta) + y\cos\theta = -x\sin\theta + y\cos\theta$$

It is reasonable that any physically significant vector should transform in the same way; so we shall now include this rotational transformation in our definition of a vector:

$$A'_x = A_x\cos\theta + A_y\sin\theta \tag{1.35}$$

$$A'_y = -A_x\sin\theta + A_y\cos\theta$$

The vector must not change form under such a rotation: If $A_x = f(x, y)$, then $A'_x = f(x', y')$ in terms of the same function f. (Note that a vector is unaffected by a translation; in other words, you can move a vector around, parallel to itself.)

▼ EXAMPLE 1-4

(a) Is $\mathbf{A} = y\hat{\mathbf{x}} + x\hat{\mathbf{y}}$ a vector? **(b)** Is $\mathbf{A} = -y\hat{\mathbf{x}} + x\hat{\mathbf{y}}$ a vector?

ANSWERS **(a)** From Equation 1.35,

$$A'_x = A_x\cos\theta + A_y\sin\theta$$
$$= y\cos\theta + x\sin\theta$$

Now since $A_x = y$, we must find that $A'_x = y'$. However, from Equation 1.34,

$$y' = -x\sin\theta + y\cos\theta$$

which is not the same as A'_x: The sign of the $x \sin \theta$ term is wrong. So this vector candidate fails the rotation-transformation test.

(b) Again from Equation 1.35,

$$A'_x = A_x \cos \theta + A_y \sin \theta$$
$$= - y \cos \theta + x \sin \theta$$
$$= - y'$$

in view of Equation 1.34. So far so good. Also,

$$A'_y = - A_x \sin \theta + A_y \cos \theta$$
$$= y \sin \theta + x \cos \theta$$
$$= x'$$

from Equation 1.34. So this is an acceptable vector.

Now we generalize to three dimensions. To facilitate notation, we number the coordinates: $x \rightarrow x_1$, $y \rightarrow x_2$, $z \rightarrow x_3$. The transformations of Equation 1.35 become

$$A'_1 = A_1 \cos(x'_1, x_1) + A_2 \cos(x'_1, x_2) + A_3 \cos(x'_1, x_3) \qquad \textbf{(1.36)}$$
$$A'_2 = A_1 \cos(x'_2, x_1) + A_2 \cos(x'_2, x_2) + A_3 \cos(x'_2, x_3)$$
$$A'_3 = A_1 \cos(x'_3, x_1) + A_2 \cos(x'_3, x_2) + A_3 \cos(x'_3, x_3)$$

Here the angle between axes x'_1 and x_1, for example, is given as (x'_1, x_1). Thus, in terms of the angle θ in Figure 1.11, we have $\cos \theta = \cos(x'_1, x_1)$. These "direction cosines" will now be abbreviated as a_{11} and so forth. Then Equation 1.35 can be written in the brief form:

$$A'_i = \sum_{j=1}^{3} a_{ij} A_j$$

This represents three equations, one each for $i = 1$, 2, and 3. The following **summation convention** (introduced by Einstein) is commonly employed: If an index is repeated in a term, it is summed over. With this, Equation 1.36 boils down to

$$A'_i = a_{ij} A_j \qquad \textbf{(1.37)}$$

which again represents three equations, one each for the three possible values of i. Each of these equations involves a summation over the three values of j.

Other parts of the summation convention are that no index may occur three times in any term, and if an index occurs only once in a term, then the equation is valid for each possible value of that index.

The inverse transformation is

$$A_i = a'_{ij} A'_j \qquad \textbf{(1.38)}$$

These a'_{ij} coefficients have the relationship

$$a'_{ij} = a_{ji}$$

since this is just the cosine of the same angle taken in the opposite direction.

The direction cosines may also be written as

$$a_{ij} = \frac{\partial x'_i}{\partial x_j}$$

If for example we differentiate $x'_i = a_{ij}x_j$ with respect to x_j, we obtain $\partial x'_i/\partial x_j = a_{ij}$. (This is called "contravariant." In other coordinate systems there is a distinction between "covariant" and contravariant transformations, but the distinction disappears in Cartesian coordinates.)

For the scalar S, the corresponding definition would involve simply $S' = S$. There is no transformation involved; a scalar just sits there.

As mentioned, a vector equation doesn't mix coordinates. For example, a force in the x direction produces an acceleration that is also in the x direction. However, in physics it often happens that there is an interdependence, and we need a structure that will mix coordinates. Here is a simple example from Section 9.2. Since anisotropic crystals polarize more easily in some directions than others, an electric E field in the x direction may produce a polarization P component in other directions. This may be represented using a tensor with components χ_{ij}:

$$P_x = \epsilon_0(\chi_{xx} E_x + \chi_{xy} E_y + \chi_{xz} E_z)$$
$$P_y = \epsilon_0(\chi_{yx} E_x + \chi_{yy} E_y + \chi_{yz} E_z)$$
$$P_z = \epsilon_0(\chi_{zx} E_x + \chi_{zy} E_y + \chi_{zz} E_z)$$

We abbreviate it as

$$P_i = \epsilon_0 \sum \chi_{ij} E_j$$
$$\rightarrow \epsilon_0 \chi_{ij} E_j \qquad \text{using the summation convention}$$

The χ_{ij}'s constitute a tensor of second rank.

By definition, the components T_{ij} of a second-rank tensor **T** transform according to the form:

$$T_{ij} = \sum\sum a_{ik}a_{jl} T_{kl} \rightarrow a_{ik}a_{jl} T_{kl} \qquad \text{(1.39)}$$

using the summation convention for each repeated index. This is like applying the vector transformation *twice* (Eq. 1.37).

These transformations are "orthogonal," which means that

$$a_{ij}a_{kj} = \delta_{ik} \qquad \text{(1.40)}$$

where the Kronecker delta, δ_{ik}, is one if $i = k$ and zero otherwise. This may be seen as follows. If we start with the transformation

$$A'_i = a_{ij}A_j$$

and insert the inverse transformation,

$$A_j = a'_{jk}A'_k$$

along with $a'_{jk} = a_{kj}$, then we get

$$A'_i = a_{ij} a_{kj} A'_k$$

and so $a_{ij} a_{kj}$ must be one if $i = k$ and zero otherwise; therefore, $a_{ij} a_{kj} = \delta_{ik}$. Similarly, using inverse transformations we can easily show that $a_{ij} a_{ik} = \delta_{jk}$ (but not $a_{ij} a_{jk} = \delta_{ik}$).

Note that the direction cosines a_{ij} are nine in number and can be assembled into a matrix that resembles a tensor, yet they do not constitute a tensor. If we apply the transformation to them, we find

$$a'_{ij} = a_{ik} a_{jl} a_{kl} = a_{ik}(a_{jl} a_{kl}) = a_{ik} \delta_{jk} = a_{ij}$$

but that's wrong: $a'_{ij} \neq a_{ij}$ (instead, $a'_{ij} = a_{ji}$).

There are several interesting vector operations corresponding to multiplication. If we take the direct product $A_i B_j$ and set the indices equal, we obtain $A_i B_i$, which represents the three terms of the scalar product or dot product, $\mathbf{A} \cdot \mathbf{B}$. (This is called "contraction.") The vector or cross product is an antisymmetric tensor:

$$\mathbf{A} \times \mathbf{B} = (A_y B_z - A_z B_y) \,\hat{\mathbf{x}} + (A_z B_x - A_x B_z) \,\hat{\mathbf{y}} + (A_x B_y - A_y B_x) \,\hat{\mathbf{z}}$$

which may be written in matrix form (with j's for columns displayed for your convenience):

$$A_i B_j - A_j B_i = \begin{array}{c} j = \\ \\ \\ \\ \end{array} \begin{vmatrix} 1 & 2 & 3 \\ 0 & A_x B_y - A_y B_x & -A_z B_x + A_x B_z \\ -A_x B_y + A_y B_x & 0 & A_y B_z - A_z B_y \\ A_z B_x - A_x B_z & -A_y B_z + A_z B_y & 0 \end{vmatrix} \qquad \textbf{(1.41)}$$

Since there are only three independent components in this case, it is convenient to ascribe each component to one of the three coordinate axes; for example, $A_x B_y - A_y B_x$ is the z component of the cross product, $(\mathbf{A} \times \mathbf{B})_z$. The Lorentz force, mentioned at the beginning of this section, is an example of such a quantity. This "pseudovector" or axial vector behaves in most ways like a true or polar vector; however, under coordinate inversion its components don't change sign, whereas those of a true vector do.

Similarly, the curl is a pseudovector:

$$\nabla \times \mathbf{A} \rightarrow \frac{\partial A_j}{\partial x_i} - \frac{\partial A_i}{\partial x_j}$$

but we usually don't mind, because it behaves like a vector.

1.7. Orthogonal Curvilinear Coordinates

So far we have done our analysis using Cartesian (rectangular) coordinates where appropriate. Many of our results are good for other coordinate systems. However, the Cartesian expression for ∇ has no elementary counterpart in other coordinates, so we will be needing explicit expressions for gradients, divergences, curls, and Laplacians in the two most common alternative systems, namely spherical (or spherical polar) and cylindrical (or circular cylindrical).

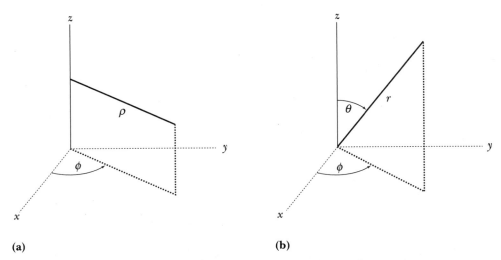

Figure 1.12   **(a)** Cylindrical coordinates. **(b)** Spherical coordinates.

We shall begin the development in a general way. Our general coordinates for a point are q_1, q_2, q_3. These correspond to x, y, z, or in cylindrical coordinates to ρ, ϕ, z (Fig. 1.12a) or in spherical coordinates to r, θ, ϕ (Fig. 1.12b). We restrict ourselves to orthogonal coordinates: that is, the unit vectors $\hat{\mathbf{q}}_1$, $\hat{\mathbf{q}}_2$, $\hat{\mathbf{q}}_3$ are mutually perpendicular at any point.

In the coordinate systems under consideration, an infinitesimal displacement $d\mathbf{l}$ in an arbitrary direction may be written in the following ways:

$$
\begin{array}{llll}
\text{Cartesian} & d\mathbf{l} = dx\,\hat{\mathbf{x}} & + \, dy\,\hat{\mathbf{y}} & + \, dz\,\hat{\mathbf{z}} & \textbf{(1.42)} \\
\text{Cylindrical:} & d\mathbf{l} = d\rho\,\hat{\boldsymbol{\rho}} & + \, \rho\,d\phi\,\hat{\boldsymbol{\phi}} & + \, dz\,\hat{\mathbf{z}} \\
\text{Spherical:} & d\mathbf{l} = dr\,\hat{\mathbf{r}} & + \, r\,d\theta\,\hat{\boldsymbol{\theta}} & + \, r\sin\theta\,d\phi\,\hat{\boldsymbol{\phi}} \\
\text{General:} & d\mathbf{l} = h_1\,dq_1\,\hat{\mathbf{q}}_1 + h_2\,dq_2\,\hat{\mathbf{q}}_2 + h_3\,dq_3\,\hat{\mathbf{q}}_3
\end{array}
$$

For example, in cylindrical coordinates, if we move by $d\phi$ in the $\hat{\boldsymbol{\phi}}$ direction, keeping ρ and z fixed, the distance traveled is $\rho\,d\phi$. These equations display the meanings of the h's and q's in the various systems; for example, in cylindrical coordinates, $q_2 = \phi$ and $h_2 = \rho$.

Two notes on cylindrical and spherical coordinates: **(1)** The unit vectors need not retain any particular direction in space; for example, in Figure 1.12a, when ϕ changes, $\hat{\rho}$ changes direction. **(2)** If ρ or r is zero, then ϕ is undefined.

Now we can develop the gradient, divergence, curl, and Laplacian in turn.

Gradient

For an arbitrary scalar function $f(q_1, q_2, q_3)$, we have, from the chain rule,

$$
df = \frac{\partial f}{\partial q_1}\,dq_1 + \frac{\partial f}{\partial q_2}\,dq_2 + \frac{\partial f}{\partial q_3}\,dq_3 \tag{1.43}
$$

Also, from Section 1.2, Equation 1.14,

$$df = (\nabla f) \cdot dl$$

So in view of (from Eq. 1.42)

$$dl = h_1 dq_1 \,\hat{\mathbf{q}}_1 + h_2 dq_2 \,\hat{\mathbf{q}}_2 + h_3 dq_3 \,\hat{\mathbf{q}}_3$$

it must be that we can write gradient as

$$\nabla f = \frac{1}{h_1} \frac{\partial f}{\partial q_1} \hat{\mathbf{q}}_1 + \frac{1}{h_2} \frac{\partial f}{\partial q_2} \hat{\mathbf{q}}_2 + \frac{1}{h_3} \frac{\partial f}{\partial q_3} \hat{\mathbf{q}}_3 \tag{1.44}$$

because then $(\nabla f) \cdot dl$ satisfies the chain rule equation for df, Equation 1.43.

When we substitute h and q from Equation 1.42, we find for cylindrical coordinates

$$\nabla f = \frac{\partial f}{\partial \rho} \hat{\boldsymbol{\rho}} + \frac{1}{\rho} \frac{\partial f}{\partial \phi} \hat{\boldsymbol{\phi}} + \frac{\partial f}{\partial z} \hat{\mathbf{z}} \tag{1.45}$$

and for spherical coordinates

$$\nabla f = \frac{\partial f}{\partial r} \hat{\mathbf{r}} + \frac{1}{r} \frac{\partial f}{\partial \theta} \hat{\boldsymbol{\theta}} + \frac{1}{r \sin \theta} \frac{\partial f}{\partial \phi} \hat{\boldsymbol{\phi}} \tag{1.46}$$

as displayed inside the front cover

◥ EXAMPLE 1-5

Find the gradient of $2r$ in spherical coordinates. Then do it in rectangular coordinates.

ANSWER Spherical coordinates:

$$\nabla (2r) = \frac{\partial (2r)}{\partial r} \hat{\mathbf{r}} = 2\hat{\mathbf{r}}$$

there being no angular dependence. So the scalar field $2r$ is increasing in the positive radial direction.

Rectangular coordinates are more complicated here because that system is less well suited to the problem. To begin with, the vector $\mathbf{r}$ is

$$\mathbf{r} = r\hat{\mathbf{r}} = x\hat{\mathbf{x}} + y\hat{\mathbf{y}} + z\hat{\mathbf{z}}$$

and its magnitude is

$$r = \sqrt{x^2 + y^2 + z^2}$$

For the x coordinate of the gradient,

$$\frac{\partial r}{\partial x} \hat{\mathbf{x}} = \frac{\partial \sqrt{x^2 + y^2 + z^2}}{\partial x} \hat{\mathbf{x}} = \frac{x\hat{\mathbf{x}}}{\sqrt{x^2 + y^2 + z^2}}$$

So, combining the corresponding results for x, y, and z, we find

$$\nabla(2r) = 2\frac{x\hat{\mathbf{x}} + y\hat{\mathbf{y}} + z\hat{\mathbf{z}}}{\sqrt{x^2 + y^2 + z^2}} \qquad \text{in rectangular coordinates}$$

$$= 2\frac{r\hat{\mathbf{r}}}{r} = 2\hat{\mathbf{r}} \qquad \text{as before, in spherical coordinates}$$

Divergence

Proceeding as in Section 1.5, we integrate the vector function $\mathbf{A}\,(q_1, q_2, q_3)$ over the surface of an infinitesimal six-sided figure analogous to a cube (but no longer necessarily a cube because we employ curvilinear coordinates). In the q_1 direction, the contribution is

$$(A_1 h_2 h_3)(q_1 + \Delta q_1, q_2, q_3)\,\Delta q_2 \Delta q_3 - (A_1 h_2 h_3)(q_1, q_2, q_3)\,\Delta q_2 \Delta q_3 \qquad (1.47)$$

The notation is peculiar, so watch closely. The distance element in the direction of q_2 is $h_2 \Delta q_2$, and so the area element is $\Delta a = h_2 h_3 \Delta q_2 \Delta q_3$. However, in curvilinear coordinates not only A_1 but also h_2 and h_3 may change when we go from q_1 to $q_1 + \Delta q_1$. So we have lumped the product of A_1, h_2, and h_3 into one function of the three coordinates: $(A_1 h_2 h_3)(q_1, q_2, q_3)$. When we differentiate, we have to do the whole thing, namely, the function $(A_1 h_2 h_3)$ considered as a unit.

Upon multiplying and dividing by Δq_1, we obtain a partial derivative form as before. Thus, we find that the q_1 contribution is

$$\int_1 \mathbf{A}\cdot d\mathbf{a} = \frac{\partial}{\partial q_1}\,(h_2 h_3 A_1)\,\Delta q_1 \Delta q_2 \Delta q_3$$

$$= \frac{1}{h_1 h_2 h_3}\frac{\partial}{\partial q_1}\,(h_2 h_3 A_1)\,\Delta v \qquad (1.48)$$

where the volume element is $\Delta v = h_1 h_2 h_3 \Delta q_1 \Delta q_2 \Delta q_3$ (as we remarked before, the distance element in the direction of q_1 is $h_1 \Delta q_1$ and so on). Adding the contributions from the q_2 and q_3 sides, we obtain

$$\oint \mathbf{A}\cdot d\mathbf{a} = \frac{1}{h_1 h_2 h_3}\left[\frac{\partial}{\partial q_1}(h_2 h_3 A_1) + \frac{\partial}{\partial q_2}(h_3 h_1 A_2) + \frac{\partial}{\partial q_3}(h_1 h_2 A_3)\right]\Delta v$$

$$= \int \nabla\cdot\mathbf{A}\,dv$$

by the divergence theorem. This implies that the divergence is

$$\nabla\cdot\mathbf{A} = \frac{1}{h_1 h_2 h_3}\left[\frac{\partial}{\partial q_1}(h_2 h_3 A_1) + \frac{\partial}{\partial q_2}(h_3 h_1 A_2) + \frac{\partial}{\partial q_3}(h_1 h_2 A_3)\right] \qquad (1.49)$$

When we substitute the appropriate q's and h's from Equation 1.42, we find:

Cylindrical: $\nabla \cdot \mathbf{A} = \dfrac{1}{\rho}\dfrac{\partial(\rho A_\rho)}{\partial\rho} + \dfrac{1}{\rho}\dfrac{\partial A_\phi}{\partial\phi} + \dfrac{\partial A_z}{\partial z}$ (1.50)

Spherical: $\nabla \cdot \mathbf{A} = \dfrac{1}{r^2}\dfrac{\partial(r^2 A_r)}{\partial r} + \dfrac{1}{r\sin\theta}\dfrac{\partial(A_\theta\sin\theta)}{\partial\theta} + \dfrac{1}{r\sin\theta}\dfrac{\partial A_\phi}{\partial\phi}$ (1.51)

◥ EXAMPLE 1-6

Given that $\mathbf{A} = 2\rho\hat{\rho}$, find $\nabla \cdot \mathbf{A}$.

ANSWER Cylindrical coordinates are implied.

$$\nabla \cdot \mathbf{A} = \frac{1}{\rho}\frac{\partial(\rho A_\rho)}{\partial\rho} + 0 + 0$$

$$= \frac{1}{\rho}\frac{\partial(2\rho^2)}{\partial\rho} = 4$$

so there is a constant source of flux per unit volume (Section 1.5).

Curl

Proceeding as in Section 1.5, we integrate the vector function $\mathbf{A}(q_1, q_2, q_3)$ over the edge of an infinitesimal four-sided loop analogous to a square (but no longer necessarily square because we employ curvilinear coordinates). In the $q_1 = $ constant plane, the contribution from the four sides is

$$\oint \mathbf{A}_1 \cdot d\mathbf{l} = (A_2 h_2)(q_1, q_2, q_3)\,\Delta q_2 - (A_2 h_2)(q_1, q_2, q_3 + \Delta q_3)\,\Delta q_2 \qquad (1.52)$$
$$+ (A_3 h_3)(q_1, q_2 + \Delta q_2, q_3)\,\Delta q_3 - (A_3 h_3)(q_1, q_2, q_3)\,\Delta q_3$$

The distance element in the q_2 direction is $h_2\Delta q_2$. However, as before, both A_2 and h_2 can change as we move in the q_3 direction; so, we combine the product of A_2 and h_2 into one function of the coordinates, $(A_2 h_2)(q_1, q_2, q_3)$. Multiplying and dividing by the appropriate increments, this expression becomes (in the limit)

$$\oint \mathbf{A}_1 \cdot d\mathbf{l} = -\frac{(A_2 h_2)(q_1, q_2, q_3 + \Delta q_3) - (A_2 h_2)(q_1, q_2, q_3)}{\Delta q_3}\,\Delta q_2 \Delta q_3$$

$$+ \frac{(A_3 h_3)(q_1, q_2 + \Delta q_2, q_3) - (A_3 h_3)(q_1, q_2, q_3)}{\Delta q_2}\,\Delta q_2 \Delta q_3 \qquad (1.53)$$

$$\rightarrow \left(\frac{\partial(A_3 h_3)}{\partial q_2} - \frac{\partial(A_2 h_2)}{\partial q_3} \right)\frac{\Delta a_1}{h_2 h_3}$$

where a_1 is the 1 component of the area. Cyclic rotation of coordinates gives us the other components; the total is

$$\oint \mathbf{A} \cdot d\mathbf{l} = \left[\frac{1}{h_2 h_3} \left(\frac{\partial (A_3 h_3)}{\partial q_2} - \frac{\partial (A_2 h_2)}{\partial q_3} \right) + \frac{1}{h_3 h_1} \left(\frac{\partial (A_1 h_1)}{\partial q_3} - \frac{\partial (A_3 h_3)}{\partial q_1} \right) \right.$$

$$\left. + \frac{1}{h_1 h_2} \left(\frac{\partial (A_2 h_2)}{\partial q_1} - \frac{\partial (A_1 h_1)}{\partial q_2} \right) \right] \cdot \Delta \mathbf{a} \tag{1.54}$$

$$= \int \nabla \times \mathbf{A} \cdot d\mathbf{a} \qquad \text{by Stokes' theorem}$$

This implies that the bracketed quantity represents the curl. Substituting the appropriate h's and q's from Equation 1.42, we obtain the results listed inside the front cover:

Cylindrical:

$$\nabla \times \mathbf{A} = \left(\frac{1}{\rho} \frac{\partial A_z}{\partial \phi} - \frac{\partial A_\phi}{\partial z} \right) \hat{\boldsymbol{\rho}} + \left(\frac{\partial A_\rho}{\partial z} - \frac{\partial A_z}{\partial \rho} \right) \hat{\boldsymbol{\phi}} + \frac{1}{\rho} \left(\frac{\partial (\rho A_\phi)}{\partial \rho} - \frac{\partial A_\rho}{\partial \phi} \right) \hat{\mathbf{z}} \tag{1.55}$$

Spherical: $$\nabla \times \mathbf{A} = \frac{1}{r \sin \theta} \left(\frac{\partial (A_\phi \sin \theta)}{\partial \theta} - \frac{\partial A_\theta}{\partial \phi} \right) \hat{\mathbf{r}}$$

$$+ \frac{1}{r} \left(\frac{1}{\sin \theta} \frac{\partial A_r}{\partial \phi} - \frac{\partial (r A_\phi)}{\partial r} \right) \hat{\boldsymbol{\theta}} + \frac{1}{r} \left(\frac{\partial (r A_\theta)}{\partial r} - \frac{\partial A_r}{\partial \theta} \right) \hat{\boldsymbol{\phi}} \tag{1.56}$$

◤ EXAMPLE 1-7

Given that $\mathbf{A} = \dfrac{\hat{\mathbf{r}}}{r}$, find $\nabla \times \mathbf{A}$.

ANSWER

$$\nabla \times \mathbf{A} = (0 + 0) \, \hat{\mathbf{r}} + (0 + 0) \, \hat{\boldsymbol{\theta}} + (0 + 0) \hat{\boldsymbol{\phi}} = 0$$

(except at exactly $r = 0$, where it is undefined). $\mathbf{A}$ is strictly radial and conveys no tendency toward rotation.

Laplacian

The Laplacian of a scalar function is the divergence of the gradient:

$$\nabla^2 f = \nabla \cdot \nabla f \tag{1.57}$$

We start with the expression for the divergence, Eq. 1.49, and for example for A_1 we substitute the corresponding component of the gradient, namely, $\dfrac{1}{h_1} \dfrac{\partial f}{\partial q_1}$. The result is

$$\nabla^2 f = \frac{1}{h_1 h_2 h_3} \left[\frac{\partial}{\partial q_1} \left(\frac{h_2 h_3}{h_1} \frac{\partial f}{\partial q_1} \right) + \frac{\partial}{\partial q_2} \left(\frac{h_3 h_1}{h_2} \frac{\partial f}{\partial q_2} \right) \right.$$

$$\left. + \frac{\partial}{\partial q_3} \left(\frac{h_1 h_2}{h_3} \frac{\partial f}{\partial q_3} \right) \right] \tag{1.58}$$

Substituting q's and h's from Equation 1.42, we find:

Cylindrical: $$\nabla^2 f = \frac{1}{\rho} \frac{\partial}{\partial \rho} \left(\rho \frac{\partial f}{\partial \rho} \right) + \frac{1}{\rho^2} \frac{\partial^2 f}{\partial \phi^2} + \frac{\partial^2 f}{\partial z^2}$$ **(1.59)**

Spherical: $$\nabla^2 f = \frac{1}{r^2} \frac{\partial}{\partial r} \left(r^2 \frac{\partial f}{\partial r} \right) + \frac{1}{r^2 \sin \theta} \frac{\partial}{\partial \theta} \left(\sin \theta \frac{\partial f}{\partial \theta} \right)$$
$$+ \frac{1}{r^2 \sin^2 \theta} \frac{\partial^2 f}{\partial \phi^2}$$ **(1.60)**

 EXAMPLE 1-8

Given that $f = 1/r$, find $\nabla^2 f$.

ANSWER

$$\nabla^2 f = \frac{1}{r^2} \frac{\partial}{\partial r} \left(r^2 \frac{\partial f}{\partial r} \right) = 0$$

(except at exactly $r = 0$, where it is undefined).

The Laplacian of a vector cannot be found as above because there is no appropriate general expression for ∇; but rather, it is defined using the relationship we have already found for Cartesian coordinates in Eq. 1.25:

$$\nabla^2 \mathbf{A} \equiv \nabla(\nabla \cdot \mathbf{A}) - \nabla \times (\nabla \times \mathbf{A})$$

EXAMPLE 1-9

Given that $\mathbf{A} = \hat{\mathbf{r}}/r$, find $\nabla^2 \mathbf{A}$.

ANSWER From the curl example presented in Example 1-7, $\nabla \times \mathbf{A} = 0$. For the divergence,

$$\nabla \cdot \left(\frac{\hat{\mathbf{r}}}{r} \right) = \frac{1}{r^2} \frac{\partial (r^2/r)}{\partial r} = \frac{1}{r^2}$$

so

$$\nabla^2 \mathbf{A} = \nabla(\nabla \cdot \mathbf{A}) - \nabla \times (\nabla \times \mathbf{A})$$
$$= \nabla \left(\frac{1}{r^2} \right) - \nabla \times (0)$$

$$\nabla^2 \mathbf{A} = -2 \frac{1}{r^3} \hat{\mathbf{r}}$$

1.8 SUMMARY

We have presented sufficient vector analysis to enable us to develop Maxwell's equations and their immediate consequences. The necessary formulas are collected inside the front cover.

Much of the formalism is based on the "del" differential operator; in Cartesian coordinates:

$$\nabla = \hat{\mathbf{x}} \frac{\partial}{\partial x} + \hat{\mathbf{y}} \frac{\partial}{\partial y} + \hat{\mathbf{z}} \frac{\partial}{\partial z} \tag{1.12}$$

In other coordinate systems, this operator has no unique representation.

In terms of del, we described the gradient, divergence, and curl in Section 1.2 and the Laplacian in Section 1.3.

$$\text{Gradient of } f = \text{grad } f = \nabla f \tag{1.13}$$

$$\text{Divergence of } \mathbf{A} = \text{div } \mathbf{A} = \nabla \cdot \mathbf{A} \tag{1.15}$$

$$\text{Curl of } \mathbf{A} = \text{curl } \mathbf{A} = \nabla \times \mathbf{A} \tag{1.16}$$

$$\text{Laplacian of } f = \nabla^2 f \tag{1.21}$$

$$\text{Laplacian of } \mathbf{A} = \nabla^2 \mathbf{A} \tag{1.25}$$

We developed the corresponding formulas in various coordinate systems using general properties of curvilinear coordinates in Section 1.7.

In rectangular coordinates, the Laplacian of a vector is simply the result of applying the scalar Laplacian operator to each component. In other systems, it must be defined as

$$\nabla^2 \mathbf{A} = \nabla(\nabla \cdot \mathbf{A}) - \nabla \times (\nabla \times \mathbf{A}) \tag{1.25}$$

We demonstrated a uniqueness theorem in Section 1.4: The curl and divergence suffice to determine a vector field in a region where the field has specified values on the boundary.

We developed two particularly useful integral relationships involving del:

$$\int \nabla \cdot \mathbf{A} \, dv = \oint \mathbf{A} \cdot d\mathbf{a} \qquad \text{(Divergence theorem)} \tag{1.31}$$

$$\int (\nabla \times \mathbf{A}) \cdot d\mathbf{a} = \oint \mathbf{A} \cdot d\mathbf{l} \qquad \text{(Stokes' theorem)} \tag{1.33}$$

◥ PROBLEMS

1-1 Calculate dot product, cross product, and angle between:

$$\mathbf{A} = -\hat{\mathbf{x}} + 7\hat{\mathbf{y}} + 6\hat{\mathbf{z}} \quad \text{and} \quad \mathbf{B} = -\hat{\mathbf{x}} - 6\hat{\mathbf{y}} + 6\hat{\mathbf{z}}$$

1-2 Find the angle between any two of the four body diagonals of a cube using a dot product.

1-3 Show that $\mathbf{A} \times (\mathbf{B} \times \mathbf{C}) + \mathbf{B} \times (\mathbf{C} \times \mathbf{A}) + \mathbf{C} \times (\mathbf{A} \times \mathbf{B}) = 0$.

1-4 Show that $(\mathbf{A} \times \mathbf{B}) \cdot (\mathbf{C} \times \mathbf{D}) = (\mathbf{A} \cdot \mathbf{C})(\mathbf{B} \cdot \mathbf{D}) - (\mathbf{A} \cdot \mathbf{D})(\mathbf{B} \cdot \mathbf{C})$.

1-5 Using the dot product, derive the law of cosines:

$$C^2 = A^2 + B^2 - 2\,AB\cos\theta.$$

1-6 Using the cross product, derive the law of sines, Figure 1.13:

$$\frac{\sin\alpha}{A} = \frac{\sin\beta}{B} = \frac{\sin\gamma}{C}$$

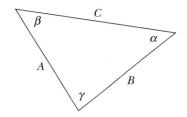

Figure 1.13

1-7 Show that the vectors $\mathbf{A}$, $\mathbf{B}$, and $\mathbf{C}$ are coplanar if $\mathbf{A} \cdot \mathbf{B} \times \mathbf{C} = 0$.

1-8 The vertices of a triangle ABC are at the points $(1, -1, 0)$, $(0, 1, -1)$, and $(-1, 0, 2)$, respectively. Place point D so that the figure $ABCD$ forms a plane parallelogram.

1-9 For constant vector $\mathbf{A}$, show that $(\mathbf{r} - \mathbf{A}) \cdot \mathbf{A} = 0$ is the equation of a plane.

*1-10 For constant vector $\mathbf{A}$, show that $(\mathbf{r} - \mathbf{A}) \cdot \mathbf{r} = 0$ is the equation of a sphere.

1-11 The vectors $\mathbf{A}$, $\mathbf{B}$, and $\mathbf{C}$, with tails together, define a tetrahedron. Find the total vector area (taking outward-pointing as positive).

1-12 Prove: (a) $\nabla \cdot (f\mathbf{A}) = \mathbf{A} \cdot \nabla f + f\nabla \cdot \mathbf{A}$.
  (b) $\nabla \cdot (\mathbf{A} \times \mathbf{B}) = \mathbf{B} \cdot (\nabla \times \mathbf{A}) - \mathbf{A} \cdot (\nabla \times \mathbf{B})$.
  (c) $\nabla \times (\mathbf{A} \times \mathbf{B}) = (\mathbf{B} \cdot \nabla)\mathbf{A} - (\mathbf{A} \cdot \nabla)\mathbf{B} + \mathbf{A}(\nabla \cdot \mathbf{B}) - \mathbf{B}(\nabla \cdot \mathbf{A})$.

1-13 If $\mathbf{A}$ and $\mathbf{B}$ are each irrotational, show that $\mathbf{A} \times \mathbf{B}$ is solenoidal.

1-14 If f satisfies Laplace's equation, show that ∇f is both solenoidal and irrotational.

1-15 Find the gradient of $f = x - xy + 3z^2$.

1-16 Find the Laplacian of $f = x - xy + 3z^2$.

1-17 Find the curl and the divergence of $\mathbf{A} = x\hat{\mathbf{x}} - xy\hat{\mathbf{y}} + 3z^2\hat{\mathbf{z}}$.

1-18 Find the Laplacian of $\mathbf{A} = x\hat{\mathbf{x}} - xy\hat{\mathbf{y}} + 3z^2\hat{\mathbf{z}}$ by direct application of ∇^2 and by Equation 1.25.

1-19 For fields of the form $r^n\,\hat{\mathbf{r}}$, show that (except at $r = 0$) the divergence is zero only for $n = -2$.

1-20 For fields of the form $\rho^n\hat{\boldsymbol{\phi}}$, show that (except at $\rho = 0$) the curl is zero only for $n = -1$.

1-21 Find a pair of fields having equal divergences in some region, having the same values on the boundary of that region, and yet having different curls.

*1-22 Find a pair of fields having equal curls in some region, having the same values on the boundary of that region, and yet having different divergences.

1-23 Demonstrate the divergence theorem for the vector $\mathbf{A} = y\hat{\mathbf{y}}$, for a cube of side s having a corner at the origin and edges lying along the positive x, y, z axes.

1-24 Demonstrate Stokes' theorem for the vector $\mathbf{A} = x\hat{\mathbf{y}}$, for a square of side s having a corner at the origin and edges lying along the positive x, y axes.

1-25 Show that for a closed surface, $\oint d\mathbf{a} = 0$, where $\mathbf{a}$ is area. (See inside front cover.)

1-26 Prove that the area of a plane closed curve is

$$\mathbf{a} = \frac{1}{2} \oint \mathbf{r} \times d\mathbf{l}$$

for arbitrary origin.

1-27 In Cartesian coordinates, find $\nabla \cdot (x\hat{\mathbf{x}})$. In cylindrical coordinates, find $\nabla \cdot (\rho\hat{\boldsymbol{\rho}})$. In spherical coordinates, find $\nabla \cdot (r\hat{\mathbf{r}})$.

1-28 Find the curl and the divergence of $\hat{\boldsymbol{\theta}}$. Sketch the field and comment.

1-29 **(a)** Calculate $\int_a^b \mathbf{A} \cdot d\mathbf{l}$ for the vector $\mathbf{A} = y\hat{\mathbf{y}}$, for a path around two sides of a square in the xy plane, in the following order: from point $(0, 0)$ to point $(1, 0)$ to point $(1, 1)$. Compare with the result obtained by going around the other way: point $(0, 0)$ to $(0, 1)$ to $(1, 1)$. Are they the same?
(b) Repeat for $\mathbf{A} = x\hat{\mathbf{y}}$. Are the two results the same?

(c) Why do parts (a) and (b) give different answers?

1-30 Show that $\cos^2\alpha + \cos^2\beta + \cos^2\gamma = 1$ for the direction cosines of an arbitrary vector $\mathbf{r}$, Figure 1.14.

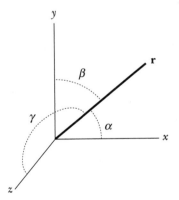

Figure 1.14

1-31 Show that length of a vector doesn't change in a rotational transformation: $A'^2 = A^2$.

*1-32 **(a)** Show that if the curl and divergence of a vector field are specified everywhere in space, the field itself is determined up to the gradient of a scalar function that satisfies Laplace's equation. **(b)** Then show that the field is completely determined within a region if (in addition) the value of the field is specified on the surface of the region. (Don't use Section 1.4.)

C H A P T E R

2

ELECTRIC FIELD E—GAUSS'S LAW

W e live in an electronic age, of course. But in another sense, people have always existed and acted within an electromagnetic framework: Our bodies and our surroundings are entirely constructed of electrically charged particles. Electrical phenomena were noticed by the ancient Greeks, and the work "electricity" comes from their word "elektron" for amber, a substance that readily displays electrostatic effects. The second half of the word "**electrostatic**" is Greek for "stationary": In this chapter we will restrict ourselves to charges that don't move. (The motion of charges can result in magnetism and radiation, which we will describe in later chapters.)

So we shall now embark upon a study of this thing called electricity: What is electrical charge, and how does it behave? Let us begin, not with an equation, but with a simple experiment.

2.1. Coulomb's Law for Electrostatic Force *F* and Charge *Q*

We already know that ordinary matter consists of negative electrons, in states of various energies, surrounding heavier positively charged nuclei. So if we rub a hard rubber rod with fur, it is not surprising that some electrons are rubbed off the fur, the rod becomes negatively charged, and the fur is left positively charged. With care we can detect that the fur and rod attract each other. If we then hold the rod near an uncharged small, light pith ball, suspended by a thread, they are attracted at first, Figure 2.1a, because of induced polarization. (This is discussed in Sections 8.2 and 9.1. Instead of a pith ball you could use a lightly crumpled ball of aluminum foil.) Then the rod touches the pith ball and some electrons jump across, leaving both rod and ball negatively charged, Figure 2.1b, at which point they repel each other.

This is perhaps the simplest and best-known of electrostatics demonstrations; yet, like all such demonstrations, it is superficially complex and ambiguous, and it requires explanation and interpretation. The electrons we've been manipulating are completely invisible. It is the duty and pleasure of the physicist to see the ideal relationship that lies behind the curtain imposed by real materials: "**Like charges repel**," and unlike charges attract.

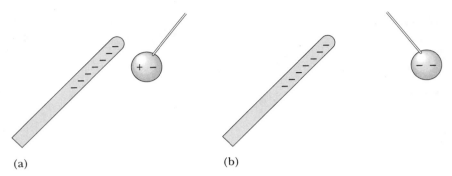

(a)   (b)

Figure 2.1   **(a)** Charged rod, uncharged pith ball. **(b)** Charged rod and pith ball.

The electrostatic force may be calculated using **Coulomb's law,** published in 1785. In SI units (see Appendix D on Units):

$$F = \frac{qQ}{4\pi\epsilon_0 r^2}$$ (2.1)

See Figure 2.2. F is force in newtons (N), Q and q are electrical charges in coulombs (C), r is their separation in meters, and $\epsilon_0 = 8.85 \times 10^{-12}$ C^2/N · m^2, a constant given inside the back cover, called "permittivity of free space" for reasons that will become clear later on (Section 9.2). It is an inverse-square law, like Newton's law of gravitation, which came a century earlier.

Limitations on Coulomb's law: Q and q are "point" charges, that is, small in size compared with their separation. They are "quasi-static": They must be more or less stationary (velocity v $\ll$ c, where c is the speed of light), and not strongly acceler-ated (acceleration $a \ll c^2/r$); we will encounter the more general case as we go along, and especially in Chapter 18, on Relativity.

(For Coulomb's law in this simple scalar form, a positive F implies repulsion; but it turns out that we will have little use for this rule.)

◥ EXAMPLE 2-1

Find the force F between two equal charges, $q = Q = 1$ C, separated by $r = 1$m.

ANSWER For convenience, we can set $1/4\pi\epsilon_0 \approx 9 \times 10^9$ N · m^2/C^2. Plugging these numbers into Coulomb's law, we find:

$$F = \frac{qQ}{4\pi\epsilon_0 r^2}$$

$$= 9 \times 10^9 \text{ N} \cdot \text{m}^2/\text{C}^2 \frac{1\text{C} \times 1\text{ C}}{(1\text{m})^2} = 9 \times 10^9 \text{ N} \qquad \text{repulsion}$$

This is an astonishing force! A kilogram weighs $mg = 9.8$ N (we are using mass $m = 1$ kg and gravitational acceleration $g = 9.8$ m/s^2), so the force represents the weight of about 10^9 kg, or a million metric tons. Remember that a car weighs only

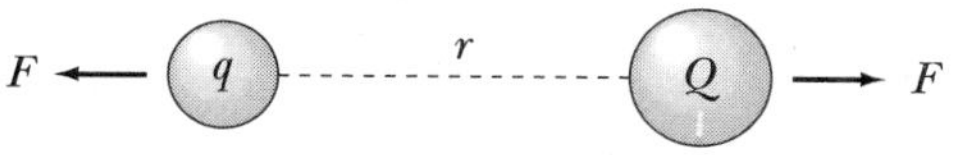

$$F \longleftarrow \, q \,\cdots\cdots \overset{r}{\cdots\cdots}\, Q \, \longrightarrow F$$

Figure 2.2   Coulomb's law.

a ton or two. One might at first suspect that somehow the coulomb happens to be a very large unit, but that is not the best explanation: The mass of a piece of copper containing one coulomb of electrons is only

$$1\,\text{C} \times \frac{1\ \text{electron}}{1.6 \times 10^{-19}\text{C}} \times \frac{1\ \text{atom}}{29\ \text{electrons}} \times \frac{1\ \text{mole}}{6 \times 10^{23}\ \text{atoms}} \times \frac{63.5\ \text{gram}}{1\ \text{mole}}$$

$$= 2.3 \times 10^{-5}\ \text{gram}$$

which is a microscopic speck; it would take 100,000 of them to make a penny. So, we have plenty of coulombs around.

Notice, by the way, the form of the preceding calculation, in which we have arranged the quantities so that it is obvious that the units cancel out. This represents a useful approach to derivations of this type, and you can be guided toward the correct result simply by observing the units. We have used Avogadro's number and the charge on an electron, from the table inside the back cover, and the atomic weight and number of copper, from a periodic table.

To get a better feel for the relatively enormous size of the electrostatic force, let's try an analogous gravitational calculation. The gravitational force between two 1-kilogram masses separated by 1 meter is:

$$F = G\frac{mM}{r^2}$$

$$= 6.7 \times 10^{-11}\,\text{N} \cdot \text{m}^2/\text{kg}^2\,\frac{1\ \text{kg} \times 1\ \text{kg}}{(1\ \text{m})^2} = 6.7 \times 10^{-11}\,\text{N} \tag{2.2}$$

which is smaller by a factor of about 10^{20}.

Why, then, is our life dominated by gravity rather than by electrostatic forces? Well, in the first place, on a microscopic scale, our life *is* dependent on electrostatic forces; they are what hold our molecules together, and gravity is usually negligible for interactions between small particles.

In the second place, there exist both positive and negative charges, and on a large scale they tend to be almost equal in number, so the electrostatic forces nearly cancel out. Whereas there are no negative gravitational masses, so the gravitational forces, however weak, just keep adding up. Because of this, gravity appears much more important to creatures the size of human beings.

Now that we have Coulomb's law, all electrostatic problems are in principle solved. If we have more than two charges, we can just add the forces as vectors. In vector form, Coulomb's law becomes

$$\boxed{\mathbf{F} = \frac{qQ\hat{\mathbf{r}}}{4\pi\epsilon_0 r^2}} \tag{2.3}$$

where now **F** is a **vector, in boldface,** having both direction and magnitude; and $\hat{\mathbf{r}}$ is a dimensionless **unit vector** pointing in the direction of the separation vector **r**:

$$\hat{\mathbf{r}} = \frac{\mathbf{r}}{r}$$

(For example, if both charges are positive, the force on Q is directed away from q.)

In **superposition** we trust: usually forces, fields, and so forth, due to different sources simply add up as vectors. (However, we should warn you here that there are important exceptions, in particular in the interaction of fields with materials. In Sections 9.3 and 10.8 we describe materials that "saturate" at high field magnitude, so that addition of sources does not produce a proportionate increase of field.) The following example will illustrate the use of these concepts.

 EXAMPLE 2-2

Suppose there are three charges located as follows, Figure 2.3a: 1 μC (microcoulomb, 10^{-6}C) located at the origin; -2 μC at 4 cm, 0 (on the x axis); and 3 μC at 0, 5 cm (on the y axis). Find the force on the 3 μC charge.

ANSWER The angle shown is arctan $(5/4) = 51.3°$, and the hypotenuse is $\sqrt{4^2 + 5^2} = 6.403$ cm. When we apply Coulomb's law between the 3 μC charge and the 1 μC charge, we find (remember to convert to coulombs and meters):

$$F_{1,3} = \frac{qQ}{4\pi\epsilon_0 r^2}$$

$$= 9 \times 10^9 \frac{3 \times 10^{-6} \times 1 \times 10^{-6}}{0.05^2} = 10.80 \text{ N}$$

in the $+y$ direction, Figure 2.3b; and for the -2 μC charge, we obtain 13.17 N directed down at 51.3° from the 3 μC.

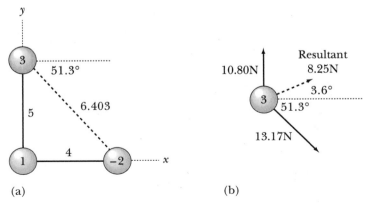

(a) (b)

Figure 2.3 **(a)** Example: charges. **(b)** Example: vectors.

Adding these vectors by components:

x components		y components	
10.80 cos 90°	= 0	10.80 sin 90°	= 10.80
13.17 cos − 51.3°	= 8.23	13.17 sin − 51.3°	= − 10.28
	8.23 = F_x		0.52 = F_y

Then we find the resultant vector **F** having the components F_x and F_y:

$$F = \sqrt{F_x^2 + F_y^2} \qquad\qquad \theta = \arctan\left(\frac{F_y}{F_x}\right)$$

$$= \sqrt{8.23^2 + 0.52^2} = 8.25 \text{ N} \qquad\qquad = \arctan\left(\frac{0.52}{8.23}\right) = 3.6°$$

in magnitude up from the + x axis

2.2.   Field E and Charge Density ρ

Yes, our electrostatic problems are solved, in principle. However, we have a number of reasons to introduce an electric field as an intermediary in electrostatics.

(a) Newton himself remarked in connection with his law of gravitation: "That one body may act upon another at a distance through a vacuum, without the mediation of any thing else, . . . is to me so great an absurdity, that I believe no man, who has in philosophical matters a competent faculty of thinking, can ever fall into it."

(b) Coulomb's law suggests an instantaneous interaction, but it would not be reasonable for the interaction between two bodies to be instantaneous, and in fact it would be contrary to the theory of relativity (Chapter 18). Indeed, it is found experimentally that when one body is moved, the effect cannot be felt on another body until, at the earliest, a later "retarded" time that is the separation of the bodies divided by c, the speed of light (Section 14.4).

(c) Further developments, in later chapters, will suggest that the proposed electric field has considerable objective reality, and in particular that it possesses mass and energy. For example, it is by means of electric and magnetic fields that the sun's energy reaches us.

(d) Even as a pure mathematical fiction the electric field would be useful, because it enables us to break what might be a complicated problem into two easier problems. If we wish to determine the force between two charge distributions, we can first find the field due to one, and then find the effect of that field on the other.

So we now introduce the **electric field E.** For two charges, as above, we have a force of

$$F = \frac{qQ}{4\pi\epsilon_0 r^2}$$

Figure 2.4 Electric field.

Dividing by q, we obtain the defining equation for the magnitude of the **E** field:

$$\frac{F}{q} = E \tag{2.4}$$

(as mentioned before, q is stationary) and the Coulomb's-law field for a single stationary charge Q:

$$E = \frac{Q}{4\pi\epsilon_0 r^2}$$

This is shown in vector form in Equation 2.5 and Figure 2.4:

$$\boxed{\mathbf{E} = \frac{Q\hat{\mathbf{r}}}{4\pi\epsilon_0 r^2}} \tag{2.5}$$

The **E** field is a vector that points from + to −, and the force on a + charge is in the direction of the field. Its units are newtons per coulomb (N/C).

At 1 meter from a charge of 1 coulomb, the field would be 9×10^9 N/C. Ordinary materials break down in such a large field (air sparks at about a million N/C). So, it is not practical to confine a whole coulomb of excess + or − charge in a volume the size of a cubic meter. The electric field near the earth's surface is around 100 N/C, directed downward; the necessary negative charge is maintained by atmospheric phenomena such as thunderstorms.

 EXAMPLE 2-3

In the three-charge problem shown in Example 2-2 and Figure 2.3, find the field at the 3 μC charge.

ANSWER

$$E = \frac{F}{q}$$

$$= \frac{8.25 \text{ N}}{3 \times 10^{-6}\text{ C}} = 2.75 \times 10^6 \text{ N/C}$$

at 3.6° above the $+x$ axis.

 EXAMPLE 2-4

Now replace the 3 μC charge with a charge of -4 μC and find the force and the field there.

ANSWER The field is the same, regardless of the charge there:

$$E = 2.75 \times 10^6 \, \text{N/C}$$

Since we already know the field, the force is most easily calculated as

$$F = qE$$
$$= -4 \times 10^{-6}\text{C} \times 2.75 \times 10^6 \, \text{N/C} = -11\text{N}$$

The force points in the opposite direction now, at 3.6° below the $-x$ axis.

Instead of discrete point charges, we may face a continuous charge distribution. Then we can just add up the vector contributions of the various parts of the distribution; that is, we can integrate over the charge elements dQ. Suppose for example we want to know the field **E** at a field point of vector location **r**, Figure 2.5, caused by a continuous source distribution of charge density ρ given by $\rho(\mathbf{r}') = dQ/dv'$. Then Equation 2.5 must be generalized to the form:

$$\mathbf{E}(\mathbf{r}) = \int \frac{\rho(\mathbf{r}')\,\hat{\mathbf{r}}''}{4\pi\epsilon_0 r''^2}\,dv'$$

where $\mathbf{r}'' = \mathbf{r} - \mathbf{r}'$. It's a little ungainly in this form, and so it is usually abbreviated as

$$\mathbf{E} = \int \frac{\hat{\mathbf{r}}\,\rho\,dv}{4\pi\epsilon_0 r^2} \tag{2.6}$$

So we may have a volume charge density ρ, where ρ is the charge dQ per unit volume dv; or a surface charge density σ per area da; or a linear charge density λ per length dl; then the charge elements dQ become respectively:

Volume charge density	Surface charge density	Linear charge density
$dQ = \rho\,dv$	$dQ = \sigma\,da$	$dQ = \lambda\,dl$

Correspondingly we can write Equation 2.6 as

$$\mathbf{E} = \int \frac{\hat{\mathbf{r}}\,\rho\,dv}{4\pi\epsilon_0 r^2} \quad \text{or} \quad \mathbf{E} = \int \frac{\hat{\mathbf{r}}\,\sigma\,da}{4\pi\epsilon_0 r^2} \quad \text{or} \quad \mathbf{E} = \int \frac{\hat{\mathbf{r}}\,\lambda\,dl}{4\pi\epsilon_0 r^2} \tag{2.7}$$

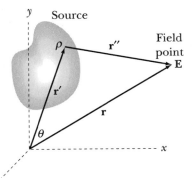

Figure 2.5   Charge distribution.

It's possible that the charge distribution may include the origin, where $r = 0$; but that doesn't give any trouble as long as ρ is well behaved (integrable) because the amount of charge that is precisely at the origin is precisely zero, and any limiting procedure you dream up will lead to a well-defined field.

The integral for **E** is more complicated than it may look: It is a vector relationship, and you have to integrate each vector component separately. This is shown in the next example.

 EXAMPLE 2-5

A line charge of constant linear charge density λ coulombs per meter extends along the positive y axis from 0 to L. Find the field at position a along the positive x axis.

ANSWER See Figure 2.6a. In this case, the charge element is $dQ = \lambda \, dy$. For the x component, we multiply by $\cos \theta$:

$$E_x = \int_0^L \frac{\cos \theta \, \lambda \, dy}{4\pi\epsilon_0 r^2}$$

We have three variables: θ, y, and r; so we must get rid of two. We shall keep the θ variable. From Figure 2.6a,

$$\tan \theta = y/a, \text{ so } dy = a \sec^2 \theta \, d\theta \qquad \text{and} \qquad \cos \theta = a/r, \text{ so } r = a/\cos \theta = a \sec \theta$$

This substitution yields a trivial integral. For the x component:

$$E_x = \frac{\lambda}{4\pi\epsilon_0} \int_0^{y=L} \frac{\cos \theta \, a \sec^2\theta \, d\theta}{a^2 \sec^2\theta}$$

$$= \frac{\lambda}{4\pi\epsilon_0} \int_0^{y=L} \frac{\cos \theta \, d\theta}{a} = \frac{\lambda}{4\pi\epsilon_0 a} \sin \theta \Big]_0^{y=L}$$

$$E_x = \frac{\lambda}{4\pi\epsilon_0 a} \left(\frac{L}{\sqrt{a^2 + L^2}} - 0 \right) \qquad \text{(Fig. 2.6b)} \qquad \qquad \textbf{(2.8)}$$

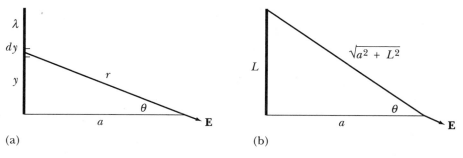

Figure 2.6 (a) **E** field of finite line charge. (b) Upper limit of $\sin \theta$ for $y = L$.

For the y component, we use $-\sin\theta$ instead of $\cos\theta$; then a similar calculation yields

$$E_y = \frac{\lambda}{4\pi\epsilon_0} \int_0^{y=L} \frac{-\sin\theta\,d\theta}{a} = \frac{\lambda}{4\pi\epsilon_0 a} \cos\theta \Big]_0^{y=L}$$

$$E_y = \frac{\lambda}{4\pi\epsilon_0 a}\left(\frac{a}{\sqrt{a^2+L^2}} - 1\right) \qquad \text{(Fig. 2.6b)} \qquad \textbf{(2.9)}$$

Magnitude and direction of the vector **E** are given by

$$E = \sqrt{E_x^2 + E_y^2} \qquad \text{and} \qquad \theta = \arctan\frac{E_y}{E_x}$$

Are our results correct? We can check by looking at the case in which $a \gg L$: Then the field should approach that of a point charge. We find for the x component:

$$E_x \rightarrow \frac{\lambda}{4\pi\epsilon_0 a}\frac{L}{\sqrt{a^2}} = \frac{\lambda L}{4\pi\epsilon_0 a^2} = \frac{Q}{4\pi\epsilon_0 a^2}$$

which is indeed the field of a point charge; and for the y component:

$$E_y = \frac{\lambda}{4\pi\epsilon_0 a}\left(\frac{a}{\sqrt{a^2+L^2}} - 1\right) = \frac{\lambda}{4\pi\epsilon_0 a}\left(\frac{1}{\sqrt{1+(L/a)^2}} - 1\right)$$

$$\approx \frac{\lambda}{4\pi\epsilon_0 a}\left(\frac{1}{1 + \frac{1}{2}(L/a)^2} - 1\right) \approx \frac{\lambda}{4\pi\epsilon_0 a}(1 - \tfrac{1}{2}(L/a)^2 - 1) = -\frac{L}{2a}E_x$$

which is negligible compared to E_x for $a \gg L$. And correspondingly, the angle θ goes to zero. So, that gives us one check on the validity of our result.

Another check is this: For $a \ll L$, the field should approach that for a semi-infinite wire. We shall treat that case as another example.

 EXAMPLE 2-6

A semi-infinite line charge of linear charge density λ extends along the positive y axis. Find the field at position a along the positive x axis.

ANSWER See Figure 2.7. We proceed just as in the preceding example, except that instead of the limit $y \rightarrow \infty$ it is easier to use $\theta = \pi/2$. We arrive at

$$E_x = \frac{\lambda}{4\pi\epsilon_0 a}\sin\theta \Big]_0^{\pi/2} = \frac{\lambda}{4\pi\epsilon_0 a} \qquad \textbf{(2.10)}$$

and

$$E_y = \frac{\lambda}{4\pi\epsilon_0 a}\cos\theta \Big]_0^{\pi/2} = -\frac{\lambda}{4\pi\epsilon_0 a} \qquad \textbf{(2.11)}$$

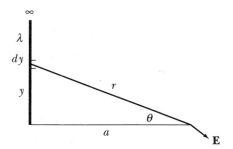

Figure 2.7   E field of semi-infinite line charge.

The magnitude of the field is then

$$E = \sqrt{E_x{}^2 + E_y{}^2} = \frac{\lambda\sqrt{2}}{4\pi\epsilon_0 a}$$

and the direction is given by

$$\theta = \arctan\frac{E_y}{E_x} \Rightarrow 45° \qquad \text{below the positive } x \text{ axis}$$

Now: Does this indeed represent the limiting case for the preceding example, for $a \ll L$, as we suggested above? The x component of that example is (Eq. 2.8)

$$E_x = \frac{\lambda}{4\pi\epsilon_0 a}\frac{L}{\sqrt{a^2 + L^2}} \to \frac{\lambda}{4\pi\epsilon_0 a}$$

and the y component is (Eq. 2.9)

$$E_y = \frac{\lambda}{4\pi\epsilon_0 a}\left(\frac{a}{\sqrt{a^2 + L^2}} - 1\right) \to \frac{\lambda}{4\pi\epsilon_0 a}(0 - 1)$$

Those are the correct components, so it all looks okay.

◀ EXAMPLE 2-7

Now repeat the calculation for a line charge extending from $-\infty$ to ∞, Figure 2.8.

ANSWER The integrations are formally the same as above. But this time the y components cancel out, by symmetry, for the line charge sections of positive and negative y, so $E_y = 0$; and the corresponding x components are equal, so we get twice the previous x component. The result is

$$E_x = \frac{\lambda}{4\pi\epsilon_0 a}\sin\theta\Big]_{-\pi/2}^{\pi/2} = \frac{\lambda}{2\pi\,\epsilon_0 a}$$

so

$$E = \frac{\lambda}{2\pi\epsilon_0 a} \qquad\qquad (2.12)$$

along the positive x axis.

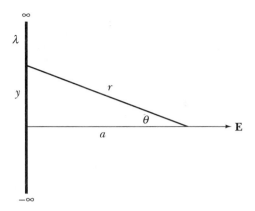

Figure 2.8   **E** field of infinite line charge.

Now we'll show a famous geometrical construction that enables us to find the field inside a spherical shell with very little calculation. (After we learn Gauss's law, in the next section, we'll do it even more easily.)

 **EXAMPLE 2-8**

Find *E* inside a spherical shell of radius *R* and uniform surface charge σ.

ANSWER   Direct integration of the Coulomb's-law field is left for Problem 2-21. The field at the arbitrary inside point P, Figure 2.9, is zero because of cancellation of the fields of charges on opposite sides, such as Q_1 and Q_2, as we shall now show. We draw a narrow cone through P. The two parts of the cone are similar: The opening angles are the same, and the angles where the cone hits the sphere are the same. So the areas of the cone–sphere intersections are proportional to the squares of the distances:

$$\frac{(\text{area})_1}{r_1^2} = \frac{(\text{area})_2}{r_2^2}$$

The charges are also proportional to the areas:

$$\frac{(\text{area})_1}{Q_1} = \frac{(\text{area})_2}{Q_2}$$

So the charges are proportional to the squares of the distances:

$$\frac{Q_1}{r_1^2} = \frac{Q_2}{r_2^2} \implies \frac{Q_1}{4\pi\epsilon_0 r_1^2} = \frac{Q_2}{4\pi\epsilon_0 r_2^2}$$

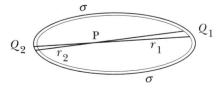

Figure 2.9   **E** inside spherical shell.

This implies that the fields due to these two charges are equal and opposite. The entire spherical surface can be mapped in this way: For any charge Q_1 on one side, there is a corresponding Q_2 on the other side that cancels its effect. Consequently the total field is zero:

$$E = 0$$

everywhere inside the spherical shell. (Similarly, the gravitational field is zero inside a uniformly massive spherical shell.)

2.3.   Electric Field Lines—Gauss's Law in Integral Form

We are now describing the electrostatic interaction in terms of an electric field **E** that fills all space. This is a vector field, which means, in effect, there are three numbers associated with every point in space. That may be easy to *say,* but it's not easy to *see,* especially since the field itself is invisible. We can show representative **E** vectors as in Figure 2.10a for a single charge; however, it's still unsatisfactory. In the first place, the dynamic range is limited: We can't draw the arrows close to the charge because they would be too long. More important, a good model makes significant concepts stand out rather than hiding them; but the vector picture gives little indication of the kind of underlying continuity of the electric field that we are about to study in this section. A better way is to show the corresponding **electric field lines,** or lines of flux, Figure 2.10b. (They were once called "lines of force," but that is a misnomer; a field is not a force.)

Looking at the lines for a single charge, we observe that the field points away from a positive charge, and so do the field lines; also, the field falls off inversely as the square of the distance, and so in principle does the density of the field lines. It is easy to see that the field lines don't disappear into thin air; they begin on posi-

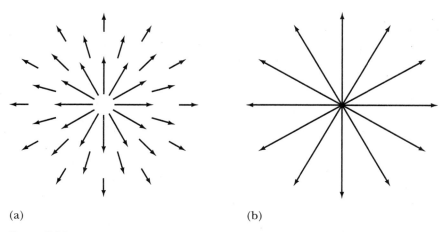

(a)   (b)

Figure 2.10   **(a)** Single charge, **E** vectors. **(b)** Single charge, field lines.

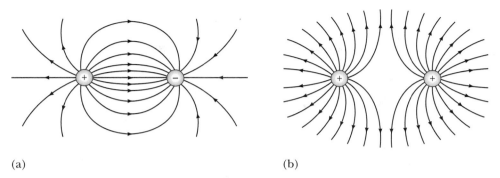

Figure 2.11   **(a)** Opposite charges. **(b)** Equal charges.

tive charges and terminate on negative ones, or else they go all the way to infinity. The idea goes back to Faraday, who imagined that they are like rubber bands that pull in the direction of their length and that somehow repel each other sideways. More field lines are shown in Figure 2.11.

There are two **caveats:** First, these field lines are lacking in objective reality, and we will find, when we learn about relativity, that field lines which are useful in one frame of reference can simply disappear altogether in another frame. Secondly, the density of field lines shown in the simple-charge figure actually falls off inversely as the first power of the distance, rather than the square, because the picture is two-dimensional whereas the field fills three dimensions. So, the field lines can faithfully represent field direction, but their two-dimensional representation on the printed page may give only qualitative information regarding their actual three-dimensional density. (See Appendix C for calculation and plotting of field lines.)

Ordinary charged bodies are usually complicated structures, with all kinds of field lines running in and out; however, at large distance the irregularities get smoothed out, as illustrated in Figure 2.12, and you are left with a simple charge that is just the net total of the positive and negative charges present.

There is no actual "number of field lines" coming out of a charge, Figure 2.13a, but the **electric flux** is related to the field lines and is readily defined quantitatively. The flux through an area is the integral of the (normal) field over the area,

$$\text{flux} = \int \mathbf{E} \cdot d\mathbf{a} \qquad (2.13)$$

Thus, flux is a scalar. The field **E** is the flux density, meaning the flux per unit area. We count only the field perpendicular to the surface, because to the extent that the field is parallel to the surface, it doesn't pass through it. The direction of the area element *d***a** is taken to be the outward-pointing normal to the surface if it is closed, as shown in Figure 2.13b. This means we are in effect finding the component of **E** that is parallel to *d***a**. The reason we take *d***a** to be normal (perpendicular to the surface) is that the normal has a well-defined direction, whereas a line tan-

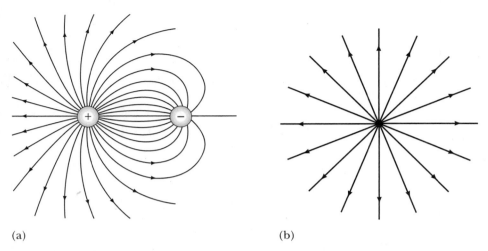

(a)   (b)

Figure 2.12   (a) Unequal opposite charges. Some field lines stop at the other charge and others continue to infinity. **(b)** Same as Figure 2.12a, from far away. It looks like a single charge equal to the sum of the two.

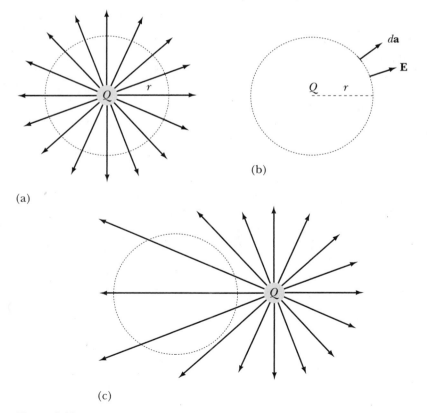

(a)

(b)

(c)

Figure 2.13   (a) Field lines. **(b)** Flux. **(c)** For charge outside, flux in cancels flux out.

gent to the surface could point in many different directions. And here is a reminder about the scalar **dot product** of two vectors:

$$\mathbf{A} \cdot \mathbf{B} = |\mathbf{A}|\,|\mathbf{B}| \cos \theta = A_x B_x + A_y B_y + A_z B_z$$

where $|\mathbf{A}|$ means the magnitude of the vector $\mathbf{A}$, always positive, and θ is the angle between the vectors.

The total flux is easily calculated for a sphere of radius r with a charge Q at the center, Figure 2.13. E is constant over the surface of the sphere, so we can take it outside of the integral sign. Also, $\mathbf{E}$ is perpendicular to the surface, so it is parallel to $d\mathbf{a}$; so $\cos \theta = 1$. The area of the sphere is $4\pi r^2$. So the flux is

$$\text{flux} = 4\pi r^2 E$$

$$= \frac{4\pi r^2 Q}{4\pi \epsilon_0 r^2}$$

So finally

$$\text{flux} = \frac{Q}{\epsilon_0} \tag{2.14}$$

which may also be written as

$$\boxed{\oint \mathbf{E} \cdot d\mathbf{a} = \int \frac{\rho}{\epsilon_0}\, dv} \tag{2.15}$$

where the integral on the left covers the entire (closed) surface of the sphere, and v is its volume. This is a general result and it is called **Gauss's law in integral form.** It doesn't depend on the radius of the sphere: As the radius gets bigger, the field E gets smaller, but the area gets bigger in the same proportion. If you have a surface other than a sphere, and $\mathbf{E}$ is not perpendicular to the surface, then the $\cos \theta$ factor cancels the increased area; see below. What this means is, the electric flux begins and ends on electric charge (or it keeps going to infinity); it doesn't just appear or disappear.

And if there are several charges inside the surface, or a charge distribution, we can just add up the contributions, invoking superposition. The net flux out (positive) will be the net charge divided by ϵ_0; if it turns out negative, then the field points inwards. It is remarkable, by the way, that charges outside the surface are simply ignored: as Figure 2.13c suggests, the flux in cancels the flux out for such charges. (Remember, flux is scalar, but it can be positive or negative.)

We can use Gauss's law more generally using the concept of solid angle Ω. To remind you about Ω: An ordinary angle element $d\theta$ is given by $\hat{\boldsymbol{\theta}} \cdot d\mathbf{l}/r$ in radians, Figure 2.14a, where l is length; and correspondingly, a solid angle element $d\Omega$ is given in steradians by $\hat{\mathbf{r}} \cdot d\mathbf{a}/r^2$, Figure 2.14b. The circumference of a circle is $2\pi r$, and so there are $2\pi r/r = 2\pi$ radians in a circle; correspondingly, the area of a sphere is $4\pi r^2$, and so there are $4\pi r^2/r^2 = 4\pi$ steradians in a sphere. As we have mentioned, $d\mathbf{a}$ points outward.

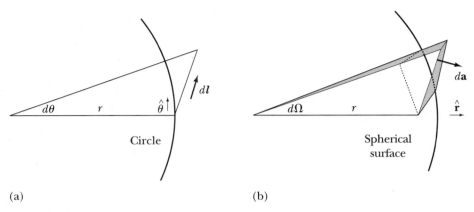

Figure 2.14   **(a)** Angle. **(b)** Solid angle.

Another approach is to set $d\Omega = d\theta \sin \theta\, d\phi$. Then integration over all angles gives

$$\int d\Omega = \int_0^{2\pi} \int_0^{\pi} \sin \theta\, d\theta\, d\phi = 4\pi$$

Now we begin with E in vector form:

$$\mathbf{E} = \frac{Q\,\hat{\mathbf{r}}}{4\pi\epsilon_0 r^2}$$

Multiply both sides by $d\mathbf{a}$ (a dot product) and integrate over an arbitrary closed surface S containing Q, shown in Figure 2.15:

$$\text{Electric flux} = \int \mathbf{E} \cdot d\mathbf{a} = \int \frac{Q\,\hat{\mathbf{r}} \cdot d\mathbf{a}}{4\pi\epsilon_0 r^2}$$

$$= \frac{Q}{4\pi\epsilon_0} \int \frac{\hat{\mathbf{r}} \cdot d\mathbf{a}}{r^2}$$

$$= \frac{Q}{4\pi\epsilon_0} \int d\Omega$$

So, as before (Eq. 2.14),

$$\text{flux} = \frac{Q}{\epsilon_0} \tag{2.16}$$

since the total solid angle is 4π.

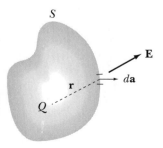

Figure 2.15   Gauss's law.

Note that the inverse-square $(1/r^2)$ character of the electric field was essential in carrying through this proof. For a simple charge, this implies that only the inverse-square field can be represented with continuous field lines.

It is surprising how much Gauss's simple formula, based on an equally simple concept, helps in calculating some fields. The reason is that we can draw the arbitrary surface S anywhere we want, so we can take advantage of the geometry. For example, go back to the infinite line charge in the preceding section, Figure 2.8, where we found $E = \lambda/2\pi\epsilon_0 a$. We had to carry through an integration there. We will now find the same result much more easily using Gauss's law. (We will write it as $E = \lambda/2\pi\epsilon_0 r$, with r instead of a, because we are using a for area here.)

We place a **Gaussian surface** around the line, in this case a cylinder of radius r and length y, shown in Figure 2.16. (The field lines shown in Figure 2.16a actually come out three-dimensionally, but it's hard to show that here, on 2-dimensional paper.) E is constant on the curved surface of Figure 2.16b, by symmetry. The area of the curved surface is $2\pi r y$, and all of the lines of flux that come out of length y must pass through that surface. The field is obviously perpendicular to the wire (symmetry again), so no field lines pass through the flat ends of the Gaussian tin can. The total charge inside the surface is $Q = \lambda y$.

Starting with Gauss's law,

$$\text{flux} = \frac{Q}{\epsilon_0}$$

$$\int \mathbf{E} \cdot d\mathbf{a} = \frac{\lambda y}{\epsilon_0}$$

$$E \, 2\pi r y = \frac{\lambda y}{\epsilon_0}$$

so

$$E = \frac{\lambda}{2\pi r \epsilon_0} \tag{2.17}$$

as before. But this time we didn't have to integrate; the symmetry of the problem took care of that for us.

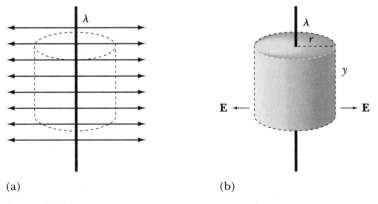

(a)   (b)

Figure 2.16   (a) Field lines of line charge. (b) $\mathbf{E}$ field of line charge.

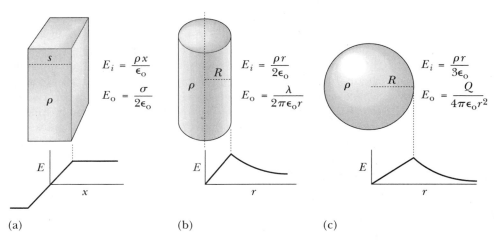

Figure 2.17   **(a)** Infinite plane (slab). **(b)** Infinite cylinder. **(c)** Sphere.

However, you must recognize the limitations as well as the power of Gauss's law. It's always true, but it's not always helpful. For an infinite line charge the calculation of the field E is trivial; but for the semi-infinite line, as in Figure 2.7, from the origin to infinity, Gauss's law is of no assistance, because the necessary symmetry is missing: The field is *not* perpendicular to the curved surface, and field lines *do* pass through the flat ends of the tin can. In this case you either carry through the integration or accept an approximation.

The symmetrical geometries for which Gauss's law is useful, along with the corresponding **E** fields, are shown in Figure 2.17. E_i means inside, E_o is outside, and all have uniform ρ. For the sphere, the appropriate Gaussian surface is a sphere; for the cylinder, it is a cylinder; and for the plane, a cylinder will do.

The following example involves a volume charge distribution.

EXAMPLE 2-9

Find **E**, inside and outside, at distance x from the center of an infinite plane of thickness s and uniform charge density ρ, Figures 2.17a and 2.18.

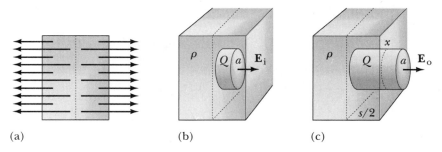

Figure 2.18   **(a)** Field lines. **(b)** **E** inside. **(c)** **E** outside.

ANSWER We employ a cylindrical Gaussian surface of area a and length x. (It needn't be round; we just need the area a.) By symmetry, $E = 0$ at $x = 0$. Only the end on the right, of area a, contributes to the electric flux, because, for the end on the left, $\mathbf{E} = 0$; for the sides, $\mathbf{E}$ is parallel to the sides and thus perpendicular to the area element $d\mathbf{a}$. (This is unlike the cylinder for the line charge in Figure 2.16, in which the flux comes out the curved sides and not the ends.)

Inside, the charge density is ρ inside the entire Gaussian surface of volume $v = ax$, Figure 2.18b, so $Q = \rho\, ax$. Then,

$$\text{Flux} = E_i a = \rho a x / \epsilon_0$$

$$E_i = \rho\, x / \epsilon_0 \tag{2.18}$$

Outside, the charge density is ρ only up to $x = s/2$, as in Figure 2.18c, so $Q = \rho\, as/2$. Then,

$$\text{Flux} = E_o a = \rho a(s/2)/\epsilon_0$$

$$E_o = \rho\, s/2\epsilon_0$$

and

$$E_o = \sigma/2\epsilon_0 \tag{2.19}$$

in terms of charge per unit area $\sigma = \rho s$.

So the field increases linearly inside the plane, as shown in Figure 2.17a, and is constant outside.

It is a consequence of Gauss's law that there is **no stable equilibrium** in electrostatics (Earnshaw's theorem, Section 12.1). Figure 2.19 provides a particular example of this general rule. It consists of eight identical positive charges at the corners of a cube. One might expect that a ninth positive charge might be trapped in the middle of the cube, at the point marked "x." However, if a positive charge were trapped there, it would imply that the electric field was pointing inward from all directions. Gauss's law would then imply presence of a negative charge at the center, but there is none. In fact, the positive charge has an escape route from the center of the cube through the center of any face: The field points outward (see Problem 2-35).

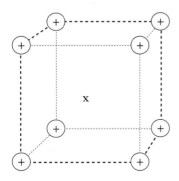

Figure 2.19 No stable configuration.

And if we were to place a negative charge near the center too, just to hold the positive charge, then the positive charge would accelerate toward the negative one, and the situation would not be static.

Then how does an ordinary atom retain its stability? Well, as for the negative electrons, classically they are accelerating toward the positive nucleus, and that does not constitute static equilibrium; furthermore, if it were not for Bohr's postulate and/or quantum mechanics, the electrons would quickly radiate away their energy and combine somehow with the nucleus. As for the nucleus, the protons are indeed repelled electrostatically, but they are held together by the strong (nuclear) interaction, which is not subject to Gauss's law as it is not inverse-square. We will revisit this idea in connection with Laplace's equation, Section 12.1.

◤ EXAMPLE 2-10

Find the pressure (force per unit area) between two oppositely charged parallel planes of charge densities $\pm\sigma$.

ANSWER From the preceding example and Figure 2.20, the field at plane **b**, caused by plane **a**, is $E_a = \sigma/2\epsilon_0 = E/2$ (E being the total field due to both planes). That is the field which pulls on plane **b**.

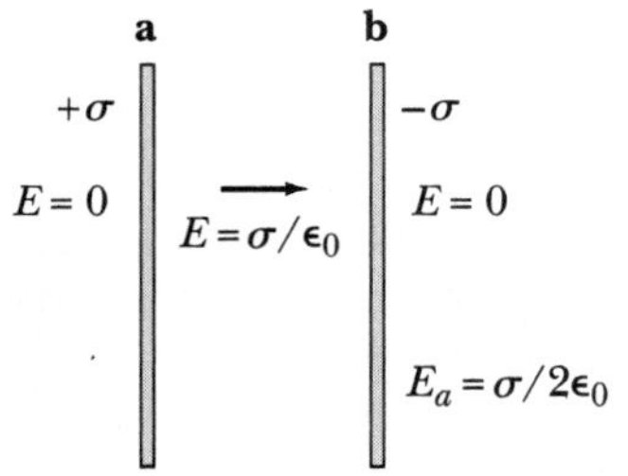

a   **b**

$+\sigma$ $-\sigma$

$E = 0$ $E = 0$

$\xrightarrow{}$
$E = \sigma/\epsilon_0$

$E_a = \sigma/2\epsilon_0$

Figure 2.20   Parallel planes.

This can be seen from another standpoint. If we consider **b** to be a slab of finite thickness and uniform charge density, then each small charge Δq of **b** responds to the E_b field of the rest of **b**, as well as to E_a. A Δq on the left edge of **b** experiences the field E, and one on the right edge sees zero field. The average field on the Δq elements of **b** is again $E/2$. The argument is independent of **b**'s thickness and so it holds in the limit of an infinitesimally thick surface charge distribution: The effective field on such a charge is the average of the fields on the two sides.

So anyway, for area a of plane **b**, the charge is $Q = \sigma a$, and the force is $F = QE_a = a\,\sigma^2/2\epsilon_0$, and so the pressure P is

$$P = \frac{F}{a} = \frac{\sigma^2}{2\epsilon_0}$$

$$= \frac{\epsilon_0 E^2}{2}$$

(We examine this pressure further in Sections 6.1 and 14.2.)

2.4.   Gauss's Law in Differential Form—Divergence of E, $\nabla \cdot$ E

Gauss's law in integral form is very good as far as it goes, but it is a little lacking in generality. A surface is implied, over which one integrates. However, in the first place, there seems something arbitrary about this surface. Also, what if the observer changes reference frames? The surface changes, or the result is no longer valid. It would be nice if there were a simple law of absolute generality (within the realm of classical electromagnetism), relativistically correct, correct in every conceivable way.

Needless to say, there is such a law.

We begin by examining Gauss's law microscopically. Figure 2.21 shows a small cube, in a region of charge density ρ, with an **E** field, too. We are going to find the difference between flux in and flux out of this cube; we'll call it Δ flux for short. By Gauss's law, that difference will be equal to Q/ϵ_0, where Q is the charge in the cube.

First we look at the difference in flux due to E_x, as shown. After we understand this, the y and z directions will be easy. The net flux involving E_x is

$$E_x(x + \Delta x, y, z)\, \Delta y\, \Delta z - E_x(x, y, z)\, \Delta y\, \Delta z$$

To simplify the notation, we have placed the point (x, y, z) in the center of the left-hand face of the cube, and we assume that the component $E_x(x, y, z)$ represents a suitable average over that face. (We do this because y and z are not going to change in a significant way in this first calculation.) When we multiply and divide by Δx, the thing flips itself into a partial derivative form:

$$\Delta \text{ flux in } x \text{ direction} = \frac{E_x(x + \Delta x, y, z)}{\Delta x} \Delta x\, \Delta y\, \Delta z - \frac{E_x(x, y, z)}{\Delta x} \Delta x\, \Delta y\, \Delta z \qquad (2.20)$$

$$= \left(\frac{E_x(x + \Delta x, y, z) - E_x(x, y, z)}{\Delta x} \right) \Delta x\, \Delta y\, \Delta z$$

$$= \frac{\partial E_x}{\partial x} \Delta v \qquad (2.21)$$

in the limit as $\Delta x \to 0$, where the volume element is $\Delta v = \Delta x\, \Delta y\, \Delta z$.

For y and z the story is similar. We can just add them up:

$$\Delta \text{ flux} = \left(\frac{\partial E_x}{\partial x} + \frac{\partial E_y}{\partial y} + \frac{\partial E_z}{\partial z} \right) \Delta v$$

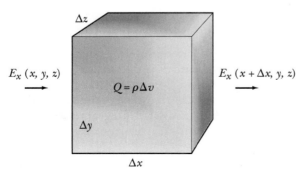

Δz

$E_x(x, y, z)$

$Q = \rho \Delta v$

$E_x(x + \Delta x, y, z)$

Δy

Δx

Figure 2.21   Small charged cube.

At this point it is convenient to define a vector differential operator, sometimes called **del,** to make all this easier to write (Section 1.2):

$$\nabla \equiv \frac{\partial}{\partial x} \hat{\mathbf{x}} + \frac{\partial}{\partial y} \hat{\mathbf{y}} + \frac{\partial}{\partial z} \hat{\mathbf{z}}$$

(2.22)

The operator is incomplete as it stands: It is waiting for something to operate on. Meanwhile, it does *not* operate on the included unit vectors, which are just constants. With this del, the above equation becomes

$$\Delta \text{ flux} = \nabla \cdot \mathbf{E} \, \Delta v$$

We treat del somewhat like a vector. Writing it out in detail,

$$\nabla \cdot \mathbf{E} = \left(\frac{\partial}{\partial x} \hat{\mathbf{x}} + \frac{\partial}{\partial y} \hat{\mathbf{y}} + \frac{\partial}{\partial y} \hat{\mathbf{z}} \right) \cdot (E_x \, \hat{\mathbf{x}} + E_y \, \hat{\mathbf{y}} + E_z \, \hat{\mathbf{z}})$$

When we multiply this out, three terms survive, just as for any dot product of two vectors:

$$\nabla \cdot \mathbf{E} = \frac{\partial E_x}{\partial x} \hat{\mathbf{x}} \cdot \hat{\mathbf{x}} + \frac{\partial E_y}{\partial y} \hat{\mathbf{y}} \cdot \hat{\mathbf{y}} + \frac{\partial E_z}{\partial z} \hat{\mathbf{z}} \cdot \hat{\mathbf{z}}$$

The decedents are the six cross terms, such as $\partial E_x / \partial y \, \hat{\mathbf{x}} \cdot \hat{\mathbf{y}}$, with $\hat{\mathbf{x}} \cdot \hat{\mathbf{y}} = 0$, because $\hat{\mathbf{x}}$ is perpendicular to $\hat{\mathbf{y}}$. The dot products such as $\hat{\mathbf{x}} \cdot \hat{\mathbf{x}}$ are 1, because the unit vector $\hat{\mathbf{x}}$ is parallel to itself; so what's left is

$$\nabla \cdot \mathbf{E} = \frac{\partial E_x}{\partial x} + \frac{\partial E_y}{\partial y} + \frac{\partial E_z}{\partial z}$$

(2.23)

which is what we wanted. Now, gathering it together: Gauss's law is (Eq. 2.14)

$$\Delta \text{ flux} = \frac{Q}{\epsilon_0} = \frac{\rho \, \Delta v}{\epsilon_0}$$

By analyzing the cube, above, we found

$$\Delta \text{ flux} = \nabla \cdot \mathbf{E} \, \Delta v$$

So, cancelling the Δv, we get

$$\nabla \cdot \mathbf{E} = \frac{\rho}{\epsilon_0}$$

(2.24)

This is **Gauss's law in differential form.** It is completely true and general, within the realm of classical electromagnetism, and it is one of Maxwell's four equations, all of which we will get to in due course. Compared with the integral form of Gauss's law, the differential form is true at every point and not just when integrated over some surface. We have derived it, using the integral form, for the particular case of Coulomb's law, but Coulomb is only for the static case. Later we shall simply postulate Gauss's law for the general case.

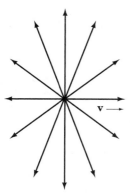

Figure 2.22   Charge at high speed.

It is also true that Coulomb's law contains information that Gauss's law does not. Consider for example the field of a rapidly moving (relativistic) charge, Figure 2.22. It obeys Gauss's law, since the field lines are continuous, but not Coulomb's law, because Coulomb's law implies a field that doesn't change with angle. (We will discover later, especially in Equation 3.32, that a Coulomb's-law field has zero curl; but Gauss's law imposes no such restriction because it only involves the divergence.)

When we integrate both sides of Gauss's law over an enclosed volume v, we arrive at the **divergence theorem** (which is implicit in the above derivation):

$$\int \nabla \cdot \mathbf{E} \, dv = \int \frac{\rho}{\epsilon_0} \, dv = \frac{Q}{\epsilon_0}$$

So,

$$\boxed{\int \nabla \cdot \mathbf{E} \, dv = \oint \mathbf{E} \cdot d\mathbf{a}} \qquad (2.25)$$

(This was once called Gauss's theorem, but that led to some confusion with Gauss's law.) Both sides of this equation represent electric flux. The expression $\nabla \cdot \mathbf{E}$ is called the **divergence of E** and is sometimes written div **E**. (See Section 1.2.) It is a scalar, like the dot product of two vectors $\mathbf{A} \cdot \mathbf{B}$. Consider a volume v bounded by a surface S, Figure 2.23. If there is positive charge Q inside v, then there is a divergence of **E**, $\nabla \cdot \mathbf{E}$, in the volume, and field lines are coming into exis-

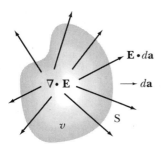

Figure 2.23   Divergence theorem.

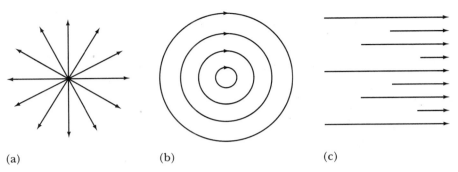

(a)   (b)   (c)

Figure 2.24   **(a)** Divergence at center point. **(b)** No divergence. **(c)** Divergence in volume.

tence there. These field lines find their way out through the surface S and produce a flux of **E**, **E** · **da**, over the surface area. As before, **da** points outward.

More briefly: **If there is a positive divergence, then field lines are being created;**
that is, there is a net positive flux through a surface enclosing the positive divergence.

We use the divergence theorem to pass directly from the differential to the integral form of Gauss's law:

$$\nabla \cdot \mathbf{E} = \frac{\rho}{\epsilon_0}$$

$$\int \nabla \cdot \mathbf{E} \, dv = \int \frac{\rho}{\epsilon_0} \, dv$$

$$\oint \mathbf{E} \cdot d\mathbf{a} = \frac{Q}{\epsilon_0}$$

Figure 2.24 illustrates fields with and without divergence. You should learn to
recognize the presence of a divergence by inspection. In Figure 2.24a there is a positive point charge at the center, so the divergence is large there and zero elsewhere—
the field lines just keep going out. In Figure 2.24b the field lines don't begin or end,
so there is no divergence. Figure 2.24c represents a volume distribution of charge
density ρ; field lines begin throughout the volume and proceed toward the right.

The "del" operator given above works only in Cartesian coordinates. In other
coordinate systems, there are corresponding vector relationships that are given inside the front cover; for derivations, see Section 1.7.

◤ EXAMPLE 2-11

What is the divergence of the field $\mathbf{E} = r\,\hat{\mathbf{r}}$?

ANSWER See Figure 2.25. In spherical coordinates, $\nabla \cdot \mathbf{E}$ is shown inside the front
cover:

$$\nabla \cdot \mathbf{E} = \frac{1}{r^2} \frac{\partial(r^2 E_r)}{\partial r} + \frac{1}{r \sin \theta} \frac{\partial E_\theta \sin \theta}{\partial \theta} + \frac{1}{r \sin \theta} \frac{\partial E_\phi}{\partial \phi}$$

$$= \frac{1}{r^2} \frac{\partial(r^2 E_r)}{\partial r} + 0 + 0$$

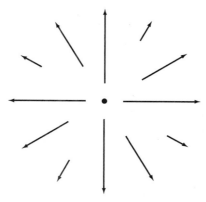

Figure 2.25   Divergence in volume.

there being no angular component or dependence. Since $E_r = r$, the result is

$$\nabla \cdot \mathbf{E} = \frac{1}{r^2} \frac{\partial(r^3)}{\partial r} = 3$$

We can also do it in rectangular coordinates:

$$r\hat{\mathbf{r}} = x\hat{\mathbf{x}} + y\hat{\mathbf{y}} + z\hat{\mathbf{z}}$$

Then,

$$\nabla \cdot \mathbf{E} = \frac{\partial x}{\partial x} + \frac{\partial y}{\partial y} + \frac{\partial z}{\partial z} = 1 + 1 + 1 = 3$$

as before.

 EXAMPLE 2-12

In the preceding example, what is the charge density?

ANSWER From Gauss's law,

$$\rho = \epsilon_0 \, \nabla \cdot \mathbf{E}$$
$$= 3 \, \epsilon_0$$

In this case, field lines are coming into being everywhere at a constant rate per unit volume.

 EXAMPLE 2-13

What is the divergence of $\mathbf{E} = 3 \, x\hat{\mathbf{x}}$ (see Fig. 2.24c)?

ANSWER

$$\nabla \cdot \mathbf{E} = \frac{\partial(3x)}{\partial x} + 0 + 0 = 3$$

So it too is 3, just like the preceding example with $\mathbf{E} = r\hat{\mathbf{r}}$, and the corresponding charge density is again $\rho = 3\,\epsilon_0$.

So when you have a uniform charge density everywhere, what field do you really get? Figure 2.24c shows one possibility, and Figure 2.25 shows another. In such a case you need additional information, such as how the field behaves at large distances. Simply specifying the charge density does not uniquely determine the field. (See Sections 1.4 and 12.1 on Uniqueness.)

The divergence is not affected by mere coordinate translations or additive constants. Obviously

$$\nabla \cdot (\mathbf{E} + \mathbf{E_o}) = \nabla \cdot \mathbf{E} \tag{2.26}$$

where $\mathbf{E_o}$ is a constant. And if we move the coordinate system by the constant x_0, the corresponding derivative could be written

$$\frac{\partial E_x}{\partial (x - x_0)} = \frac{\partial E_x}{\partial x}$$

For a point charge at the origin, Figure 2.26, what is the divergence of $\mathbf{E}$? We start with the Coulomb's law field and take the divergence:

$$\nabla \cdot \mathbf{E} = \nabla \cdot \frac{Q\hat{\mathbf{r}}}{4\pi\epsilon_0 r^2} = \frac{Q}{4\pi\epsilon_0} \nabla \cdot \frac{\hat{\mathbf{r}}}{r^2}$$

and in spherical coordinates (see inside front cover)

$$\nabla \cdot \frac{\hat{\mathbf{r}}}{r^2} = \frac{1}{r^2} \frac{\partial}{\partial r} \left(\frac{r^2}{r^2} \right)$$

Elsewhere than the origin, this gives $\dfrac{\partial}{\partial r} \dfrac{r^2}{r^2} = 0$, so the divergence is zero: Field lines simply continue on out into space. But at the origin it goes infinite: A lot of field lines come into existence all at once. To find $\nabla \cdot \dfrac{\hat{\mathbf{r}}}{r^2}$ at the origin we employ the divergence theorem. Our Gaussian surface is a sphere about the origin:

$$\int \nabla \cdot \frac{\hat{\mathbf{r}}}{r^2}\, dv = \oint \frac{\hat{\mathbf{r}}}{r^2} \cdot d\mathbf{a} = \frac{4\pi r^2}{r^2} = 4\pi \tag{2.27}$$

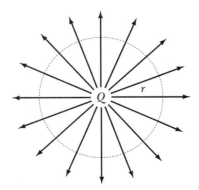

Figure 2.26   Divergence for point charge.

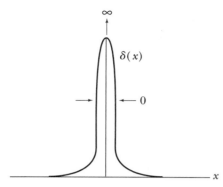

Figure 2.27   Delta function limit.

regardless of radius. So aside from the origin the divergence is zero, yet its integral is 4π.

This implies that $\nabla \cdot \dfrac{\hat{\mathbf{r}}}{r^2}$ is a bizarre kind of function that is zero everywhere except at the origin, and yet goes to infinity in such a way that its integral remains finite, namely, 4π. It may be regarded as the limit of a function such as

$$\delta(x) = \lim \frac{1}{\epsilon \sqrt{\pi}} \exp(-x^2/\epsilon^2) \qquad (2.28)$$
$$\text{as } \epsilon \to 0$$

shown in Figure 2.27; as it gets taller and narrower, its area remains constant. It is called a **Dirac delta function.** (There are other mathematical representations.) In one dimension, $\delta(x) = 0$ except at $x = 0$, and at the origin it goes to infinity in such a way that

$$\boxed{\int_{-\infty}^{\infty} \delta(x) \, dx = 1} \qquad (2.29)$$

◢ EXAMPLE 2-14

Evaluate $\displaystyle\int_{-\infty}^{\infty} (x + 7) \, \delta(x) \, dx$.

ANSWER

$$\int_{-\infty}^{\infty} (x + 7) \, \delta(x) \, dx = \int_{-\infty}^{\infty} (0 + 7) \, \delta(x) \, dx = 7 \int_{-\infty}^{\infty} \delta(x) \, dx = 7$$

since the integrand is significant only at $x = 0$.

The delta function in three dimensions is $\int \delta(\mathbf{r}) \, dv = 1$, or $\int \delta^3(r) \, dv = 1$; either way, it is the product of one-dimensional δ's, and its dimensions are those of volume^{-1}. This indicates that

$$\nabla \cdot \frac{\hat{\mathbf{r}}}{r^2} = 4\pi\delta(\mathbf{r}) \tag{2.30}$$

since if we integrate either side we get 4π:

$$\int \nabla \cdot \frac{\hat{\mathbf{r}}}{r^2} \, dv = \int 4\pi\delta(\mathbf{r}) \, dv$$

$$\oint \frac{\hat{\mathbf{r}}}{r^2} \cdot d\mathbf{a} = 4\pi \int \delta(\mathbf{r}) \, dv$$

$$\frac{4\pi r^2}{r^2} = 4\pi$$

Then to answer the original question: for a point charge Q at the origin, the charge density is

$$\rho = Q\delta(\mathbf{r}) \tag{2.31}$$

and so

$$\nabla \cdot \mathbf{E} = \nabla \cdot \frac{Q\hat{\mathbf{r}}}{4\pi\epsilon_0 r^2} = \frac{Q}{\epsilon_0}\delta(\mathbf{r}) \tag{2.32}$$

which is reasonable after you understand the meaning of divergence and of the delta function.

2.5. SUMMARY

This chapter begins with the interaction between stationary charges, and it introduces the electric field as an intermediary in that interaction. Its ultimate purpose is to provide experimental and conceptual support for the general and fundamental law of Gauss.

The force of repulsion between two stationary like charges is given by Coulomb's law:

$$F = \frac{qQ}{4\pi\epsilon_0 r^2} \tag{2.1}$$

or, in vector form,

$$\mathbf{F} = \frac{qQ\,\hat{\mathbf{r}}}{4\pi\epsilon_0 r^2} \tag{2.2}$$

The corresponding electric field is

$$\mathbf{E} = \frac{\mathbf{F}}{q} \tag{2.4}$$

The electric field may be visualized using lines of flux. Electric flux is

$$\text{flux} = \int \mathbf{E} \cdot d\mathbf{a} \tag{2.13}$$

Gauss's law in integral form facilitates calculation of fields of appropriate symmetry:

$$\oint \mathbf{E} \cdot d\mathbf{a} = \int \frac{\rho}{\epsilon_0} \, dv \qquad \text{or} \qquad \text{flux} = \frac{Q}{\epsilon_0} \tag{2.14, 2.15}$$

The differential vector operator del behaves somewhat like a vector:

$$\nabla = \frac{\partial}{\partial x} \hat{\mathbf{x}} + \frac{\partial}{\partial y} \hat{\mathbf{y}} + \frac{\partial}{\partial z} \hat{\mathbf{z}} \tag{2.22}$$

A positive divergence $\nabla \cdot \mathbf{E}$ indicates a source of electric flux.
Gauss's law in differential form is one of Maxwell's four equations:

$$\nabla \cdot \mathbf{E} = \frac{\rho}{\epsilon_0} \tag{2.24}$$

A point charge may be described in terms of the Dirac delta function:

$$\rho = Q \, \delta(\mathbf{r}) \tag{2.31}$$

 **PROBLEMS**

2-1 Charges of 7 nC and -3 nC are separated by 2 cm. Find the force. Is it attractive or repulsive? (n means nano, which is 10^{-9}.)

2-2 Identical charges separated by 1 cm experience a repulsion of 0.1 N. What is the magnitude of each charge?

2-3 If the field near the earth's surface is 100 N/C directed downward everywhere, what is the charge Q on the whole earth? (Idealize: The charge is at the center of the earth.) If the charge is actually on the earth's surface, what is the surface charge density σ?

2-4 If the field near the earth's surface is 100 N/C directed downward everywhere, and the gravitational field is 9.8 N/kg, also downward, find the acceleration of an electron there; is it up or down?

2-5 A charge of 3 nC is located at the origin, and 5 nC is located at 0.3 m along the x axis. Where is the electric field zero?

2-6 A charge of 3 nC is located at the origin, and -5 nC is located at 0.3 m along the x axis. Where is the electric field zero?

2-7 A charge of 4 nC is located at the origin, and -5 nC is at $x = 0.3$ m; find the field at $y = 0.2$ m on the y axis.

2-8 A charge of 4 nC is located at the origin, and 5 nC is at $x = 0.3$ m; find the field at $y = 0.2$ m on the y axis.

2-9 Identical charges of 1 μC are located at the corners of a square: At the origin, at $(0, 0.1)$, at $(0.1, 0)$, and at $(0.1, 0.1)$. Find the field at the center, $(0.05, 0.05)$.

2-10 For Problem 2-9, replace the charge at the origin with -1 μC and repeat.

2-11 Find the electric field magnitude at position z on the axis of a ring of radius a and of linear charge density λ, lying in the xy plane and centered on the z axis, Figure 2.28. Check it for $z \rightarrow 0$ and for $z \gg s$.

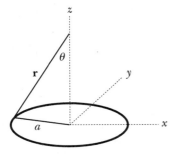

Figure 2.28

2-12 Find the electric field magnitude at position z on the axis of a square of side s and of linear charge density λ, lying in the xy plane and centered on the z axis. Check it for $z \rightarrow 0$ and for $z \gg a$.

2-13 Find **E** at 3 cm from a long line charge of 4 nC/m.

2-14 Find **E** at 3 cm from a large plane of surface charge 4 nC/m^2.

2-15 Starting with Gauss's law in differential form, find the field inside and outside of a sphere of radius R and constant charge density ρ; then re-express in terms of charge Q. (Use a spherical Gaussian surface, Fig. 2.29.)

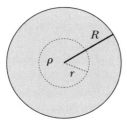

Figure 2.29

2-16 Find the field inside and outside of an infinite cylinder of radius R and uniform charge density ρ; then re-express in terms of linear charge density λ. (Use length l and a cylindrical Gaussian surface, Fig. 2.30.)

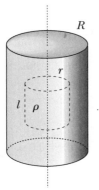

Figure 2.30

2-17 Find the field inside and outside of an infinite cylinder of radius R and charge density $\rho(r) = \sigma/r$, where σ is a constant; then, re-express in terms of linear charge density λ.

2-18 Find the field inside and outside of a sphere of radius R and charge density $\rho(r) = \sigma/r$, where σ is a constant; then, re-express in terms of total charge Q.

2-19 A hemispherical surface of radius R has surface charge σ. Find the field at the center (at distance R from the surface).

2-20 A thin disk of uniform surface charge density σ and radius a lies in the xy plane, centered on the z axis, Figure 2.28. Find the field at point z along the z axis. Check for $z \to 0$ and for $z \gg a$.

*2-21 Find E, inside and outside, at distance z from the center of a uniformly charged spherical shell of radius R and charge Q, by direct integration of Coulomb's law and by Gauss's law.

2-22 Find the field E_i between, and E_o outside of, two infinite parallel planes of uniform surface charge density σ, separated by distance s, Figure 2.31. For $E_i = 10^6$ N/C, find σ.

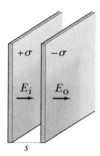

Figure 2.31

2-23 Find the field E_i between, and E_o outside of, two infinite coaxial cylinders, Figure 2.32 of inner and outer radii r_1 and r_2 and of linear charge density $+\lambda$ and $-\lambda$, respectively. For $E_i = 10^6$ N/C at $r = 1$ cm, find λ.

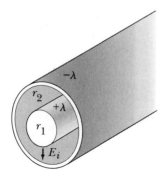

Figure 2.32

2-24 Find the field E_i between, and E_o outside of, two concentric spheres, Figure 2.33, of inner and outer radii r_1 and r_2 and of charge $+Q$ and $-Q$, respectively. Evaluate at $r = r_1 = 1$ cm for $Q = 2 \times 10^{-9}$C.

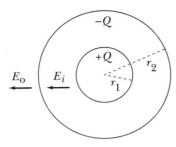

Figure 2.33

2-25 Superposition: An infinite cylinder of uniform charge density ρ and radius R contains a cylindrical cavity of radius $R/2$ as shown, Figure 2.34. Find the field at point p on the surface.

2-26 Superposition: A sphere of radius R and uniform charge density ρ contains a spherical cavity of radius $R/2$ as shown, Figure 2.34. Find the field at point p on the surface.

2-27 Superposition: A sphere of uniform charge density ρ and radius R contains a spherical cavity of radius $R/2$ as shown, Figure 2.34. Find the field in the cavity.

2-28 Superposition: An infinite cylinder of radius R and uniform charge density ρ contains a cylindrical cavity of radius $R/2$ as shown, Figure 2.34. Find the field in the cavity.

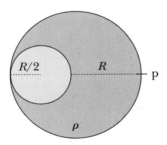

Figure 2.34

2-29 Find the divergence of the field $\mathbf{E} = x\,\hat{\mathbf{x}}$. Sketch such a field. Find the charge density ρ.

2-30 Find the divergence of the field $\mathbf{E} = x\,\hat{\mathbf{y}}$. Sketch such a field. Find the charge density ρ. (Such a field cannot exist in electrostatics.)

2-31 A cube of side 1 unit exists with the origin and (1, 1, 1) on its diagonal. There is a charge of q at the origin. Find the flux through each face.

2-32 Two particles of identical mass m and charge q are suspended by threads of length l from a common point; find the charge q in terms of m, l, and θ, which is half of the angle between the threads.

2-33 Find the force between two hemispheres of a uniformly charged spherical shell of surface charge density σ and radius R, Figure 2.35.

2-34 Find the force between two hemispheres of a uniformly charged sphere of charge density ρ and radius R.

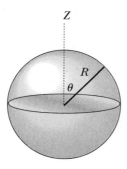

Figure 2.35

*2-35 Eight identical positive charges Q occupy the corners of a cube of side s, Figure 2.36. Show that the field on a line from the center to one face points away from the center (except at the center). Numerical results will do.

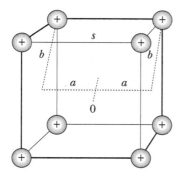

Figure 2.36

2-36 Two infinite planes of uniform surface charge density $-\sigma$ and $+2\sigma$ respectively are: **(a)** parallel and separated by distance s. Find E everywhere. Sketch the planes and field lines. **(b)** mutually perpendicular. Find E everywhere. Sketch the planes and field lines.

2-37 The step function is $\theta(x) = 0$ for $x < 0$, $= 1$ for $x \geq 0$ (Fig. 2.37). Find $d\theta(x)/dx$.

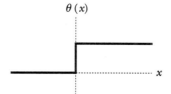

Figure 2.37

CHAPTER

3

MAGNETIC FIELD B—
AMPERE'S LAW

I n the preceding chapter we kept the charges stationary. Now we will relax
that restriction: We are going to let them move a little. We will find that the
electrostatic force is still there, but also something new arises: We enter the
realm of **magnetism.** ("Magnet" comes from the ancient Greek name of Magnesia,
a city in Asia Minor where magnetic rocks were found.)

3.1. Magnetic Force for Current I

Charge in motion constitutes electrical current **I**. The unit of current is the ampere (amp, A), and an ampere of current transports charge of a coulomb per second from one location to another (Appendix D). The direction of the current is
that of the motion of positive charges; if the charge carriers are negative, they go
the other way. Currents may exert forces on each other, as we can demonstrate: If
we run $I = 50$ amps through two parallel wires separated by $r = 1$ cm, Figure 3.1a,
we find that the wires attract each other feebly. This is a small effect: From Chapter
2 we know that there are many thousands of coulombs of electrons available, and
the separation is small, yet the force barely moves the wires (we calculate it in the

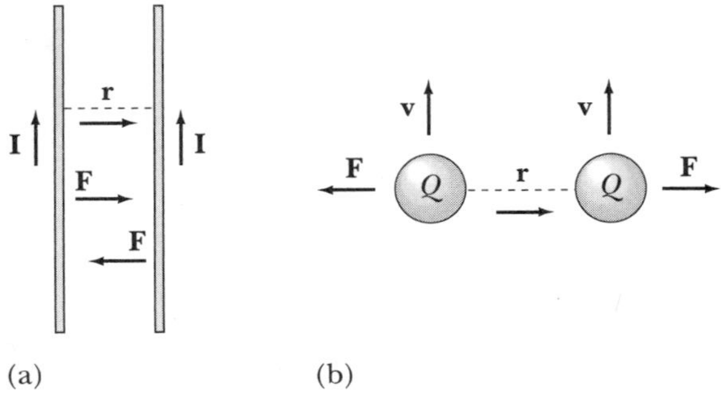

(a)   (b)

Figure 3.1   (a) Parallel currents. (b) Charges moving parallel.

next section). As mentioned in Section 2.1, all such demonstrations are complicated by the actual invisibility of the ideal phenomena being considered. In this
case we might suspect that what we are seeing is some sort of electrostatic effect,
but after connecting the wires in various ways to various power sources, we find
that the force depends only on the currents and their separation; so it is something new, and we call it magnetic force. Thus we find: **Parallel currents attract.**

Consequently, it is only natural to suspect that like charges moving parallel, as
in Figure 3.1b, will display a reduction in their mutual repulsion. And it turns out
to be true. When they are stationary, their repulsion is given by Coulomb's law; but
when they are moving, the magnetic attraction springs up, reducing the net repulsion. At what speed do the electrostatic repulsion and magnetic attraction cancel
to give zero force? Answer: at the speed of light!

Now in the relativity chapter later on, we will find that when you move perpendicular to a force, the force is reduced in such a way as to reach zero at the speed
of light. But that's exactly what the magnetic force contributes. So we will observe
that magnetism is fundamentally a relativistic phenomenon. The electrons drift
very slowly, less than one millimeter per second; however, there are lots of them
(see Problems 3.1 and 3.2) and their electrostatic repulsion is relatively large, so a
small relativistic effect is detectable. We shall not pursue this idea further here, because magnetic phenomena are far easier to describe in the historical conceptual
framework that goes back to Faraday and Oersted, who knew nothing of Einstein's
relativity. We shall take a relativistic approach to electromagnetism in Chapter 18.

Considering the success of Coulomb's law, and of Newton's law of gravitation,
it would be reasonable to seek to express the magnetic force in terms of an inverse-square law. Early attempts were based on the concept of magnetic north and
south poles, which were analogous to plus and minus charges: The corresponding
forces would obey an inverse-square law, and like poles would repel. However, such
magnetic monopoles do not appear to exist (see Section 5.2). It appears instead
that, classically at any rate, magnetic forces are due to electrical currents, that is, to
moving charges and to changing electric fields. (Poles are nonetheless useful conceptually, and we return to them in Section 3.4.)

An inverse-square law for current elements was proposed soon after Oersted
demonstrated that a current deflects a compass needle, in 1820. In integral form,
the force $\mathbf{F}_{ab}$ of circuit a on circuit b, Figure 3.2a, is found to be rather complicated:

$$\mathbf{F}_{ab} = \frac{\mu_0}{4\pi} I_a I_b \oint_a \oint_b dl_b \times \frac{dl_a \times \hat{\mathbf{r}}}{r^2} \tag{3.1}$$

but fortunately this expression is seldom employed in practical calculations. The
constant μ_0 is $4\pi \times 10^{-7}$ in appropriate SI units: N/A^2 in this case. The length
vectors l_a and l_b are in meters and point in the directions of the corresponding
currents $\mathbf{I}_a$ and $\mathbf{I}_b$ (in amperes). The unit vector $\hat{\mathbf{r}}$ points from dl_a to dl_b, and r is
the separation in meters. Each line integral is carried out over the corresponding
closed circuit.

As with Coulomb's law (Section 2.1), this force law is limited to quasi-static situations, with circuits moving slowly and with currents not strongly accelerated.

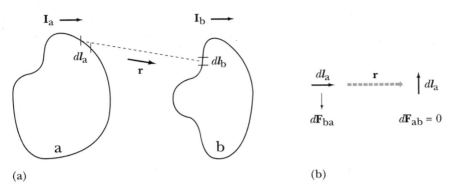

Figure 3.2   **(a)** Circuits. **(b)** Current elements.

The force law just given for magnetism is cumbersome partly because it involves integrals over two complete circuits. Why can't we just find the force between two tiny circuit elements? That's similar to what we did with Coulomb's law, to start with: We employed two point charges rather than extended charge distributions, and the result was quite simple.

Well, the Coulomb's-law force between two circuit elements of length dl (scalar) and constant linear charge density λ would be (Sections 2.1, 2.2)

$$d\mathbf{F}_{ab} = \frac{1}{4\pi\epsilon_0} \lambda_a \lambda_b \frac{dl_a dl_b\, \hat{\mathbf{r}}}{r^2}$$

That's not bad. The analogous magnetic force would be, from the above expression for the magnetic force,

$$d\mathbf{F}_{ab} = \frac{\mu_0}{4\pi} I_a I_b \frac{dl_b \times (dl_a \times \hat{\mathbf{r}})}{r^2} \tag{3.2}$$

There are similarities: Each expression is inverse square in r, and each involves the product of the charges or currents, that is, λ or I. However, there is an asymmetry in the magnetic force between the circuit elements. If we look at two particular line elements, Figure 3.2b, we find that $dl_a \times \hat{\mathbf{r}} = 0$ because the vectors are parallel, so $d\mathbf{F}_{ab} = 0$. But for $d\mathbf{F}_{ba}$, reversing the direction of $\hat{\mathbf{r}}$, $dl_b \times \hat{\mathbf{r}}$ points up out of the paper, and so $dl_a \times (dl_b \times \hat{\mathbf{r}})$ points down as shown. The force elements $d\mathbf{F}_{ab}$ and $d\mathbf{F}_{ba}$ are not equal and opposite. This indicates that Newton's third law is obeyed only after the integration is completed about the two closed circuits.

The problem remains when we replace $I_a dl_a$ and $I_b dl_b$ with moving simple charges. In that case, there is no question of integrating over a complete circuit. Rather, as we will find later, the electromagnetic field is contributing to the overall momentum balance; we will encounter field momentum in Chapter 14 (Poynting's vector).

Aside from the asymmetry we have just mentioned, the above magnetic force law is much less useful than Coulomb's law for the following reason. In Coulomb's law there are only two directions (vectors) and they are parallel: the force $\mathbf{F}$ and the separation $\mathbf{r}$. But in the magnetic case there are two additional directions, namely, those of the two current elements ($I\,dl$), and no one is necessarily parallel or perpendicular to anyone else. The situation becomes much more complicated. So we shall go over to a field representation without delay.

3.2.   Biot-Savart Law for Field B and Current Density J

The justifications we gave in Section 2.2 for introducing the electric field **E** are also operative for the **magnetic field B.** The preceding integral for **F** is quite unwieldy and we need to break it up into simpler pieces. First we rearrange it:

Analogous electrostatic formulas:

$$\mathbf{F}_{ab} = I_b \oint dl_b \times \left(\frac{\mu_0}{4\pi} I_a \oint \frac{dl_a \times \hat{\mathbf{r}}}{r^2} \right) \quad (3.3) \qquad \mathbf{F} = \lambda_a \oint dl_a \frac{1}{4\pi\epsilon_0} \lambda_b \oint \frac{\hat{\mathbf{r}}\, dl_b}{r^2}$$

Then we simply split it in two. One part,
the **Biot-Savart law,** is shown in Figure 3.3a:

$$\boxed{\mathbf{B} = \frac{\mu_0}{4\pi} I \oint \frac{dl \times \hat{\mathbf{r}}}{r^2}} \quad (3.4) \qquad \mathbf{E} = \frac{1}{4\pi\epsilon_0} \lambda \oint \frac{\hat{\mathbf{r}}}{r^2}\, dl$$

The other part is shown in Figure 3.3b:

$$\boxed{\mathbf{F} = I \oint dl \times \mathbf{B}} \quad (3.5) \qquad \mathbf{F} = \lambda \oint \mathbf{E}\, dl$$

The preceding formulas for **B** and **F** apply to different circuits, each with its own current I. Here circuit "a" produces a field **B**, and circuit "b" reacts to it. The magnetic field **B** is measured in N/A·m, or tesla (T). The constant μ_0 is $4\pi \times 10^{-7}$ SI units, T·m/A here (or N/A², Section 3.1. It is called the permeability of free space for reasons that become clearer in Chapter 10, Magnetic

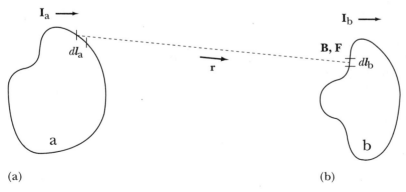

(a)   (b)

Figure 3.3   (a) Circuit producing **B**. (b) Circuit affected by **B**.

Materials). The line integrals are carried over closed current loops, and they are path dependent: You have to follow the wire. The limitations on the Biot-Savart law are similar to those on Coulomb's law, Section 2.1: It works in the quasi-static case, but when things happen fast it can't keep up.

EXAMPLE 3-1

Find the force between two 20 cm sections of wire, separated by 1 cm, each carrying an electric current of 50 amperes(A), in parallel.

ANSWER These wires are not closed loops, of course, but the approximation is okay; the end effects are relatively small. We shall do the calculation in two parts: First we calculate the field **B** caused by one wire, and then we find the force **F** on the other. For simplicity, we employ the approximation that the first wire is infinitely long; this makes only a small difference in **B** at the second wire. (Without this approximation the calculation of **F** becomes formidable.)

For $d\mathbf{l}$ we use $d\mathbf{y}$, Figure 3.4a. The right-hand rule indicates that $d\mathbf{y} \times \hat{\mathbf{r}}$ points into the page; this direction is sometimes represented by $\otimes$, representing the tail feathers of an arrow. As a cross product, its magnitude is $|d\mathbf{y}|\,|\hat{\mathbf{r}}|\sin\theta = dy\sin\theta$, since the magnitude of the unit vector $\hat{\mathbf{r}}$ is one. Then,

$$B = \frac{\mu_0 I_1}{4\pi}\int \frac{dy\sin\theta}{r^2} = \frac{\mu_0 I_1}{4\pi}\int \frac{dy\cos\varphi}{r^2}$$

We go over to the φ variable, obtaining a trivial integral (compare with Section 2.2):

$$r = a/\cos\varphi \qquad \text{and} \qquad dy = a\sec^2\varphi\,d\varphi$$

$$B = \frac{\mu_0 I_1}{4\pi}\int_{-\pi/2}^{\pi/2} \frac{\cos\varphi\,a\sec^2\varphi\,d\varphi}{a^2\sec^2\varphi}$$

$$= \frac{\mu_0 I_1}{4\pi a}\int_{-\pi/2}^{\pi/2}\cos\varphi\,d\varphi = \frac{\mu_0 I_1}{4\pi a}\sin\varphi\Big]_{-\pi/2}^{\pi/2} = \frac{\mu_0 I_1}{2\pi a} \qquad (3.6)$$

$$= \frac{4\pi\times10^{-7}\,\text{T·m/A}\times50\,\text{A}}{2\pi\times0.01\,\text{m}} = 10^{-3}\,\text{T}$$

For comparison, the earth's magnetic field is around 0.5×10^{-4} T, and a strong magnet can give 0.5 tesla. (In terms of the commonly used Gaussian unit, corresponding values are 0.5 gauss and 5000 gauss.)

For techniques involving a finite straight wire, refer to analogous calculations in Section 2.2.

Now we find the force on the other wire, Figure 3.4b.

$$\mathbf{F} = I_2 \oint d\mathbf{l} \times \mathbf{B} \qquad (3.7)$$

$$F = BI_2 l = 10^{-3}\,\text{T}\times50\,\text{A}\times0.2\,\text{m} = 0.010\,\text{N}$$

This corresponds to the weight of only one gram, but it is readily detectable. The **right-hand rule** for the force says: when we push $d\mathbf{l}$ into **B** with our right fin-

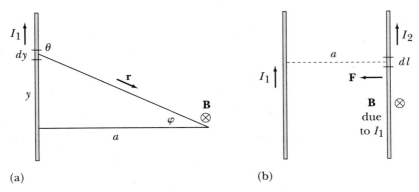

(a)   (b)

Figure 3.4   (a) B field of straight wire. **(b)** Force on wire.

gers, our thumb points in the direction of **F**. Another useful right-hand rule is shown in Figure 3.5: Point your straight fingers in the direction of the magnetic field **B**. Stick your thumb out in the direction of $I\,dl$. Then your palm pushes in the direction of **F**. This is handy when all three variables are involved: B, F, and I. (I_2 is also sitting in its own magnetic field, of course, but that contributes nothing to the force on I_2—otherwise the wire would start moving all by itself.)

The magnetic force between two infinite parallel wires is used for defining the ampere (Appendix D). Combining F and B from above, we obtain

$$B = \frac{\mu_0 I_1}{2\pi a} \quad \text{and} \quad F = BI_2 l$$

so

$$\frac{F}{l} = \frac{\mu_0 I_1 I_2}{2\pi a} \tag{3.8}$$

Now if $I_1 = I_2$, and $a = 1$ m, and $F/l = 2 \times 10^{-7}$ N/m, then $I_1 \equiv 1$ ampere, by definition. The coulomb is then defined as an ampere second. (By contrast, in the

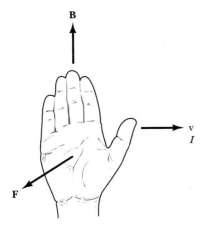

Figure 3.5   A right-hand rule.

Gaussian system of units, the unit of charge is first defined using Coulomb's law, and then the unit of current is defined in terms of charge per unit time. Appendix D gives further information on the Gaussian system.)

EXAMPLE 3-2

Find **B** at distance z above the center of a circular loop of radius a carrying current I. Evaluate for $z = 0$, $a = 1$ cm, and $I = 1$ A.

ANSWER See Figure 3.6; a, z, r, and dB lie in the plane of the paper, and I is perpendicular to r. Consequently the $\sin\theta$ of the cross product $= 1$ and θ disappears from the calculation. However, because of the cross product, the field element dB caused by a given current line element is perpendicular to $\mathbf{r}$, so it makes an angle φ with the z axis. By symmetry, **B** will point in the z direction, so we need only the z component of dB, that is, $\cos\varphi\, dB$. φ and r are constant, so

$$B = \frac{\mu_0 I}{4\pi} \oint \frac{\cos\varphi\, dl}{r^2} = \frac{\mu_0 I}{4\pi} \frac{\cos\varphi}{r^2} \oint dl$$

$$= \frac{\mu_0 I}{4\pi} \frac{\cos\varphi\, 2\pi a}{r^2} = \frac{\mu_0 I a^2}{2(z^2 + a^2)^{3/2}} \tag{3.9}$$

Note that here the $\cos\varphi$ comes from taking the z component of **B**, whereas in the preceding example it arose out of a cross product.

At the center of the loop, $z = 0$ and the expression becomes

$$B = \frac{\mu_0 I}{2a}$$

$$= \frac{4\pi \times 10^{-7} \times 1}{2 \times 0.01} = 6.3 \times 10^{-5} \text{ T}$$

which is about the size of the earth's magnetic field. To get large magnetic fields, one needs many turns of wire at high current; for most purposes, appropriate placement of magnetic iron is also quite helpful. We examine such matters in Chapter 10.

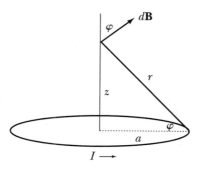

Figure 3.6   **B** field of current loop.

The current **I** encompasses all of the charge flowing from one location to another, for example, through a wire. If we are concerned with the rate of charge flow at different points in a volume, we may need the concept of volume **current density J**, in amperes/m^2: The current dI passing through a small surface $d\mathbf{a}$ is **J** $\cdot$ $d\mathbf{a}$. (The vector **a** is normal to the surface.) Similarly, when current flows on a surface, then we can draw a line element dl on the surface and express the **surface current density K,** in amperes/m, using current element $dI = |\mathbf{K} \times d\mathbf{l}|$. (For both **J** and **K**, we are interested in the flow *across* the corresponding area or length element.) In terms of these currents, the expression for the Biot-Savart law, Equation 3.4, becomes one of the following (compare with Eq. 2.7):

$$\mathbf{B} = \frac{\mu_0}{4\pi} \oint \frac{I\,d\mathbf{l} \times \hat{\mathbf{r}}}{r^2} \text{ or } \boxed{\mathbf{B} = \frac{\mu_0}{4\pi} \int \frac{\mathbf{J} \times \hat{\mathbf{r}}}{r^2}\,dv} \text{ or } \mathbf{B} = \frac{\mu_0}{4\pi} \int \frac{\mathbf{K} \times \hat{\mathbf{r}}}{r^2}\,da \quad \textbf{(3.10)}$$

The length l integration is carried over a complete closed loop that follows the path of the current I. The volume v integration is carried out over all space, or at least where **J** is not zero; anyway, it is not just over a closed loop, and there is no path dependence. Like dv, the area differential da is a scalar.

(It happens that the **J** given here may not be complete: Later, Section 3.5, we will find that the total volume current density can more generally be written **J** + $\epsilon_0 \partial \mathbf{E}/\partial t$; however, the difference is unimportant in the quasi-static case for which the Biot-Savart law is valid. We examine this point further in Section 3.8.)

If we introduce $Q\mathbf{v}$, where **v** is velocity, in place of the current element $I\,dl$, we would expect to obtain the B field due to a single charge Q moving slowly:

$$\boxed{B = \frac{\mu_0}{4\pi} \frac{Q\mathrm{v}\sin\theta}{r^2}} \quad \textbf{(3.11)}$$

as shown in Figure 3.7, with **B** pointing up out of the page (the symbol represents the head of an arrow pointing towards you). And it is indeed true, but that is perhaps partly accidental. After all, this is no electrostatic or magnetostatic situation; **E** and **B** fields are changing in nontrivial ways. Also, not surprisingly, at high speed the field scrunches up, somewhat like the electrostatic field in Figure 2.22. We will return to this formula several times in later chapters.

3.3.   Lorentz Force—Trajectories

As we have seen, a current passing through a magnetic field experiences a force. However, that current consists of many individual moving charges. Now, what is

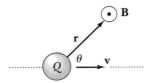

Figure 3.7   **B** field of moving charge.

the magnetic force on those elementary charges? Equation 3.5 suggests that the force **F** on a current element Il is $Il \times \mathbf{B}$. The current I is the charge Q per unit time t. If the current element is represented as, say, a line segment of charge Q and length l, moving at speed $\mathbf{v} = l/t$, then

$$\mathbf{F} = Il \times \mathbf{B} = \frac{Q}{t} l \times \mathbf{B} = Q\mathbf{v} \times \mathbf{B}$$

So, when we include both electric and magnetic fields in the expression for the force on a charge Q, moving at velocity $\mathbf{v}$, we obtain the fundamental **Lorentz force:**

$$\boxed{\mathbf{F} = Q\mathbf{E} + Q\mathbf{v} \times \mathbf{B}} \qquad (3.12)$$

If the charge is stationary, the force depends on **E**; if it moves, there is an additional force proportional to **v** and **B**. (If it accelerates, then there may be additional forces, but they are usually neglected). The direction of $\mathbf{v} \times \mathbf{B}$ is given by the usual right-hand rule: The fingers push **v** into **B**, and the thumb points in the direction of $\mathbf{v} \times \mathbf{B}$. (Or see Fig. 3.5.)

 EXAMPLE 3-3

A magnetic field of 0.2 T points in the $+x$ direction, and a charge of 1 μC moves in the xy plane at 300 m/s, 30° up from the x axis. Find the force.

ANSWER The Lorentz force is

$$\mathbf{F} = Q\mathbf{v} \times \mathbf{B}$$
$$F = 10^{-6}\,\text{C} \times 300\text{ m/s} \times 0.2\text{ T} \times \sin 30° = 3 \times 10^{-5}\text{ N}$$

The force is directed down into the paper, along the negative z axis, Figure 3.8.

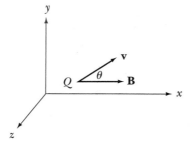

Figure 3.8   Lorentz force.

In the preceding example, the force turns out to be rather small, despite the fact that the charge of 1 μC is pretty big for an electrostatic charge; the field is about what you would get from a good permanent magnet; and v is nearly the

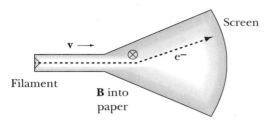

Figure 3.9   Cathode ray tube (CRT).

speed of sound. To get large magnetic forces, you need to enlist many coulombs. In a wire you typically have 10^4 C or more to work with, instead of only 10^{-6} C (Problems 3-1 and 3-2), so you can get significant forces despite the low drift velocity of the electrons.

You can also get substantial effects with small forces if the mass is small. For example, the narrow electron beam in your television tube (a cathode ray tube, CRT) is deflected back and forth across the screen at a frequency of 15,750 Hz using magnetic fields. The process works as follows, shown in Figure 3.9. The electron beam is born when a red-hot filament heats a cathode and electrons come boiling out. The electrons are then accelerated to a speed of nearly 10^8 m/s using an electrostatic field. After they have reached that speed, magnetic fields are usually more effective, for the following reason. The ratio of electric to magnetic force is E/vB (for B perpendicular to v). Large fields would be $E = 10^6$ N/C and $B = 1$ T; so for such fields the forces would be equal at v = 10^6 m/s, and at 10^8 m/s the magnetic force would be 100 times larger. As it turns out, a magnetic field of around 0.01 T is adequate to guide the beam over your TV screen.

The $\mathbf{v} \times \mathbf{B}$ force is always perpendicular to the velocity, and so it does no work and does not change the speed of a charged particle. Instead, it tends to make the particle go in a circle, or in a helix—see below. If $\mathbf{v}$ is perpendicular to $\mathbf{B}$, we get a circular orbit, Figure 3.10, in a uniform $\mathbf{B}$ field. The centripetal force is the magnetic force:

$$\frac{m\mathbf{v}^2}{r} = QvB$$

so

$$\boxed{m\mathbf{v} = QrB} \tag{3.13}$$

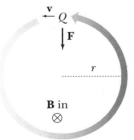

Figure 3.10   Charge orbit in $\mathbf{B}$ field.

This means that a singly charged elementary particle's trajectory in a given transverse magnetic field depends on its momentum. In a transverse electrostatic field of given length, the deflection depends on momentum times speed, as you may easily verify. Consequently, in crossed E and B fields, you can separate charged particles by velocity: If there is no deflection, then

$$\mathbf{F} = 0 = Q\mathbf{E} + Q\mathbf{v} \times \mathbf{B}$$

so

$$\mathbf{v} = -E/B$$

regardless of mass.

A particle trapped, going in circles, in a magnetic field displays a characteristic **cyclotron frequency** usually given as an angular frequency:

$$\omega = QB/m \qquad (3.14)$$

where $\omega = v/r$; this follows immediately from $mv = QrB$, Equation 3.13. (Note that, nonrelativistically, that is, for $v \ll c$, ω is independent of speed v.)

◤ EXAMPLE 3-4

Find the cyclotron frequency for electrons, speed 10^5 m/s, in a field of 100 gauss.

ANSWER As mentioned previously, the gauss is the commonly used Gaussian unit of magnetic field (see Appendix D); a tesla is 10^4 gauss. The electron speed is irrelevant.

$$\omega = QB/m$$
$$= 1.6 \times 10^{-19}\,\text{C} \times 0.01\,\text{T}/9.1 \times 10^{-31}\,\text{kg} = 1.8 \times 10^9\,\text{rad/s}$$

This cyclotron frequency is fundamental to operation of cyclotron particle accelerators (Problem 6-37), and it can be of interest in studies of elementary particles, materials, and interstellar space.

If the charged particle's velocity is neither parallel nor perpendicular to **B**, the particle describes a helix. In Figure 3.11 the **B** field is parallel to the z axis, and the initial velocity is parallel to the xz plane. The particle moves uniformly in the z direction with speed v_z, while it continues to go in circles of radius mv_x/QB.

If there is a region of crossed **E** and **B** fields, we can get cycloids, Figure 3.12, if the speed is not too great. The particle starts from rest, and it tends to migrate in the direction of the vector $\mathbf{E} \times \mathbf{B}$. Analysis of this case is simplified by relativity, so we defer discussion to Section 18.7.

. . . Still . . . There's something funny about this force $\mathbf{F} = Q\mathbf{v} \times \mathbf{B}$. Suppose we have Q moving at **v** through **B**, with no **E** field, and we have the Lorentz force, and the particle path proceeds to curve because of the force. Now,

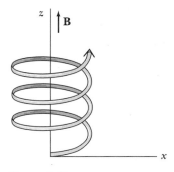

Figure 3.11   Helix in **B** field.

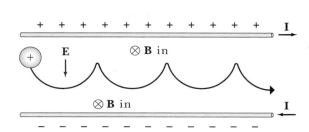

Figure 3.12   Crossed **E** and **B**, with cycloid.

suppose we jump aboard the charge. Now there's no **v**, and so there's no $Q\mathbf{v} \times \mathbf{B}$. Yet, the force must still be there. So the

QUESTION: is, where does this force come from?

ANSWER: It comes from relativity. The relativistic transformation of **B** at speed **v** yields an electric field $\mathbf{E} = \mathbf{v} \times \mathbf{B}$ in the new reference frame, in this case. Charge distributions can change if necessary to produce the required **E** field. We will examine how this happens in more detail in Section 18.7. As we mentioned in Section 3.1, magnetism is a relativistic phenomenon. However, we shall continue to use the usual concepts and notation, developed before relativity, because they are far easier to deal with.

However, you should remember that if you move, relative to a frame having a **B** field, you may find a corresponding **E** field springing up. The important point for the Lorentz force is, for speed not too fast, the force will be the same: $Q\mathbf{E} \Longleftrightarrow Q\mathbf{v} \times \mathbf{B}$, which means: $Q\mathbf{E}$ in one reference frame can turn into $Q\mathbf{v} \times \mathbf{B}$ in another, and vice versa.

3.4.   Magnetic Field Lines and Poles—Divergence of B

The magnetic field may be visualized in **magnetic field lines.** As in the case of electric field lines, Section 2.3, Faraday imagined them to pull in the direction of their length and to repel each other sideways. But unlike *E* lines, these lines of *B* do not begin on the field sources, but instead they tend to form closed loops. For a *straight wire,* Figure 3.13a, they form circles, centered on the wire, with direction given by a **right-hand rule:** Put your thumb in the direction of the current, and the fingers will tend to curl in the direction of the **B** field. For a cylindrical hollow **solenoid** wrapped with conducting wire, Figure 3.13b, the **right-hand rule** is this: Put your thumb in the direction of **B**, and your fingers will tend to curl in the direction of the current.

In addition to the two **caveats** mentioned in connection with Figures 2.10 and 2.11, for real magnetic field lines there is a third: The loops do not actually close

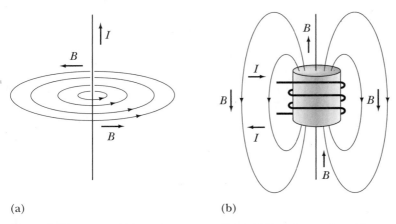

(a)   (b)

Figure 3.13   (a) Field lines for a wire. (b) Field lines for a solenoid.

precisely. For example, if we add a small magnetic field in the direction of the wire in Figure 3.13a, the circles become helices (spirals) that do not close on themselves after one turn around the wire. A more detailed description (Slepian) would be both tedious and unnecessary here, however, since the field lines are used only for illustration of the direction and magnitude of the field, and not for quantitative calculation.

It already seems clear from the pictures that lines of B don't begin or end. Comparing with Section 2.4, this implies that the divergence of **B** is always zero:

$$\boxed{\nabla \cdot \mathbf{B} = 0} \qquad (3.15)$$

which is our second Maxwell equation. In view of the divergence theorem, Sections 1.5 and 2.4, this means that

$$\Phi = \oint \mathbf{B} \cdot d\mathbf{a} = 0 \qquad (3.16)$$

where the integral is carried over any closed surface, and Φ is the net **magnetic flux** out of the surface (units: weber, Wb, or $T \cdot m^2$). The equation is analogous to Gauss's law, but it turns out to be of little use in practical field calculations. We will deal with $\nabla \cdot \mathbf{B}$ further in Section 3.7.

In the case of the **E** field, divergence implies sources of flux, namely, electrical charges. In magnetism, since $\nabla \cdot \mathbf{B} = 0$, it appears that there are no such sources: **no magnetic monopoles.** Their possible existence has tantalized both theoreticians and experimentalists for generations, but as yet nothing has come either of speculation or of experiment; see Section 5.2 for more on this.

The utility of the concept of **magnetic poles** is scarcely diminished by the evident nonexistence of monopoles. Magnetic *dipoles do* exist. In other words, magnetic poles always come in pairs. In the vicinity of a magnet or solenoid, the direction of **B** is indicated simply by labeling the magnet **North,** or "N," where flux

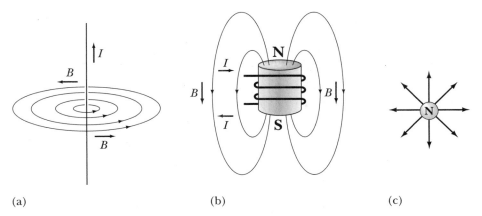

(a)   (b)   (c)

Figure 3.14   **(a)** For a wire, no poles. **(b)** Poles for a solenoid. **(c)** North magnetic monopole.

comes out, and **South,** or "S," where it goes in. Figure 3.14b repeats Figure 3.13b, but with poles shown. In the case of the field lines of a wire, Figure 3.14a, there is no particular place where the lines go "in" or "out" of anything, so the pole concept is not applicable. Figure 3.14c shows a North magnetic monopole.

Like magnetic poles repel, and unlike attract. This is consistent with the rule given in Section 3.1: Parallel currents attract. In Figure 3.15 we have apposed unlike poles of two identical solenoids. They attract, because the currents in the wires are parallel (rather than antiparallel).

"North poles" are so named because solenoids or bar magnets tend to align themselves in the earth's magnetic field, with the north pole of the magnet pointing more or less toward the north pole of the earth. This of course implies that the earth itself is like a large magnet, Figure 3.16, with its geographical north pole near a magnetic south pole. At the geographical north pole the field **B** points downward into the earth and has magnitude 0.62×10^{-4} T; at the equator it is

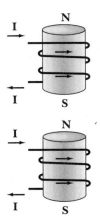

Figure 3.15   Unlike poles attract.

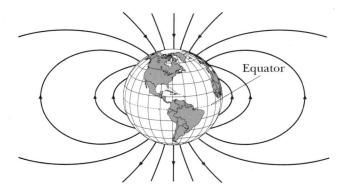

Figure 3.16   Earth's magnetic field.

around 0.5×10^{-4} T and is horizontal and directed more or less north. It is probably due to some kind of internal dynamo; see Section 4.3. It is not caused by magnetic iron, because the earth's interior is above the Curie temperature (Section 10.2). The earth's magnetic field reverses itself every few hundred thousand years; what we will do then remains to be seen. (The sun's magnetic field reverses itself every 11 years in conjunction with the sunspot cycle.)

3.5.   Ampère's Law in Integral Form

It's important of course to realize that there are no magnetic monopoles, but it doesn't help much in calculating **B**. Ampère's law does help, a great deal in fact. It's kind of like Gauss's law in its usefulness, and in its limitations.

For the straight wire of Figure 3.4, we have already shown that $B = \mu_0 I/2\pi r$ (we use r in place of a here). When we do a line integral of this expression about a closed circular loop centered on the wire, Figure 3.17a, we get

$$\oint \mathbf{B} \cdot d\mathbf{l} = \oint \frac{\mu_0 I}{2\pi r} \, dl$$

$$= \frac{\mu_0 I}{2\pi r} \oint dl = \frac{\mu_0 I}{2\pi r} 2\pi r$$

so

$$\oint \mathbf{B} \cdot d\mathbf{l} = \mu_0 I \tag{3.17}$$

This is part of Ampère's law. We have found it for a circle centered on a straight wire. However, the radius doesn't matter, suggesting that the law may have more generality. Later, in Section 3.7, we will carry through a mathematical development linking this law with the Biot-Savart law, but even that approach is not completely satisfactory in view of the fact that Biot-Savart is limited to the quasi-static case. Ultimately one must simply postulate Ampère's law for the general case, based on its agreement with experiment.

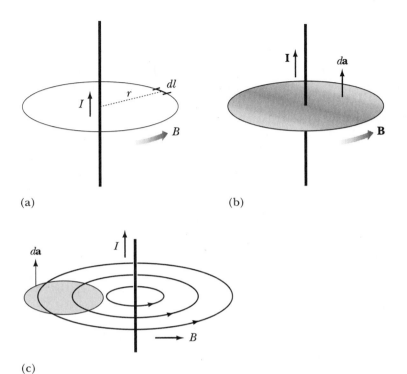

Figure 3.17   (a) Infinite wire. (b) Amperian loop. (c) Ignore current outside loop.

More importantly, the preceding form of Ampère's law is missing a term. **B** depends also on the rate of change of **E**, the electric field. The effect was discovered theoretically by Maxwell and later verified experimentally. As we will remark again in connection with Einstein's discovery of time dilation, it is pretty good to simply *explain* an observation; however, to *predict* an effect that has never been seen before is fantastic, almost magical. The effect is not readily demonstrated in quasi-static cases, but it is often of crucial importance, especially for radiation. The new term is μ_0 times the **displacement current,**

$$\frac{d}{dt} \int \epsilon_0 \mathbf{E} \cdot d\mathbf{a} \tag{3.18}$$

(for **a**, see subsequent text) and for magnetic fields it behaves very much like the **conduction current density J** that consists of moving charges. We will discuss it further in this section and also in Sections 3.8 and 5.1. (In particular: Why doesn't the Biot-Savart law need a displacement current term?) In Section 9.1 we will add another term, involving materials, to the displacement current.

So, the general form of **Ampère's law in integral form** looks like this:

$$\oint \mathbf{B} \cdot d\mathbf{l} = \mu_0 I + \mu_0 \frac{d}{dt} \int \epsilon_0 \mathbf{E} \cdot d\mathbf{a} \tag{3.19}$$

(It was once called "Ampère's circuital law" to avoid confusion with another law.)

The area **a** is bounded by the **"Amperian loop"** on which **B** is integrated, for example the circle in Figure 3.17b, and the direction of *d***a** is given by the **right-hand rule:** thumb in direction of *d***a**, fingers in direction taken on the contour. Current **I** and changing **E** field pass through the contour. The area needn't be flat; see the last example in this section.

As with Gauss's law, Figure 2.13c, Ampère's law simply ignores current passing outside of the Amperian loop, Figure 3.17c. Field lines tend to make opposite contributions on opposite sides of the loop, and thus their effect is cancelled out.

◢ EXAMPLE 3-5

Find the magnetic field at distance r from an infinite straight wire carrying constant current I. (Neglect E.) Evaluate for 1 cm and 10 A.

ANSWER We draw an Amperian loop (or Stokesian loop, Section 3.6) about the wire, a circle, Figure 3.17a, analogous to the Gaussian surface of Gauss's law. **B** is constant on this loop of circumference $2\pi r$, and current I passes through the loop (that is, through the surface bounded by the loop; the current is in the wire, and the wire passes through the loop). Then,

$$\oint \mathbf{B} \cdot d\mathbf{l} = \mu_0 I$$

$$= B \oint dl = 2\pi r B$$

so

$$B = \frac{\mu_0 I}{2\pi r} \tag{3.20}$$

as before, but this time we didn't have to integrate. Evaluating:

$$B = \frac{4\pi \times 10^{-7} \times 10}{2\pi \times 0.01} = 2 \times 10^{-4} \, \text{T}$$

It won't erase your magnetic tape.

The reasoning above is somewhat circular, of course, because it was the **B** field for a straight wire that suggested Ampère's law to us in the first place. So let's try something new:

◢ EXAMPLE 3-6

Find the magnetic field of an infinite solenoid having radius R and N' turns per meter of wire carrying current I. Evaluate for $R = 3$ cm, $N' = 10^4$ turns/m, and $I = 1$ A.

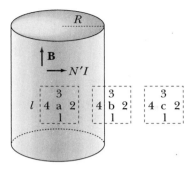

Figure 3.18   Infinite solenoid.

ANSWER Figure 3.18 shows our solenoid, which we display in finite size because of environmental considerations. Symmetry dictates that **B** must be parallel to the axis of the solenoid. The right-hand rule says: If our fingers point to the right, then **B** points up, as shown. Three Amperian loops are shown; we will consider $\oint \mathbf{B} \cdot d\mathbf{l}$ for each of these loops.

Loop a: Sides 1 and 3 are perpendicular to the field and so contribute nothing to the line integral. There is no current through the loop, so the line integral must be zero, and so the field at side 2 must equal that at line 4. This means that the field is constant inside the solenoid.

Loop c: By the same reasoning, the field is a constant outside the solenoid. But what is that constant? Since the divergence of **B** is zero, in a finite solenoid all of the field lines that go up the solenoid inside must return outside the solenoid. However, the area outside the solenoid is much larger than inside, so the field must be correspondingly smaller. In the limit of infinite length, it is reasonable to assume that the field outside is zero. A formal proof requires integrating the Biot-Savart formula over the entire solenoid surface; it's not so hard, and it's in Problem 3-17.

Loop b: Since the field outside is zero, only side 4 contributes to the line integral. The current through the loop is $l\,N'I$, where l is the length of the loop. So Ampère's law says:

$$\oint \mathbf{B} \cdot d\mathbf{l} = \mu_0 I \text{ (total)}$$

$$Bl = \mu_0 l N' I$$

$$\boxed{B = \mu_0 N' I} \tag{3.21}$$

The radius R doesn't matter. In fact, the solenoid doesn't even have to be circular; as long as we know the field is parallel to the axis, any cross section will do. Now we evaluate B in the above example:

$B = \mu_0 N' I$
$\quad = 4\pi \times 10^{-7}\,\text{T} \cdot \text{m/A} \times 10^4/\text{m} \times 1\,\text{A} = 1.26 \times 10^{-2}\,\text{T}\ (= 126\ \text{gauss})$

It's not much of a magnetic field; we need to put in some iron to get a more respectable value, but that comes later.

Instead of $N'I$, the current could be given in terms of a surface current density, K amperes per meter; then we would have

$$B = \mu_0 K \qquad (3.22)$$

As with Gauss's law, Ampère's law is always true but not always useful. It is limited to a few geometries; you can't use it for example to find the field of a *semi*-infinite

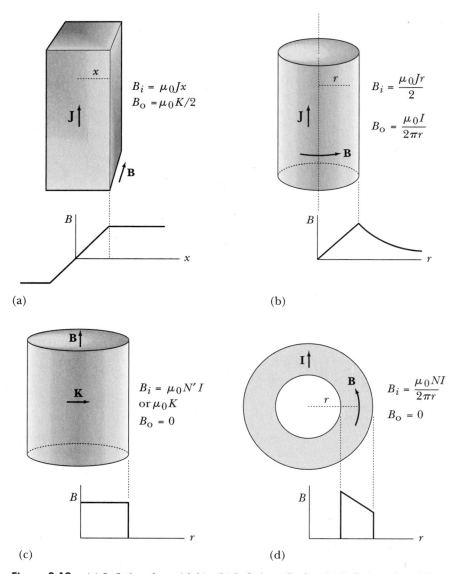

$$B_i = \mu_0 J x$$
$$B_o = \mu_0 K/2$$

(a)

$$B_i = \frac{\mu_0 J r}{2}$$

$$B_o = \frac{\mu_0 I}{2\pi r}$$

(b)

$$B_i = \mu_0 N' I$$
or $\mu_0 K$

$$B_o = 0$$

(c)

$$B_i = \frac{\mu_0 N I}{2\pi r}$$

$$B_o = 0$$

(d)

Figure 3.19　(a) Infinite plane (slab). (b) Infinite cylinder. (c) Infinite solenoid. (d) Toroid.

wire, or a ring, because those fields are curved in nonobvious ways. Figure 3.19 shows the symmetrical geometries for which Ampère's law is useful, with their B fields.

For the plane and the solenoid, the appropriate Amperian loop is a rectangle; for the cylinder and the toroid it is a circle. To be useful, these loops must be in the plane of **B** and perpendicular to **J**.

The following example involves a volume current distribution.

 **EXAMPLE 3-7**

Find **B**, inside and outside, at distance x from the center of an infinite plane of thickness s and current density **J**, Figure 3.19a and 3.20. (It is analogous to Fig. 2.18.)

ANSWER We employ an Amperian loop of sides l and x. By symmetry, $B = 0$ at $x = 0$. Only the side on the right, of length l, contributes to the line integral of **B**, because, for the side on the left, $B = 0$; and for the top and bottom, **B** is perpendicular to the length element x. *Inside,* the current density is **J** inside the entire Amperian loop of area lx, Figure 3.17b, so $I = Jlx$. Then from Ampère's law,

$$\oint \mathbf{B} \cdot d\mathbf{l} = \mu_0 I$$

$$Bl = \mu_0 Jl\,x$$

$$B = \mu_0 Jx \qquad\qquad\qquad (3.23)$$

Outside, the current density is **J** only up to $x = s/2$, Figure 3.17c, so $I = Jl\,s/2$. Then:

$$\oint \mathbf{B} \cdot d\mathbf{l} = \mu_0 I$$

$$B = \mu_0 Js/2 = \mu_0 K/2 \qquad\qquad (3.24)$$

where current per unit area is $K = Js$. So the field increases linearly inside the plane, as shown in Figure 3.19a, and is constant outside.

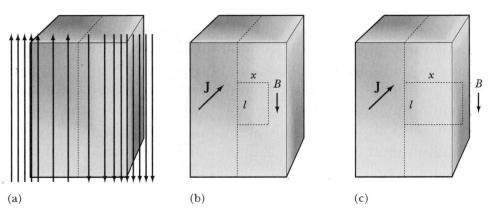

(a) (b) (c)

Figure 3.20   (a) Field lines. **(b)** **B** inside. **(c)** **B** outside.

The following example will help clarify the role of the displacement current and the flexibility of the surface spanning the Amperian loop.

◥ EXAMPLE 3-8

Two large parallel plates of area a and uniform charge density σ are perpendicular to an otherwise-continuous wire carrying constant current I, Figure 3.21. Find the field at distance r from the wire in two different ways: using the conduction current, and using the displacement current.

ANSWER Ampère's law is

$$\oint \mathbf{B} \cdot d\mathbf{l} = \mu_0 I + \mu_0 \frac{d}{dt} \int \epsilon_0 \mathbf{E} \cdot d\mathbf{a}$$

In each case we employ a circular Amperian loop of radius r, Figure 3.21.

Conduction current, Figure 3.21a: We bend the surface so that it passes through the wire. The E field outside the parallel plates is negligible, so we have

$$\oint \mathbf{B} \cdot d\mathbf{l} = \mu_0 I$$

$$2\pi r B = \mu_0 I$$

$$B = \mu_0 I / 2\pi r \qquad (3.25)$$

Displacement current, Figure 3.21b: We bend the surface so that it passes between the parallel plates. No conduction current passes through the surface, so we have

$$\oint \mathbf{B} \cdot d\mathbf{l} = \mu_0 \frac{d}{dt} \int \epsilon_0 \mathbf{E} \cdot d\mathbf{a}$$

$$2\pi r B = \mu_0 \epsilon_0 a \frac{dE}{dt}$$

From Problem 2-22, $E = \sigma / \epsilon_0$, so $dE/dt = d\sigma / \epsilon_0 dt$. As current flows, charge accumulates on the plates, and so $Q = \sigma a$ and $I = dQ/dt = a d\sigma/dt$. Then,

$$2\pi r B = \mu_0 \epsilon_0 \, a d\sigma / \epsilon_0 dt = \mu_0 dQ/dt = \mu_0 I$$

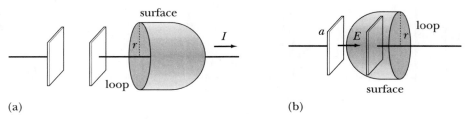

(a)   (b)

Figure 3.21   (a) Conduction current. **(b)** Displacement current.

So,

$$B = \mu_0 I / 2\pi r$$

as before.

This shows that Maxwell's displacement term must be present in Ampère's law, in principle, so that we get the same **B** field regardless of where we choose to place the surface spanning the Amperian loop. However, it is also true that this term happens to be insignificant in the usual quasi-static cases, so it can usually be neglected (see Section 3.8 for more).

3.6.   Ampère's Law in Differential Form—Curl of B

As with Gauss's law, there is a more elegant form of Ampère's law waiting to find expression. We begin by examining the law microscopically. Figure 3.22 shows a small stationary square loop in the xy plane, in a region of field **B** and total current density $\mathbf{J}_T$, which includes displacement current density $\epsilon_0 \partial \mathbf{E} / \partial t$ for now, to simplify the equations; we shall separate it out at the end. After we deal with the xy plane, we can easily add the xz and yz planes. We will calculate the line integral of **B** around the loop and set it equal to μ_0 times the z component of the current through the area enclosed by the loop, J_z.

Current I_T flows through the area of the loop; it is directed up out of the page, and **B** points counterclockwise, as shown. (However, it doesn't always have to point in the same direction as the contour we choose.) We begin at the lower left corner and proceed counterclockwise around the square, multiplying each side by the corresponding field component. (As in Figure 1.10, we assume that $B_x(x, y, z)$ represents a suitable average over the side.)

From the preceding section, Ampère's law says

$$\mu_0 I_T = \oint \mathbf{B} \cdot d\mathbf{l} \qquad (3.26)$$

Total current I_T is current density J_z times area $\Delta x \Delta y$; on the right side of the equa-

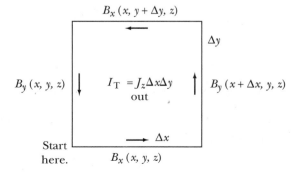

Figure 3.22   Small square in current density.

tion we place the results from making one complete round trip about the contour:

$$\mu_0 J_z \Delta x \Delta y = B_x(x, y, z)\Delta x + B_y(x + \Delta x, y, z)\Delta y - B_x(x, y + \Delta y, z)\Delta x - B_y(x, y, z)\Delta y$$

When we regroup, and we multiply and divide by Δx or Δy, the result turns into a pair of partial derivative forms:

$$\mu_0 J_z \Delta x \Delta y = \frac{B_y(x + \Delta x, y, z) - B_y(x, y, z)}{\Delta x}\Delta x \Delta y$$
$$- \frac{B_x(x, y + \Delta y, z) - B_x(x, y, z)}{\Delta y}\Delta x \Delta y$$

(3.27)

Canceling $\Delta x \Delta y$,

$$\mu_0 J_z = \frac{\partial B_y}{\partial x} - \frac{\partial B_x}{\partial y}$$

Correspondingly, in the yz and xz planes,

$$\mu_0 J_x = \frac{\partial B_z}{\partial y} - \frac{\partial B_y}{\partial z} \quad \text{and} \quad \mu_0 J_y = \frac{\partial B_x}{\partial z} - \frac{\partial B_z}{\partial x}$$

To simplify the writing of these expressions, we once again turn to the del operator ∇ (Sections 1.2 and 2.4). Again we treat it like a vector, and in this case the above partial derivatives fit neatly into a cross product type of format:

$$\nabla \times \mathbf{B} = \left(\frac{\partial}{\partial x}\hat{\mathbf{x}} + \frac{\partial}{\partial y}\hat{\mathbf{y}} + \frac{\partial}{\partial z}\hat{\mathbf{z}}\right) \times (B_x\hat{\mathbf{x}} + B_y\hat{\mathbf{y}} + B_z\hat{\mathbf{z}})$$

(3.28)

When we multiply this cross product out there are nine terms, of which three are zero. The six survivors look like this: $\partial B_y/\partial x\,\hat{\mathbf{x}} \times \hat{\mathbf{y}}$, where $\hat{\mathbf{x}} \times \hat{\mathbf{y}} = \hat{\mathbf{z}}$. The three that don't make it look like this: $\partial B_y/\partial y\,\hat{\mathbf{y}} \times \hat{\mathbf{y}}$, where $\hat{\mathbf{y}} \times \hat{\mathbf{y}} = 0$.

The six surviving terms are exactly the same as the partial derivative terms we found by going around the contour. For example, from above, two of them are found in the z component:

$$\mu_0 J_z = \frac{\partial B_y}{\partial x} - \frac{\partial B_x}{\partial y}$$

So putting all three components of $\mathbf{J}$ together, we find

$$\mu_0 \mathbf{J}_T = \left(\frac{\partial B_z}{\partial y} - \frac{\partial B_y}{\partial z}\right)\hat{\mathbf{x}} + \left(\frac{\partial B_x}{\partial z} - \frac{\partial B_z}{\partial x}\right)\hat{\mathbf{y}} + \left(\frac{\partial B_y}{\partial x} - \frac{\partial B_x}{\partial y}\right)\hat{\mathbf{z}}$$

which can be written as

$$\mu_0 \mathbf{J}_T = \nabla \times \mathbf{B}$$

Separating out the displacement current, we arrive at **Ampère's law in differential form:**

$$\boxed{\nabla \times \mathbf{B} = \mu_0 \mathbf{J} + \mu_0 \epsilon_0 \frac{\partial \mathbf{E}}{\partial t}}$$

(3.29)

This is another Maxwell equation. It is complete, correct, and general. Compared with the integral form, the differential form of Ampère's law is true at every point and not just when integrated over some line contour. We have discussed its significance in some particular cases; later we shall simply postulate it for the general case.

When we integrate both sides of Ampère's law over a surface bounded by a contour,

$$\int \nabla \times \mathbf{B} \cdot d\mathbf{a} = \int \left(\mu_0 \, \mathbf{J} + \mu_0 \epsilon_0 \frac{\partial \mathbf{E}}{\partial t} \right) \cdot d\mathbf{a}$$

we arrive at **Stokes' theorem** (which is implicit in the above derivation):

$$\boxed{\int \nabla \times \mathbf{B} \cdot d\mathbf{a} = \oint \mathbf{B} \cdot d\mathbf{l}} \qquad (3.30)$$

illustrated in Figure 3.23. (See inside front cover.)

The expression $\nabla \times \mathbf{B}$ is called the **curl of B** and is sometimes written curl **B**. (See Section 1:2.) It is a vector, like the cross product $\mathbf{A} \times \mathbf{B}$. Consider an area a bounded by an arbitrary contour, a **"Stokesian loop"** (or Amperian loop), Figure 3.23. If current **J** passes through the loop, then there is a curl of **B**, $\nabla \times \mathbf{B} = \mu_0 \mathbf{J}$, inside. This generates field lines that make their presence felt at the contour, producing a line integral of **B**, $\oint \mathbf{B} \cdot d\mathbf{l}$, around the contour. The right-hand rule here is: fingers in direction of $d\mathbf{l}$, thumb in direction of $\nabla \times \mathbf{B}$ (which is also the direction of **J**).

More briefly: **if there is a curl, then field lines tend to involve circulation;** see below.

We can use Stokes' theorem to pass directly from the differential to the integral form of Ampère's law:

$$\nabla \times \mathbf{B} = \mu_0 \mathbf{J} + \mu_0 \epsilon_0 \frac{\partial \mathbf{E}}{\partial t}$$

$$\int \nabla \times \mathbf{B} \cdot d\mathbf{a} = \int \left(\mu_0 \mathbf{J} + \mu_0 \epsilon_0 \frac{\partial \mathbf{E}}{\partial t} \right) \cdot d\mathbf{a}$$

$$\oint \mathbf{B} \cdot d\mathbf{l} = \mu_0 I + \mu_0 \int \epsilon_0 \frac{\partial \mathbf{E}}{\partial t} \cdot d\mathbf{a}$$

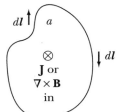

Figure 3.23   An illustration of Stokes' theorem—a "Stokesian" or Amperian loop.

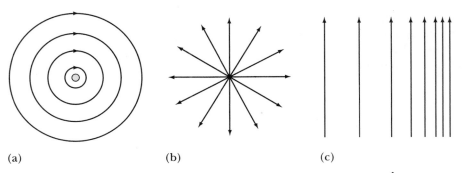

Figure 3.24   **(a)** Curl at center; not elsewhere if field is proportional to $\hat{\phi}/\rho$. **(b)** No curl. **(c)** Curl everywhere.

Figures 3.24 and 3.25 illustrate fields with and without curl. One conceptual test for the presence of curl is this: Imagine that the lines represent fluid flow, and put in a paddle wheel, Figure 1.7b; if the wheel goes around, there's a curl. Another test: If you can draw a closed contour that has a nonzero line integral around it, there's a curl. If $\nabla \times \mathbf{B} = 0$, the field is called **irrotational.** If $\nabla \cdot \mathbf{B} = 0$, it's **solenoidal.**

◢ EXAMPLE 3-9

Find the curl of the field $\mathbf{B} = \rho\hat{\phi}$ in cylindrical coordinates.

ANSWER   See Figure 3.25. From inside the front cover,

$$\nabla \times (\rho\hat{\phi}) = \frac{1}{\rho} \frac{\partial}{\partial\rho} (\rho\rho)\, \hat{\mathbf{z}} = 2\hat{\mathbf{z}}$$

It is constant, and it points in the z direction.

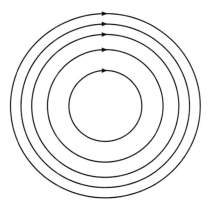

Figure 3.25   Curl everywhere.

 **EXAMPLE 3-10**

In the preceding example, what is the current density?

ANSWER From Ampère's law,

$$\mathbf{J} = \nabla \times \mathbf{B}/\mu_0$$

$$J = 2/\mu_0$$

The same **J** can produce a field of $\mathbf{B} = 2x\hat{\mathbf{y}}$, Figure 3.24c. As with Figures 2.24c and 2.25, we need to know the boundary conditions to specify the field uniquely (Section 1.4).

 EXAMPLE 3-11

Find the curl of the field of an infinite thin wire, $\mathbf{B} = \dfrac{\mu_0 I}{2\pi\rho}\,\hat{\phi}$, Figure 3.24a.

ANSWER From inside the front cover,

$$\nabla \times (\hat{\phi}/\rho) = \frac{1}{\rho}\frac{\partial}{\partial\rho}(\rho/\rho)\hat{\mathbf{z}} = 0$$

the curl is zero everywhere except at $\rho = 0$. Yet somehow there is a field anyway. Also, there is a finite line integral around the wire; for a circular Stokesian loop:

$$\oint(\hat{\phi}/\rho)\cdot dl = (1/\rho)2\pi\rho = 2\pi$$

$$= \int\nabla \times (\hat{\phi}/\rho)\cdot d\mathbf{a} \qquad \text{by Stokes' theorem}$$

So $\nabla \times (\hat{\phi}/\rho)$ must involve a two-dimensional Dirac delta function (see Section 2.4), whose value is zero except at the origin, and yet whose integral is one:

$$\int\delta(\rho)\,da = 1$$

We evidently have

$$\nabla \times (\hat{\phi}/\rho) = 2\pi\delta(\rho)\hat{\mathbf{z}}$$

since

$$\int 2\pi\delta(\rho)\hat{\mathbf{z}}\cdot d\mathbf{a} = 2\pi$$

(By the right-hand rule, $\hat{\mathbf{z}}$ is parallel to **a**, the area vector.) So we have found that

$$\nabla \times \mathbf{B} = \mu_0 I\,\delta(\rho)\hat{\mathbf{z}} \qquad (3.31)$$

for the **B** field of a long thin wire, with the current I flowing in the z direction. By way of comparison, in Section 2.4 we obtained

$$\nabla \cdot \mathbf{E} = Q\delta(\mathbf{r})/\epsilon_0$$

(Eq. 2.32) for the E field of a point charge.

 EXAMPLE 3-12

Find the curl of a Coulomb's-law field, Figure 3.24b.

ANSWER From Section 2.2,

$$\mathbf{E} = Q\,\hat{\mathbf{r}}/4\epsilon_0\pi r^2$$

There is no θ or ϕ component or dependence, so an inspection of the expression for $\nabla \times \mathbf{E}$ in spherical coordinates, inside the front cover, reveals that each of its six terms is zero. So, in this case

$$\nabla \times \mathbf{E} = 0 \tag{3.32}$$

Since any electrostatic field is built up of the fields of individual elementary charges, this carries the important corollary that **any electrostatic field has zero curl.**

3.7. $\nabla \cdot \mathbf{B}$, $\nabla \times \mathbf{B}$, and Biot-Savart

Now we are ready to follow more-or-less standard mathematical developments linking the Biot-Savart law with the divergence and curl of **B**. We might hesitate to call these derivations: Biot-Savart is actually limited to the quasi-static case, whereas the differential forms of Maxwell's equations are completely general. In that sense, what we do here could be regarded as constituting a partial justification of the Biot-Savart form. However, we don't disparage Biot-Savart, despite its limitations, because it provides a way of calculating the field in many asymmetrical quasi-static cases where Ampère's law would be much more difficult.

We begin by taking the **divergence** of the Biot-Savart expression, Section 3.2, with the intention of showing that it is zero.

Referring to Figure 3.26, the current source of the **B** field at **r** is located at **r'**. The distance from source point to field point is $\mathbf{r''} = \mathbf{r} - \mathbf{r'}$. The current density is $\mathbf{J'} = \mathbf{J}(\mathbf{r'})$. So, the Biot-Savart law is

$$\mathbf{B} = \frac{\mu_0}{4\pi}\int \frac{\mathbf{J'} \times \hat{\mathbf{r}}''}{r''^2}dv' \tag{3.33}$$

We take the divergence at the field point, at **r**:

$$\nabla \cdot \mathbf{B} = \nabla \cdot \frac{\mu_0}{4\pi}\int \frac{\mathbf{J'} \times \hat{\mathbf{r}}''}{r''^2}dv' \tag{3.34}$$

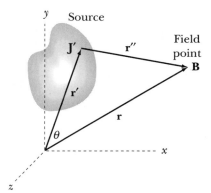

Figure 3.26   Current distribution.

Are we permitted to slip the del operator ∇ past the integral sign? Not necessarily; we have to rationalize that sort of thing on a case-by-case basis. We *can* do it if we are using different variables for the differentiation and integration. For the differentiation, we are evaluating the divergence at the field point,

$$\mathbf{r} = x\hat{\mathbf{x}} + y\hat{\mathbf{y}} + z\hat{\mathbf{z}}$$

so we are differentiating with respect to x, y, and z, while keeping x', y', and z' constant. For the integration we are evaluating the integrand at the source point,

$$\mathbf{r}' = x'\hat{\mathbf{x}} + y'\hat{\mathbf{y}} + z'\hat{\mathbf{z}}$$

and the distance from source point to field point is

$$\mathbf{r}'' = (\mathbf{r} - \mathbf{r}') = (x - x')\mathbf{x} + (y - y')\hat{\mathbf{y}} + (z - z')\hat{\mathbf{z}} \tag{3.35}$$

In the integration, it is x', y', and z' that vary, and the x, y, and z are constant. Thus, we are indeed using different variables for the differentiation and integration, so the order of differentiation and integration may be interchanged:

$$\nabla \cdot \mathbf{B} = \frac{\mu_0}{4\pi} \int \nabla \cdot \left(\mathbf{J}' \times \frac{\hat{\mathbf{r}}''}{r''^2} \right) dv' \tag{3.36}$$

Next, our hardy del operator has to get past the parenthesis. In general, for vectors **A** and **B**, we can write

$$\nabla \cdot (\mathbf{A} \times \mathbf{B}) = \mathbf{B} \cdot (\nabla \times \mathbf{A}) - \mathbf{A} \cdot (\nabla \times \mathbf{B}) \tag{3.37}$$

(see inside front cover for the identity, and see Chapter 1 for such derivations). Now we apply the identity:

$$\nabla \cdot \mathbf{B} = \frac{\mu_0}{4\pi} \int \nabla \cdot \left(\mathbf{J}' \times \frac{\hat{\mathbf{r}}''}{r''^2} \right) dv' \tag{3.38}$$

$$= \frac{\mu_0}{4\pi} \int \frac{\hat{\mathbf{r}}''}{r''^2} \cdot \nabla \times \mathbf{J}' \, dv' - \frac{\mu_0}{4\pi} \int \mathbf{J}' \cdot \left(\nabla \times \frac{\hat{\mathbf{r}}''}{r''^2} \right) dv'$$

In this equation, $\nabla \times \mathbf{J}' = 0$ because $\mathbf{J}'$ depends on primed (source) variables but not on unprimed (field) ones, whereas the differentiation is with respect to field variables. Also, $\nabla \times \dfrac{\hat{\mathbf{r}}''}{r''^2} = 0$ because the curl of an inverse-square field is zero, as observed in Example 3-12. Thus, we arrive at

$$\boxed{\nabla \cdot \mathbf{B} = 0} \tag{3.39}$$

as already suggested in Section 3.4.

Now we tackle the **curl** of Biot-Savart:

$$\nabla \times \mathbf{B} = \frac{\mu_0}{4\pi} \int \nabla \times \frac{\mathbf{J}' \times \hat{\mathbf{r}}''}{r''^2} \, dv' \tag{3.40}$$

As before, we are integrating over source variables at $\mathbf{r}'$, Figure 3.26, and differentiating at the field point at $\mathbf{r}$, so we have exchanged the order of differentiation and integration. We can expand the integrand using another identity from the front cover:

$$\nabla \times (\mathbf{A} \times \mathbf{B}) = (\mathbf{B} \cdot \nabla)\mathbf{A} - (\mathbf{A} \cdot \nabla)\mathbf{B} + \mathbf{A}(\nabla \cdot \mathbf{B}) - \mathbf{B}(\nabla \cdot \mathbf{A}) \tag{3.41}$$

so

$$\nabla \times \left(\mathbf{J}' \times \frac{\hat{\mathbf{r}}''}{r''^2} \right) = 0 - (\mathbf{J}' \cdot \nabla) \frac{\hat{\mathbf{r}}''}{r''^2} + \mathbf{J}' \left(\nabla \cdot \frac{\hat{\mathbf{r}}''}{r''^2} \right) - 0$$

Here $\mathbf{A}$ means $\mathbf{J}'$, and its derivatives are zero because $\mathbf{J}'$ doesn't depend on field variables. So, the expression for $\nabla \times \mathbf{B}$ becomes

$$\nabla \times \mathbf{B} = \frac{\mu_0}{4\pi} \int \mathbf{J}' \left(\nabla \cdot \frac{\hat{\mathbf{r}}''}{r''^2} \right) dv' - \frac{\mu_0}{4\pi} \int (\mathbf{J}' \cdot \nabla) \frac{\hat{\mathbf{r}}''}{r''^2} \, dv' \tag{3.42}$$

Now we will get rid of the second integral. The differential operator ∇ operates on $\mathbf{r}$, and $\mathbf{r}'' = \mathbf{r} - \mathbf{r}'$. If we use instead ∇', which operates on $\mathbf{r}'$, the $\mathbf{r}''$ derivatives will change sign. So, the second integral becomes

$$+ \frac{\mu_0}{4\pi} \int (\mathbf{J}' \cdot \nabla') \frac{\hat{\mathbf{r}}''}{r''^2} \, dv'$$

The x component is, throwing in another vector identity,

$$+ \frac{\mu_0}{4\pi} \int (\mathbf{J}' \cdot \nabla') \frac{x''}{r''^3} \, dv' = - \frac{\mu_0}{4\pi} \int \nabla' \cdot \left(\frac{x''}{r''^3} \mathbf{J}' \right) dv' - \frac{\mu_0}{4\pi} \int (\nabla' \cdot \mathbf{J}') \frac{x''}{r''^3} \, dv' \tag{3.43}$$

We assume steady-state conditions, so $\nabla' \cdot \mathbf{J}' = 0$, which gets rid of the second integral. We can apply the divergence theorem to the first integral; it becomes

$$\frac{\mu_0}{4\pi} \int \nabla' \cdot \left(\frac{x''}{r''^3} \mathbf{J}' \right) dv' = \frac{\mu_0}{4\pi} \int \frac{x''}{r''^3} \mathbf{J}' \cdot d\mathbf{a}' \tag{3.44}$$

If we make the volume of integration very large, so that all currents are inside, then $\mathbf{J}' = 0$ on the surface, so this area integral vanishes. (Nonetheless, we will not shrink from applying the result to infinite current distributions.) The same holds for the y and z components.

So, now we are left with the first integral above, Eq. 3.43:

$$\nabla \times \mathbf{B} = \frac{\mu_0}{4\pi} \int \mathbf{J}' \left(\nabla \cdot \frac{\hat{\mathbf{r}}''}{r''^2} \right) dv' = \frac{\mu_0}{4\pi} \int \mathbf{J}' \, 4\pi \delta(\mathbf{r}'') \, dv' \tag{3.45}$$

using the Dirac delta function from Section 2.4, which is different from 0 only where $\mathbf{r}'' = 0$, so $\mathbf{r} = \mathbf{r}'$. Then $\mathbf{J}'$ is effectively constant and $\mathbf{J}' = \mathbf{J}(\mathbf{r})$; it comes outside of the integral, so

$$\nabla \times \mathbf{B} = \frac{\mu_0}{4\pi} \mathbf{J} \int 4\pi \delta(\mathbf{r}'') \, dv' \tag{3.46}$$

and finally

$$\boxed{\nabla \times \mathbf{B} = \mu_0 \mathbf{J}} \tag{3.47}$$

as in Section 3.6. Thus, we "derive" part of Ampère's law starting with Biot-Savart, but remember what we said in the first paragraph of this section, concerning the significance of this derivation.

3.8. $\partial E / \partial t$ and Biot-Savart

The displacement current density $\epsilon_0 \partial \mathbf{E} / \partial t$ is an essential part of Ampère's law: See Section 3.5, and especially the last example in that section; yet it plays no role in the Biot-Savart law, Section 3.2. Why is this?

We can provide an intuitive answer immediately: In the volume integral of Biot-Savart, $\mathbf{E}$ points in all different directions, so its contribution to $\mathbf{B}$ tends to cancel out; whereas in the area integral implied by Ampère's law, $\mathbf{E}$ tends to point in more or less the same direction across the area, so there is not the same kind of cancellation. Now of course an intuitive idea is not a proof, but it may provide encouragement and guidance in establishing a proof.

We will now present a formal demonstration that $\epsilon_0 \partial \mathbf{E} / \partial t$ has negligible effect in the Biot-Savart law for the quasi-static case. We begin by restating Biot-Savart with the presumed displacement term:

$$\mathbf{B} = \frac{\mu_0}{4\pi} \int \frac{(\mathbf{J}' + \epsilon_0 \partial \mathbf{E}' / \partial t) \times \hat{\mathbf{r}}''}{r''^2} \, dv' \tag{3.48}$$

Figure 3.26 explains the various primes and double primes. We wish to show that the additional term in $\partial \mathbf{E}' / \partial t$ is zero:

$$\int \frac{\partial \mathbf{E}' / \partial t \times \hat{\mathbf{r}}''}{r''^2} \, dv' = \frac{\partial}{\partial t} \int \mathbf{E}' \times \frac{\hat{\mathbf{r}}''}{r''^2} \, dv' \tag{3.49}$$

should be zero.

Now, we have seen the del operator acting as a vector in a dot product, the divergence, Section 2.4, and in a cross product, the curl, Section 3.6. There remains the possibility that del may act upon a scalar, f:

$$\nabla f = \frac{\partial f}{\partial x}\,\hat{\mathbf{x}} + \frac{\partial f}{\partial y}\,\hat{\mathbf{y}} + \frac{\partial f}{\partial z}\,\hat{\mathbf{z}}$$

The physical significance of such an operation will be discussed later, in Section 6.2. Here we are interested in it merely as a formal operation. When we perform this operation on the scalar $1/r$, using the information inside the front cover for spherical coordinates, we find

$$\nabla'\frac{1}{r''} = \frac{\hat{\mathbf{r}}''}{r''^2} \tag{3.50}$$

(positive because we are differentiating with respect to source variables, $\mathbf{r}' = \mathbf{r} - \mathbf{r}''$, Fig. 3.26). So, we can rewrite the integral in question, Eq. 3.49, as

$$\frac{\partial}{\partial t}\int \mathbf{E}' \times \nabla'\frac{1}{r''}\,dv'$$

Now we introduce a vector identity:

$$\int \mathbf{E}' \times \nabla'\frac{1}{r''}\,dv' = \int \frac{1}{r''}\,\nabla' \times \mathbf{E}'\,dv' - \int \nabla' \times \frac{\mathbf{E}'}{r''}\,dv' \tag{3.51}$$

(We pluck these vector identities from inside the front cover.)

For the first integral, we remind you, from Example 3-12, that the curl of an electrostatic field is zero, so for the quasi-static case $\nabla \times \mathbf{E} = 0$, and the first integral vanishes.

For the second integral we use an integral formula from the front cover which is neither Stokes' theorem nor the divergence theorem:

$$\int \nabla \times \mathbf{A}\,dv = -\oint \mathbf{A} \times d\mathbf{a} \tag{3.52}$$

so in this case

$$\int \nabla' \times \frac{\mathbf{E}'}{r''}\,dv' = \oint \frac{\mathbf{E}'}{\mathbf{r}''} \times d\mathbf{a}' \tag{3.53}$$

Now at large distances $\mathbf{E}'$ is radial and behaves as $1/r^2$ (for any finite-size nonzero net charge). So the integral vanishes over a surface at large r.

Thus, we dispose of both parts of the integral in Equation 3-51. So the Biot-Savart law does not require a term in $\partial\mathbf{E}/\partial t$. We repeat, this conclusion holds for the quasi-static case, with $\nabla \times \mathbf{E} = 0$, which is sufficient, because Biot-Savart makes no pretense to complete generality.

We must also mention here that there exist other currents, in particular the polarization current density, $\partial\mathbf{P}/\partial t$, Section 9.1, and the magnetization current density, $\nabla \times \mathbf{M}$, Section 10.1. For these the curl of $\mathbf{B}$ does not necessarily vanish, and they *may* figure, in principle, in the Biot-Savart law.

3.9 SUMMARY

When charges move, they may be affected by magnetic fields by means of the Lorentz force. Such fields are produced only by other moving charges; there are no magnetic monopoles. Further analysis brings us to the fundamental Ampère's law.

The magnetic field may be calculated using the Biot-Savart law:

$$\mathbf{B} = \frac{\mu_0 I}{4\pi} \int \frac{dl \times \hat{\mathbf{r}}}{r^2} \tag{3.4}$$

For a long straight wire, this becomes

$$B = \frac{\mu_0 I}{2\pi r} \tag{3.6}$$

The force on a wire is

$$\mathbf{F} = I \oint dl \times \mathbf{B} \tag{3.5}$$

For a straight wire segment perpendicular to $\mathbf{B}$,

$$F = BIl \tag{3.7}$$

The Lorentz force on a charge moving in electric and magnetic fields is

$$\mathbf{F} = Q\mathbf{E} + Q\mathbf{v} \times \mathbf{B} \tag{3.12}$$

For a charge moving in a circle in a magnetic field,

$$m\mathbf{v} = QrB \tag{3.13}$$

The magnetic field may be visualized using magnetic field lines. Magnetic flux is

$$\Phi = \int \mathbf{B} \cdot d\mathbf{a} \tag{3.16}$$

One of Maxwell's equations says that there are no magnetic monopoles:

$$\nabla \cdot \mathbf{B} = 0 \tag{3.15}$$

Ampère's law facilitates calculation of fields of appropriate symmetry:

$$\oint \mathbf{B} \cdot dl = \mu_0 I + \mu_0 \frac{d}{dt} \int \epsilon_0 \mathbf{E} \cdot d\mathbf{a} \tag{3.19}$$

Ampère's law in differential form is another of Maxwell's four equations:

$$\nabla \times \mathbf{B} = \mu_0 \mathbf{J} + \mu_0 \epsilon_0 \frac{\partial \mathbf{E}}{\partial t} \tag{3.29}$$

The **B** field is always solenoidal: $\nabla \cdot \mathbf{B} = 0$. In the electrostatic case, **E** is irrotational:

$$\nabla \times \mathbf{E} = 0 \tag{3.32}$$

 PROBLEMS

3-1 Assuming that there is one free electron per atom in copper, find the drift velocity of conduction electrons in a wire of radius 0.5 mm carrying a current of 3 A.

3-2 For Problem 3-1, find the charge of the conduction electrons in a 10-cm segment of wire.

3-3 Find the force between a square loop of side s and an infinite wire at distance s, Figure 3.27, each carrying current I. Evaluate for 1 A and 1 cm.

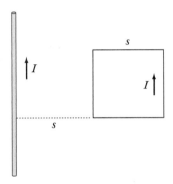

Figure 3.27

3-4 A copper wire oriented East–West carries current density **J** sufficient to levitate the wire in the earth's magnetic field, which is 0.5×10^{-4} T directed North, and gravitational field $g = 9.8$ m/s^2 downward. Find **J**. Is it reasonable? Does **J** point East or West?

3-5 Current I flows upward along the y axis from 0 to L, Figure 3.4a. Show that, at distance a on the x axis,

$$B = \frac{\mu_0 I}{2\pi a} \frac{L}{\sqrt{L^2 + a^2}}$$

in the negative z direction. Evaluate for $a = L = 10$ cm and $I = 1.5$ A.

3-6 Find B at distance z on the axis of a square loop of side s, lying in the xy plane, carrying current I, Figure 3.28. Evaluate at $z = 0$ for $I = 1$ A and $s = 1$ cm.

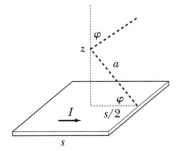

Figure 3.28

3-7 Find B at the center of the semicircle in Figure 3.29. Evaluate for $I = 1$ A and $r = 1$ cm. Compare with the earth's magnetic field.

Figure 3.29

3-8 Find B at the center of the semicircle in Figure 3.30. Evaluate for $I = 1$ A and $r = 1$ cm. Compare with the earth's magnetic field.

Figure 3.30

3-9 A charge of -2 nC moves East with speed 2000 m/s, in a magnetic field of 5×10^{-5} T directed Northward. Find the force.

3-10 A proton traveling across a magnetic field of 1.5 T describes a circle of radius 7 cm. What is its speed?

3-11 If at the earth's surface the electric field is 100 N/C pointed down, and the magnetic field is 5×10^{-5} T pointed northward, at what minimum velocity must an electron travel to experience no Lorentz force?

3-12 Starting with the Lorentz force, derive the expression for the cyclotron frequency ω for a particle of charge Q and mass m moving across (perpendicular to) a magnetic field B. Calculate the cyclotron frequency for a proton, and for an electron, in a magnetic field of 1.2 T.

3-13 Find B at 3 cm from a large plane of surface current 4 A/m.

3-14 Find B at 3 cm from a long wire of current 4 A.

3-15 Starting with Ampère's law in differential form, find the magnetic field at r from the axis, inside and outside of a circular toroid, Figure 3.31, of major radius R and minor radius a, wrapped with N turns of wire carrying current I. Evaluate for $r = R = 5$ cm, $a = 2$ cm, $N = 1000$, $I = 3$ A. (The Amperian loop is a circle in the plane of the toroid.)

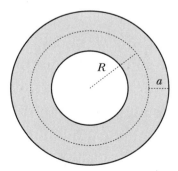

Figure 3.31

3-16 Find B inside and outside of an infinite cylinder of radius R and current density $J(r) = K/r$, where K is a constant, Figure 3.32; then reexpress in terms of current I. (Compare with Problem 2-17. Use a circular Amperian loop.)

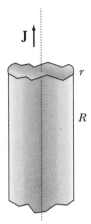

Figure 3.32

3-17 Integrate the field of a loop, Section 3.2, from $-\infty$ to ∞ to find the field on the axis of an infinite solenoid of circular cross section and of surface current density $K = N'I$. What does the result imply for the field outside the solenoid (in view of Ampère's law)?

*3-18 Find the B field at z (from the end) on the axis of a finite solenoid of radius a and length L, Figure 3.33, carrying surface current density $K = N'I$. Show that it is proportional to the solid angle subtended by the surface current.

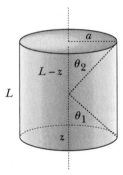

Figure 3.33

3-19 Find the B field between parallel planes, Figure 3.34, separated by s, carrying opposite surface current densities of K. Evaluate for $K = 1000 \text{ A/m}$.

Figure 3.34

3-20 Find the B field between coaxial cylinders of radii r_1 and r_2, Figure 3.35, carrying opposite currents of I. Evaluate at $r = 1$ cm for $r_1 = 0.5$ cm, $r_2 = 1.5$ cm, and $I = 10$ A.

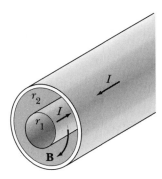

Figure 3.35

3-21 Superposition: An infinite cylinder of uniform (axial) current density $\mathbf{J}$ and radius R contains a cylindrical cavity of radius $R/2$, Figure 3.36. Find the $\mathbf{B}$ field at point p on the surface.

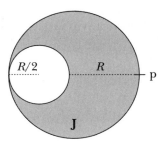

Figure 3.36

*3-22 Superposition: For Problem 3-21, find $\mathbf{B}$ inside the cavity.

3-23 Find the curl of the field $\mathbf{B} = x\,\hat{\mathbf{x}}$. Sketch such a field. Find the current density $\mathbf{J}$ (for $\partial \mathbf{E}/\partial t = 0$). Is such a field possible?

3-24 Find the curl of the field $\mathbf{B} = x\,\hat{\mathbf{y}}$. Sketch such a field. Find the current density $\mathbf{J}$ (for $\partial \mathbf{E}/\partial t = 0$). Is such a field possible?

3-25 Helmholtz coils: Two identical coaxial coils of N turns and of radius R, carrying current I in the same sense, are separated by distance R along the z axis, Figure 3.37. At the center, show that $B = 8\mu_0 NI/5^{3/2} R$. Evaluate for $N = 100$, $I = 2$ A, $R = 0.1$ m. (It

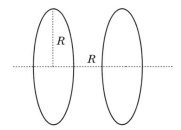

Figure 3.37

is not hard to show that at the center $\partial B/\partial z = 0$ and $\partial^2 B/\partial z^2 = 0$; consequently, this presents a simple way of producing a rather uniform yet accessible magnetic field.)

3-26 Magnetic focusing in a cyclotron: In a region of cylindrical geometry, Figure 3.38, where $B = B_z$ and $\partial B_z/\partial\rho < 0$, use $\nabla \cdot \mathbf{B}$ and $\nabla \times \mathbf{B}$ to show that the **B** field bows in such a way as to push orbiting particles toward the median plane. Sketch such a field.

Figure 3.38

3-27 A disk of radius R and uniform surface charge density σ lies in the xy plane with its axis coinciding with the z axis, Figure 3.39. It rotates about its axis with angular velocity ω. Find B at position z on the z axis. Evaluate for $R = z = 5$ cm, $\sigma = 2\ \mu C/m^2$, at 7200 rpm (rev/min).

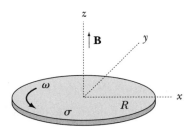

Figure 3.39

3-28 A uniform ion beam of radius R consists of N' ions per meter of mass m, speed v, and charge q. Find the Lorentz force (due to both E and B) on an ion at the edge of the beam. At what speed would the force become zero?

3-29 An electron travels in a circular orbit about a fixed proton with angular momentum $\hbar = h/2\pi$ (Bohr model). Calculate the average B, in teslas, at the proton due to the electron.

3-30 A long straight wire at $x = a$ carries current I in the negative y direction, and another at $x = -a$ carries I in the positive y direction. Find the B_z field along the x axis. Sketch it.

3-31 Superposition: Show that the field on the axis at the end of a semi-infinite solenoid is half of the field well inside of the solenoid, Figure 3.40. Then show that half of the flux inside of the solenoid leaves through the walls, and half leaves through the open end.

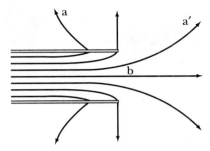

Figure 3.40

3-32 Superposition: Show that the field line passing through the edge of a semi-infinite solenoid is straight and perpendicular to the solenoid axis, Figure 3.40. Then compare the fields in regions marked a and a'.

3-33 Show that there is no net force on a rigid current loop in a uniform magnetic field. (But the net torque need not be zero.)

3-34 A circular current loop of current I and radius R is partially inserted into a region of uniform magnetic field **B**, Figure 3.41. The plane of the loop is perpendicular to **B**. Show that the force on the loop equals BIl, where l is the chord. Find the direction in this case.

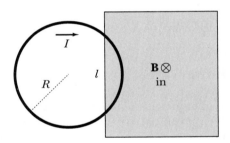

Figure 3.41

3-35 Uniform surface current density **K** exists on the xy plane, in a thin strip from $x = -w/2$ to $+w/2$, extending to $y = \pm\infty$. Find **B** at a on the $+x$ axis outside of the strip. Check for large a.

*3-36 Show that the force between two loops, Eq. 3.1, can be written in the symmetrical form:

$$\mathbf{F} = \frac{\mu_0}{4\pi} \oint a \oint b \, \frac{\hat{\mathbf{r}}}{r^2} \, dl_a \cdot dl_b$$

C H A P T E R

4

ELECTROMAGNETIC INDUCTION – FARADAY'S LAW

As we have seen, stationary charge is described using electrostatics, and uniform motion of charge brings in magnetism. In this chapter we will for the first time be concerned with the effects of the acceleration of charge, and we will see that it can produce electric fields. However, we will find that there are also other ways whereby one may induce electricity using magnetic fields, by exploiting uniform relative motion of charges. These diverse processes may be conveniently summarized in Faraday's law of electromagnetic induction.

4.1. Faraday's Law in Integral Form; emf

If you pass a current through a wire, you get a magnetic field. It is reasonable to guess that there might be a reciprocal effect: If you pass a wire through a magnetic field, you can get a current. Such effects are readily demonstrated. Referring to Figure 4.1a, if we pass a current of a few amperes through a wire in the field of a permanent magnet, a noticeable Lorentz force is observed, $\mathbf{F} = Q\mathbf{v} \times \mathbf{B}$ (Section 3.3). The direction is given by the corresponding right-hand rule: $\mathbf{B}$ points into the paper, and if we press $\mathbf{v}$ into $\mathbf{B}$ with our fingers, our thumb points in the direction of $\mathbf{F}$ (or see Fig. 3.5). In Figure 4.1b we move the same wire through the magnetic field, and using an ammeter between the ends we might find a current of a milliamp or so in the direction of the "emf." Both forces $\mathbf{F}$ shown here represent the Lorentz force on the positive charge carriers in the wire (in Fig. 4.1b there is also a force, not shown, that we must exert to move the wire). The speed of the charge carriers would be v in the absence of the Lorentz force. "Emf" stands for "electromotive force," but it is not really a force; so it is sometimes called "electromotance." Roughly speaking, it involves the product of electric field E and distance l, and it tends to make electric current flow; we'll discuss it further in the next page or two.

There are other ways to generate current using magnetism, and in fact our modern electrical civilization depends on them. They were discovered in 1832 by Michael Faraday, and independently by Joseph Henry, in the course of investiga-

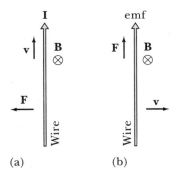

Figure 4.1   (a) Pass current through wire. (b) Pass wire through magnetic field.

(a)   (b)

tions into the nature of electricity, which had little practical application at that time. Indeed, after a public lecture and demonstration, a member of the audience asked Faraday, "This is all very interesting, but what use is it?" His famous reply: "What use is a newborn baby?"

Here are four different geometries, shown in Figures 4.2a–4.2d, each of which yields an induced emf. Current flows in the conducting rectangular loop in the direction of the emf.

Looking at Figures 4.2a and 4.2b, clearly the emf should be the same; it shouldn't matter whether you move the loop or the wire. Yet at a fundamental level the physical description is different. In Figure 4.2a the emf is due to the Lorentz force $Q\mathbf{v} \times \mathbf{B}$ on the charges in the wire; it is greater on the left side of the loop than the right because of the $1/r$ falloff of the B field. In Figure 4.2b the charges in the loop aren't moving (if the current is small), so there is no Lorentz force; we will touch upon the reason for the emf in the chapter on Relativity, Section 18.9.

Fields don't move. An electric or magnetic field has a magnitude at a given point, and it has space and time derivatives there. Indeed, waves may travel through such a field. However, it has no characteristic corresponding to the speeds of its sources. The idea of an E or B field, of some given magnitude and direction, "moving West at 20 m/s," is physically meaningless. Thus, in Figure 4.2b, the magnetic field lines are *not* being carried along with the wire. It is sometimes intuitively attractive to imagine field lines sweeping through space or cutting through conductors; for example, in plasmas it may be useful to imagine that the magnetic field lines are carried along with the plasma. But you should not forget that fields don't actually move.

Looking at Figures 4.2b and 4.2c, clearly an emf in one implies an emf in the other: The loop can't tell whether the wire is approaching or the current is increasing. Yet Figure 4.2c for the first time brings in an effect caused by **acceleration** of charge, which we shall examine in this and later chapters. For example, Figure 14.24 (Section 14.6) suggests that when we accelerate a charge toward the right, neighboring charges will (later) experience a force pointing towards the left, that is, antiparallel to the acceleration of the first charge.

Is there some generalization covering all of these emf's? In the three preceding cases, we find B increasing in one reference frame or another; in particular, in

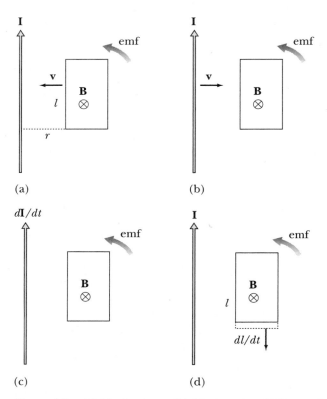

(a)   (b)

(c)   (d)

Figure 4.2   (a) Moving loop. (b) Moving wire. (c) Increasing *I*. (d) Expanding loop.

Figure 4.2a an observer sitting on the moving loop finds an increasing *B*. But in the expanding loop of Figure 4.2d, there is no question of *B* changing at all. Faraday understood what it is that all four cases have in common: The magnetic flux through the loop is increasing. There is no other law of physics comparable to the law of **electromagnetic induction** in its extraordinary ability to summarize several fundamentally different phenomena in one very simple relationship, **Faraday's law in integral form:**

$$\text{emf} = -\frac{d\Phi}{dt} = -\dot{\Phi} \tag{4.1}$$

(the dot implies time derivative). This is sometimes called the **flux rule.** It implies that when the magnetic flux through a circuit changes, an emf is induced that tends to drive charge around the circuit. In this context, **emf** involves an integral about a closed circuit:

$$\text{emf} = \oint \mathbf{E} \cdot d\mathbf{l}$$

(emf is often represented by the symbol $\mathcal{E}$. Sometimes we will permit a contribution to emf involving a path that is not closed, $\int_a^b \mathbf{E} \cdot d\boldsymbol{l}$; see Section 6.4 for more.)

As in the preceding chapter, the magnetic flux is given by

$$\Phi = \int \mathbf{B} \cdot d\mathbf{a}$$

where the surface **a** is bounded by the wire, and the vector **a** is normal to the surface itself. Of course Φ is a scalar, but we will sometimes think of it as "pointing" in the direction of **B**.

There are obvious exceptions to the flux rule. For example, in Figure 4.3, if we change the position of the switch from a to b, the flux in the circuit changes, but there is no corresponding emf. An old-timer would say: No wire is cutting flux lines. In Figure 4.2d, the lower wire is actually moving through a B field, and the emf is due to the Lorentz force; but in Figure 4.3, neither lower wire is moving.

Combining the equations for Φ and emf, we can write Faraday's law as

$$\boxed{\text{emf} = \oint \mathbf{E} \cdot d\boldsymbol{l} = -\frac{d}{dt} \int \mathbf{B} \cdot d\mathbf{a}} \qquad \textbf{(4.2)}$$

Units: The electric field E is force per unit charge, N/C; and dl is distance in meters, m. Also, force times distance is work in joules: $\text{N} \cdot \text{m} = \text{J}$. This means that the units of emf are work per unit charge, J/C, which we define as V, **volt.** Consequently, the units of the electric field E can be given as volts per meter, V/m, just as well as in the units we have been using up until now, N/C. Also, in view of Equation 4.1, a **weber** is a volt-second (as well as a tesla-meter^2, Section 3.4).

The usual **right-hand rule** applies: Place the thumb in the direction of $d\mathbf{a}$, and the fingers curl in the direction of dl.

We shall now derive Faraday's law for Figure 4.2a, using the Lorentz force, which results in what is sometimes called **motional emf,** caused by motion of a wire through a B field. We shall simply postulate that it holds for the other cases; of course it's verified experimentally.

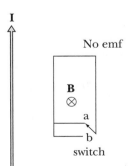

Figure 4.3   Failure of flux rule.

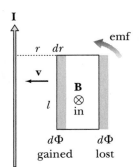

Figure 4.4   Flux rule.

In time dt, the side marked l moves a distance dr and sweeps through an area of $l\,dr$, Figure 4.4. The flux $d\Phi = Bl\,dr$ thus enters the loop. So for the side marked l:

$$\frac{d\Phi}{dt} = Bl\frac{dr}{dt}$$

$$\text{emf} = Bl\text{v}$$

If B entered at angle θ, this would obviously become

$$\boxed{\text{emf} = Bl\text{v}\sin\theta}$$

There is no E field in Figure 4.2a; the force on the charges in the loop is simply $Q\text{v} \times \mathbf{B}$. But in their rest frame, Figure 4.2b, there is no v for them, so no $\text{v} \times \mathbf{B}$; instead, in this frame there is an $\mathbf{E}$ field accompanying the changing $\mathbf{B}$ field, and the force is $Q\mathbf{E}$. We mentioned this phenomenon in Section 3.3, and a fuller discussion of how fields come and go in different frames of reference is deferred to Chapter 18, on Relativity. The force on the charges must be the same in the two cases (or nearly so, for small speed): $Q\mathbf{E} = Q\text{v} \times \mathbf{B}$. When a wire moves across a $\mathbf{B}$ field in the laboratory frame of reference, the effective $\mathbf{E}$, in its rest frame, is the same as $\text{v} \times \mathbf{B}$ in the lab frame. So for the side marked l:

$$\text{emf} = \int \mathbf{E}\cdot d\mathbf{l} = -\text{v}Bl = -\frac{d\Phi}{dt} \qquad \text{from above}$$

Sign: If $d\mathbf{a}$ points into the paper, so that $\mathbf{B}\cdot d\mathbf{a}$ is positive, then by the right-hand rule $\mathbf{B}$ and $d\mathbf{l}$ point in opposite directions. So for this side, we indeed find that the emf $= -d\Phi/dt$. We can then carry through the same argument for the flux leaving the loop on the other side. As for the top and bottom sides, no flux crosses them. So for the emf for the whole contour, and for the net flux gained, we find

$$\text{emf} = -\frac{d\Phi}{dt}$$

so we have arrived at Faraday's law for motional emf, starting with the Lorentz force in Figure 4.2a.

◥ EXAMPLE 4-1

A rod slides at speed v along parallel rails, which are connected electrically and are separated by distance l, in a uniform magnetic field **B** directed out of the paper, Figure 4.5. Find the emf around the circuit. Evaluate for $B = 0.1$ T, v $= 3$ m/s, $l = 70$ cm.

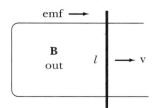

Figure 4.5   Rod and rails.

ANSWER As above,

$$\text{emf} = Blv$$
$$= 0.1 \text{ T} \times 0.7 \text{ m} \times 3 \text{ m/s} = 0.21 \text{ V} \qquad \text{clockwise}$$

Although the direction is contained in principle in the right-hand rule, in practice the easiest and most reliable way to find the direction of the emf is **Lenz's law:** The induced current flows in such a direction as to tend to oppose the *change* in B flux (not necessarily **B** itself). In this case the flux in the circuit is increasing *out of* the paper, so the current wants to flow in such a way as to create magnetic flux *into* the paper; this means clockwise.

In connection with Lenz's law it is worth noting that Nature does not always react in such a way as to *oppose* change; for example, when iron is placed in a magnetic field, "bound currents" arise which *enhance* the magnetic field that produces them (see Chapter 10).

The emf given in Faraday's law is for one complete contour. If there are N loops connected together in series, then obviously we get more emf:

$$\text{emf} = -N\frac{d\Phi}{dt} \qquad\qquad (4.4)$$

The product $N\Phi$ may be called the **flux linkage.**

◥ EXAMPLE 4-2

A loop of $N = 100$ turns of wire and of radius 2 cm lies in the plane of the paper. A magnetic field of 0.15 T is introduced, pointing into the paper, in time 0.3 s. Find the emf. Does current tend to flow clockwise or counterclockwise?

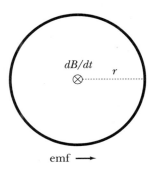

Figure 4.6   Changing flux.

ANSWER See Figure 4.6. For each loop we have an emf of $-d\Phi/dt$. The area is πr^2. So, we have

$$\text{emf} = -\pi r^2 N \, dB/dt$$
$$= -3.14 \times (0.02 \text{ m})^2 \times 100 \times 0.15 \text{ T}/0.3 \text{ s}$$
$$= -0.063 \text{ volt} \qquad \text{counterclockwise}$$

The flux is increasing *into* the paper, so by Lenz's law, the induced emf must tend to produce flux *out of* the paper.

It is remarkable that the *B* field need not touch a wire at all in order that an emf be induced there. In Figure 4.7 we see an infinite solenoid with increasing current. A circular Stokesian loop is drawn outside of the solenoid. The *B* field outside of an infinite solenoid is always zero, Section 3.5; however, there is flux passing through the loop, even though the flux doesn't touch the loop. So an emf is induced in the direction shown by *E* (see Section 6.3 for more).

4.2. Faraday's Law in Differential Form

Faraday's law in integral form, Equation 4.2, seems to invite Stokes' theorem (inside front cover, and Sections 1.5 and 3.6), so let's apply it:

$$\oint \mathbf{E} \cdot d\boldsymbol{l} = -\frac{d}{dt} \int \mathbf{B} \cdot d\mathbf{a}$$

$$\int \nabla \times \mathbf{E} \cdot d\mathbf{a} = -\frac{d}{dt} \int \mathbf{B} \cdot d\mathbf{a}$$

This is true for arbitrary areas, so the urge to set integrands equal is almost irresistible. If only we could sneak the time derivative past the integral sign . . . .

Unfortunately, the case of motional emf stands in the way, Figures 4.2a and 4.2d. There the flux in the loop is changing, but **B** is not changing at any point, so

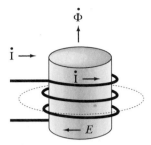

Figure 4.7   Long solenoid.

we can't simply put a $d\mathbf{B}/dt$ inside the integral sign. What we *can* do, however, is to set that case aside, and to write Faraday's law for the other cases, Figures 4.2b and 4.2c, where the loop is stationary. In these cases the **B** field really is changing at a given point. Now we can make the time derivative a partial derivative, keeping the other variables constant:

$$\int \nabla \times \mathbf{E} \cdot d\mathbf{a} = -\frac{\partial}{\partial t} \int \mathbf{B} \cdot d\mathbf{a}$$

$$= -\int \frac{\partial}{\partial t} \mathbf{B} \cdot d\mathbf{a} \tag{4.5}$$

It's okay to put it inside, because the integration and differentiation involve different variables. This being true for arbitrary contours, the integrands must be equal:

$$\boxed{\nabla \times \mathbf{E} = -\frac{\partial \mathbf{B}}{\partial t}} \tag{4.6}$$

This is the other Maxwell equation, **Faraday's law in differential form;** now we've seen all four. The equation is true at every point and under all circumstances.

However, you should understand that the integral form of Faraday's law is more inclusive than the differential form because it contains the case of motional emf, Figures 4.2a and 4.2d.

◢ EXAMPLE 4-3

Describe the magnetic field associated with $\mathbf{E} = x\hat{\mathbf{y}}$.

ANSWER If we take the curl, $\nabla \times \mathbf{E} = \hat{\mathbf{z}}$ so $\partial \mathbf{B}/\partial t = -\hat{\mathbf{z}}$. Integrating, $\mathbf{B} = -t\hat{\mathbf{z}} + \mathbf{B}_0$. B is increasing with time in the negative z direction, but we don't know what $\mathbf{B}_0$ is.

Now that Faraday's law has been shorn of motional content, we can apply it to the remark made near the beginning of the chapter, in connection with Figure 4.2c: Accelerating charge induces an emf in the opposite direction in its neighbor-

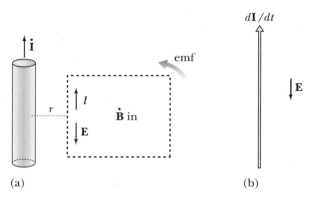

Figure 4.8   **(a)** Infinite wire. **(b)** Antiparallel **E**.

hood. In Figure 4.8a, an infinite straight wire carries increasing current. We start with Faraday's law, integrate it over the loop, and apply Stokes' theorem.

$$\nabla \times \boldsymbol{E} = -\frac{\partial \boldsymbol{B}}{\partial t}$$

$$\int \nabla \times \boldsymbol{E} \cdot d\mathbf{a} = -\int \frac{\partial}{\partial t} \boldsymbol{B} \cdot d\mathbf{a}$$

$$\oint \boldsymbol{E} \cdot d\boldsymbol{l} = -\frac{\partial}{\partial t} \Phi$$

Now, by the right-hand rule, $\partial \boldsymbol{B}/\partial t$ points into the paper through the loop. Then we put our fingers in the direction of l around the loop, and our thumb points in the direction of $\partial \Phi/\partial t$, indicating that that quantity is positive. That means that $\boldsymbol{E} \cdot d\boldsymbol{l}$ is predominantly negative; in particular, $\boldsymbol{E}$ must be antiparallel to l on the side of the loop closest to the wire, because the change in $\boldsymbol{B}$ field is greatest there. The same result is obtained more easily using Lenz's law: The wire produces flux that increases into the paper, so the emf must be such as to tend to produce a current whose flux points out of the paper. And that is what we wanted to see: The E field associated with accelerating charge tends to push neighboring charge in the direction opposite to the acceleration, Figure 4.8b. (See Fig. 14.24 for an intuitive explanation of this.)

4.3.   Generators and Motors

The electric currents governed by Faraday's law provide the means whereby ordinary people get power in all corners of the globe. Figure 4.9 shows an electrical **generator** in its simplest form: A rotating rectangular loop of wire in a magnetic field B. Angular velocity is $\omega = d\theta/dt$, where θ is the angle between the loop (normal) and the field. The area of B subtended by the loop is proportional to sin θ, so the output is a sine wave of voltage or emf. The electrical power output is provided

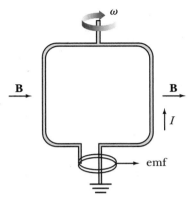

Figure 4.9   AC generator.

by the mechanical input power required to keep the rotating loop turning. For the emf produced, see Problem 4-15.

A generator becomes a **motor** when external emf is applied. Indeed, there have been automobiles that used the same device for starting-motor and generator; but the practice has been abandoned for reasons that are physically trivial but financially significant. As a motor, the device still acts as a generator in that it produces a "back emf." If there is no load, the device spins freely, and the wires cutting across magnetic field lines produce an emf that cancels out the imposed voltage, thus conserving energy. For the torque produced, see Problem 4-16.

When we rotate a conducting cylinder in a magnetic field, a motional emf arises. This is a Faraday disk, or a **homopolar generator.** The disk can even be a cylindrical permanent magnet rotating about its own N–S axis, Figure 4.10; the effect is due not to any imagined rotation of the field but to the motion of the electrically conducting magnet through its own (stationary) field. An emf arises be-

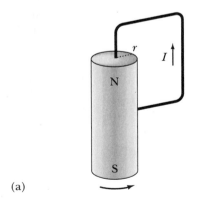

(a)

(b)

Figure 4.10   **(a)** Homopolar generator.
(b) Area swept out.

tween the axis and the outside surface of the magnet, because of the Lorentz force on the free electrons. If we connect a stationary wire from axis to surface as shown, a current I flows. If the wire rotates along with the magnet, there is no effect; differential rotation is required. (So if we hold the magnet and move the wire, the current flows the other way.)

▽ EXAMPLE 4-4

In Figure 4.10, for average $B = 0.5$ T at the top of the magnet, at 1800 rpm (revolutions per minute), with $r = 1.0$ cm, estimate the induced emf.

ANSWER As the magnet surface passes under the stationary dotted line in Figure 3.10a, an area of $r^2 d\theta / 2$ is swept out in time dt, Figure 4.10b. Then,

$$\text{emf} = -d\Phi/dt = B\frac{r^2 d\theta}{2\,dt} = B\omega\, r^2/2 \tag{4.7}$$

$$= 0.5 \text{ T} \times \frac{1800 \text{ rev/min} \times 2\pi \text{ rad/rev}}{60 \text{ s/min}} \times (0.01 \text{ m})^2/2 = 4.7 \text{ mV}$$

from axis to edge. A nearly equal flux passes out the side of the cylinder (Problem 3-31), so the total will be around 9.4 mV. Homopolar generators are of limited practical importance because the emf tends to be small.

If we cause current to flow in the wire in Figure 4.10, using a battery for example, then the magnet tends to rotate, and we have a **homopolar motor** (see Problem 4-17).

It is possible to use the current produced by a homopolar generator to provide its own field, Figure 4.11. The copper disk rotates with angular velocity ω, as shown, and current I flows in the stationary wire loop. With relative rotation in the

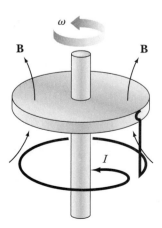

Figure 4.11   Dynamo.

correct sense, a *B* field arises spontaneously, inducing current *I*, which in turn increases magnetic field *B*. Rotation in the other direction reduces any *B* field originally present. This kind of dynamo is probably responsible for magnetic fields in bodies such as the earth and sun (Section 3.4). Differential motion is provided by internal convection due to temperature differences. (See Problem 10-34 for analysis.)

QUESTION: Why are practical motors and generators based on magnetic effects, rather than electrostatic ones?

ANSWER: Because there are no magnetic monopoles. As mentioned in Section 3.1, classically all magnetic forces are due to electrical currents (including changing electric fields; in quantum mechanics we also find that some particles have intrinsic magnetic moments not explicable in terms of classical current flow). Ordinary materials are composed of electric monopoles, namely electrons and positive nuclei. When electric fields are generated which are intense enough to provide useful forces, these electric monopoles rush in to cancel them out; that is, the materials "break down" electrically. Imagine magnets in a world of plentiful magnetic monopoles: As soon as large magnetic fields were produced, the magnetic monopoles would rush in and cancel them. In that sense we appear to be fortunate that magnetic monopoles are, to say the least, very scarce.

4.4 SUMMARY

To produce electricity using a magnetic field, we may move a conductor across the field or we may change the field. This information is contained in Faraday's law in various ways.

When the magnetic flux through a loop changes, the induced emf is given by Faraday's law in integral form:

$$\text{emf} = -\frac{d\Phi}{dt} \tag{4.1}$$

which is the same as

$$\oint \mathbf{E} \cdot d\mathbf{l} = -\frac{d}{dt} \int \mathbf{B} \cdot d\mathbf{a} \tag{4.2}$$

For a moving conductor this may become:

$$\text{emf} = Blv \sin \theta \tag{4.3}$$

For *N* loops in series,

$$\text{emf} = -N\frac{d\Phi}{dt} \tag{4.4}$$

where $N\Phi$ is the flux linkage.

The direction of the emf is provided by Lenz's law: Induced current tends to flow in such a direction as to produce a magnetic flux opposing the imposed change in flux.

Fields don't move; instead, they have space and time derivatives.

Faraday's law in differential form is one of Maxwell's equations:

$$\nabla \times \mathbf{E} = -\frac{\partial \mathbf{B}}{\partial t} \tag{4.6}$$

It is not entirely equivalent to the integral form above.

 PROBLEMS

4-1 A square loop of side 60 cm and of 17 turns lies in the horizontal plane, Figure 4.12. A magnetic field, directed 25° away from vertically downward, decreases from 0.13 T to 0.07 T in 4 seconds. Find the induced emf. Is it clockwise or counterclockwise as seen from above?

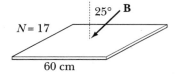

Figure 4.12

4-2 A loop of radius 3 cm and of 170 turns lies in the plane of the paper. A magnetic field is introduced in 0.2 second, producing a clockwise emf of 0.025 volt. What is the vertical component of the field?

4-3 A vertical wire of length 5 cm is passed across a horizontal field of 0.2 T at a speed of 20 m/s and at an angle of 30° relative to the field. Find the emf.

4-4 An airplane of wingspan 20 m flies northward at 300 m/s through the earth's magnetic field, assumed to be downward at 20° north of vertical, and of magnitude 5×10^{-5} T. Find the emf between the wing tips (including direction). What current flows?

4-5 A circular loop of radius 2 cm and of 57 turns is flipped in 0.1 s in a magnetic field of 1.3 T; that is, its axis is initially parallel and finally antiparallel to the field. The resistance of the loop is $R = 16$ ohms (Ω). Using Ohm's law, emf = IR, find the average current.

4-6 A square loop of side 1.25 cm and of 137 turns is flipped in 0.1 s in a magnetic field of 2.1 T. The resistance of the loop is 13 Ω. Find the total charge that flows.

4-7 An infinite solenoid of radius $R = 2$ cm carries $N' = 1700$ turns/m. The current increases from 0 to 1.2 A in 0.4 s. Find the emf in a single loop at $r = 3$ cm, outside the solenoid, Figure 4.7.

4-8 Repeat the preceding problem for a single loop at $r = 1$ cm, inside the solenoid.

4-9 For Figure 4.13a, assume that $I = 3$ A, $r = 4$ cm, v $= 20$ cm/s, $l = 5$ cm, and the other side of the loop is 2.5 cm. Find the emf about the loop without assuming that B changes at any point.

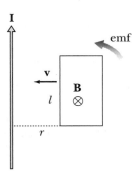

Figure 4.13   (a) Moving loop.

4-10 Repeat the preceding problem for Figure 4.13b without assuming that the loop is moving.

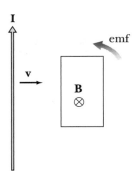

Figure 4.13   (b) Moving wire.

4-11 For Figure 4.13c, find the value of dI/dt that yields the same emf as that in Problem 4-9.

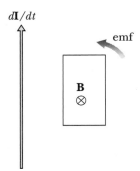

Figure 4.13   (c) Increasing I.

4-12 For Figure 4.13d, find the speed v = dl/dt that gives the same emf as that in Problem 4-9.

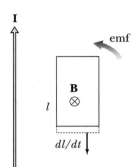

Figure 4.13   (d) Expanding loop.

4-13 $\mathbf{E} = 17\hat{\phi}/\rho$ in cylindrical coordinates; describe **B**. Sketch **E**.

4-14 $\mathbf{E} = 17\hat{\phi}$ in cylindrical coordinates; describe **B**. Sketch **E**.

4-15 A loop of *N* turns of area *a* rotates with angular velocity *ω* normal to a magnetic field **B**, Figure 4.9. At $t = 0$ the plane of the loop is parallel to **B**, as shown. Find the induced emf. Find the maximum emf for $N = 17$, $a = 10$ cm^2, $B = 0.40$ T, $f = 60$ Hz. $(\omega = 2\pi f)$

4-16 A plane loop of *N* turns of current *I* and area *a* sits with vector **a** normal to a magnetic field **B**, Figure 4.14. Find the torque. Evaluate for $N = 17$, $a = 10$ cm^2, $B = 0.40$ T, $I = 2$ A.

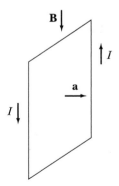

Figure 4.14

4-17 A current of $I = 3$ A flows $r = 1.0$ cm across a field of 0.5 T in a stationary homopolar motor, Figure 4.15; find the torque.

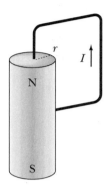

Figure 4.15

4-18 **(a)** A rod of length L rotates with angular velocity ω about an axis that is perpendicular to the rod at its center and that is parallel to the magnetic field **B**. Find the emf for the length of the rod. Evaluate for $B = 2$ T, $L = 30$ cm, rotational rate 7200 rpm (revolutions per minute). **(b)** Repeat for axis at one end.

4-19 Satellite tether: Two earth satellites are connected by a 20-km conductive tether that moves perpendicular to the earth's field of 5×10^{-5} T. Their altitude is about 200 miles. Find the induced emf along the tether. (The electrical circuit is closed by ion guns at the satellites; the current may be several amperes.)

*4-20 Betatron: In this electron accelerator, the electrons are not only accelerated but also held in orbit, of constant r, by the same increasing **B** field, directed into the page, Figure 4.16. Show that the average B_i inside the electron orbit is twice the B_0 at the orbit. (This works even relativistically, so the betatron can accelerate electrons up to speeds exceeding $0.99c$, where c is the speed of light; it is limited by the tendency of the electrons to radiate away their energy when accelerated.) Would this work for accelerating protons?

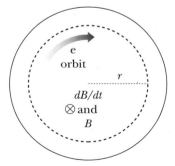

Figure 4.16

C H A P T E R
5

MAXWELL'S EQUATIONS IN VACUUM

Everything we have done so far has built toward the complete and correct formulation of the fields of classical electromagnetism that is contained in Maxwell's equations. To this end, in preceding chapters it has been necessary to provide some experimental verification of the fundamental relationships, as well as to buttress the intuition with necessary concepts. Our further development of electromagnetic theory in subsequent chapters will be both simpler and sounder, being based on Maxwell's firm foundation.

5.1. Maxwell's Equations in Differential and Integral Form

Now we are ready to present all four Maxwell equations on one page:

- Gauss's law; electric flux begins and ends on charge (or at infinity):

$$\oint \mathbf{E} \cdot d\mathbf{a} = \frac{Q}{\epsilon_0} \qquad\qquad \nabla \cdot \mathbf{E} = \frac{\rho_t}{\epsilon_0} \qquad (5.1)$$

- No magnetic monopoles; magnetic field lines don't begin or end:

$$\oint \mathbf{B} \cdot d\mathbf{a} = 0 \qquad\qquad \nabla \cdot \mathbf{B} = 0 \qquad (5.2)$$

- Faraday's law of electromagnetic induction; a changing $\mathbf{B}$ produces an $\mathbf{E}$:

$$\oint \mathbf{E} \cdot d\mathbf{l} = -\frac{d}{dt} \int \mathbf{B} \cdot d\mathbf{a} \qquad\qquad \nabla \times \mathbf{E} = -\frac{\partial \mathbf{B}}{\partial t} \qquad (5.3)$$

- Ampère's law; $\mathbf{B}$ is produced by current and by changing $\mathbf{E}$:

$$\oint \mathbf{B} \cdot d\mathbf{l} = \mu_0 I + \mu_0 \frac{d}{dt} \int \epsilon_0 \mathbf{E} \cdot d\mathbf{a} \qquad \nabla \times \mathbf{B} = \mu_0 \mathbf{J}_t + \mu_0 \epsilon_0 \frac{\partial \mathbf{E}}{\partial t} \qquad (5.4)$$

The fields interact with charge mainly by means of the Lorentz force:

$$\mathbf{F} = Q\mathbf{E} + Q\mathbf{v} \times \mathbf{B} \qquad (5.5)$$

Only the "differential forms" on the right are properly called "Maxwell's equations." When we do numerical calculations, the "integral forms" on the left are usually the ones we need. However, it is the differential forms that are complete, correct, general, elegant, and that will form the basis of our further study of electromagnetism.

The "t" ("total") subscript implies inclusion of all electric charge, even that which is locked inside atoms. Materials are included, in principle, since classically materials may be regarded as nothing but complicated configurations of charge and current. (In practice, however, inclusion of materials will be facilitated by the introduction of special fields designed for that purpose, Chapters 9–11.) In vacuum we can leave the "t" off if we wish.

We have included a brief colloquial description of each equation, but these things must be taken with a grain of salt. For one thing, a cause-and-effect relationship may be implied; however, if a curl of $\mathbf{E}$ is always accompanied by a changing $\mathbf{B}$, then which one can be called the cause and which the effect? We will follow the convenient conventional usage: What is on the left side of the equation is "due to" what is on the right side.

Here is a quick summary of how we arrived at Maxwell's equations:

Coulomb's law gave us Gauss's law, $\nabla \cdot \mathbf{E} = \rho/\epsilon_0$ (Section 2.4) and also $\nabla \times \mathbf{E} = 0$ (Section 3.6), which is missing a term. But Coulomb is static, whereas Gauss is more general.

The *Biot-Savart law* yielded $\nabla \cdot \mathbf{B} = 0$ (Section 3.7) and Ampère's law, $\nabla \times \mathbf{B} = \mu_0\mathbf{J}$ (Section 3.7), which is also missing a term. But again, Ampère is more general.

Faraday's law produced $\nabla \times \mathbf{E} = -\partial\mathbf{B}/\partial t$ (Section 4.2), which supplies the missing term mentioned in connection with Coulomb's law. Here we find that Faraday's law as expressed by Faraday is *more* general than the differential form found in Maxwell's equation.

Ampère's law, as completed by Maxwell, says $\nabla \times \mathbf{B} = \mu_0\mathbf{J} + \mu_0\epsilon_0\partial\mathbf{E}/\partial t$ (Section 3.6), providing the other missing term.

Maxwell's equations summarize the behavior of classical electromagnetic fields. They cannot be derived in their generality; rather, we have guessed at them, and postulated them, based on some simple experiments in special cases. In other words, we have been setting the conceptual foundation for Maxwell's equations. Proof comes with their agreement with experiment.

This point merits further discussion. Feynman asserts (II 26-2): "It is sometimes said, by people who are careless, that all of electrodynamics can be deduced solely from the Lorentz transformation and Coulomb's law. Of course, that is completely false. . . . There are several additional tacit assumptions. . . . (Whenever you see (such) a sweeping statement, you always find that it is false.)" And Jackson says (page 578): "At present it is popular in undergraduate texts and elsewhere to attempt to derive magnetic fields and even Maxwell's equations from Coulomb's law of electrostatics and the kinematics of special relativity. It should be immediately apparent that without additional assumptions this is impossible."

The alternative view is frequently expressed; for example: ". . . many authors have obtained Maxwell's equations from Coulomb's law and special relativity"

(Neuenschwander). After examining the alternatives, this book will side with Feynman and Jackson, as far as rigorous development is concerned. On the other hand, the relativistic approach to electrodynamics is frequently attractive and enlightening. Einstein has remarked (Autobiographical Notes, 9–11), ". . . it was quite sufficient for me if I could peg proofs upon propositions the validity of which did not seem to me to be dubious. . . . If it thus appeared that it was possible to get certain knowledge of the objects of experience by means of pure thinking, this rested upon an error." So we point out relativistic concepts as the need arises, and we deal with the whole question in more detail in Chapter 18, on Relativity.

Maxwell's equations are statements involving the curls, divergences, and time derivatives of the **E** and **B** fields, and in that sense they contain all there is to know about the nature and behavior of the fields, for given boundary conditions (Section 1.4; real-world fields approach zero at infinity, and this in itself can provide satisfactory boundary conditions for completing the specification of the fields). Roughly speaking, the divergence provides useful information on how the field changes as you move in the direction of the field, and the curl does the same for motion across the field.

Maxwell's equations are not all entirely independent. If we take the divergence of Faraday's equation, we find:

$$\nabla \cdot (\nabla \times \mathbf{E}) = -\nabla \cdot \frac{\partial \mathbf{B}}{\partial t}$$

$$0 = -\frac{\partial}{\partial t} \nabla \cdot \mathbf{B}$$

because the divergence of a curl is identically zero (see front cover). This means that $\nabla \cdot \mathbf{B}$ is necessarily a constant because of Faraday's law. If it was ever zero at a given point, it must still be zero there. But who knows what it was in the past? The second equation is still necessary to specify the value of the constant, namely zero.

If we take the divergence of Ampère's equation, we obtain:

$$\nabla \cdot (\nabla \times \mathbf{B}) = \nabla \cdot \mu_0 \mathbf{J} + \nabla \cdot \mu_0 \epsilon_0 \frac{\partial \mathbf{E}}{\partial t}$$

$$0 = \nabla \cdot \mathbf{J} + \nabla \cdot \epsilon_0 \frac{\partial \dot{\mathbf{E}}}{\partial t}$$

$$\boxed{0 = \nabla \cdot \mathbf{J} + \frac{\partial \rho}{\partial t}} \tag{5.6}$$

in view of Gauss's law, $\nabla \cdot \mathbf{E} = \rho/\epsilon_0$. This is the **equation of charge continuity,** that is, charge conservation. It says that if there is a positive divergence of **J** in some region, then charge is leaving the region, and so the charge density ρ must be getting smaller there. So conservation of charge is not a separate assumption; it is built into the Maxwell equations. Even if there were no independent demonstration of Maxwell's displacement current $\epsilon_0 \partial \mathbf{E}/\partial t$, it or something equivalent would be required simply by charge conservation.

What happens when we apply the del operator to Maxwell's equations in other ways? The answer is, we tend to wind up with wave equations. For example, let's take the curl of Faraday's law:

$$\nabla \times \mathbf{E} = -\frac{\partial \mathbf{B}}{\partial t}$$

Take the curl:

$$\nabla \times (\nabla \times \mathbf{E}) = -\frac{\partial \nabla \times \mathbf{B}}{\partial t}$$

Add Ampère's law:

$$\nabla(\nabla \cdot \mathbf{E}) - \nabla^2 \mathbf{E} = -\frac{\partial}{\partial t}\left(\mu_0 \mathbf{J} + \mu_0 \epsilon_0 \frac{\partial \mathbf{E}}{\partial t}\right)$$

and Gauss's law:

$$\nabla^2 \mathbf{E} - \mu_0 \epsilon_0 \frac{\partial^2 \mathbf{E}}{\partial t^2} = \mu_0 \frac{\partial \mathbf{J}}{\partial t} + \nabla \frac{\rho}{\epsilon_0}$$

and we find the second time and space derivatives that characterize wave equations. These wave equations describe electromagnetic radiation, including light. All of this is enormously important for reasons historical, theoretical, and practical. But patience: It will take us a little time to get there (Chapter 14).

What happens if we take curls and divergences of the first two Maxwell equations,

$$\nabla \cdot \mathbf{E} = \frac{\rho_t}{\epsilon_0} \quad \text{and} \quad \nabla \cdot \mathbf{B} = 0$$

Well, we just can't do it, because we only take curls and divergences of vectors (Chapter 1). We can however take their gradients; and when we do, and we employ the other Maxwell equations as before, we again end up with wave equations.

Now we've at least looked at various combinations of the del operator with Maxwell's equations. "There's gold in them thar hills," and much of the remainder of this textbook will be devoted to the exploitation of relationships to be found in Maxwell's equations.

At this point we should have a look at Maxwell's role in development of the equations that now bear his name. Maxwell began his study of electromagnetism with a thorough examination of Faraday's *Experimental Researches in Electricity*. (Einstein: "The pair Faraday–Maxwell has a most remarkable inner similarity with the pair Galileo–Newton—the former of each pair grasping the relationships intuitively, and the second one formulating these relationships exactly and applying them quantitatively.") In 1861 he extended Faraday's field lines (Sections 2.3 and 3.4) into a mechanical "vortex" model that filled all space with what amounted to gear wheels and idler wheels, Figure 5.1. The electric current flows through a wire

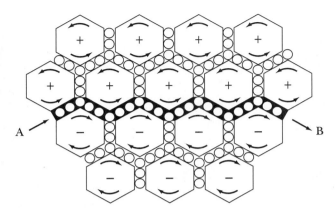

Figure 5.1 Maxwell's mechanical model.

left to right, from A to B. The little circles are the idler wheels: Translational motion of the idlers constitutes electric current, and they are free to move in a conductor, but in an insulator they can only sit and spin. The larger hexagons are cross sections of magnetic field tubes, which rotate as shown, and rotation of the tubes represents magnetic field. Because of the idlers in the wire, the tubes above the wire rotate counterclockwise, and those below the wire rotate clockwise. The magnetic field propagates outwards though space because of the rotation of the tubes produced by the idlers. If the field encounters another wire, it can induce a current there.

This lovely picture did not survive until Maxwell's 1873 book, *A Treatise on Electricity and Magnetism*. He kept only the abstract mathematical equations that had been suggested by the model; he removed the scaffolding, so to speak, after the building was constructed. But he retained his interest in visualization of the fields, and his book included numerous plates such as Figure 5.2, meticulously executed, showing field lines and equipotentials (see Section 6.3). For budding theoreticians we doubt that there is a moral here somewhere. On the one hand, the greatest theoretician of the 19th century relied on pictures and models. On the other hand, the 20th century theoretician Dirac has remarked that it is more important that one's equations be beautiful mathematically than that they correspond to currently accepted concepts.

Maxwell's equations as written by Maxwell are scarcely recognizable to the modern reader. As an explorer in a new field, he was burdened by a clumsy notation. In his "Recapitulation" (Articles 604–619) the divergence of **B**, $\nabla \cdot \mathbf{B}$, is given as:

$$\frac{da}{dx} + \frac{db}{dy} + \frac{dc}{dz} = 0$$

and we find these three equations:

$$4\pi u = \frac{d\gamma}{dy} - \frac{d\beta}{dz} \qquad 4\pi v = \frac{d\alpha}{dz} - \frac{d\gamma}{dx} \qquad 4\pi w = \frac{d\beta}{dx} - \frac{d\alpha}{dy}$$

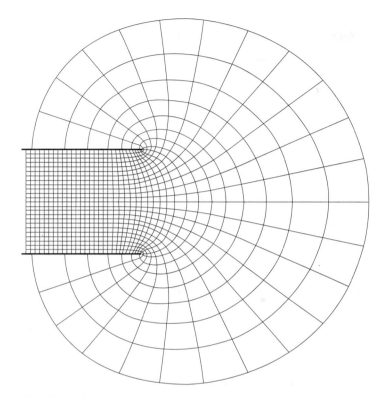

Figure 5.2   A plate from Maxwell.

instead of the one equivalent modern equation $4\pi\mathbf{J}_T = \nabla \times \mathbf{H}$, where $\mathbf{J}_T$ is "true current," including displacement current, and $\mathbf{H}$ is analogous to $\mathbf{B}$ (Chapter 10). Moreover, his selection of equations is somewhat different from ours; for example, $\mathbf{B} = \nabla \times \mathbf{A}$ is included (Section 6.2), but $\nabla \times \mathbf{E} = -\partial\mathbf{B}/\partial t$ is not, although it is derivable from other relationships that are given. He remarked: "These may be regarded as the principal relations among the quantities we have been considering. They may be combined so as to eliminate some of these quantities, but our object at present is not to obtain compactness in the mathematical formulae, but to express every relation of which we have any knowledge." Maxwell died, at 48, during the preparation of his second edition, and it remained for Hertz and Heaviside to put his equations into their present elegant form. (In fact, they were for a time called the "Hertz–Heaviside" equations.) (Nahin)

5.2.   Magnetic Monopoles

We have mentioned (Sections 3.1 and 4.3) that there are no magnetic monopoles, and it's a good thing too. However, there is an intriguing symmetry in Maxwell's equations; in particular, a changing $\mathbf{B}$ field makes an $\mathbf{E}$ field, and a changing $\mathbf{E}$

field makes a **B** field. And there are holes that seem to beg to be filled in. With magnetic monopoles denoted by the subscript m here, Maxwell's equations assume the following form:

$$\nabla \cdot \mathbf{E} = \frac{\rho}{\epsilon_0}$$

$$\nabla \cdot \mathbf{B} = \rho_{\mathrm{m}}$$

$$\nabla \times \mathbf{E} = -\mathbf{J}_{\mathrm{m}} - \frac{\partial \mathbf{B}}{\partial t} \qquad (5.7)$$

$$\nabla \times \mathbf{B} = \mu_0 \mathbf{J} + \mu_0 \epsilon_0 \frac{\partial \mathbf{E}}{\partial t}$$

Now the symmetry is complete. A flow of electric charge makes a magnetic field, and a flow of magnetic charge makes an electric field. The vacancies in Maxwell's equations are all filled.

This is so attractive that there are physicists even now, at this very moment, somewhere in the world, planning and executing attempts at detection of the magnetic monopole. Indeed, there have been several published reports of success (Price, Cabrera), but they have always faded away with time. Up until now.

But even though we have never seen a magnetic monopole, we know that its magnetic charge must be rather large, on quantum mechanical grounds. The reasoning is simple and instructive, so we will go through it here. The idea originated in 1931 with Dirac, who also predicted the positron, which does exist.

We place a positron $+e$ and a North magnetic monopole Q_{m} as shown in Figure 5.3. We move the positron in a semicircular arc about the monopole. Because of the magnetic field, the system experiences a torque, so angular momentum changes. In quantum mechanics, angular momentum is quantized; it cannot change by less than $h/4\pi$, where h is Planck's constant.

The magnetic field due to Q_{m} is

$$B = \frac{Q_{\mathrm{m}}}{4\pi r^2}$$

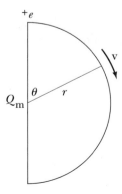

Figure 5.3

and it points away from Q_m. The force on the positron is $F = evB$, directed up out of the page. The corresponding torque points downward in the figure (by the right-hand rule), and its magnitude is

$$dL/dt = F r \sin \theta$$

where L is angular momentum. The total change in L is

$$\Delta L = \int F r \sin \theta \, dt \qquad (5.8)$$

$$= \int evB \, r \, \sin \theta \, dt$$

pointing downwards. For a line element $v \, dt$ we can substitute $r \, d\theta$, so

$$\Delta L = \int_0^\pi eB \, r^2 \sin \theta \, d\theta$$

$$= 2 \, eB \, r^2$$

$$= 2 \, e \frac{Q_m}{4\pi r^2} \, r^2 = 2 \, e \frac{Q_m}{4\pi}$$

We started with $+e$ at the top of Figure 5.3, and after the movement it is at the bottom. So in principle all we have done is to turn the system upside down and displace it by r. If the angular momentum has changed, then the fields themselves must have an intrinsic angular momentum L_0; this will be explained later, Section 14.1, in connection with Poynting's vector. It is significant that ΔL is independent of r; it suggests that it is a real physical quantity and not just some sort of mathematical artifact, and it carries the remarkable corollary that if a magnetic monopole exists, then it has this sort of angular momentum in conjunction with each and every electron in the universe.

The intrinsic angular momentum L_0 evidently points up at the beginning of the movement in Figure 5.3, and afterwards it points down, so that the change ΔL is twice L_0. We set L_0 equal to $h/4\pi$ as a minimum:

$$L_0 = \frac{\Delta L}{2} = e \frac{Q_m}{4\pi} = \frac{h}{4\pi}$$

So,

$$\boxed{Q_m = \frac{h}{e}} \qquad (5.9)$$

$$= \frac{6.6 \times 10^{-34} \, \text{J} \cdot \text{s}}{1.6 \times 10^{-19} \, \text{C}} = 4.1 \times 10^{-15} \, \text{Wb}$$

(Some authors use instead: $g = h/\mu_0 e$.)

The force between two monopoles is

$$F = \frac{Q_m B}{\mu_0} = \frac{Q_m^2}{4\pi\mu_0 r^2} \qquad (5.10)$$

which is 4700 times larger than the force between two electrons. The monopole's mass is unknown, but a monopole should be readily detected in flight by its heavy ionization (see Problem 5-16).

Absence (or dearth) of magnetic monopoles is of great significance for our technology. Our electric motors usually employ magnetic forces, rather than electrostatic forces, because whenever you try to create a large electric field, materials ionize and the "electric monopoles," that is, the ions, come and cancel out your field. If we were surrounded by numerous magnetic monopoles, then magnets would suffer the same fate, and electric motors would be limited to comparatively weak forces.

5.3 SUMMARY

The fields of classical electromagnetism are governed by Maxwell's equations. We have developed these relationships on both experimental and conceptual bases; however, they are presented at this point as postulates, suitable as a foundation for further development of electromagnetic theory:

$$\nabla \cdot \mathbf{E} = \frac{\rho}{\epsilon_0} \tag{5.1}$$

$$\nabla \cdot \mathbf{B} = 0 \tag{5.2}$$

$$\nabla \times \mathbf{E} = -\frac{\partial \mathbf{B}}{\partial t} \tag{5.3}$$

$$\nabla \times \mathbf{B} = \mu_0 \mathbf{J} + \mu_0 \epsilon_0 \frac{\partial \mathbf{E}}{\partial t} \tag{5.4}$$

The fields interact with charge mainly via the Lorentz force:

$$\mathbf{F} = Q\mathbf{E} + Q\mathbf{v} \times \mathbf{B} \tag{5.5}$$

Maxwell's equations imply the conservation of charge, as expressed by the equation of charge continuity:

$$0 = \nabla \cdot \mathbf{J} + \frac{\partial \rho}{\partial t} \tag{5.6}$$

Maxwell's equations do not contain the magnetic monopole explicitly, but neither do they preclude its existence. It would have magnetic charge that is a multiple of

$$Q_{\mathrm{m}} = \frac{h}{e} = 4.1 \times 10^{-15} \text{ Wb} \tag{5.9}$$

PROBLEMS

5-1 A wave traveling in free space (no ρ, no J) in the positive x direction at speed c is described by:

$$\mathbf{E} = E_0 \cos k(x - ct) \,\hat{\mathbf{y}}$$

$$\mathbf{B} = B_0 \cos k(x - ct) \,\hat{\mathbf{z}}$$

where k, E_0, and B_0 are constants. Under what circumstances do these $\mathbf{E}$ and $\mathbf{B}$ fields satisfy all of Maxwell's equations?

5-2 Repeat the preceding problem for:

$$\mathbf{E} = E_0 \cos k(x - ct) \,\hat{\mathbf{x}}$$

$$\mathbf{B} = B_0 \cos k(x - ct) \,\hat{\mathbf{x}}$$

5-3 See Problem 2-30: $\mathbf{E} = x\,\hat{\mathbf{y}}$. Is it possible now? How?

5-4 See Problem 3-23: $\mathbf{B} = x\,\hat{\mathbf{x}}$. Is it possible now? How?

5-5 In cylindrical coordinates, how would you produce an $\mathbf{E}$ field of the form: **(a)** $\rho\,\hat{\rho}$; **(b)** $\rho\,\hat{\phi}$; **(c)** $\rho\,\hat{\mathbf{z}}$? (Here ρ is a coordinate and not a charge density.)

5-6 In cylindrical coordinates, how would you produce a $\mathbf{B}$ field of the form: **(a)** $\rho\,\hat{\rho}$; **(b)** $\rho\,\hat{\phi}$; **(c)** $\rho\,\hat{\mathbf{z}}$?

5-7 If current density $\mathbf{J} = J_0\hat{\mathbf{r}}$ everywhere, find the rate of change of charge density ρ. Then, for a sphere of radius R centered at the origin, find the total current I through the surface of the sphere, and find the rate of change of charge Q inside by integrating $\partial\rho/\partial t$ over the volume.

5-8 Repeat the preceding problem for $\mathbf{J} = I_0 \,\hat{\mathbf{r}}/r^2$.

*5-9 For a charge Q moving at speed v ($\ll c$) along the z axis, write $\mathbf{B}$ at distance r and angle θ starting from the Biot-Savart law. Then find the same result using Maxwell's displacement current density $\epsilon_0\partial\mathbf{E}/\partial t$. (For the latter, set up a Stokesian loop of constant radius ρ with its axis on the path of the charge, Figure 5-4, and calculate the rate of change of electric flux through the loop.)

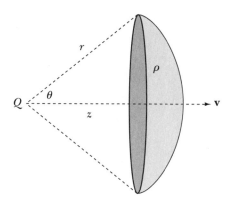

Figure 5.4

5-10 Demonstrate that the Coulomb field $\mathbf{E} = \dfrac{Q\hat{\mathbf{r}}}{4\pi\epsilon_0 r^2}$, for a stationary point charge, follows from Maxwell's equations.

5-11 A sphere of radius R has decreasing uniform charge density $\rho = \rho_0 e^{-t/\tau}$ with radial velocity distribution. At the surface, find J, displacement current density $\epsilon_0 \partial E/\partial t$, E, and B.

5-12 A circular parallel-plate capacitor of radius R and spacing s is being charged by constant current I. Find the magnetic field between the plates at distance r from the axis.

5-13 Show that Maxwell's equations including the magnetic monopole, Section 5.2, imply conservation of magnetic charge.

5-14 Find the ratio between: **(a)** the force on a magnetic monopole in a field of $B = 1$ T; and **(b)** the force on an electron in a field of $E = 10^6$ V/m.

5-15 Find the ratio between: **(a)** the Lorentz force between a magnetic monopole and an electron moving by at speed nearly c, and **(b)** the Coulomb force between two electrons. (This is one way of explaining the large ionization of rapidly moving magnetic monopoles.)

*5-16 High-energy particle beams may be focused using groups of magnetic quadrupole lenses, Figure 5.5. The magnetic field B is zero at the xy origin, which is the center of the path of the particle beam, and B changes linearly with distance from that origin. Observe that in this case, for protons coming out of the paper (in the direction of the $+z$ axis), the lens is converging in the xz plane: For example, a proton at some position on the $+x$ axis experiences a force in the $-x$ direction.

    (a) Find the focal length f in the xz plane, for magnet thickness s, in terms of $\partial B_y/\partial x$, for particles of charge q and momentum p.

    (b) Show that Maxwell's equations imply that, for B fields in vacuum, if the simple quadrupole lens is converging in the xz plane, it is diverging in the yz plane.

    (c) Show that for two successive thin lenses, of equal power, one converging and one diverging, spaced much closer than the focal length, the net result is focusing in both planes. Find the total focal length f_t of such a pair of lenses of spacing z. (This is "strong focusing." Figure 3.38 involves weak focusing.)

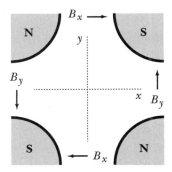

Figure 5.5

C H A P T E R

6

ENERGY AND POTENTIALS

Energy is the fundamental commodity. Although it was discovered a scant 150 years ago, it dominates our lives economically, politically, and, in the present instance, in the realm of physics. For example, the energy we get from the sun arrives in the form of electromagnetic fields. It is fair to say that electromagnetic fields are of interest, and of practical importance, primarily because they contain and transport energy. In this chapter we shall examine electromagnetic energies and the potentials from which they may be calculated.

6.1. Energy Density in E and B Fields

To evaluate the **E** field energy, we begin by breaking an arbitrary field up into more easily analyzed components, Figure 6.1. Looking at the microscopic portion of the field intersected by the two parallel lines, we shall pretend that it is produced instead by the parallel, uniformly charged planes of Figure 6.2. In this we are employing the fundamental assumption that the nature of an **E** or **B** field does not depend on its source. The **E** field in Figure 6.1 could even be caused by a changing **B** field, but in any case we shall treat it in terms of local electric charges.

From Section 2.3, each plane contributes the field $E = \sigma/2\,\epsilon_0$, so the field between the planes is

$$E = \sigma/\epsilon_0 \tag{6.1}$$

and $E = 0$ outside. We are going to move one plane by a small (virtual) displacement x and find out how much work is done on the small area yz.

The charge on yz is σyz. The effective field is $E/2$ (Section 2.3), so the magnitude of the resulting force on the charge is

$$F = \sigma yz\, E/2$$

The work we do during displacement x is

$$W = Fx = \sigma xyz\, E/2$$

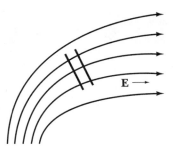

Figure 6.1 Field lines.

So the work per unit volume $v = xyz$, which is swept out by yz, is

$$\frac{W}{xyz} = \frac{\sigma E}{2}$$

which we can put in the form, with $\sigma = \epsilon_0 E$:

$$\boxed{\frac{dW}{dv} = \frac{1}{2}\,\epsilon_0 E^2} \tag{6.2}$$

The result is true in empty space even for nonstatic cases, including radiation, because we can always make the parallel planes small enough to reduce transient effects. (However, in materials the formula is a little different, as we shall see in Section 9.6.)

Now in this model we have expended energy, and we have created a volume of field. The question is, is the energy really in the field? Or is it perhaps associated with the charges instead? In electrostatics this question is physically meaningless, which is to say, it doesn't matter as long as you don't let the charges move; in fact, in Section 6.5 we calculate the energy as if it were associated with the charges, for the static case. However, when you let the charges move arbitrarily, you find that

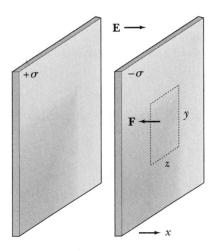

Figure 6.2 Parallel charged planes.

the **E** and **B** fields can carry energy away to great distances, as radiation, independently of the charges. It would be unthinkable that the energy resides on the charges while they are standing still and suddenly transfers itself to the fields when they start to move. So it is more fruitful to believe that the energy is located in the fields.

 **EXAMPLE 6-1**

Find the energy density in a field of 10^6 V/m.

ANSWER

$$\frac{dW}{dv} = \epsilon_0 E^2/2$$

$$= 8.85 \times 10^{-12} \, C^2/N \cdot m^2 \times (10^6 \, N/C)^2/2 = 4.4 \, J/m^3$$

which isn't much, given that the field is rather strong for an electrostatic field.

Now we shall do the same thing for the **B** field. In Figure 6.3 we have two parallel planes carrying surface current K in opposite directions. From Section 3.5, each plane contributes the field $B = \mu_0 K/2$, so the field between the planes is

$$B = \mu_0 K \tag{6.3}$$

and $B = 0$ outside. The force on a wire of length l is $F = BIl$ (Section 3.2); in this case, the effective field is $B/2$, the current I across the small area yz is Ky, and the length is z, so

$$F_x = Kyz \, B/2$$

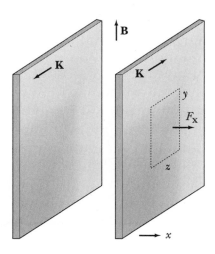

Figure 6.3   Plane surface currents.

The planes repel each other, so you get energy out when you separate them (if the current is held constant). The work the area yz puts out during displacement x is

$$W_x = -F_x x = -KxyzB/2$$

W_x is no vector component; its subscript is for identification only. So the work per unit volume v is, with $K = B/\mu_0$:

$$\frac{W_x}{v} = -\frac{KB}{2} = -\frac{B^2}{2\mu_0} \tag{6.4}$$

Now the field volume has gone up, and yet the system has done work; so who is paying for all this?

The answer is, it takes energy to keep the current going, because there is a back emf due to Lenz's law, Section 4.1. Width l is z here, and speed v is dx/dt, so (Eq. 4.3)

$$\text{emf} = Blv \implies Bz\frac{x}{t}$$

The effective field at a plane is $B/2$ (as with $E/2$ in the preceding case), so half of the emf comes from one plane and half from the other. This emf is associated with a force F_z on the moving charge and an energy input W_z required to keep the current flowing:

$$W_z = F_z z = Q\,\text{emf} = Q\,Bz\frac{x}{t}$$

$$= Bxz\,I = \frac{Bxyz\,I}{y} = Bv\frac{I}{y} = BvK$$

So,

$$\frac{W_z}{v} = BK = \frac{B^2}{\mu_0}$$

The net energy density, which appears in the field, is

$$\frac{W}{v} = \frac{W_z}{v} + \frac{W_x}{v} = \frac{B^2}{\mu_0} - \frac{B^2}{2\mu_0}$$

which implies that

$$\boxed{\frac{dW}{dv} = \frac{B^2}{2\mu_0}} \tag{6.5}$$

Again, this result holds in vacuum even for radiation (but it must be modified in materials; see Section 10.2).

 **EXAMPLE 6-2**

Find the energy density in a field of 1 tesla.

ANSWER

$$W/v = B^2/2\mu_0$$
$$= (1 \text{ T})^2/2 \times 4\pi \times 10^{-7} \text{ T} \cdot \text{m/A}$$
$$= 4.0 \times 10^5 \text{ T} \cdot \text{A/m or J/m}^3$$

This is a hundred thousand times greater than the energy density found before in a readily attainable electric field. The reason for the disparity is that a strong electric field breaks matter into its constituent electric monopoles, that is, electrons and positive ions. But, as was remarked in Section 5.2, one may create large magnetic fields without fear that the surrounding matter will break up into magnetic monopoles, because there aren't any.

Looking back at these two parallel developments for *E* and *B*, we find that in each case the first equation for force *F* can be divided through by area *yz*, yielding **pressure.** Thus, in both cases, numerically and in terms of units,

$$\text{Pressure} = \text{energy density}$$

so the pressure *P* is

$$P = \frac{\epsilon_0 E^2}{2} \quad \text{or} \quad \frac{B^2}{2\mu_0} \tag{6.6}$$

a useful result but not a general one (see Section 14.2). As we remarked before in connection with field lines (sections 2.3 and 3.4), the pressure may be described as a pull in the direction of the field or a push sideways.

 EXAMPLE 6-3

For $E = 10^6$ V/m, the electrostatic pressure is only $\epsilon_0 E^2/2 = 4.4$ N/m^2, or 4.4 Pa (pascals); for oppositely charged parallel plates it is directed inwards. This is not an impressive number; the personal danger from sparking is much greater. On the other hand, for $B = 1$ T, the magnetic pressure is $B^2/2\mu_0 = 4.0 \times 10^5$ Pa, or 4 atmospheres, directed outwards on the current-carrying coils that produce the field. Such a pressure may result in substantial forces, and larger magnetic fields can be as dangerous as explosives. The highest steady field attained so far is about 35 T, for which the pressure is 4800 atmospheres, or around 35 tons per square inch. Experiments at higher fields may be done in pulsed mode, up to 70 T for 10 ms, or even in explosive mode, 200 T for 10 μs, the measurements being taken quickly by a remote observer (*Physics Today,* June 1996).

6.2. Scalar Potential *V* and Vector Potential A—Gradient of *V*

We could in principle calculate electromagnetic energies using the energy densities developed in the preceding section (see the Problems). But it is often easier to calculate them directly in terms of the geometrical relationships between charges or currents. For this purpose, we define the **scalar potential** V and **vector potential** **A** as follows:

$$E = -\nabla V - \frac{\partial A}{\partial t} \qquad (6.7)$$

and

$$B = \nabla \times A \qquad (6.8)$$

Why do we select these particular forms? The most important answer, but not an immediately satisfying one, is that further study will reveal their usefulness. For now, we are going to provide two significant points on their behalf: **(1)** these equations are not merely consistent with Maxwell's equations, they *contain* two of them; **(2)** the dimensions and units of **A** and V suggest that these quantities have something to do with energy.

But first: This is our first exposure to the gradient, written grad V or ∇V (see Section 1.2), in an electromagnetic context, so we will pause and examine it for a moment.

We have already used the del operator "∇" on vectors. If we do what looks like a dot product, we get a *divergence*, like $\nabla \cdot E$, Sections 1.2 and 2.4, which represents a source or sink of flux. And for a cross product, we get a *curl*, $\nabla \times B$, Sections 1.2 and 3.6, which has to do with a kind of rotational effect of the field lines. Now if we apply the del operator to the scalar V, we get the *gradient*, ∇V:

$$\nabla V = \frac{\partial V}{\partial x}\hat{x} + \frac{\partial V}{\partial y}\hat{y} + \frac{\partial V}{\partial z}\hat{z} \qquad (6.9)$$

Its physical significance is this: It tells us in what direction V is changing, and its slope. Analogously, if h is height on a hill, then ∇h is a vector pointing uphill, and its magnitude is the slope, rise over run.

◢ EXAMPLE 6-4

Find the gradient of $V = 2x$ (in volts), and state its significance.

ANSWER

$$\nabla V = \nabla(2x)$$

$$= \frac{\partial 2x}{\partial x}\hat{x} + 0\,\hat{y} + 0\,\hat{z} = 2\,\hat{x} \qquad \text{volts/meter}$$

Now $\hat{x}$ is a unit vector in the x direction, and so every time you go one meter in the positive x direction, V increases by two volts; however, if you go in the y or z direction, you get nothing. According to Equation 6.7, this implies a contribution to **E** of 2 V/m in the negative x direction, but of course $\partial A/\partial t$ may also be contributing to **E**.

Now we deliver the two supporting points promised above.

(1) Maxwell's equations: if we take the curl of the first of the two potential equations, Equation 6.7, we find:

$$\mathbf{E} = -\nabla V - \frac{\partial \mathbf{A}}{\partial t}$$

$$\nabla \times \mathbf{E} = -\nabla \times (\nabla V) - \frac{\partial}{\partial t} \nabla \times \mathbf{A} \tag{6.10}$$

$$\nabla \times \mathbf{E} = \qquad 0 \qquad - \frac{\partial \mathbf{B}}{\partial t}$$

because the curl of any gradient is identically zero (see inside front cover) and the curl of **A** is **B**. So we retrieve Faraday's law in differential form. And if we take the divergence of the second equation, Equation 6.8:

$$\mathbf{B} = \nabla \times \mathbf{A}$$
$$\nabla \cdot \mathbf{B} = \nabla \cdot (\nabla \times \mathbf{A}) = 0 \tag{6.11}$$

because the divergence of a curl is identically zero (inside front cover). This is Maxwell's equation for the divergence of **B**. These two Maxwell equations are contained in the definitions of the potentials, Equations 6.7 and 6.8.

(2) Units of energy: Equation 6.7 requires that units of E must be the same as those of V/x. E is volts per meter, and volts is joules per coulomb (Section 4.1), so the units of the scalar potential V are joules per coulomb. Thus the units suggest that V can tell us how much energy it takes to put a charge somewhere. But the force on the charge depends also on changing magnetic flux (Faraday's law), so the potential doesn't always tell the whole story about the energy.

Equation 6.7 also indicates that the units of E are those of A/t; so the units of A are $V \cdot s/m$, which can be written as $J \cdot s/C \cdot m$, or $J/A \cdot m$, joules per ampere-meter. This implies that **A** may involve the energy required to place a current of given length somewhere. The units of **A** can also be written in terms of momentum per unit charge, kg $\cdot$ m/s $\cdot$ C, which suggests that **A** can serve as a kind of "momentum potential" (see Problem 6-18; in quantum mechanics, momentum is $mv + qA$).

Although these V and **A** definitions, Equations 6.7 and 6.8, are sufficient to enable one to determine the fields from the potentials, they do not completely determine the potentials themselves. You can add constants to V and **A** without changing **E** and **B**; these constants are usually so chosen that the potentials are zero at infinity. Furthermore, the curl of **A** is defined by Equation 6.8, but to complete **A**'s description we need its divergence; and for the scalar V we would like an expression involving the time derivative. Both may be combined in the **Lorentz condition:**

$$\nabla \cdot \mathbf{A} = -\mu_0 \epsilon_0 \frac{\partial V}{\partial t} \qquad (6.12)$$

(See Section 14.3 for more on this; for gauge transformation, see Problem 6-14.)

In connection with the potential equations, it is worth noting: If the curl of a vector is zero, then it may always be represented as the gradient of a scalar, like $\mathbf{E} = -\nabla V$ in electrostatics, since the curl of a gradient is zero:

$$\text{If} \quad \nabla \times \mathbf{E} = 0, \quad \text{then} \quad \mathbf{E} = -\nabla V \qquad (6.13)$$

And if the divergence of a vector is zero, then it may be represented as the curl of a vector, like $\mathbf{B} = \nabla \times \mathbf{A}$ in general, since the divergence of a curl is zero:

$$\text{If} \quad \nabla \cdot \mathbf{B} = 0, \quad \text{then} \quad \mathbf{B} = \nabla \times \mathbf{A}$$

We shall provide concrete examples of these potentials, and especially of the energy associated with them, in Sections 6.3 to 6.5.

6.3.   Calculation of Potentials

There are various useful ways of figuring out what the scalar and vector potentials are in particular cases, based on equations from the preceding section. We shall gather some examples together in this section, for your convenience.

 EXAMPLE 6-5

Find the scalar potential V at 3 cm from a stationary charge of 2 nC (in electrostatics).

ANSWER We begin with

$$\mathbf{E} = -\nabla V - \partial \mathbf{A}/\partial t$$

This is a static problem, so $\partial \mathbf{A}/\partial t = 0$.

$$\mathbf{E} = -\nabla V$$

For the simple charge,

$$\mathbf{E} = \frac{Q \hat{\mathbf{r}}}{4\pi\epsilon_0 r^2}$$

In spherical coordinates, $\nabla V = \dfrac{\partial V}{\partial r} \hat{\mathbf{r}}$, there being no angular dependence. So,

$$\frac{\partial V}{\partial r} = -\frac{Q}{4\pi\epsilon_0 r^2}$$

Integrating,

$$V = - \int_r^\infty \frac{Q \, dr}{4\pi\epsilon_0 r^2}$$

since, as mentioned above, V is usually taken to be zero at infinity. So,

$$\boxed{V = \frac{Q}{4\pi\epsilon_0 r}} \qquad \text{for a point charge} \qquad \textbf{(6.14)}$$

$$= 9 \times 10^9 \; (\text{N} \cdot \text{m}^2/\text{C}^2) \times 2 \times 10^{-9} \, \text{C}/0.03 \, \text{m}$$
$$= 600 \, \text{N} \cdot \text{m}/\text{C} \quad \text{or} \quad \text{volts}$$

It would take 600 joules of energy to bring a coulomb of charge to this point. So even a nanocoulomb is a substantial charge, in terms of the voltage it can produce. For comparison, the voltage available at an ordinary household outlet is only 120 V, and a battery typically provides 12 V or less.

Electrostatic fields may be represented by **equipotentials** or by electric field lines, or even by both. Equipotentials are surfaces of constant voltage. Figure 6.4a shows a single charge with its radial field lines and its spherical equipotentials; it corresponds to the case of the preceding example. Figure 6.4b represents two long line charges perpendicular to the paper. Equipotentials are always perpendicular to E field lines in the static case (this follows from $\mathbf{E} = -\nabla V$). Caveats concerning field lines were presented in Section 2.3; in particular, a two-dimensional representation on paper can falsify the magnitude of a three-dimensional field. For equipotentials this problem does not arise. See Section 5.1 for a Maxwell picture of equipotentials and field lines, and see Appendix C for a discussion of how to calculate them.

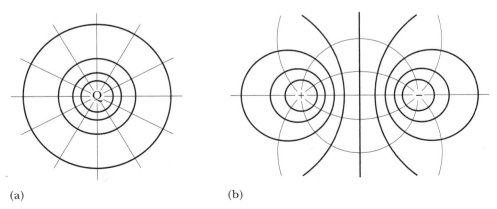

(a)   (b)

Figure 6.4   (a) Point charge. (b) Opposite line charges. (Equipotentials, dark lines; $\mathbf{E}$ field lines, light lines.)

In the preceding example we found it convenient in the static case to calculate V starting with a known $\mathbf{E}$:

$$\mathbf{E} = -\nabla V \Longrightarrow \boxed{\nabla V = V_b - V_a = -\int_a^b \mathbf{E} \cdot d\boldsymbol{l}} \tag{6.15}$$

(In practice the sign in front of the integral may vary, depending on the direction of integration.) Analogously, one may often calculate $\mathbf{A}$ starting with a known $\mathbf{B}$:

$$\mathbf{B} = \nabla \times \mathbf{A}$$

$$\int \mathbf{B} \cdot d\mathbf{a} = \int \nabla \times \mathbf{A} \cdot d\mathbf{a}$$

so,

$$\boxed{\Phi = \oint \mathbf{A} \cdot d\boldsymbol{l}} \tag{6.16}$$

using Stokes' theorem and the definition of magnetic flux, $\Phi = \int \mathbf{B} \cdot d\mathbf{a}$.

EXAMPLE 6-6

Find the vector potential $\mathbf{A}$ at 7 cm from the axis of an infinite solenoid of radius 2 cm whose B field is 0.3 T. (Remember, $B = 0$ outside of the infinite solenoid.)

ANSWER We draw a circular loop about the solenoid, Figure 6.5, and find the flux through it:

$$\Phi = \oint \mathbf{A} \cdot d\boldsymbol{l}$$

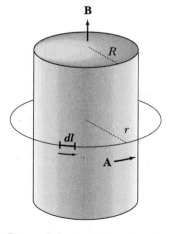

Figure 6.5   Infinite solenoid.

By symmetry, **A** is constant, so

$$\pi R^2 B = 2\pi r A$$
$$A = R^2 B/2r$$
$$= 0.02^2 \, \text{m}^2 \times 0.3 \, \text{T}/2 \times 0.07 \, \text{m} = 8.6 \times 10^{-4} \, \text{T} \cdot \text{m}$$

If we put our right thumb in the direction of **B**, then our fingers curl in the direction of *dl* and of **A**.

This value of *A* is typical for small magnets, but we don't have much intuitive basis for judging it, unlike the case for *V*: Voltmeters are common, but nobody has a tesla-meter meter.

QUESTION: If *B* = 0 always outside of an infinite solenoid, then how do the electrons in the loop know what to do when the flux changes?

ANSWER: They respond to $-\partial \mathbf{A}/\partial t$. (Compare with Fig. 4.7.) Remember that $\mathbf{E} = -\nabla V - \partial \mathbf{A}/\partial t$, and so in a sense the connection of *E* and *B* is only indirect.

For *V* for a single stationary charge, we have noted previously that

$$V = \frac{Q}{4\pi\epsilon_0 r}$$

For a charge distribution, with superposition, we would have

$$\boxed{V = \int \frac{\rho \, dv}{4\pi\epsilon_0 r}}$$ **(6.17)**

EXAMPLE 6-7

A uniform line charge λ exists on the *y* axis from 0 to *L*; find the potential *V* on the *x* axis at *a* (Fig. 6.6).

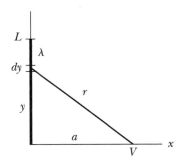

Figure 6.6   Line charge.

ANSWER For the charge element, we use $\lambda\, dy$ for $\rho\, dv$.

$$V = \int_0^L \frac{\lambda\, dy}{4\pi\epsilon_0 r}$$

$$= \frac{\lambda}{4\pi\epsilon_0} \int \frac{dy}{\sqrt{a^2 + y^2}} = \frac{\lambda}{4\pi\epsilon_0} \ln(y + \sqrt{y^2 + a^2})\Big]_0^L$$

So,

$$V = \frac{\lambda}{4\pi\epsilon_0} \ln\left(\frac{L + \sqrt{L^2 + a^2}}{a}\right) \tag{6.18}$$

For small a this goes to infinity, and for large a it goes to zero, as expected. For small L it becomes $Q/4\pi\epsilon_0 a$, the point-charge potential with $Q = \lambda L$. However, for large L it goes to infinity, logarithmically but inexorably. We can check this result for an infinite line using the Gauss's-law field:

$$E = \lambda/2\pi\epsilon_0 r = -\nabla V \qquad \text{(for the quasi-static case)}$$

We integrate the E field from a to some distance R and then find the limit as R goes to infinity:

$$V = \int_a^R \lambda\, dr/2\pi\epsilon_0 r = \lambda \ln(R/a)/2\pi\epsilon_0$$

Sure enough, it diverges as $R \to \infty$. If you construct an infinitely long line of uniform linear charge density, you produce a thing of infinite voltage, not to mention infinite energy, with dire consequences for everything. So don't do that. You can, however, produce parallel-line charges of opposite charge, or concentric cylinders of opposite charge; see the Problems.

The situation for the vector potential **A** of an infinite current-carrying wire is analogous to that for V, above. Because of this, **A** is infinite for a solenoid such as that of Figure 6.5 if the wire of which it is wound goes from top to bottom: It is in effect the sum of a solenoid plus an infinite wire. However, if we wrap the solenoid with wire in such a way that the wire enters and leaves at the same point, we avoid this difficulty.

We can also avoid the problem of infinite **A** for infinite conductors if we have more than one of them. For example, for concentric cylinders, or a coaxial cable, Figure 6.7, we don't encounter the infinity, mentioned above, if we can fix the potential on one conductor.

 EXAMPLE 6-8

Find **A** at radius r in the space between concentric cylinders of radii r_1 and r_2 carrying opposite currents I, Figure 6.7.

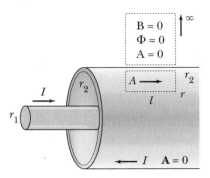

Figure 6.7   Coaxial cylinders.

ANSWER We begin with $\mathbf{B} = \nabla \times \mathbf{A}$ and apply Stokes' theorem (Sections 1.5, 3.6) for the dashed contours shown: $\oint \mathbf{A} \cdot d\mathbf{l} = \Phi$. Outside, there is no B, so no Φ and no $\mathbf{A}$. One side of the lower contour is on the outside cylinder, where $\mathbf{A} = 0$, and for the sides on the left and right, $\mathbf{A}$ is perpendicular to $d\mathbf{l}$; so only the side nearest the central wire contributes to the line integral.

$$\Phi = \int \mathbf{B} \cdot d\mathbf{a} = \oint \mathbf{A} \cdot d\mathbf{l}$$

$$\int_r^{r_2} \frac{\mu_0 I}{2\pi\rho} \, l \, d\rho = l\mathrm{A}$$

so,

$$\frac{\mu_0 I}{2\pi} \ln\left(\frac{r_2}{r}\right) = A \tag{6.19}$$

in the direction of the current I in the central conductor. Check: When $r = r_2$, $A = \ln 1 = 0$, as was specified for the outside conductor. Like V in the preceding example, A is logarithmic, and it would go to infinity if the outer conductor weren't there to stop it.

In Example 6-7 and Figure 6.6, we applied the expression for V, Equation 6.17:

$$V = \int \frac{\rho \, dv}{4\pi\epsilon_0 r}$$

We shall now demonstrate that for constant current the analogous expression,

$$\boxed{\mathbf{A} = \int \frac{\mu_0 \mathbf{J} \, dv}{4\pi r}} \tag{6.20}$$

is consistent with $\mathbf{B} = \nabla \times \mathbf{A}$ in the quasi-static case.

We begin by taking its curl. We can interchange the order of integration and differentiation, because the integration is over source variables and the differentia-

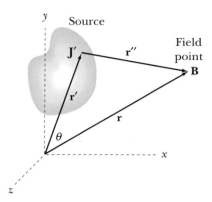

Figure 6.8   Current distribution.

tion is over field variables, Figure 6.8 (compare with Fig. 3.26). With $\mathbf{r}'' = \mathbf{r} - \mathbf{r}'$, we have:

$$\nabla \times \mathbf{A} = \nabla \times \int \frac{\mu_0 \mathbf{J}' \, dv'}{4\pi r''} = \int \nabla \times \frac{\mu_0 \mathbf{J}' \, dv'}{4\pi r''}$$

We employ a vector identity on this to get

$$\nabla \times \mathbf{A} = \int \frac{\mu_0 \nabla \times \mathbf{J}' \, dv'}{4\pi r''} + \int \frac{\mu_0 \nabla \left(\frac{1}{r''}\right) \times \mathbf{J}' \, dv'}{4\pi}$$

Now in the first term, $\nabla \times \mathbf{J}'$ is zero because the curl involves field coordinates and $\mathbf{J}'$ involves source variables. In the second term, $\nabla \frac{1}{r''} = -\frac{\hat{r}''}{r''^2}$. So, the result is

$$\nabla \times \mathbf{A} = \int \frac{\mu_0 \mathbf{J}' \times \hat{r}'' \, dv'}{4\pi r''^2} = \mathbf{B} \tag{6.21}$$

which is the Biot-Savart law, so Equation 6.20 is okay.

Note that the expression for $\mathbf{A}$,

$$\mathbf{A} = \int \frac{\mu_0 \mathbf{J} \, dv}{4\pi r}$$

implies that $\mathbf{A}$ tends to point in the direction of current flow. For a closed wire loop we can write the current element as $I \, dl$, and the expression becomes

$$\mathbf{A} = \oint \frac{\mu_0 I \, dl}{4\pi r} \tag{6.22}$$

 **EXAMPLE 6-9**

A square wire loop of side s carries current I. Find $\mathbf{A}$ on a line passing through the center of the loop, parallel to a side, at a distance s from the center.

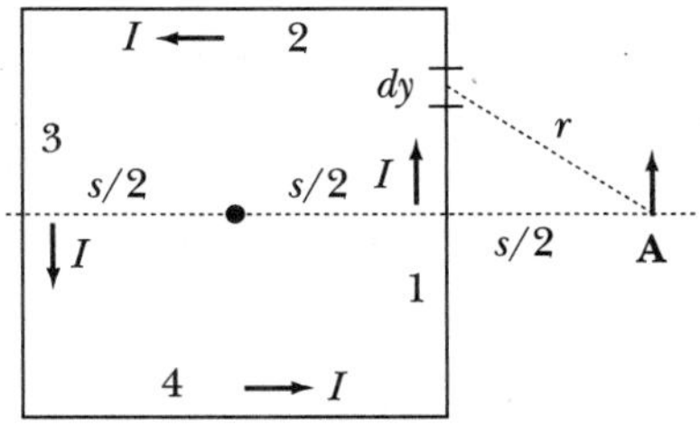

Figure 6.9 Current loop.

ANSWER See Figure 6.9. **A** will point in the positive y direction, because side "1" is closest to it. Set $s/2 = a$. For the side labeled "1," we have

$$A_1 = \int_{-a}^{a} \frac{\mu_0 I\, dy}{4\pi r}$$
$$= \frac{\mu_0 I}{4\pi}\int \frac{dy}{\sqrt{a^2 + y^2}} = \frac{\mu_0 I}{4\pi}\ln(y + \sqrt{y^2 + a^2})\Big]_{-a}^{a}$$

So,

$$A_1 = \frac{\mu_0 I}{4\pi}\ln\left(\frac{\sqrt{2}+1}{\sqrt{2}-1}\right) = \frac{\mu_0 I}{2\pi}\ln(\sqrt{2}+1)$$

in the $+y$ direction.

Sides 2 and 4 cancel, by symmetry. Side 3 involves the same integral as side 1, with $3a$ inside the radical instead of a:

$$A_3 = \frac{\mu_0 I}{4\pi}\ln(y + \sqrt{y^2 + (3a)^2}\Big]_{-a}^{a}$$

So,

$$A_3 = \frac{\mu_0 I}{4\pi}\ln\left(\frac{\sqrt{10}+1}{\sqrt{10}-1}\right) = \frac{\mu_0 I}{2\pi}\ln\left(\frac{\sqrt{10}+1}{3}\right) \quad \text{in the } -y \text{ direction}$$

and finally,

$$A_1 - A_3 = \frac{\mu_0 I}{2\pi}\ln\left(\frac{3(\sqrt{2}+1)}{\sqrt{10}+1}\right)$$

(It doesn't involve s.)

EXAMPLE 6-10

Find **A** at the center of the same current loop (Fig. 6.9).

ANSWER **A** = 0 by symmetry. Contributions from opposite sides cancel.

We have now found two ways of calculating each of **E**, **B**, V, and **A** (listed in Eq. 6.23). Of these two ways, one is easy if appropriate symmetry exists, and the other is useful in more general geometries but usually more difficult. Most of our applications have been limited to quasi-static phenomena; radiation will be dealt with later. With this in mind, we omit $\partial \mathbf{E}/\partial t$ and $\partial \mathbf{A}/\partial t$.

Asymmetric		Symmetric		
$\mathbf{E} = \displaystyle\int \frac{\rho\, \hat{\mathbf{r}}\, dv}{4\pi\epsilon_0 r^2}$	(2.7)	$\displaystyle\oint \mathbf{E}\cdot d\mathbf{a} = Q/\epsilon_0$	(2.15)	
$\mathbf{B} = \displaystyle\int \frac{\mu_0 \mathbf{J} \times \hat{\mathbf{r}}\, dv}{4\pi\, r^2}$	(3.10)	$\displaystyle\oint \mathbf{B}\cdot d\mathbf{l} = \mu_0 I$	(3.19)	(6.23)
$V = \displaystyle\int \frac{\rho\, dv}{4\pi\epsilon_0 r}$	(6.17)	$V = -\displaystyle\int \mathbf{E}\cdot d\mathbf{l}$	(6.15)	
$\mathbf{A} = \displaystyle\int \frac{\mu_0 \mathbf{J}\, dv}{4\pi\, r}$	(6.20)	$\displaystyle\oint \mathbf{A}\cdot d\mathbf{l} = \Phi$	(6.16)	

Two further remarks: **(1)** All of these formulas are true in both symmetric and non-symmetric cases; the distinction merely involves convenience in practical calculations. **(2)** It often happens that it is easier to approach the **E** and **B** fields indirectly, by calculating the potentials first and then differentiating (Eqs. 6.7 and 6.8); this is done, for example, in the case of dipoles in the next chapter, Sections 7.1 and 7.2.

   All the equations in Equation 6.23 represent an impressive panoply of formulae; nevertheless, if you stray from an axis of symmetry, you may find yourself in a mathematical thicket. For example, here is a problem that you might expect to be easy.

EXAMPLE 6-11

Find the potential V in the plane of a ring of radius R and linear charge density λ, at distance $a < R$ from the center, Figure 6.10a.

ANSWER

$$V = \int \frac{dq}{4\pi\epsilon_0 r} = \frac{1}{4\pi\epsilon_0} 2 \int_0^\pi \frac{\lambda R\, d\theta}{\sqrt{a^2 + R^2 + 2aR\cos\theta}}$$

It isn't in my table of integrals. We execute a half-angle substitution:

$$\phi = \theta/2 \quad \text{and} \quad \cos\theta = 1 - 2\sin^2\phi$$

so,

$$V = \frac{\lambda}{4\pi\epsilon_0} 4 \int_0^{\pi/2} \frac{R\, d\phi}{\sqrt{a^2 + R^2 + 2aR - 4aR\sin^2\phi}}$$

$$= \frac{\lambda R}{\pi\epsilon_0(a + R)} \int_0^{\pi/2} \frac{d\phi}{\sqrt{1 - m\sin^2\phi}}$$

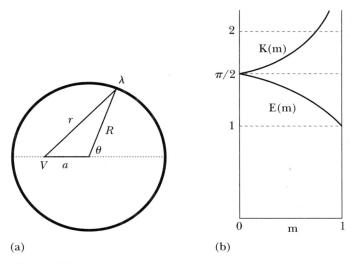

Figure 6.10   **(a)** Charged ring. **(b)** Complete elliptic integrals.

where

$$m = \frac{4aR}{(a + R)^2}$$

Now we introduce K(m), the complete **elliptic integral** of the first kind, "complete" because the upper limit is $\pi/2$:

$$K(m) = \int_0^{\pi/2} \frac{d\phi}{\sqrt{1 - m \sin^2\phi}} \qquad \text{with } 0 \leq m \leq 1$$

This function is plotted in Figure 6.10b. In terms of K(m), V becomes

$$V = \frac{\lambda R}{\pi\epsilon_0(a + R)} K(m)$$

The complete elliptic integral of the second kind would be

$$E(m) = \int_0^{\pi/2} \sqrt{1 - m \sin^2\phi} \, d\phi$$

And the point here is, these elliptic integrals *can not* be expressed in terms of the elementary functions: algebraic, trig and inverse trig, log, and exponential. They are a new kind of function, and you have to look up their values in a table, just as you would for, say, the $\sin^{-1}$ function. (Be aware: The "elliptic function" is not the same as the elliptic integral.)

As Figure 6.10b suggests, in this example

$$V \to \infty \quad \text{as} \quad a \to R \quad \text{and} \quad V = \lambda/2\epsilon_0 \quad \text{at} \quad a = 0$$

6.4.   Particle Energy; Conservative Field

Both scalar potential V and vector potential $\mathbf{A}$ can contribute to the energy of a charged particle by means of the $\mathbf{E}$ and $\mathbf{B}$ fields they are associated with:

$$\mathbf{E} = -\nabla V - \frac{\partial \mathbf{A}}{\partial t}$$

and

$$\mathbf{B} = \nabla \times \mathbf{A}$$

For the magnetic field, the force Lorentz $Q\mathbf{v} \times \mathbf{B}$ is always perpendicular to the displacement, so *no* work is done by a constant $\mathbf{B}$. The $\mathbf{E}$ field *can* do work on a charge, with or without the assistance of $\partial \mathbf{A}/\partial t$. However, if we are limited to electrostatics, with $\partial \mathbf{A}/\partial t = 0$, so that $\mathbf{E}$ is derived only from ∇V, then $\mathbf{E}$ is called **conservative:** When we carry a charge about a closed contour, no energy is gained or lost. To demonstrate this: Energy is force QE times distance l; so for a charge carried about a closed contour, the energy W put into the system is

$$
\begin{aligned}
W &= -\oint Q\mathbf{E} \cdot d\mathbf{l} \\
&= -\int Q\nabla \times \mathbf{E} \cdot d\mathbf{a} \qquad \text{using Stokes' theorem} \\
&= -\int Q\nabla \times \left(-\nabla V - \frac{\partial \mathbf{A}}{\partial t} \right) \cdot d\mathbf{a} \\
&= -\int Q\nabla \times (-\nabla V) \cdot d\mathbf{a} = 0
\end{aligned}
\tag{6.24}
$$

because we are assuming $\partial \mathbf{A}/\partial t = 0$ and because the curl of a gradient is always zero (see inside front cover). Note that if $\nabla \times \mathbf{E} = 0$ everywhere, then $W = 0$ and the field is conservative.

So we have several equivalent **tests for a conservative E field** $(W = 0)$:

- It has no curl: $\nabla \times \mathbf{E} = 0$.
- It is the gradient of a scalar: $\mathbf{E} = -\nabla V$ (because the curl of a gradient is zero).
- Any closed contour integral $\oint_a^b \mathbf{E} \cdot d\mathbf{l}$ is zero (Stokes' theorem).
- The line integral from one point to another $\int \mathbf{E} \cdot d\mathbf{l}$ is independent of the path.

The last of the above four tests for a conservative field is an obvious corollary of the next-to-last one: If any closed contour gives zero, then it doesn't matter how you get from one point to another. For example, in Figure 6.11a, if the line integral all the way around from point **a** back to point **a** is zero, then whatever energy we gain in carrying a test charge on path 1, from **a** to **b**, we must give back on the return path 2 from **b** back to **a**. So if we traverse path 2 in the other direction, from **a** to **b**, the energy gain must be the same as along path 1.

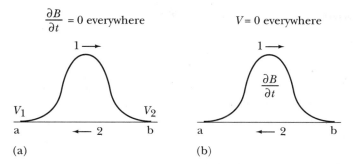

Figure 6.11   (a) Path independence. (b) Path dependence.

By contrast, if there is a $\partial \mathbf{B}/\partial t$ as in Figure 6.11b, then by Faraday's law there is a curl of **E**. The line integral about the loop is not zero, and there is a path dependence. Energy is transferred to a charge traveling a complete circuit along this path, even though $V = 0$ everywhere in Figure 6.11b. So it's not conservative. The energy comes from whatever provides the $\partial \mathbf{B}/\partial t$.

The closed-contour test is illustrated for three cases in Figure 6.12. For the closed contour **a** in Figure 6.12a, there is no $\partial \mathbf{A}/\partial t$, so $\mathbf{E} = -\nabla V$ and it is conservative. There is no curl, and no net work is done in carrying a test charge completely around the closed contour labeled **a**. That is the situation in electrostatics.

In Figure 6.12b, there is a solenoid at the center with a changing **B**. Outside of the solenoid, at loop **a**, $\partial \mathbf{A}/\partial t$ is proportional to $1/r$, and in such a field $\nabla \times \mathbf{E} = 0$ (Problem 6-17); so there is no energy to be gained by going around loop **a**. But there is a curl of **E** in the center, due to Faraday's law, $\nabla \times \mathbf{E} = -\partial \mathbf{B}/\partial t$, and so a charge may gain or lose energy in going around the closed loop **b**. This energy is supplied by whatever agency is providing the current in the solenoid.

In Figure 6.12c, showing a rapidly moving charge, the loop **b** also has a finite line integral, because the E field is greater for the vertical segment than the horizontal one. There is also a $\partial \mathbf{B}/\partial t$ in the vicinity, because the charge is moving. If a

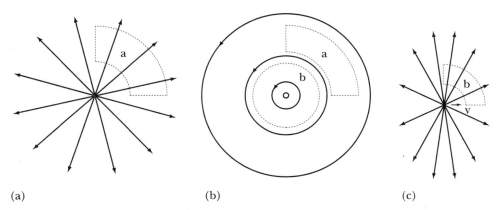

(a)   (b)   (c)

Figure 6.12   (a) **E** due to Q at center. (b) **E** due to $\partial \mathbf{B}/\partial t$ at center. (c) **E** of moving Q.

small test charge goes around the loop, it can gain energy at the expense of Q's kinetic energy.

QUESTION: Can the B magnetic field be conservative too?

ANSWER: In a sense, yes, if its curl is zero. There are no magnetic monopoles to take advantage of it, but it does permit one to define a magnetic scalar potential, Section 12.1, which can be useful in computations.

Now, in Section 4.1 we introduced a voltage: **emf** $= \oint \mathbf{E} \cdot d\mathbf{l}$. Here and in the preceding section we have introduced another voltage: **potential** $V = -\int \mathbf{E} \cdot d\mathbf{l}$. What is the distinction between these voltages?

The answer is, since $\nabla \times \mathbf{E} = -\partial \mathbf{B}/\partial t$ in general, then (Stokes)

$$\oint \mathbf{E} \cdot d\mathbf{l} = -\int \frac{\partial \mathbf{B}}{\partial t} \cdot d\mathbf{a}$$

So emf is due to changing magnetic $\mathbf{B}$ fields, and V contributes nothing to emf. To put it another way: also, in general,

$$\mathbf{E} = -\nabla V - \frac{\partial \mathbf{A}}{\partial t}$$

so $\mathbf{E}$ consists of two terms, both expressible as volts per meter. But the ∇V term contributes no energy to a charge carried about a closed contour (since $\nabla \times \nabla V = 0$), and emf is due to the $\partial \mathbf{A}/\partial t$ term alone.

So far we have simply limited emf (or ε) to the definition:

$$\mathrm{emf} = \oint \mathbf{E} \cdot d\mathbf{l}$$

However, the following more general concept is often useful: emf is "any influence that causes charge to circulate around a closed path" (Purcell). The energy required may come from electromagnetic sources ($\partial \mathbf{A}/\partial t$) or others (e.g., chemical processes). For example, electromotive force is often ascribed to a battery, even though there is no $\oint \mathbf{E} \cdot d\mathbf{l}$. One may colloquially attribute emf to a component that is not part of a closed circuit at all, for example $\int_a^b \mathbf{E} \cdot d\mathbf{l}$ along path 1 in Figure 6.11b. It may even be referred to as a potential difference, but of course this is inaccurate because the potential V is zero everywhere in the figure. And we may be doing the same occasionally, but do bear in mind the distinction between emf and potential alluded to above.

Now we're going to actually calculate a charged particle's energy. As we mentioned before, energy input W is minus force QE times distance l:

$$W = -\int Q\mathbf{E} \cdot d\mathbf{l}$$

so,

$$\frac{W}{Q} = -\int \mathbf{E} \cdot d\mathbf{l} = \Delta V \qquad (6.25)$$

in the static case. In Section 6.2 we pointed out that V includes an arbitrary constant that is usually so chosen that $V = 0$ at infinity. So the energy required to place a charge Q at a point r of pre-existing V is

$$W = \int_{\infty}^{r} Q\mathbf{E} \cdot d\mathbf{l}$$

$$= QV$$

Once the charge Q is there, of course, it changes V by its mere presence, but that "self-energy" doesn't affect how much energy it takes to get the charge there. (However, if Q causes other charges to change position, that's a different story.)

Energies of small particles are often given in terms of **electron volts,** eV, rather than joules. One eV is the energy an electron gains in rising through one volt:

$$W = QV$$

$$= 1.6 \times 10^{-19} \, C \times 1 \, V = 1.6 \times 10^{-19} \, J$$

So the conversion factor between joules and electron volts is numerically equal to the electron charge.

 EXAMPLE 6-12

In a color TV tube, an electron is accelerated by 25 kilovolts. Find its kinetic energy in eV and in J.

ANSWER

$$W = \frac{1}{2} \, mv^2 = QV$$

$$= 1.6 \times 10^{-19} \times 25 \times 10^{3} = 4.0 \times 10^{-15} \, J \qquad \textbf{(6.26)}$$

$$= 25 \text{ keV}$$

This calculation is correct because it involves only QV, which is okay even in relativity (if $\partial \mathbf{A} / \partial t = 0$). However, the expression $\frac{1}{2} \, mv^2$, which was not actually employed, is only a nonrelativistic approximation for the kinetic energy. Carrying it through to find the speed,

$$\frac{1}{2} \, mv^2 \approx 4.0 \times 10^{-15} \, J$$

$$v = \sqrt{2 \times 4.0 \times 10^{-15} / 9.1 \times 10^{-31}} = 9.4 \times 10^{7} \, m/s$$

The fact that the speed found is a significant fraction of the speed of light, $c = 3 \times 10^{8}$ m/s, warns us that nonrelativistic kinematics is likely to be inadequate. We'll do it better in Chapter 18; the relativistically correct answer is 9.1×10^{7} m/s. We have gone out of our way to do the calculation here just to alert you to the fact that, for small particles and high voltages, some of the usual nonrelativistic calculations may be inaccurate.

6.5. Energy Density for V and A

We already have the correct expressions for energy density, static or otherwise, namely $\epsilon_0 E^2/2$ and $B^2/2\mu_0$, from Section 6.1. However, in the static case there are expressions involving the potentials that are often easier to calculate.

For the total electrostatic energy, integrating over all space,

$$
\begin{aligned}
W &= \frac{\epsilon_0}{2} \int E^2 \, dv \\[2mm]
&= -\frac{\epsilon_0}{2} \int \nabla V \cdot \mathbf{E} \, dv \qquad (\text{if } \partial \mathbf{A}/\partial t = 0) \\[2mm]
&= \frac{\epsilon_0}{2} \int V \nabla \cdot \mathbf{E} \, dv - \frac{\epsilon_0}{2} \int \nabla \cdot (V\mathbf{E}) \, dv \qquad \text{using a vector identity} \\[2mm]
&= \int \frac{V\rho}{2} \, dv - \frac{\epsilon_0}{2} \oint V\mathbf{E} \cdot d\mathbf{a} \qquad (\text{divergence theorem and } \epsilon_0 \nabla \cdot \mathbf{E} = \rho)
\end{aligned}
\tag{6.27}
$$

Now the second integral is an integral over a surface enclosing all space. Then it will be zero if we get far enough away from all charges, by the following reasoning: At large distance, the charge distribution looks like a point charge, so V goes to zero as $1/r$, and E as $1/r^2$. The area $\mathbf{a}$ only increases as r^2, like the area of a sphere. So the integral tends to get smaller as $1/r$, and it can be discarded at large r. Consequently, we can write

$$
W = \int \frac{V\rho}{2} \, dv
$$

so we find the energy density to be, in the static case,

$$
\boxed{\frac{dW}{dv} = \frac{\rho V}{2}}
\tag{6.28}
$$

If the charge Q is all at potential V, then $W = QV/2$.

Why is there a factor of $1/2$ in this equation? If we bring in a charge Q from infinity (where $V = 0$) and place it at a location of pre-existing potential V, the energy is $W = QV$, as we saw in the preceding section, without any $1/2$. However, if there is no pre-existing V, then the first little bit ΔQ of charge arrives when $V = 0$; the last one arrives at full V; and the average ΔQ arrives at potential $V/2$.

The two energy densities themselves, $\epsilon_0 E^2/2$ and $\rho V/2$, are not the same, and in fact $\rho V/2$ may even be negative in some regions. Their integrals over all space happen to be equal in the static case, but not necessarily in the dynamic case. For example, it is possible that the charge distribution accelerated at some previous time, gave off radiation, and then became quiescent. In such a case, the integral of $\epsilon_0 E^2/2$ would still be correct, but the integral of $\rho V/2$ would be incomplete, because it does not include the radiated energy.

◤ EXAMPLE 6-13

Find the electrostatic energy of a uniformly charged spherical shell of radius $R = 5$ cm, having a field of $E = 10^6$ V/m at its surface, Figure 6.13.

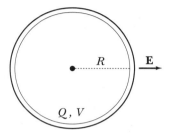

Figure 6.13   Charged shell.

ANSWER The energy is

$$W = \int \rho V \, dv / 2 = QV/2$$

Outside of the shell, the field behaves as if all of the charge were concentrated at the center, so

$$V = Q/4\pi\epsilon_0 R = RE \quad (= 5 \times 10^4 \text{ V}) \qquad \text{and} \qquad Q = 4\pi\epsilon_0 R^2 E \quad (= 278 \text{ nC})$$

So,

$$\begin{aligned} W &= QV/2 \\ &= 4\pi\epsilon_0 R^3 E^2/2 \\ &= \frac{(0.05 \text{ m})^3 \times (10^6 \text{ V/m})^2}{9 \times 10^9 \text{ N} \cdot \text{m}^2/\text{C}^2} \times 2 = 0.0069 \text{ J} \end{aligned}$$

The energy is very small; so this baseball-size object, charged to 50,000 volts ($= ER$), could make a small spark but would pose no threat to human life or limb. Here we have assumed that the energy is entirely contained in the infinitely thin spherical shell where the charge density is located. In Problem 6-8, we will repeat the calculation by integrating $\epsilon_0 E^2/2$, and we will find energy everywhere *except* in that sphere; yet the result will be the same.

The preceding example dealt with the self-energy of one charge, which might be written $W_s = Q_1 V_1/2$ for a spherical shell, where the voltage V_1 is due to Q_1 itself. However, when you have two small charges as in Figure 6.14, you often care only about the interaction energy W_i between the charges, because the individual charges come ready made and you don't change them. W_i is the energy it takes to bring one charge from infinity up to a distance r from the other (which is held sta-

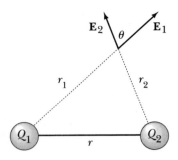

Figure 6.14   Interaction energy.

tionary). That energy is easy to calculate using potentials; for two simple charges:

$$W_i = Q_1 V_2 = \frac{Q_1 Q_2}{4\pi\epsilon_0 r}$$

On the other hand, for the same two simple charges, the calculation in terms of fields is much harder. The total energy is

$$W_t = \frac{\epsilon_0}{2} \int (\mathbf{E}_1 + \mathbf{E}_2)^2 \, dv$$

$$= \frac{\epsilon_0}{2} \int E_1{}^2 \, dv + \frac{\epsilon_0}{2} \int E_2{}^2 \, dv + \epsilon_0 \int \mathbf{E}_1 \cdot \mathbf{E}_2 \, dv \qquad (6.29)$$

The first two terms are self-energies again, and the third is W_i. That integral turns out to be difficult even for two simple charges — try it and see. This illustrates the fact, mentioned at the beginning of this section, that the potentials may be more useful than the field in calculating energies.

For the magnetic energy density of steady currents, we can proceed as before for the electrostatic case. Integrating the field energy density over all space.

$$W = \frac{1}{2\mu_0} \int B^2 \, dv$$

$$= \frac{1}{2\mu_0} \int \mathbf{B} \cdot \nabla \times \mathbf{A} \, dv \qquad \text{(since } \mathbf{B} = \nabla \times \mathbf{A}) \qquad (6.30)$$

$$W = \frac{1}{2\mu_0} \int \mathbf{A} \cdot \nabla \times \mathbf{B} \, dv + \frac{1}{2\mu_0} \int \nabla \cdot (\mathbf{A} \times \mathbf{B}) \, dv \qquad \text{(using a vector identity)}$$

The second integral is zero as before: We use the divergence theorem to change it to $\int \mathbf{A} \times \mathbf{B} \cdot d\mathbf{a}$, and we argue again that the product $\mathbf{A} \times \mathbf{B}$ gets small faster than the area $\mathbf{a}$ gets big. In the first integral, from Ampère's law, $\nabla \times \mathbf{B} = \mu_0 \mathbf{J}$ if $\partial \mathbf{E}/\partial t = 0$. So, in the static case

$$\boxed{\frac{dW}{dv} = \frac{\mathbf{A} \cdot \mathbf{J}}{2}} \qquad (6.31)$$

▼ EXAMPLE 6-14

A long solenoid of length l and radius R carries current I in N' turns per meter, Figure 6.15. Find the energy using $\mathbf{A} \cdot \mathbf{J}/2$. Evaluate for $l = 10$ cm, $R = 1$ cm, $I = 2$ A, $N' = 10^4$ turns/meter.

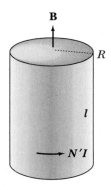

Figure 6.15   Finite solenoid.

ANSWER $\mathbf{A}$ and $\mathbf{J}$ are parallel, so we have

$$W = \int \frac{\mathbf{A} \cdot \mathbf{J}}{2} \, dv$$

$$W = \int \frac{A}{2} J \, dv \Longrightarrow \int \frac{A}{2} K \, da$$

$$= \frac{A}{2} K 2\pi Rl$$

where the surface current density is $K = N'I$, and of course $B = \mu_0 N' I = \mu_0 K$. As before, we find $\mathbf{A}$ thus:

$$\Phi = \int \mathbf{B} \cdot d\mathbf{a} = \oint \mathbf{A} \cdot dl$$

$$\mu_0 K\pi R^2 = 2\pi RA$$

$$A = \mu_0 KR/2$$

So,

$$W = (\mu_0 KR/4) K 2\pi Rl$$

$$= \mu_0 K^2 \pi R^2 l/2$$

Evaluating,

$$W = 4\pi \times 10^{-7} (10^4 \times 2)^2 \pi (0.01)^2 \, 0.1/2 = 7.9 \times 10^{-3} \text{J}$$

a rather small energy because of the small volume involved. Here we have assumed that all of the energy is contained in the infinitely thin cylindrical shell where the current density $\mathbf{J}$ is located. In Problem 6-1, we will repeat the calculation by inte-

Figure 6.16   Electron.

grating $B^2/2\mu_0$, and we will find that all of the energy is inside the cylinder, rather than just on the surface; yet the result will be the same. (In Problem 10-32 we do it again using the concept of inductance.)

How big is an electron (Fig. 6.16)? Surprisingly, we can at least set a lower limit using Einstein's celebrated formula $E = mc^2$ (Fig. 6.17), which we will have to write as $W = mc^2$ to avoid confusion with the electric field E. The formula says that energy has mass. The electron has energy in its field, and this contributes to its mass. If we suppose the electron is a spherical shell of radius r_e, then its energy is

Figure 6.17 $E = mc^2$.

$$\frac{QV}{2} = \frac{1}{2} \frac{e^2}{4\pi\epsilon_0 r_e}$$

If it's a different shape, the leading factor of $\frac{1}{2}$ may change. Since we don't know what shape it is, and we just want an approximation, we will leave out the $\frac{1}{2}$. Certainly this energy accounts for some of the electron mass. For our lower limit, we will assume that *all* of the mass of the electron is due to its field energy:

$$W = mc^2 = \frac{e^2}{4\pi\epsilon_0 r_e}$$

So,

$$r_e = \frac{e^2}{4\pi\epsilon_0 mc^2} = 2.82 \times 10^{-15} \text{ m} \tag{6.32}$$

ALRIGHT RUTH,   I ABOUT GOT THIS ONE RENORMALIZED.

Figure 6.18   Renormalized.

This is called the **classical electron radius.** It is about the size of an atomic nucleus. The electron cannot be smaller than this, because if it were its mass would be larger.

Unfortunately, the electron *is* smaller, and yet its mass is *not* larger. In fact, in high-energy scattering experiments the electron behaves like a point charge, which would then have infinite mass. This and related infinities are "swept under the rug" in a process called "renormalization" (Fig. 6.18), in a discipline called "quantum electrodynamics," or QED for short. This theory has been amazingly successful in predicting and explaining modern experimental results. Many of its proponents are no more satisfied with renormalization than you or I, but it seems to work. We will return to this matter when we do dipoles (how can a point charge have a magnetic dipole moment?).

6.6 SUMMARY

The most important single characteristic of a field is the energy it contains. Energy calculation is often facilitated by using a potential as an intermediary. In a conservative field, a test charge does not gain or lose energy over a closed trajectory.

The energy density of E and B fields is

$$\frac{dW}{dv} = \frac{1}{2}\epsilon_0 E^2 + \frac{1}{2}B^2/\mu_0 \tag{6.2, 6.5}$$

The scalar and vector potentials are given by:

$$\mathbf{E} = -\nabla V - \frac{\partial \mathbf{A}}{\partial t} \tag{6.7}$$

and

$$\mathbf{B} = \nabla \times \mathbf{A} \tag{6.8}$$

The gradient ∇V gives the direction and magnitude of increasing V.
In the static case, potentials are

$$V = \int \frac{\rho \, dv}{4\pi\epsilon_0 r} \tag{6.17}$$

and

$$\mathbf{A} = \int \frac{\mu_0 \mathbf{J} \, dv}{4\pi r} \tag{6.20}$$

We have found two ways each to calculate $\mathbf{E}$, $\mathbf{B}$, V, and $\mathbf{A}$ for the quasi-static case: $\hspace{4cm}$ (6.23)

Asymmetric: $\quad \mathbf{E} = \int \frac{\rho \hat{r} \, dv}{4\pi\epsilon_0 r^2} \quad \mathbf{B} = \int \frac{\mu_0 \mathbf{J} \times \hat{r} \, dv}{4\pi r^2} \quad V = \int \frac{\rho \, dv}{4\pi\epsilon_0 r} \quad \mathbf{A} = \int \frac{\mu_0 \mathbf{J} \, dv}{4\pi r}$

Symmetric: $\quad \oint \mathbf{E} \cdot d\mathbf{a} = Q/\epsilon_0 \quad \oint \mathbf{B} \cdot d\mathbf{l} = \mu_0 I \quad V = -\int \mathbf{E} \cdot d\mathbf{l} \quad \oint \mathbf{A} \cdot d\mathbf{l} = \Phi$

A conservative field has no curl and may be expressed as the gradient of a scalar. In the static case, the energy density is

$$\frac{dW}{dv} = \frac{1}{2} \rho V + \frac{1}{2} \mathbf{A} \cdot \mathbf{J} \qquad \text{(6.28, 6.31)}$$

The classical electron radius is

$$r_e = \frac{e^2}{4\pi\epsilon_0 mc^2} = 2.82 \times 10^{-15} \text{ m} \qquad \text{(6.32)}$$

◢ PROBLEMS

6-1 Find the approximate energy in the field of a solenoid of radius 1.0 cm, length 10 cm, and $N' = 10^4$ turns per meter of current 2 A, by integrating $B^2/2\mu_0$. Compare with Example 6-14, which involved $\mathbf{A} \cdot \mathbf{J}/2$.

6-2 For the preceding problem, find the pressure on the coils. Is it inward or outward?

6-3 Find $\mathbf{A}$ at 1 cm from the axis of an infinite solenoid of radius 2 cm whose B field is 0.3 T. (Assume all current is azimuthal.)

6-4 Find V at 2 cm from a charge of -3 nC.

6-5 Find $\mathbf{E}$ and V, at distance r from the center, inside and outside of a uniformly charged sphere of radius R and charge Q. Sketch E and V versus r.

6-6 Find $\mathbf{B}$ and $\mathbf{A}$, at distance r from the axis, inside and outside of an infinite solenoid of radius R and of N' turns carrying current I. Sketch B and A versus r.

6-7 Find the energy of a uniformly charged sphere of radius R and charge Q. Evaluate for $R = 5$ cm, $Q = 278$ nC.

6-8 Find the energy of a uniformly charged spherical shell of radius R and charge Q by integrating E^2 over all space. Evaluate for $R = 5$ cm and surface field $E = 10^6$ V/m. Compare with Example 6-13, which involved $\rho V/2$.

6-9 Find V_i at r inside, and V_o outside, of an infinite cylinder of radius a and uniform charge density ρ, in two ways.

6-10 Find $\mathbf{A}_i$ at r inside, and $\mathbf{A}_o$ outside, of an infinite cylinder of radius a and (axial) current I, in two ways.

6-11 Find A on the surface of the inside conductor of a pair of infinite coaxial cylinders, Figure 6.19, with $I = 1.2$ A, $r_1 = 0.08$ cm, and $r_2 = 0.31$ cm.

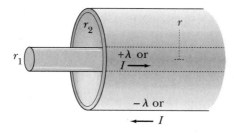

Figure 6.19

6-12 Find V on the surface of the inside conductor of a pair of infinite coaxial cylinders, Figure 6.19, with $\lambda = 0.21$ nC/m, $r_1 = 0.08$ cm, and $r_2 = 0.31$ cm.

6-13 For the solenoid of Example 6-6, assume that the field rises from 0 to 0.3 T in 0.017 s. Find the emf about the loop at r using **(a)** $d\Phi/dt$ **(b)** $\partial A/\partial t$.

6-14 Gauge transformation: Show that the **E** and **B** fields are unaffected by the following modification of **A** and V:

$$\mathbf{A}' = \mathbf{A} + \nabla\psi$$
$$V' = V - \partial\psi/\partial t$$

ψ being an arbitrary scalar.

6-15 Find **B** and **A** at the center of a circular loop of radius 1 cm and current 2 A.

6-16 Find **E** and V at the center of a circular loop of radius a and linear charge density λ. Evaluate for 1 cm and 1 nC/m.

6-17 Find $\nabla \times (\partial\mathbf{A}/\partial t)$ $(= -\nabla \times \mathbf{E})$ at distance r inside and outside of an infinite solenoid of field B and dB/dt. (So if a loop doesn't pass through or around the solenoid, it has no emf.)

6-18 A charge q is situated at a point of given vector potential **A**, caused by a current somewhere. When the current is suddenly stopped, show that the charge receives an impulse $\Delta mv = q\mathbf{A}$. (So **A** is like a "momentum potential.")

6-19 Find V on the diagonal of a square of side s and of linear charge density λ, Figure 6.20, at a distance $\sqrt{2}s$ from the corner.

6-20 Find **A** on the diagonal of a square of side s and of current I, Figure 6.20, at a distance $\sqrt{2}s$ from the corner.

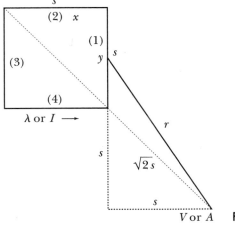

V or A **Figure 6.20**

6-21 An alpha particle (helium nucleus) of charge $+2e$ and kinetic energy 7.7 MeV collides head-on with a gold nucleus and bounces backwards. Assuming infinite mass for the gold nucleus, find the distance of closest approach. (The Rutherford experiment, 1906, suggests that the nucleus is smaller than this distance.)

6-22   Assuming a gold nucleus is a uniformly charged sphere of radius $R = 1.2 \times 10^{-15}$ $A^{1/3}$ m, where A is the number of nucleons, find the potential V at its center.

6-23   A uranium 236 nucleus splits into two equal parts. The nuclear radius is $R = 1.2 \times 10^{-15} A^{1/3}$ m, where A is the number of nucleons. Find the electrostatic energy released, in MeV. (The total energy released is less, around 200 MeV, due partly to nuclear energy.)

*6-24   For the following expressions involving the vector $\mathbf{F}$ (which could be $\mathbf{E}$ or $\mathbf{B}$): If the divergence is zero, find a possible vector potential $\mathbf{A}$ such that $\nabla \times \mathbf{A} = \mathbf{F}$; if the curl is zero, find a possible scalar potential V such that $\nabla V = \mathbf{F}$.
(a) $\mathbf{F} = (x + y)\,\hat{\mathbf{x}} + (-x + y)\,\hat{\mathbf{y}} - 2z\,\hat{\mathbf{z}}$
(b) $\mathbf{F} = (x + y)\,\hat{\mathbf{x}} + (x - y)\,\hat{\mathbf{y}} - 2z\,\hat{\mathbf{z}}$
(c) $\mathbf{F} = (x + y)\,\hat{\mathbf{x}} + (x + y)\,\hat{\mathbf{y}} - 2z\,\hat{\mathbf{z}}$
(d) $\mathbf{F} = (x + y)\,\hat{\mathbf{x}} + (-x - y)\,\hat{\mathbf{y}} - 2z\,\hat{\mathbf{z}}$

6-25   Find V at z on the axis of a disk of uniform surface charge density σ and radius R, lying in the xy plane and centered on the z axis, Figure 6.21. Evaluate σ for $V = 10^4$ V at the center of the disk and $R = 2.5$ cm.

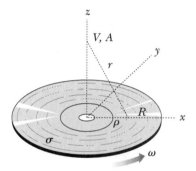

Figure 6.21

6-26   For the preceding problem, with the disk rotating at angular speed ω, find $\mathbf{A}$. Evaluate A at the center of the disk for 10^4 V and 7200 rpm.

6-27   Find the vector potential $\mathbf{A}$ near a pair of infinite wires carrying opposite currents of I, Figure 6.22, at distance r_+ from the upward current and r_- from the downward one.

6-28   Show that the scalar potential V near a pair of infinite oppositely charged lines of uniform linear charge density $\pm\lambda$, Figure 6.22, at distance r_+ from the positive wire and r_- from the negative one, is $V = \dfrac{\lambda}{2\pi\epsilon_0} \ln \dfrac{r_-}{r_+}$.

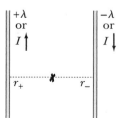

Figure 6.22

6-29 Show that the **E** field lines for two oppositely charged line charges (Fig. 6.4b) are circular arcs.

*6-30 Show that the equipotentials for two oppositely charged line charges (Fig. 6.4b) are circular cylinders.

6-31 Find the force of repulsion between the two hemispheres of a uniformly charged spherical shell of radius R and charge Q. (Think bubble.)

6-32 Create and solve a simple problem for each of the eight formulas in Equation 6.23 at the end of Section 6.3.

6-33 Show that the central field $\mathbf{E} = r^n\hat{\mathbf{r}}$ is conservative for arbitrary n.

6-34 Eight identical positive charges Q occupy the corners of a cube of side s, Figure 6.23. Show that the potential V is less at point x, at the center of one face, than at the center of the cube. (Compare with Problem 2-35.)

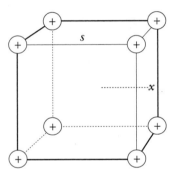

Figure 6.23

6-35 Find the potential V at the edge of a uniformly charged disk of charge density σ and radius R, Figure 6.24.

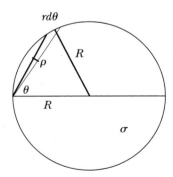

Figure 6.24

*6-36 Find the potential V in the plane of a uniformly charged disk of charge density σ and radius R, at distance $a < R$ from the center. (It involves an elliptic integral.) Check for $a = 0$ and for $a = R$.

6-37 In a cyclotron, Figure 6.25, protons p circulate between hollow "D"-shaped electrodes whose voltage alternates at the cyclotron frequency (Section 3.3). They are attracted

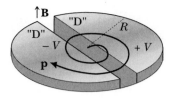

Figure 6.25

into one D, but while they are inside it the voltage is reversed, so then they are re-pelled by the first D and attracted by the other D. **B** is perpendicular to the "D's." Find the final energy, and time required, for a cyclotron of radius $R = 30$ cm, $B = 1.2$ T, and $V = \pm 1000$ volts.

CHAPTER

7

DIPOLES AND MULTIPOLES

The preceding chapters have been rather general in terms of charges and charge distributions. But now, in the next four chapters, we will be looking into the ordered charge configurations of which terrestrial molecular matter is actually comprised. Ordinary matter is more or less uncharged, but it is rich in pairs of charges called dipoles, and so we begin with the study of dipoles, which we shall regard as building blocks of dielectric and magnetic materials.

7.1. Electric Dipole p

An electric monopole is a single charge, and a dipole is two opposite charges closely spaced, or something which looks like that electrically. Such dipoles are abundant in nature. For example, in Section 2.1, when we attracted an uncharged pith ball to a charged rod, it was the little molecular dipoles which responded to the **E** field. And indeed, all dielectric materials, meaning nonconductors, may be regarded as assemblages of actual or potential dipoles.

The water molecule, for example, Figure 7.1, has a large permanent electric dipole moment; that is what makes it rather peculiar among molecules. Its positive and negative charges are not centered at the same point; it behaves like two equal opposite charges separated by a small distance. Its molecules love each other because of electrostatic attraction, and its boiling point is thereby elevated. Furthermore, its ability to dissolve salts is enhanced because its dipole molecules love the resulting electrically charged ions.

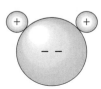

Figure 7.1   Water, H_2O.

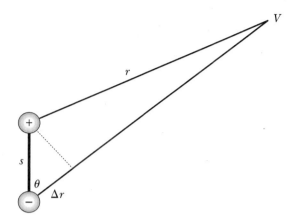

Figure 7.2   Dipole potential V.

Not surprisingly, it turns out to be economical mathematically to deal with the dipole not as just a pair of individual plus and minus charges, but as a separate object in its own right. And that is what we shall now undertake.

We suppose that opposite charges, each of magnitude Q, are separated by s, Figure 7.2. We will find the potential V at distance r and angle θ under the assumption that $r \gg s$. That is the assumption that makes our dipole approximation valid: We have to be far away, relative to the size of the dipole.

The potential at V due to the $+$ charge (Section 6.3) is

$$V_+ = \frac{Q}{4\pi\epsilon_0 r} \tag{7.1}$$

And the potential due to the $-$ charge is

$$V_- = -\frac{Q}{4\pi\epsilon_0(r + \Delta r)}$$

For the pair of charges, we add the potentials; but since one charge is $+$ and one is $-$, this amounts to taking a difference. We have

$$V = V_+ + V_- = \frac{Q}{4\pi\epsilon_0}\left(\frac{1}{r} - \frac{1}{r + \Delta r}\right) \tag{7.2}$$

When we multiply and divide by Δr, this turns into a derivative form (with $\Delta r \to 0$):

$$V = V_+ + V_- = \frac{Q}{4\pi\epsilon_0}\left(\frac{1/r - 1/(r + \Delta r)}{\Delta r}\right)\Delta r \implies -\frac{Q}{4\pi\epsilon_0}\frac{\partial(1/r)}{\partial r}\Delta r$$

$$= -\frac{\partial V_+}{\partial r}\Delta r \qquad \text{for small } \Delta r$$

That is the dipole V that we want. Differentiating V_+, with $\Delta r = s\cos\theta$, we obtain

$$V = \frac{Q}{4\pi\epsilon_0 r^2}s\cos\theta \tag{7.3}$$

As long as $s \ll r$, it doesn't matter whether we have a small Q and large s or vice versa; V depends only on their product, which we call the **dipole moment p,** with

$$\mathbf{p} = Q\mathbf{s}$$

a vector pointing in the direction from $-Q$ to $+Q$. The potential becomes

$$V = \frac{p \cos \theta}{4\pi\epsilon_0 \, r^2} = \frac{\mathbf{p} \cdot \hat{\mathbf{r}}}{4\pi\epsilon_0 \, r^2} \qquad (7.4)$$

 EXAMPLE 7-1

The dipole moment of a water molecule, H_2O, is $p = 6.2 \times 10^{-30}$ C $\cdot$ m. It is due to the two positive hydrogen ions not being centered on the negative oxygen ion. How far off center are they, effectively? That is, if $Q = 2\,e$, find s.

ANSWER

$$s = p/Q$$
$$= 6.2 \times 10^{-30}\ \text{C} \cdot \text{m}/2 \times 1.6 \times 10^{-19}\ \text{C} = 1.9 \times 10^{-11}\ \text{m}$$

which is about a fifth of the size of the water molecule.

Now we can find the **E** field of a dipole using $\mathbf{E} = -\nabla V$ (this being a static case, there is no $-\partial \mathbf{A}/\partial t$). In spherical coordinates, using gradient components from inside the front cover,

$$E_r = -\frac{\partial V}{\partial r} = \frac{2p \cos \theta}{4\pi\epsilon_0 r^3} \qquad (7.5)$$

$$E_\theta = -\frac{1}{r}\frac{\partial V}{\partial \theta} = \frac{p \sin \theta}{4\pi\epsilon_0 r^3}$$

$$E_\phi = -\frac{1}{r \sin \theta}\frac{\partial V}{\partial \phi} = 0$$

As a vector,

$$\mathbf{E} = \frac{1}{4\pi\epsilon_0 r^3}(2p \cos \theta\, \hat{\mathbf{r}} + p \sin \theta\, \hat{\boldsymbol{\theta}})$$

(A "coordinate-independent" form occurs in Problem 7–23.)

The electric field is illustrated in Figure 7.3. The field lines come up out of the dipole pointing in the direction of **p**, outside of the microscopic structure of the dipole itself; inside the dipole, as mentioned, **p** points from $-$ to $+$.

If we place a permanent dipole in a uniform external electrostatic field, it ex-

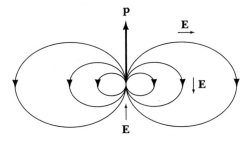

Figure 7.3 Dipole field lines.

periences a **torque.** As seen in Figure 7.4, the two equal, opposite, noncollinear forces **F** constitute a couple, and the torque τ is:

$$\tau = QE\,s\sin\theta = pE\sin\theta$$

$$\tau = \mathbf{p} \times \mathbf{E} \tag{7.6}$$

which is directed up out of the paper, by the right-hand rule for cross products. We assume that there is no changing **B** field here to contribute another, nonconservative, **E** field. When the dipole rotates, the potential **energy** $\tau\theta$ is

$$W = \int_{\pi/2}^{\theta} pE\sin\theta\,d\theta = -\mathbf{p}\cdot\mathbf{E} \tag{7.7}$$

for a permanent dipole. We start integrating at $\theta = \pi/2$ because by symmetry the energy should be zero there: The energy to remove the plus charge to infinity is QV, and for the minus charge it's $-QV$, and V is the same for both charges when $\theta = \pi/2$.

   If the dipole is not permanent but is induced by the **E** field, the energy is less, and some energy goes into the mechanical work of stretching the charge distribution. For example, this occurs when a nonpolar molecule such as CO_2 is placed in an electric field. Referring to Figure 7.5, we imagine that the negative charge is held, and the positive one is attached to it by a spring of spring constant k, at fixed θ. In this case, the separation is zero when $E = 0$. When we increase the E field from 0 to E, the final force is $\mathbf{F} = Q\mathbf{E}$, and the component of the force along the spring is

$$F\cos\theta = ks$$

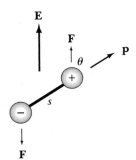

Figure 7.4 Dipole torque.

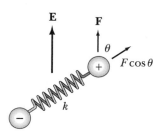

Figure 7.5   Charge on spring.

The corresponding mechanical energy is

$$W = \tfrac{1}{2} ks^2 = \tfrac{1}{2} F \cos \theta \, s = \tfrac{1}{2} QE \cos \theta \, s = \tfrac{1}{2} \mathbf{p} \cdot \mathbf{E}$$

If we somehow latch or lock the spring at this point, then we must invest $\mathbf{p} \cdot \mathbf{E}$ to remove the dipole from the field, but we get $\tfrac{1}{2} \mathbf{p} \cdot \mathbf{E}$ back when we unlatch the spring. So the net energy is

$$W = -\mathbf{p} \cdot \mathbf{E} + \tfrac{1}{2} \mathbf{p} \cdot \mathbf{E} = -\tfrac{1}{2} \mathbf{p} \cdot \mathbf{E} \tag{7.8}$$

for an induced dipole.

Since work is force times distance, a permanent dipole should align itself with an electric field and then should be attracted into the field with a force $\mathbf{F} = -\nabla W = \nabla(\mathbf{p} \cdot \mathbf{E})$. But if the $\mathbf{E}$ field is not conservative (there is a changing $\mathbf{B}$ field too), this formula is inadequate. More generally: Referring again to Figure 7.4, suppose that the field $\mathbf{E}$ points in the y direction and increases with y. Then the two forces will differ, and the net force will be

$$Q \, \Delta E_y = Q \frac{\partial E_y}{\partial y} \Delta y = Q s_y \frac{\partial E_y}{\partial y} = p_y \frac{\partial E_y}{\partial y} \tag{7.9}$$

If E_y also changes in the x direction, we get an additional $p_x \partial E_y / \partial x$. The total y component is

$$F_y = p_x \frac{\partial E_y}{\partial x} + p_y \frac{\partial E_y}{\partial y} + p_z \frac{\partial E_y}{\partial z}$$

The x and z components will be similar in an arbitrary $\mathbf{E}$ field. The total **force F** will contain nine terms, and it may conveniently be written:

$$\mathbf{F} = (\mathbf{p} \cdot \nabla) \mathbf{E} \tag{7.10}$$

This is identical to $\mathbf{F} = \nabla(\mathbf{p} \cdot \mathbf{E})$ in a conservative $\mathbf{E}$ field (no curl, Problem 7-11).

◢ EXAMPLE 7-2

A water molecule in a vacuum is attracted into a region of field $E = 10^6$ V/m. What is its kinetic energy there, in electron volts?

ANSWER Outside of the field, $\mathbf{p} \cdot \mathbf{E} = 0$; inside, $\mathbf{p}$ aligns itself with the field, and so

$$W = pE$$

$$= 6.2 \times 10^{-30}\,\text{C} \cdot \text{m} \times 10^6\,\text{V/m} = 6.2 \times 10^{-24}\,\text{joules}$$

so $6.2 \times 10^{-24}\,\text{J}/1.6 \times 10^{-19}\,\text{J/eV} = 3.9 \times 10^{-5}\,\text{eV}$

(We introduced eV in Section 6.4.) This is only $1/1000$ of thermal energy kT at room temperature T (with Boltzmann's constant k). So it is possible, but not easy, to accelerate dipoles using electric fields. By contrast, an ion having one electron charge could get a million electron volts of energy per meter in such a field.

 **EXAMPLE 7-3**

Find the force and torque on a positive charge Q and a dipole **p** that is oriented perpendicular to the line **r** joining them, Figure 7.6a.

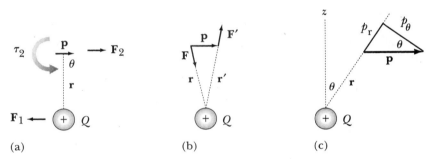

Figure 7.6   (a) Dipole and charge. (b) Forces on dipole. (c) Components.

ANSWER   The force on the charge is easy (Fig. 7.6a): from Equation 7.5,

$$\mathbf{F}_1 = QE_\theta\,\hat{\boldsymbol{\theta}} = Q\frac{p\sin\theta}{4\pi\epsilon_0 r^3}\,\hat{\boldsymbol{\theta}} \longrightarrow Q\frac{p}{4\pi\epsilon_0 r^3}\,\hat{\boldsymbol{\theta}} \qquad \text{toward the left (at } \theta = 90°)$$

The force on the dipole is $\nabla(\mathbf{p} \cdot \mathbf{E})$. Figure 7.6b shows that the force **F** on the tail of the dipole has a rightward component, and so does the force **F'** on the head, so there is a net force toward the right. But we cannot just substitute $\mathbf{p} \cdot \mathbf{E} = 0$ and get zero force. For the differentiation implied by ∇, we must allow the dipole to move in translation but not rotation, because we want force but not torque. We will use a more general location of **p**, Figure 7.6c, and then let θ become zero at the end (relative to z; it's not the same θ as for $\mathbf{F}_1$). From the figure,

$$\mathbf{p} = p_r\,\hat{\mathbf{r}} + p_\theta\,\hat{\boldsymbol{\theta}} = p\sin\theta\,\hat{\mathbf{r}} + p\cos\theta\,\hat{\boldsymbol{\theta}}$$

Then

$$\mathbf{p} \cdot \mathbf{E} = p\sin\theta\,\frac{Q}{4\pi\epsilon_0 r^2}$$

and

$$\mathbf{F}_2 = \nabla(\mathbf{p} \cdot \mathbf{E}) = \frac{pQ}{4\pi\epsilon_0}\left(\frac{-2\sin\theta}{r^3}\,\hat{\mathbf{r}} + \frac{\cos\theta}{r^3}\,\hat{\boldsymbol{\theta}}\right) \longrightarrow \frac{pQ}{4\pi\epsilon_0 r^3}\,\hat{\boldsymbol{\theta}}$$

toward the right (at $\theta = 0°$). So $\mathbf{F}_1 = -\mathbf{F}_2$, as required by Newton's third law.

The forces are not collinear; they constitute a couple whose torque is

$$\tau = \mathbf{r} \times \mathbf{F}_2 = \frac{pQ}{4\pi\epsilon_0 r^2}\,\hat{\boldsymbol{\phi}} \qquad \text{(clockwise, cw)}$$

But that's okay, because the torque on the dipole, shown in Equation 7.6, is

$$\tau_2 = \mathbf{p} \times \mathbf{E} = -p\frac{Q}{4\pi\epsilon_0 r^2}\,\hat{\boldsymbol{\phi}} \qquad \text{(counterclockwise, ccw)}$$

so the torques balance after all.

Physically the situation is as follows. If we release a dipole and a charge from rest as in Figure 7.6a, they will start moving as shown by the arrows. This overall relative motion of **p** and Q involves increasing angular momentum clockwise, and consequently the dipole itself will begin to rotate counterclockwise in such a way that the total angular momentum remains zero. Once the charge and dipole move through a significant distance and angle, of course, the situation changes. Ultimately, if there is some sort of drag or damping to get rid of kinetic energy, the charge will wind up attached to the negative end of the dipole.

7.2.   Magnetic Dipole m

Many small objects, including some atoms and elementary particles, have an intrinsic magnetic field. As with the electric dipole, it is convenient to treat these objects as infinitesimal magnetic dipoles. In this case, however, we assume classically that the field is not caused by a pair of opposite magnetic monopoles but by a current loop.

Figure 7.7 is three dimensional: The triangles are in the plane of the paper, and the square loop of side s is perpendicular to the paper. Current **I** flows around the square loop, producing a magnetic field **B** directed upwards, not shown. We will find the vector potential **A** under the assumption that $r \gg s$; as before, that is what makes the dipole approximation valid.

Of the four sides s, the ones in front and back are carrying current in opposite directions and are the same distance from the field point, so they cancel perfectly.

The side on the right contributes, into the paper (Section 6.3),

$$\mathbf{A}_+ = \frac{\mu_0 \mathbf{I}s}{4\pi r} \qquad\qquad (7.11)$$

For the pair of sides to the right and the left, we add the potentials, but since the currents are in opposite directions, this amounts to taking the difference. In the

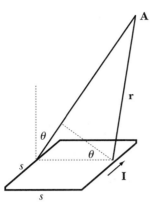

Figure 7.7 Magnetic dipole.

same fashion as for V in the preceding section, we have:

$$A = -\frac{\partial A_+}{\partial r} \Delta r$$

That is the dipole A that we want. Differentiating A_+, with $\Delta r = s \sin \theta$, we obtain

$$A = \frac{\mu_0 I s}{4\pi r^2} s \sin \theta \qquad (7.12)$$

Its direction is that of the closer current, into the paper in this case. We introduce the **magnetic moment m** as the current I times the area a of its loop,

$$m = Ia = Is^2$$

The vector **m** is normal to the loop, with direction given by the *right-hand rule:* fingers in direction of current, thumb in direction of **m**. Thus,

$$\boxed{A = \frac{\mu_0 m \sin \theta}{4\pi \, r^2} \, \hat{\phi} = \frac{\mu_0 \mathbf{m} \times \hat{\mathbf{r}}}{4\pi r^2}} \qquad (7.13)$$

◤ EXAMPLE 7-4

How big is an electron? Its magnetic moment is about one **Bohr magneton**, $\mu_B = e\hbar/2m = 9.27 \times 10^{-24}$ A $\cdot$ m^2. Assume it is a circular ring of negative charge traveling at the speed of light, Figure 7.8.

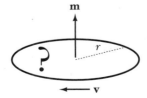

Figure 7.8 Electron.

ANSWER The time for all of the charge e to pass a point on the ring is $2\pi r/v$. So,

$$m = \mu_B = I \cdot \text{area} = \frac{e}{2\pi r/v} \cdot \pi r^2 = evv/2$$

so

$$r = 2\mu_B/ev$$
$$= 2 \times 9.3 \times 10^{-24}/1.6 \times 10^{-19} \times 3 \times 10^8 = 3.9 \times 10^{-13}\text{ m}$$

which is about 137 times r_e, the classical electron radius. The electron cannot be smaller than this and still have such a large magnetic moment. Unfortunately, the electron *is* smaller, much smaller, experimentally. In Section 6.5 we calculated the classical electron radius r_e, also wrong experimentally. Classical electromagnetism must be revised for elementary particles. Their intrinsic magnetic moment is still a meaningful quantity, but its source is not subject to classical analysis. Quantum electrodynamics (QED) deals with such matters.

In this connection, the **gyromagnetic ratio** is sometimes defined: magnetic moment divided by angular momentum. For an electron in a Bohr orbit this is

$$\frac{\pi r^2 I}{mvr} = \frac{e}{2m} \tag{7.14}$$

using $I = \dfrac{e}{2\pi r/v}$.

The field **B** of the magnetic dipole is obtained using $\mathbf{B} = \nabla \times \mathbf{A}$, in spherical coordinates, using curl components from inside the front cover:

$$B_r = \frac{1}{r\sin\theta}\frac{\partial A_\phi \sin\theta}{\partial\theta} = \frac{2\mu_0 m\cos\theta}{4\pi r^3} \tag{7.15}$$

$$B_\theta = -\frac{\partial r A_\phi}{r\partial r} = \frac{\mu_0 m\sin\theta}{4\pi r^3}$$

$$B_\phi = 0$$

The field is identical in form to that of the electric dipole, Equation 7.5. But of course at small distances their structures are quite different; see Figure 7.9 below.

The equations for **torque** and **energy** are similar to those for the electric dipole (see Problem 7–14):

$$\tau = \mathbf{m} \times \mathbf{B} \tag{7.16}$$

$$W = -\mathbf{m} \cdot \mathbf{B} \tag{7.17}$$

 EXAMPLE 7-5

What energy, in eV, is required to flip an electron in a field of 0.1 T?

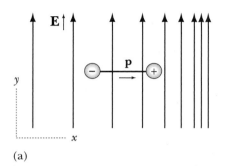

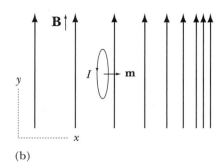

Figure 7.9   (a) Electric dipole in **E** field with curl. (b) Magnetic dipole in **B** field with curl.

ANSWER We will turn it 180°, from parallel to antiparallel with **B**, so

$$W = 2\ mB = 2\ \mu_B B$$
$$= 2 \times 9.3 \times 10^{-24}\,\text{J/T} \times 0.1\ \text{T} = 1.9 \times 10^{-24}\,\text{J} = 1.2 \times 10^{-5}\ \text{eV}$$

This small energy is easily accessible using microwaves.

The **force** on a magnetic dipole (Problem 7-14) is

$$\mathbf{F} = \nabla(\mathbf{m} \cdot \mathbf{B}) \tag{7.18}$$

unlike the electrostatic case, $\mathbf{F} = (\mathbf{p} \cdot \nabla)\,\mathbf{E}$. The difference is due to the microscopic structure of the dipoles. For example, in Figure 7.9a, an electric dipole **p** is perpendicular to an inhomogeneous **E** field. The positive end of the dipole is in a stronger field than the negative end, so there will be a net force upwards. Since $\mathbf{p} \cdot \mathbf{E} = 0$, the equation $\mathbf{F} = \nabla(\mathbf{p} \cdot \mathbf{E})$ is incorrect. (Note that the **E** field has a curl here.) And it is readily verified that $\mathbf{F} = (\mathbf{p} \cdot \nabla)\,\mathbf{E}$ does yield an upwards force:

$$F_y = p_x \frac{\partial}{\partial x} E_y \tag{7.19}$$

which is positive because E_y is increasing in the x direction.

In Figure 7.9b a magnetic dipole is perpendicular to a similar **B** field. Symmetry dictates that there is no comparable upwards force, so $\mathbf{F} = (\mathbf{m} \cdot \nabla)\,\mathbf{B}$ would be incorrect.

The elementary particles that have been checked experimentally do indeed show the magnetic structure of Figure 7.9b, a current loop, rather than being composed of a pair of magnetic monopoles arranged like the electric monopoles of Figure 7.9a (Griffiths AJP).

7.3. Electric Multipoles

In order to make an electric dipole, we started with a monopole, and we reproduced it with opposite sign, displaced from the first (Section 7.1). To make a

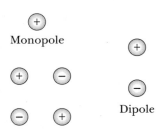

Monopole

Dipole

Quadrupole   **Figure 7.10**   Electric multipoles.

quadrupole, we can displace one dipole from another, with opposite sign; see
Figure 7.10. The next would be an octupole, and so on.

Then, to find the dipole potential V, we differentiated the monopole V. To
find the quadrupole V we would differentiate the dipole V, and so on.

Instead, we are going to take a more general approach here, one which will
yield all of the various multipole moments in one development. We begin by find-
ing the potential V at the field point $\mathbf{r}$ by integrating $\rho\,dv'$ over an arbitrary source
distribution, Figure 7.11, with $\mathbf{r}'' = \mathbf{r} - \mathbf{r}'$:

$$V = \int \frac{\rho\,dv'}{4\pi\epsilon_0 r''} \tag{7.20}$$

Then we express that integral in the form of a series. In this way we will be able to
represent the source distribution as a series of multipoles located at the origin, under the assumption that $r \gg r'$.

Using the law of cosines,

$$r'' = \sqrt{r^2 + r'^2 - 2rr'\cos\theta}$$

So,

$$\frac{1}{r''} = \frac{1}{r}\left(1 + \left(\frac{r'}{r}\right)^2 - 2\frac{r'}{r}\cos\theta\right)^{-1/2} \tag{7.21}$$

The parenthetical expression is of the form $(1 + x)^n$, with $x < 1$ and $n = -1/2$;

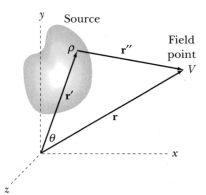

Figure 7.11   Potential at a field point.

its binomial expansion is

$$(1 + x)^n = 1 - \frac{n}{1!} x + \frac{n(n-1)}{2!} x^2 + \cdots$$

When we carry through the expansion and gather terms of like powers of r', we obtain

$$\frac{1}{r''} = \frac{1}{r}\left(1 + \left(\frac{r'}{r}\right)\cos\theta + \left(\frac{r'}{r}\right)^2\left(\frac{3\cos^2\theta - 1}{2}\right) + \cdots\right)$$

(The expressions involving $\cos\theta$ are Legendre polynomials, $P_n(\cos\theta)$, Section 12.3.)

So, the potential V becomes:

$$V = \frac{1}{4\pi\epsilon_0 r}\int \rho\, dv' + \frac{1}{4\pi\epsilon_0 r^2}\int r'\cos\theta\, \rho\, dv'$$

$$+ \frac{1}{4\pi\epsilon_0 r^3}\int r'^2\left(\frac{3\cos^2\theta - 1}{2}\right)\rho\, dv' + \cdots \tag{7.22}$$

The first term is immediately recognized as the usual monopole potential, but it is at the origin rather than at the center of charge. The integral is the **monopole moment,** which is simply the charge. The second term represents the dipole potential and may be seen to be an integral form of Equation 7.4, Section 7.1. The third term is the quadrupole, the fourth would be the octupole, and so on. Note that successive terms fall off more and more rapidly with increasing r.

Physically: Near a complicated charge distribution, the field is likely to be correspondingly complicated and to require many multipole terms for its accurate expression. As you move away from the distribution, higher-order terms become small faster than lower-order ones do, so that at large r the only significant term is the lowest order available. This point is illustrated in an elementary way in Figure 2.12.

Now, as to the second term, the dipole potential: We would like to get rid of the θ inside the integral, so that the integral doesn't depend on the position of the field point; but we can't just drag it outside, these things must be done delicately. We can write:

$$\int r'\cos\theta\, \rho\, dv' = \int \hat{\mathbf{r}}\cdot\mathbf{r}'\, \rho\, dv' = \hat{\mathbf{r}}\cdot\int \mathbf{r}'\, \rho\, dv'$$

(remembering that $\hat{\mathbf{r}}\cdot\mathbf{A} + \hat{\mathbf{r}}\cdot\mathbf{B} = \hat{\mathbf{r}}\cdot(\mathbf{A} + \mathbf{B})$). Now we define

$$\boxed{\mathbf{p} = \int \mathbf{r}'\, \rho\, dv'} \tag{7.23}$$

so the dipole becomes $\dfrac{\hat{\mathbf{r}}\cdot\mathbf{p}}{4\pi\epsilon_0 r^2}$, in agreement with Section 7.1. The expression for $\mathbf{p}$ is a generalization of $p = Qs$. For example, if we place a charge Q at $z' = s$, and

$-Q$ at the origin, the **dipole moment** is

$$\mathbf{p} = \int \mathbf{r}' \, \rho \, dv' = \int s \, \hat{\mathbf{z}}' \, \rho \, dv' = Qs \, \hat{\mathbf{z}}' \tag{7.24}$$

as before.

The third term in the potential V, Equation 7.22, is the quadrupole potential. It is complete and correct, but still we want to get rid of the θ. It's not so easy. We write

$$\int r'^2 (3 \cos^2 \theta - 1) \, \rho \, dv' = \int (3(\hat{\mathbf{r}} \cdot \mathbf{r}')^2 - r'^2) \, \rho \, dv' \tag{7.25}$$

Now we go to Cartesian coordinates and substitute

$$\mathbf{r}' = x' \, \hat{\mathbf{x}} + y' \, \hat{\mathbf{y}} + z' \, \hat{\mathbf{z}}$$

$$\hat{\mathbf{r}} = l \hat{\mathbf{x}} + m \hat{\mathbf{y}} + n \hat{\mathbf{z}}$$

where l, m, and n are the direction cosines of $\hat{\mathbf{r}}$. Writing it out, and grouping terms,

$$(3l^2 - 1) \int x'^2 \rho \, dv' + (3m^2 - 1) \int y'^2 \rho \, dv' + (3n^2 - 1) \int z'^2 \rho \, dv'$$

$$+ \, 6lm \int x'y' \rho \, dv' + 6mn \int y'z' \rho \, dv' + 6nl \int z'x' \rho \, dv'$$

These six integrals define the nine components of the **quadrupole moment** (which is a second-rank tensor). For example,

$$p_{xx} = \int x'^2 \rho \, dv' \qquad \text{and} \qquad p_{xy} = \int x'y' \, \rho \, dv'$$

The quadrupole potential then becomes

$$\frac{1}{4\pi\epsilon_0 r^3} \tag{7.26}$$

$$\left(\frac{3l^2 - 1}{2} \, p_{xx} + \frac{3m^2 - 1}{2} \, p_{yy} + \frac{3n^2 - 1}{2} \, p_{zz} + 3 \, lm \, p_{xy} + 3 \, mn \, p_{yz} + 3 \, nl \, p_{zx} \right)$$

We have achieved our goal that the quadrupole moment should not depend on the field point. It's kind of unwieldy, but fortunately it's not much used. The interesting cases, in nuclear physics and astrophysics, have cylindrical symmetry about the z axis, with $\rho(z) = \rho(-z)$, so

$$0 = p_{xy} = p_{yz} = p_{zx} \qquad \text{and} \qquad p_{xx} = p_{yy}$$

It is customary to define a single quadrupole moment in this case, a scalar; it is commonly called Q but we shall call it Q_z. Its units are $C \cdot m^2$.

$$2 \, (p_{zz} - p_{xx}) = \boxed{Q_z = \int (3 \, z'^2 - r'^2) \, \rho \, dv'} \tag{7.27}$$

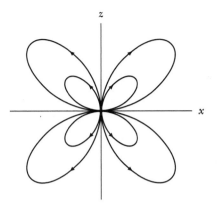

Figure 7.12   Quadrupole field lines.

Then the potential is (with $l^2 + m^2 + n^2 = 1$)

$$V = \frac{Q_z/2}{4\pi\epsilon_0 r^3}\left(\frac{3n^2 - 1}{2}\right) = \frac{Q_z}{16\pi\epsilon_0 r^3}(3\cos^2\theta_z - 1) \tag{7.28}$$

where θ_z is the angle between the z axis and $\hat{\mathbf{r}}$. For **E**, Figure 7.12, use $\mathbf{E} = -\nabla V$.

The quadrupole moment is positive for a prolate spheroid, like a football, and negative for an oblate spheroid, like the spinning earth.

(The quadrupole moment is sometimes given as our Q_z divided by e, the electronic charge, so that its units are m².)

▼ EXAMPLE 7-6

For a single charge of Q, placed at s on the z axis, find the first three multipoles.

ANSWER   The *monopole* moment is $\int \rho\, dv = Q$, located at the origin, not at s.

The *dipole* moment is $\mathbf{p} = \int \mathbf{r}\, \rho\, dv = \int z\hat{\mathbf{z}}\, \rho\, dv = Qs\, \hat{\mathbf{z}}$. For a *macroscopic* dipole centered at $s/2$, this would cancel the Q at the origin and place Q at s, which would bring us back to a single charge of Q, placed at s. But this is a *microscopic* dipole centered at the origin. Dipole moment $\mathbf{p}$ is independent of the location of the origin only if $Q = 0$.

The *quadrupole* moment is $Q_z = \int(3z^2 - r^2)\rho\, dv = 3Qs^2 - Qs^2 = 2Qs^2$. It is positive, as for a prolate spheroid. There exist a dipole and quadrupole moment only because the center of charge is not at the origin. (The octupole moment would be proportional to s^3, and so on.)

The energy of interaction may similarly be obtained by a series expansion. In general, the energy of a charge distribution $\rho(x, y, z)$ in an externally imposed potential $V(x, y, z)$ is (Section 6.4):

$$W = \int \rho V\, dv \tag{7.29}$$

We expand $V(x, y, z)$ in a Taylor's series about an origin at the center of charge:

$$V = V_0 + \frac{\partial V}{\partial x} x + \frac{\partial V}{\partial y} y + \frac{\partial V}{\partial z} z + \frac{1}{2!} \frac{\partial^2 V}{\partial x^2} x^2 + \frac{1}{2!} \frac{\partial^2 V}{\partial x \partial y} xy + \cdots$$

where the potential V_0 and the derivatives are all constants, having been evaluated at the origin.

This development appears similar to what we did above, in that it too involves a series expansion. However, note the fundamental difference in purpose: There, we were finding the potential V *caused* by a charge distribution, whereas here we are finding the energy W of a charge distribution in an *externally imposed V*.

Substituting the series in Equation 7.29, we arrive at

$$W = V_0 \int \rho \, dv + \frac{\partial V}{\partial x} \int \rho x \, dv + \frac{\partial V}{\partial y} \int \rho y \, dv + \frac{\partial V}{\partial z} \int \rho z \, dv$$

$$+ \frac{1}{2!} \frac{\partial^2 V}{\partial x^2} \int \rho x^2 \, dv + \frac{1}{2!} \frac{\partial^2 V}{\partial x \partial y} \int \rho x y \, dv \cdots$$

The first term is seen to be the monopole interaction QV, based on the approximation that the potential is that at the origin. The next three terms represent the dipole energy $- \mathbf{p} \cdot \mathbf{E}$, with $\mathbf{E} = -\nabla V$; the integrals are the x, y, and z components of the dipole moment. Then we find the quadrupole terms; the integrals are the quadrupole moments p_{xx}, p_{xy}, and so on, that we encountered above in connection with Equation 7.26.

We now restrict ourselves to cylindrically symmetrical distributions with $\rho(z) = \rho(-z)$. This eliminates cross terms in the quadrupole interaction (and incidentally it also eliminates all dipole terms; if ρ is the same at $-x$ as at x, then integration of ρx contributes nothing). The quadrupole energy becomes

$$W_Q = \frac{1}{2} p_{xx} \frac{\partial^2 V}{\partial x^2} + \frac{1}{2} p_{yy} \frac{\partial^2 V}{\partial y^2} + \frac{1}{2} p_{zz} \frac{\partial^2 V}{\partial z^2}$$

Now because of the symmetry, $p_{xx} = p_{yy}$. And the externally imposed V must obey Laplace's equation in electrostatics. This is easily seen as follows: Gauss's law says $\nabla \cdot \mathbf{E} = 0$, with zero charge density because the charge that is responsible for the externally imposed V lies elsewhere, and $\mathbf{E} = -\nabla V$ in electrostatics (since $\partial \mathbf{A}/\partial t = 0$). Putting them together, we arrive at

$$\nabla \cdot \nabla V = \nabla^2 V = \frac{\partial^2 V}{\partial x^2} + \frac{\partial^2 V}{\partial y^2} + \frac{\partial^2 V}{\partial z^2} = 0 \qquad \text{which is Laplace's equation}$$

So the quadrupole interaction energy is

$$W_Q = \frac{1}{2} (p_{zz} - p_{xx}) \frac{\partial^2 V}{\partial z^2} = \frac{Q_z}{4} \frac{\partial^2 V}{\partial z^2} \qquad (7.30)$$

Thus, generally speaking, for a monopole the energy depends on V; for a dipole, it depends on ∇V; and for a quadrupole it depends on $\nabla^2 V$. The quadrupole responds only to an inhomogeneous $\mathbf{E}$ field.

7.4 SUMMARY

Dipoles are the building blocks of ordinary dielectric and magnetic materials. In this chapter we have explored dipole fields and potentials, and we have found the forces and energies associated with dipoles in fields. Finally, we have generalized to higher-order multipoles, especially the quadrupole.

The potential for an electric dipole is

$$V = \frac{\mathbf{p} \cdot \hat{\mathbf{r}}}{4\pi\epsilon_0 r^2} \tag{7.4}$$

where the dipole moment is $p = Qs$, and the electric field is

$$\mathbf{E} = \frac{2p\cos\theta}{4\pi\epsilon_0 r^3}\hat{\mathbf{r}} + \frac{p\sin\theta}{4\pi\epsilon_0 r^3}\hat{\boldsymbol{\theta}} \tag{7.5}$$

In an external **E** field,

torque is $\qquad \tau = \mathbf{p} \times \mathbf{E}$ $\hfill$ (7.6)

energy is $\qquad W = -\mathbf{p} \cdot \mathbf{E}$ $\hfill$ (7.7)

force is $\qquad \mathbf{F} = (\mathbf{p} \cdot \nabla)\,\mathbf{E}$ $\hfill$ (7.10)

which is $\mathbf{F} = \nabla(\mathbf{p} \cdot \mathbf{E})$ if $\nabla \times \mathbf{E} = 0$.

The vector potential for a magnetic dipole is

$$A = \frac{\mu_0 \mathbf{m} \times \hat{\mathbf{r}}}{4\pi\,r^2} \tag{7.13}$$

where the magnetic dipole moment is $m = Is^2$, and the magnetic field is

$$\mathbf{B} = \frac{2\mu_0 m\cos\theta}{4\pi r^3}\hat{\mathbf{r}} + \frac{\mu_0 m\sin\theta}{4\pi r^3}\hat{\boldsymbol{\theta}} \tag{7.15}$$

In an external field,

torque is $\qquad \tau = \mathbf{m} \times \mathbf{B}$ $\hfill$ (7.16)

energy is $\qquad W = -\mathbf{m} \cdot \mathbf{B}$ $\hfill$ (7.17)

force is $\qquad \mathbf{F} = \nabla(\mathbf{m} \cdot \mathbf{B})$ $\hfill$ (7.18)

For an arbitrary electrostatic charge distribution, the dipole moment is

$$\mathbf{p} = \int \mathbf{r}\,\rho\,dv \tag{7.23}$$

and for a cylindrically symmetrical charge distribution the quadrupole moment is

$$Q_z = \int (3\,z^2 - r^2)\,\rho\,dv \tag{7.27}$$

◤ PROBLEMS

7-1 Charges of $+2$ nC and -2 nC are separated by 3 cm. Find the (electric) dipole moment. Find V and $\mathbf{E}$ at $r = 37$ cm, $\theta = 30°$.

7-2 Find the force between a dipole of 3×10^{-12} C $\cdot$ m and a charge of 4 nC separated by 0.7 m at an angle of 30° down from $\mathbf{p}$. Find the energy required to separate them.

7-3 A current of 2 mA flows about a circle of radius 3 cm. Find its magnetic (dipole) moment. Find $\mathbf{A}$ and $\mathbf{B}$ at $r = 37$ cm, $\theta = 30°$.

7-4 Find the energy required to flip a magnetic dipole of 3×10^{-4} A $\cdot$ m^2 in a magnetic field of 0.6 T.

7-5 A 3-nC charge is located at $z = 4$ cm, and a charge of -2 nC is at the origin. Find the monopole, dipole, and quadrupole (Q_z) moments.

7-6 A spherical shell of radius a has surface charge density of $\sigma = \sigma_0 \cos \theta$, Figure 7.13. Find its monopole, dipole, and quadrupole (Q_z) moments.

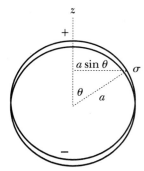

Figure 7.13

7-7 Find the potential at $x = 7$ cm, $y = 5$ cm, $z = -6$ cm, due to a dipole of $p = 3 \times 10^{-12}$ C $\cdot$ m located at the origin and pointing in the z direction.

7-8 Show that the dipole moment of a charge distribution is independent of the position of the origin if the total charge is zero.

7-9 A disk of radius a and uniform surface charge density σ spins with angular velocity ω about the z axis. Find its magnetic dipole moment.

7-10 A sphere of radius R and uniform surface charge density σ spins with angular velocity ω about the z axis. Find its magnetic dipole moment.

7-11 If $\nabla(\mathbf{p} \cdot \mathbf{E}) = (\mathbf{p} \cdot \nabla) \mathbf{E}$, then what can be said about $\mathbf{B}$?

7-12 If $\nabla(\mathbf{m} \cdot \mathbf{B}) = (\mathbf{m} \cdot \nabla) \mathbf{B}$, then what can be said about current density $\mathbf{J}$?

7-13 Find the equations for the equipotential surfaces and electric field lines of an electric dipole.

7-14 Verify that $\tau = \mathbf{m} \times \mathbf{B}$, $W = -\mathbf{m} \cdot \mathbf{B}$, and $\mathbf{F} = \nabla(\mathbf{m} \cdot \mathbf{B})$ for a small square loop of current I and side s.

7-15 The earth's magnetic field is nearly a dipole field of about 0.6 gauss (6×10^{-5} T) at a magnetic pole. **(a)** Find its magnetic moment. **(b)** What current flowing around the equator would provide this magnetic moment?

7-16 Find the energy in the earth's magnetic field (outside of the earth) in joules. Then express it in megatons of TNT. (20 kilotons, which destroyed Hiroshima, corresponds to the energy of about one gram of matter.)

7-17 Find force and torque for each case in Figure 7.14. All angles are 0 or 90°, and all separations are r.

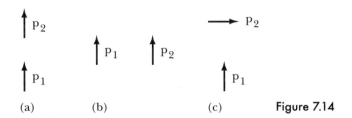

(a)   (b)   (c)   **Figure 7.14**

*7-18 For a system consisting of an arbitrarily oriented dipole **p** at distance **r** from a positive charge Q, show that the total force is zero and the total torque is zero.

7-19 In the Stern-Gerlach experiment (1922), a beam of neutral silver atoms from a 960°C oven passes through an inhomogeneous magnetic field of gradient $\partial B_z/\partial z = 50$ T/m and of length $l = 30$ cm, Figure 7.15. Assuming each atom has a magnetic moment μ_B that can point parallel or antiparallel to z, find the separation of the atoms. Would there be a force perpendicular to z? Kinetic energy in the oven is $3/2\ kT$, but in the beam it is $2\ kT$ because more energetic atoms are more likely to get into the beam.

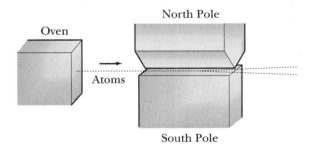

Figure 7.15   Stern-Gerlach experiment.

7-20 Repeat the preceding problem for a water molecule oriented in the z direction in an inhomogeneous electric field of gradient $\partial E_z/\partial z = 10^8$ V/m²; that is, find the size of the spread of molecules (this no longer involves quantum mechanics).

7-21 A solenoid of length 15 cm and radius 2 cm is uniformly wound with 600 turns of wire carrying 3 A. Find the magnetic field B: **(a)** in the center; **(b)** at the end, on the axis of the solenoid; **(c)** at 10 meters, on the axis; **(d)** 10 meters from the axis and the solenoid. (Good approximations will be satisfactory.)

7-22 Review Section 3.2, Figure 3.6 (round loop) and Problem 3-6 (square loop), and verify that at large distances the field is that of a magnetic dipole.

7-23 Show that the electric field of a dipole may be written in the coordinate-independent form:

$$\mathbf{E} = \frac{1}{4\pi\epsilon_0 r^3} [3\,(\mathbf{p}\cdot\hat{\mathbf{r}})\hat{\mathbf{r}} - \mathbf{p}]$$

starting with $V = \mathbf{p}\cdot\mathbf{r}/4\pi\epsilon_0 r^2$.

7-24 Show that the trajectory of a charged particle released from rest on the median plane of an electric dipole is a semicircle, like the path of a simple pendulum of $\theta = 90°$. (Consider the potential and the centripetal force. There is no gravitational field.)

7-25 **(a)** Find the quadrupole moment Q_z for a ring of charge q and radius a, in the xy plane. **(b)** Find V for $\theta_z = 0$, $r = 2a$, for the first three multipole terms. **(c)** Compare with the exact value of V.

*7-26 Find the quadrupole moment Q_z of a uniformly charged spheroid of charge Q and semiaxes a and b, Figure 7.16.

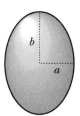

Figure 7.16

C H A P T E R

8

CONDUCTORS

T his is the second of four chapters concerned with charge and current distributions in materials. We deal here mainly with ideal conductors in which the charge carriers are free to move in whatever direction **E** beckons. We will encounter one means, the Hall effect, to assist in discovering the nature and behavior of the charge carriers.

8.1. Conductivity σ

When we apply a voltage to a resistor, current flows and energy is dissipated. For example, when we apply 10 volts to a 10-ohm, $\frac{1}{2}$-watt resistor, a current of 1 ampere flows; also, a power of 10 V × 1 A = 10 W is dissipated, and the resistor smokes. (Such resistors cost only a few cents.)

In some materials, called conductors, some of the constituent positive and negative charges move relative to each other in response to an applied electric field; these include metals, electrolytes, plasmas, and others. In one kind of ideal case, and approximately in metals, the current density is proportional to the field:

$$\boxed{\mathbf{J} = \sigma\mathbf{E}} \tag{8.1}$$

This is **Ohm's law,** and in this form it applies at every point in the material. The unit of **conductivity** σ is siemens per meter; in units, S = A/V ($=\Omega^{-1}$). Siemens was formerly called mho. (The letter σ is also customary for surface charge density; you'll have to tell the difference by the context.) The reciprocal $1/\sigma = \rho$ is called **resistivity** (another potential ambiguity).

Obedience to Ohm's law is a sometimes thing. It implies that current I is strictly proportional to voltage V. However, even for **conductors,** conditions must be imposed. A metal's conductivity decreases with temperature; for a tungsten light bulb it goes down by a factor of ten from cold to hot. Ohm's law is obeyed

Conductivities, S/m	
Aluminum	3.5×10^7
Copper	5.8×10^7
Gold	4.5×10^7
Graphite	7×10^4
Iron	1.0×10^7
Mercury	0.1×10^7
Nichrome	0.1×10^7
Sea water	5
Silicon	30
Silver	6.1×10^7
Sodium	2.2×10^7
Sulfur	1×10^{-15}

here only on the condition that temperature be held constant. For a **semiconductor** like silicon, on the other hand, the conductivity goes up with temperature, largely because of increasing numbers of available charge carriers getting shaken loose by thermal agitation. The effect is exploited in thermistors, whose resistance may be used in temperature measurement. The effective resistance of a diode (semiconductor junction) depends on the current. The resistance of an *arc* (spark) decreases with increasing current because that means that more ions are available. And for **dielectrics** or nonconductors or insulators (like sulfur), conductivities tend to be $\approx 10^{-15}$ S/m and to depend very much on impurities, temperature, and so forth.

In the case of a **superconductor,** current can continue to flow, even for years, without any voltage at all. A **superconductor** also expels magnetic flux (the Meissner effect), and at high magnetic field a superconductor goes normal. However, for example, there is an alloy of niobium and tin, Nb_3Sn, which has a critical temperature of 18 kelvin (above this it goes normal) and if cooled to 4 K it can take a magnetic field of 20 T; this is useful in practice for large magnets. Its explanation is totally quantum mechanical and outside of our purview here.

Anyway, we shall ignore all of these variations and limit ourselves to conductors that obey Ohm's law.

When we integrate Equation 8.1 over a resistor of finite size, we obtain what might be called the integral form of Ohm's law, customarily written

$$\boxed{V = IR} \tag{8.2}$$

where $V = \int \mathbf{E} \cdot dl$, and $I = \int \mathbf{J} \cdot d\mathbf{a}$, and R is the **resistance** in **ohms** (or Ω or volts per ampere). For example, for a cylindrical resistor of length l and cross sectional area a, we have

$$V = El \qquad \text{and} \qquad I = Ja$$

then from Equation 8.1,

$$J = \sigma E$$

$$\frac{I}{a} = \sigma \frac{V}{l}$$

$$V = I \frac{l}{\sigma a}$$

so the resistance may be given in terms of the conductivity σ as:

$$R = \frac{l}{\sigma a} \tag{8.3}$$

For other geometries, there are other formulas for R; see the Problems. But since **E** is proportional to **J** at every point, we always find that V is proportional to I over the whole resistor.

When charge Q is carried through an electric field **E** at velocity **v**, power P may be transferred:

$$P = \mathbf{F} \cdot \mathbf{v} = Q\mathbf{E} \cdot \mathbf{v}$$

where **F** is the force on the charge. Then the power per unit volume v is

$$\frac{dP}{dv} = \mathbf{E} \cdot \mathbf{v} \frac{dQ}{dv} = \mathbf{E} \cdot \mathbf{v} \rho$$

where ρ is charge density. So the power per unit volume may be written as

$$\boxed{\frac{dP}{dv} = \mathbf{E} \cdot \mathbf{J}} \tag{8.4}$$

at each point in the material. Units are $\dfrac{V}{m} \cdot \dfrac{A}{m^2} = W/m^3$, since an ampere is a coulomb per second, and a volt is a joule per coulomb (and a watt is a joule per second). When the power is dissipated into heat, the process is called **Joule heating.** On the other hand, when **E** and **J** point in opposite directions, then $\mathbf{E} \cdot \mathbf{J}$ is negative and electrical power is being supplied, for example, by a battery.

As with Ohm's law, there also exists what amounts to an integral form of the preceding Equation 8.4. When we carry a charge through a potential difference V, the energy required is QV (Section 6.4). If we transfer charge continuously, we find the energy per unit time is $V\, dQ/dt$, so the power is

$$\boxed{P = VI} \qquad \text{in watts} \tag{8.5}$$

Application of Ohm's law provides some convenient alternative forms: $P = I^2 R = V^2/R$.

▼ EXAMPLE 8-1

A current of 2 A flows in a copper wire of length 3 km and radius 1.1 mm. Find the voltage drop in the wire and the power dissipated.

ANSWER The conductivity of copper is 5.8×10^7 S/m $= 5.8 \times 10^7$ A/V $\cdot$ m.

$$V = IR = I\,\frac{l}{\sigma\pi r^2}$$

$$= 2\,\text{A} \times \frac{\text{V}\cdot\text{m}}{5.8 \times 10^7\,\text{A}} \times 3 \times 10^3\,\text{m} \times \frac{1}{3.14 \times (1.1 \times 10^{-3})^2\,\text{m}^2} = 27\,\text{V}$$

The units help guide us through calculations of this sort. For the power,

$$P = VI$$
$$= 27\,\text{V} \times 2\,\text{A} = 54\,\text{W}$$

For a 120 V line this would be extravagant, but perhaps not for a high-tension line.

8.2. Stationary Charge in Conductors

So far we have talked about what happens in a conductor when current flows. In a static situation, with no current flowing, $\mathbf{J} = 0$, and so, ideally at any rate,

$$\mathbf{E} = \mathbf{J}/\sigma = 0 \tag{8.6}$$

If also there is no changing vector potential $\partial\mathbf{A}/\partial t$, then there is no ∇V, in view of

$$\mathbf{E} = -\nabla V - \frac{\partial\mathbf{A}}{\partial t}$$

so the whole conductor is at one voltage: Its surface is an equipotential. If the conductor is charged, mutual repulsion drives the excess charge to the surface in such

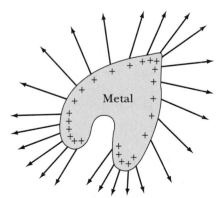

Figure 8.1   **E** field near charged conductor.

a way as to produce no **E** field inside the conductor. To achieve this, the charge will tend to accumulate at sharp points, Figure 8.1, and consequently, the external **E** field will be higher there. That is why at high voltages, sparks are more likely to occur at sharp points.

Surprisingly, when two *different* conducting materials are placed in contact, the conducting surface is *not* an equipotential; a **contact potential** arises. For example, when zinc and copper are in contact, the zinc acquires a potential of + 0.7 V relative to the copper; electrons like it better inside the copper because of its larger work function. No, this information will not help you construct your perpetual motion machine.

The equilibrium charge distribution is achieved remarkably rapidly in a good conductor. Any excess charge moves to the surface quickly, as may be seen from the following argument. We start with Gauss's law,

$$\nabla \cdot \mathbf{E} = \frac{\rho_t}{\epsilon_0}$$

This charge density ρ_t includes all excess charge. Some of the excess charge is free to roam through the conductor, while some is "bound" to atoms (Section 9.1) and is excess just because it has moved a little inside its home atoms, slightly upsetting the balance of plus and minus charge. We will make the approximation that the effect of this bound charge is small, because we are interested in an order-of-magnitude result. It turns out in fact to be an adequate approximation for this purpose. So,

$$\nabla \cdot \mathbf{E} = \frac{\rho}{\epsilon_0} \tag{8.7}$$

Substituting Ohm's law, $\mathbf{E} = \mathbf{J}/\sigma$, we find

$$\nabla \cdot \mathbf{J} = \frac{\rho\sigma}{\epsilon_0}$$

Also,

$$\nabla \cdot \mathbf{J} = -\frac{\partial \rho}{\partial t} \tag{8.8}$$

for charge continuity, Section 5.1, so,

$$-\frac{\partial \rho}{\partial t} = \frac{\rho\sigma}{\epsilon_0}$$

Integrating,

$$\rho = \rho_0 e^{-t\sigma/\epsilon_0} \tag{8.9}$$

Now in copper

$$\frac{\epsilon_0}{\sigma} = \frac{8.85 \times 10^{-12}\, \mathrm{C^2/N \cdot m^2}}{5.8 \times 10^7\, \mathrm{A/V \cdot m}} = 1.5 \times 10^{-19}\ \text{second}$$

That is the time in which the exponential will fall to $1/e = 0.368$ of its original value: its "relaxation time." The result is clearly way too small; in fact, it is less than the time for light to traverse one atom. The inaccuracy stems mainly from the fact that at very short times or correspondingly high frequencies, conductivities may decrease by a factor of 1000 or more. This is largely due to the inertia of the electrons, which is negligible at lower frequencies.

But anyway, the order of magnitude of the result makes it clear that excess charge disappears very quickly from the interior of a good conductor. For an insulator, on the other hand, conductivities can be less by a factor of 10^{20}, and corresponding relaxation times can range from seconds to years.

The field is also zero in an empty cavity inside a conductor. This is a consequence of Gauss's law (no charge, no flux), and it provides a very sensitive test of the inverse-square law. If Coulomb's law involved any power of r other than r^{-2}, then electric field lines would appear or disappear in empty space, and Gauss's law would fail (Section 2.3). So an experimenter applies a large voltage to a concentric pair of conductors; then the conductors are separated and the experimenter looks for charge on the inner one. In this way the exponent in Coulomb's law is shown to be -2 with an error of less than 1 part in 10^{10}.

What if there is a charge in the cavity? Then, Gauss's law avers, there will also be a field there. Figure 8.2 shows an uncharged spherical conductor with a cavity in which there are eight little plus charges. These induce eight minus charges on the inside surface of the metal, and eight pluses appear on the outside surface. The metal has equal numbers of plus and minus, so it remains uncharged. The lines of **E** are produced by the eight pluses in the middle; they get absorbed on the inner surface and then re-emitted on the outside surface. The field lines do not penetrate into the interior of the metal, and there is no field there.

And now if we touch the eight central pluses to the inside surface, those pluses and minuses will cancel out, and the eight pluses on the outside surface will stay put. The metal will then have a charge of eight pluses.

QUESTION: If the charge sits on the surface of a conductor, why do we need the interior of current-carrying wires? Could we use copper-coated plastic instead?

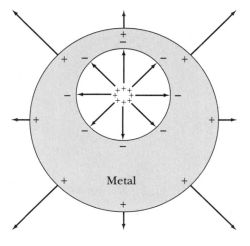

Figure 8.2   Charge in cavity in conductor.

ANSWER: From Chapter 2, we know that the excess charge is far less than a coulomb, and that is what sits on the surface; however, the total charge involved in carrying current is thousands of coulombs, which flows in the interior of the wire. So the excess charge usually makes a negligible contribution to the current. The next example illustrates this principle quantitatively. It is interesting, however, that at high frequency the current *is* confined to the surface by the "skin effect," and in such cases one can indeed use silver-plated wave guides (Section 16.3).

▼ EXAMPLE 8-2

A metal sphere of voltage $V = 10^4$ volts and radius $R = 5$ cm spins at 10,000 rpm (revolutions per minute) about a diameter, Figure 8.3. Find the current.

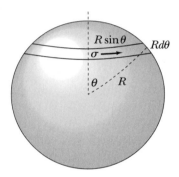

Figure 8.3   Spinning charged sphere.

ANSWER The voltage is $V = Q/4\pi\epsilon_0 R$, where Q is the charge on the sphere; and the area is $4\pi R^2$, so the surface charge density is $\sigma = \epsilon_0 V/R$. The area of the surface element shown is

$$da = 2\pi R \sin\theta \, R \, d\theta$$

and its charge is $dQ = \sigma \, da$. The sphere spins with frequency f, so the time for dQ to pass a given point is $t = 1/f$. The corresponding current element is $dI = dQ/t$. Integrating,

$$I = \int_0^\pi \frac{\sigma \, 2\pi R \sin\theta \, R \, d\theta}{1/f}$$

$$= 4\pi\epsilon_0 VRf$$

$$= 4\pi \times 8.85 \times 10^{-12} \, \text{C}^2/\text{N} \cdot \text{m}^2 \times 10^4 \, \text{V} \times 0.05 \, \text{m} \, \frac{10^4 \, \text{rev/min}}{60 \, \text{s/min}}$$

$$= 9 \, \mu\text{A}$$

which is a rather small current despite the large voltage and rapid spin. Excess charge usually contributes little current.

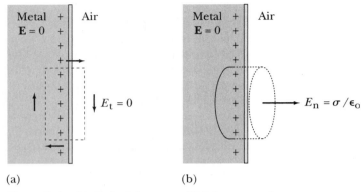

Figure 8.4   (a) Stokes' theorem. (b) Divergence theorem.

Now we shall look at the field at the surface of a charged conductor. We have, to begin with, $\nabla \times \mathbf{E} = 0$ in the static case ($\partial \mathbf{B}/\partial t = 0$). Then the line integral of $\mathbf{E}$ is zero, Figure 8.4a, by Stokes' theorem; so the tangential component E_t is continuous, the same inside as out. If in addition there is no current density $\mathbf{J}$, then

$$E_t = 0 \tag{8.10}$$

This implies that the field lines start out normal to the surface: If the tangential component E_t is zero inside it must be zero outside. This was illustrated in Figure 8.1.

By contrast, $\nabla \cdot \mathbf{E}$ is infinite at the surface of the charged conductor because there is a finite charge located in an infinitesimal volume: The area is finite, but the thickness is infinitesimal. The charge density ρ is a delta function (Section 2.4). If the surface is in the yz plane, Gauss's law says:

$$\nabla \cdot \mathbf{E} = \frac{\partial E_x}{\partial x} = \frac{\rho}{\epsilon_0} = \frac{\sigma \delta(x)}{\epsilon_0}$$

where here σ is surface charge density. (E_x is a step-function; see $\theta(x)$, Problem 2-37.) Integrating, we obtain $\Delta E_x = \sigma/\epsilon_0$, or for arbitrary surface orientation, $\Delta E_n = \sigma/\epsilon_0$. Since $E_t = 0$, E is all E_n, normal to the surface. The field inside is 0, so (Fig. 8.4b) the field outside is

$$E_n = \sigma/\epsilon_0 \tag{8.11}$$

This is a valid approximation for a curved surface if the distance from the field point to the surface is small in terms of the size and curvature of the surface.

▽ EXAMPLE 8-3

For a spherical shell of radius R and surface charge density σ, compare E at a distance of 0.1 R outside the surface with the field magnitude σ/ϵ_0 for an infinite plane of surface charge density σ (with $E = 0$ on one side).

ANSWER For the sphere,

$$E = \frac{Q}{4\pi\epsilon_0(1.1\ R)^2} = \frac{4\pi R^2\sigma}{4\pi\epsilon_0(1.1\ R)^2} = 0.83\ \frac{\sigma}{\epsilon_0}$$

so the field is already less by 17% than that given by the approximation.

Figure 8.5 compares the behavior of the electric field E and the scalar potential V for a uniformly charged sphere and a conducting sphere of unit radius, with $Q/4\pi\epsilon_0 = 1\ \mathrm{N} \cdot \mathrm{m}^2/\mathrm{C}$. The total charge on the spheres is the same, and so the graphs are identical outside the spheres. But inside, they are markedly different. There is charge density ρ inside the uniformly charged sphere, Figure 8.5a, and so there is a field E, and because of this the potential V increases toward the center. On the conductor, Figure 8.5b, all of the charge is surface charge σ, so there is no E field inside the sphere, and consequently the potential V doesn't change.

The E field in Figure 8.5a is like that of Figure 2.17c of Chapter 2, and the potential was done in Problem 6-5. No further calculation is required for Figure 8.5b.

Now we consider two separate conductors interacting through the electrostatic field. If we ground one of a pair of parallel metal plates, Figure 8.6, and if we rub a hard rubber rod with fur and touch it to the other plate, that plate may be charged to around -2000 volts, as measured by an electrometer or electroscope. The negative plate attracts positive charge into the other plate from the ground (or repels negative charge into the ground). This is an example of **electrostatic induction:** We have charged a plate positive, starting with a negative rod.

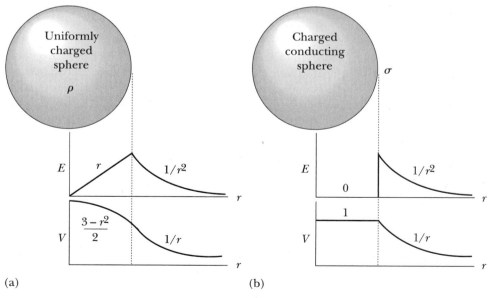

(a)   (b)

Figure 8.5   **(a)** Uniformly charged sphere. **(b)** Charged conducting sphere.

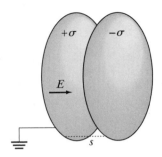

Figure 8.6   Charged metal plates.

If we then isolate (insulate) the plates and move them farther apart, increasing s in Figure 8.6, the voltage goes up! Why is that? Well, if the plates are insulated, then their charge density σ is fixed, and so $E = \sigma/\epsilon_0$ is constant. Voltage $V = \int E\, dl = Es$, so the voltage is proportional to s. This is in effect a voltage amplifier, and it was employed by Volta himself circa 1800. He applied small voltages to a pair of close-spaced silver disks that he then separated for the voltage measurements. An amplification of 1000 or more is obtainable.

By the way, how did we get up to 2000 V using fur and a rod? By the same principle. The energy required to remove an outer electron from a molecule is only a few electron volts, and so the fur and rod differ by only a few volts when they are in contact. The high voltage is generated when we pull them apart.

The field energy increases too: E is the same, but there is a larger volume of it. Of course, the requisite energy comes from our hands when we pull the oppositely charged objects apart.

Note that the left plate was still grounded, so $V = 0$, while it was already positively charged. Potential is not just a matter of what kinds of charges are present. In Figure 8.2 a conductor of constant V everywhere has positive excess charge in one place and negative in another.

It is also because of induction that a charged rod will attract an uncharged object, as in Figure 2.1a, near the beginning of the book. The negative rod attracts positive charge, and repels negative, resulting in a partial charge separation in the uncharged object. In conductors the charge is simply free to flow; in dielectrics the charge can also move, but only a little, as we will see in the next chapter under "polarization," Section 9.1. After the charge separation, the positive charges are closer than the negatives to the negative rod, so the force of attraction is stronger than that of repulsion, in view of the inverse-square law. (If the field were uniform, there would be no net attraction or repulsion.) This can also be described in terms of an induced dipole moment; Section 7.1 mentions that a dipole is attracted toward a region of increasing electrostatic field.

8.3.   Inside Metals

It is free electrons we see when we look at a metal: They gather the incident light energy and reradiate it back (Sections 14.6 and 16.3). And when we touch a metal,

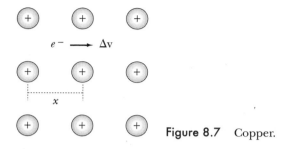

Figure 8.7 Copper.

it is the free electrons that make it seem cold, if it is at room temperature, or hot, if it is above: They provide the high thermal conductivity that carries heat toward or away from your hand much faster than most nonconductors can (but diamond conducts heat better than copper; in fact, its thermal conductivity provides the basis for a practical device to separate diamonds from imitations, using a tiny heat probe). The physical properties of a metal are largely those of the free-electron gas it contains.

In Section 8.1 we listed some electrical conductivities. Now we'll try to calculate one, say copper, from first principles, using the free-electron concept in rather naive fashion.

First Model—No Collisions Copper has one free electron per atom. The positive copper ions are suspended by electrostatic forces in a sea of negative electrons, Figure 8.7. (Copper has a face-centered cubic crystal structure, but the difference is unimportant here.) The equilibrium is stable in the sense that if a Cu^+ wanders out of place, then there are less electrons shielding it from the next Cu^+, so it gets pushed back. Taking the phrase "free electrons" at face value, we shall assume there are no collisions. When an electric field **E** is applied, the electrons of mass m and charge e undergo constant acceleration a:

$$F = ma = eE$$

Then their velocity at time t is

$$\Delta v = at = \frac{eE}{m} t$$

They just go faster and faster. Since current is flow of charge, this implies that the current goes up with time, because:

$$J = \sigma Et/t_0 \tag{8.12}$$

(t_0 being a constant). It isn't so outlandish; in fact, in an ideal plasma, Section 16.4, that's exactly what happens. But it just isn't true for a metal (except at very high frequency). So strike that model. Einstein: "The most important tool of the theoretical physicist is his wastebasket."

Second Model—Classical Electrons Going to the other extreme, we assume that the electrons collide with every positive ion, so the average distance between collisions is approximately the distance between ions, which is x, Figure 8.7. According to classical thermodynamics, the speed of the electrons is provided by the Boltzmann distribution:

$$\tfrac{1}{2}mv^2 = \tfrac{3}{2}kT \tag{8.13}$$

where k is Boltzmann's constant and T is absolute temperature, in kelvin (K). We apply the E field in the x direction, so we only want the x component of v:

$$v^2 = v_x^2 + v_y^2 + v_z^2 = 3\,v_x^2$$

Thus,

$$v_x = \sqrt{\frac{kT}{m}} \tag{8.14}$$

At room temperature, $T = 27°C$ or 300 K, so

$$v_x = \sqrt{\frac{1.38 \times 10^{-23} \times 300}{9.1 \times 10^{-31}}} = 6.7 \times 10^4 \, m/s$$

We found in Problem 3-1 that the drift velocity is much much smaller than that, to wit, around 10^{-4} m/s. Now: The electrons undergo an acceleration between collisions, and they don't lose all of their energy in each collision. They start at v_x and end at $v_x + \Delta v$. For uniform acceleration,

$$(v_x + \Delta v)^2 - v_x^2 = 2ax$$

So, neglecting the $(\Delta v)^2$,

$$2v_x \, \Delta v = 2ax$$

Now,

$$F = ma = eE \qquad \text{so} \qquad a = \frac{eE}{m}$$

$$\Delta v = \frac{ax}{v_x} = \frac{eE\,x}{mv_x} \tag{8.15}$$

The current density J is due to the drift of electrons of charge e at average speed $\Delta v/2$; and the number of free electrons is N per cubic meter. So,

$$J = Ne\frac{\Delta v}{2} = Ne\frac{eE\,x}{2mv_x}$$

It looks like we've got a winner here: J is a constant proportional to E. In fact, since $J = \sigma E$, we can write

$$\sigma = \frac{Ne^2 x}{2mv_x} \tag{8.16}$$

Let's evaluate it.

$$N = 6.02 \times 10^{23}\,\frac{atoms}{mole} \times \frac{1}{63.5}\,\frac{moles}{g} \times 8.9 \times 10^6\,\frac{g}{m^3} = 8.4 \times 10^{28}\,atoms/m^3$$

Separation of the ions is approximately

$$x = \frac{1}{\sqrt[3]{N}} = 2.3 \times 10^{-10} \, \text{m}$$

Putting it all together,

$$\sigma = \frac{Ne^2 x}{2mv_x}$$

$$= \frac{8.4 \times 10^{28} \, (1.6 \times 10^{-19})^2 \, 2.3 \times 10^{-10}}{2 \times 9.1 \times 10^{-31} \times 6.7 \times 10^4} = 0.41 \times 10^7 \, \text{S/m}$$

The measured value is 5.8×10^7. We're missing a factor of over ten. Classically we would ascribe this to the separation x, since all the other numbers are pretty firm. It would appear that the ions are small, relative to their spacing, so an electron can go past ten Cu^+ before hitting one. Qualitatively that is to some extent true (the ionic radius of Cu^+ is only 0.96×10^{-10} m). We also find that the conductivity gets smaller as the temperature increases, which is true.

How about the range of validity? Previously, we neglected $(\Delta v)^2$ with respect to $v_x \Delta v$. So we would expect the result to hold up to the point at which Δv approaches v_x in magnitude. At that point,

$$J = Ne \, \Delta v = Nev_x$$
$$= 8.4 \times 10^{28} \times 1.6 \times 10^{-19} \times 6.7 \times 10^4 = 9 \times 10^{14} \, \text{A/m}^2 \qquad \textbf{(8.17)}$$

This corresponds to about a billion amps in a 1-millimeter-diameter wire.

Such experiments are not easy to perform, although people are trying. (The wires keep exploding.) The present situation is, Ohm's law works very well in metals in all attainable situations, provided that one keeps the temperature constant.

So the preceding model seems not so bad, in terms of its results. On the other hand, quantum mechanically there is a lot more missing than a mere factor of ten, as we must now point out.

Third Model—Quantum-Mechanical Electrons Here we delve into quantum mechanics. Even if you are not familiar with the subject, you should peruse the following descriptive material anyway, to gain some appreciation of what is actually occurring in a metal. We cannot derive the results presented, because this is a classical electromagnetism textbook.

In a copper crystal, there are a large number of energy states accessible to the free electrons, the number being comparable to the number of atoms in the crystal, and so to the number of free electrons. The number is large but not infinite, and the permissible states are almost all filled. According to the Pauli exclusion principle, only one electron is permitted per state. So even at low temperature there are electrons moving at high kinetic energy because all of the lower energy states are filled. The electrons occupy what is called the Fermi sea, and only the electrons at the surface, at the highest energy, enjoy freedom to undergo a small change in energy. In copper, the Fermi energy is 7 eV, or 1.1 ×

10^{-18} J; that's "sea level" for the Fermi sea. Thermal energy kT at 300 K ($= 27°C$) is only 4.1×10^{-21} J, or 0.026 eV, or about 1/40 electron volt. So, even near absolute zero there are electrons zipping around at an effective temperature of nearly 10^5 degrees. (Celsius, Kelvin, Fahrenheit, you name it.) Obviously the Boltzmann distribution doesn't apply to the Fermi sea of electrons in a metal crystal.

So the result is: compared to the classical case, most of the electrons aren't even in the game, so the effective N is smaller; and the involved electrons are moving faster, so v_x is bigger; and as it happens, the effective distance between collisions x is larger by a factor of 100 or more; that is, in a metal the electrons can go past hundreds of ions without colliding. The fact that the classical result even comes near a factor of ten of the right σ is only because of a fortuitous cancellation of larger factors.

The simplest and clearest measurement contradicting classical physics here is probably the measurement of specific heat. Classically, the principle of equipartition of energy implies that the electrons and copper ions will share equitably in any energy that comes their way. Quantum mechanically, most of the electrons can't change energy states, because the other states are already occupied; so the electrons don't contribute much to the specific heat of the metal—and that is what is found experimentally.

Also, the high thermal conductivity of metals is due to the electrons' ability to jump past several hundred ions at once, carrying heat with them as they go.

Another measurement of the properties of conductors is provided by the Hall effect, which is discussed in the next section.

8.4. Hall Effect

Although current is conventionally described as moving from positive to negative potential, the actual direction of any charge-carrying agent in a conductor remained unknown until discovery of the **Hall effect** in 1879. It was more years before this effect was well understood, because at that time even the electron had not yet been discovered.

If we pass a current **J** through a conductor perpendicular to a magnetic field **B**, the charge carriers experience a Lorentz force proportional to their speed v. In Figure 8.8a the charge carriers are positive, and in Figure 8.8b they are negative,

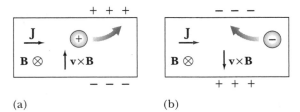

(a) (b)

Figure 8.8 **(a)** Positive carriers. **(b)** Negative carriers.

so in each case they drift upwards, as shown. Consequently the direction of the resulting Hall field $\mathbf{E}_H$ depends on the sign of the carriers; at equilibrium, there is a tiny surplus of carriers at the top. From $\mathbf{F} = q\mathbf{E} + q\mathbf{v} \times \mathbf{B} = 0$, we find that

$$E_H = vB \tag{8.18}$$

in magnitude. Also,

$$J = \rho v = vNq$$

where ρ is the charge density of the charge carriers, q is their charge, and N is their number per cubic meter. So,

$$E_H = \frac{JB}{Nq} = R_H JB \tag{8.19}$$

where the Hall coefficient is

$$R_H = \frac{E_H}{JB} = \frac{1}{Nq} \tag{8.20}$$

We will estimate R_H for copper, assuming one conduction electron per atom:

$$R_H = \frac{1}{Nq}$$

$$= \frac{63.5 \text{ g}}{\text{mole}} \times \frac{\text{mole}}{6.02 \times 10^{23} \text{ electron}} \times \frac{\text{m}^3}{8.9 \times 10^6 \text{ g}} \times \frac{\text{electron}}{-1.6 \times 10^{-19} \text{ C}}$$

$$= -7.4 \times 10^{-11} \text{ m}^3/\text{C}$$

The estimate is not so bad: 7.4 versus 5.5. But as the table shows, for many materials no such simple estimate works. For some, the carriers are "holes" in the electronic structure that behave like positive charges. The Hall effect is useful for measuring magnetic fields as well as for determining properties of conductors such as the mobility (σ/Nq) and effective mass of charge carriers. For conductors having fewer charge carriers (including semiconductors) the carriers have to move faster, giving a larger Lorentz force and a larger Hall coefficient.

Hall Coefficients, $\Omega \cdot \text{m} / \text{T}$	
Aluminum	-3.0×10^{-11}
Antimony	-200×10^{-11}
Arsenic	$+400 \times 10^{-11}$
Beryllium	$+24 \times 10^{-11}$
Bismuth	$\sim -1000 \times 10^{-11}$
Copper	-5.5×10^{-11}
Silver	-8.4×10^{-11}
Sodium	-25×10^{-11}
Zinc	$+3.3 \times 10^{-11}$

 **EXAMPLE 8-4**

For a copper strip 1 cm wide, carrying a current density of $10^7\,\text{A}/\text{m}^2$, in a field of 0.7 T, determine the Hall voltage.

ANSWER

$$V_H = E_H w = R_H JBw$$

$$= 5.5 \times 10^{-11}\,\Omega \cdot \text{m}/\text{T} \times 10^7\,\text{A}/\text{m}^2 \times 0.7\,\text{T} \times 0.01\,\text{m} = 3.8\,\mu\text{V}$$

Hall voltages tend to be pretty small.

8.5.   Eddy Currents

So far, our E field in conductors has been assumed to be due to a scalar potential difference ΔV. But changing magnetic flux $\partial\Phi/\partial t$ can be just as effective in producing a current $J = \sigma E$ via Faraday's law (Section 4.1).

If we let an aluminum rectangle fall into a magnetic field **B**, Figure 8.9a (possibly as a pendulum bob), we find that it suddenly slows down because of **eddy currents, J.** When the magnetic flux Φ through the metal changes, it produces an emf $= -d\Phi/dt$, causing current to flow. If we saw a few slots in the rectangle, Figure 8.9b, the eddy currents are greatly reduced, and the aluminum passes through the **B** field almost unaffected.

The rectangle is slowed because of Lenz's law. The conductivity σ is large, so the induced currents can be very large, on the order of hundreds of amperes; their direction is such that they oppose the imposed change in Φ. Lost energy appears as heat.

For related reasons, the iron cores of inductors and transformers are laminated: made of many insulated layers of iron, to reduce the eddy currents; other-

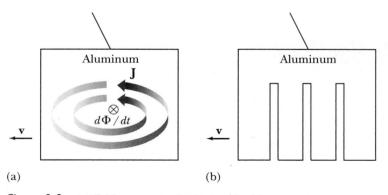

(a)   (b)

Figure 8.9   **(a)** Eddy currents. **(b)** Reduced eddy currents.

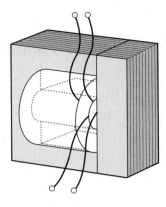

Figure 8.10   Laminated-core transformer.

wise much energy would be lost to heat. Figure 8.10 shows a sketch of a transformer with laminated-iron core. There exist magnetic dielectrics (ferrites, which are nonconductors, Chapters 9 and 10) that experience practically no eddy current; they are particularly useful at high frequencies, but at the usual 60 Hz of house current their relatively weaker magnetic performance prevents their competing successfully with laminated iron.

Eddy currents are by no means always an impediment; in many devices they are employed for constructive ends. For example, in some meters and scales, an aluminum vane in a magnetic field provides a drag that reduces unwanted oscillations. In the watt-hour meters attached to many homes, house current drives a motor that works against a calibrated drag provided by a conducting disk in a fixed magnetic field.

Automobile speedometers are electronic nowadays, but until a few years ago they exploited eddy currents. A rotating magnet was attached by a cable to the drive shaft, and it faced a more-or-less stationary metal disk on a spring attached to a pointer, Figure 8.11. The changing **B** field induced eddy currents in the disk, and the force due to the eddy currents is proportional to speed. Note that the magnet must rotate about an axis perpendicular to its field so that there is a changing field in the disk; fields don't move, and neither do they rotate. If we were

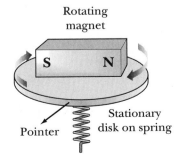

Figure 8.11   Speedometer.

simply to point the north pole of the magnet toward the disk and then rotate the magnet, there would be no effect on the disk.

We could, however, keep the magnet fixed and rotate the disk. In this case the motional emf in the disk would be caused by the Lorentz force, rather than by a changing **B** field; but the currents would be the same. That's how Faraday's law works. The currents would then exert forces on the magnet in such a direction as to tend to get it rotating in the same direction as the disk. If it were free to rotate, the magnet would rotate at the same speed as the disk, that is, it would come to rest relative to the disk.

8.6 SUMMARY

We have learned how much current flows in a conductor, and (to some extent) why. Ohm's law tells us the ratio of V/I, and Joule heating involves their product VI. The Hall effect supplies information concerning the nature of the charge carriers.

Ohm's law is

$$\mathbf{J} = \sigma \mathbf{E} \tag{8.1}$$

or for a resistor of finite size,

$$V = IR \tag{8.2}$$

The power dissipated per unit volume in a resistance is

$$\frac{dP}{dv} = \mathbf{J} \cdot \mathbf{E} \tag{8.4}$$

or for a resistor of finite size (Joule's law),

$$P = VI \tag{8.5}$$

For a good conductor with no current:

there is no internal **E** field;
the surface is an equipotential;
at the surface the tangential field is $E_t = 0$; $\tag{8.10}$

and the normal field is $E_n = \dfrac{\sigma}{\epsilon_0}$. $\tag{8.11}$

The free electrons in a metal dwell in a Fermi sea and are not subject to the classical Boltzmann distribution.

The Hall coefficient is

$$R_H = \frac{E_H}{JB} \tag{8.20}$$

Conductors may display electrostatic induction and eddy currents.

▼ PROBLEMS

8-1 A 1-megohm resistor is a cylinder of graphite-clay mixture having length 1 cm and radius 1 mm. Find the conductivity of the mixture.

8-2 A field of $E = 100$ V/m is maintained in copper. Find the current density J and the power dissipated per unit volume $\mathbf{E} \cdot \mathbf{J}$. Is it reasonable? Specific heat of copper is 387 J/kg · °C. (If you want its latent heat of fusion, you have already gone too far.)

8-3 Air may break down near 10^6 V/m. Find the corresponding charge density on a metal surface.

8-4 In the ocean, a seawater current of 0.5 m/s flows across the earth's magnetic field of 0.5×10^{-4} T. Find the electric current density.

8-5 Current of 7 A is passed through a seawater cell of dimensions 10 cm × 10 cm × 10 cm. Find the voltage. (The actual voltage includes several volts caused by "polarization" effects: elements liberated by electrolysis coat the electrodes.)

8-6 A disk of conductivity σ, radius a, and thickness s is placed in an axial magnetic field $B = B_0 \cos \omega t$, Figure 8.12. Find the power dissipated. (Thus, one may measure conductivity without electrical contact.)

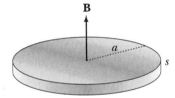

Figure 8.12

8-7 For field B, find the ratio of the Hall field to the field due to conductivity for a material of conductivity σ and Hall coefficient R_H. Evaluate for copper at 1 T.

8-8 Two metal spheres of charge Q and $-Q$ have radii of r_1 and r_2. They may be arranged concentrically or they may be separated by a large distance. Show that the difference in energy between the two arrangements equals twice the energy of the larger sphere by itself. Sketch the spheres and fields.

8-9 Of two concentric conducting spheres, the outer one, of fixed radius R, is grounded, and the inner one, of variable radius a, is maintained at potential V. Find the field E at the surface of the inner sphere, and find the radius a at which that E is minimum.

8-10 Of two coaxial conducting cylinders, the outer one, of fixed radius R, is grounded, and the inner one, of variable radius a, is maintained at potential V. Find the field E at the surface of the inner cylinder, and find the radius a at which that E is minimum.

8-11 A metal sphere of radius R and voltage V spins about a diameter with angular speed ω. Find its magnetic dipole moment. Evaluate for $V = 10^4$ volts and $R = 5$ cm at frequency $f = 10,000$ rpm (revolutions per minute; $\omega = 2\pi f$).

8-12 For the sphere of the preceding problem, find B at the center of the sphere.

8-13 A 1.5-m length of nichrome heater wire dissipates 700 W at 120 V. Find its diameter, using $\sigma = 10^6$ S/m (σ is actually smaller when hot).

8-14 Find the resistance between opposite sides of a square of silicon of thickness 1 mm. (Note that the size of the square doesn't matter.)

8-15 A sphere of radius R and uniform conductivity σ has surface voltage $V = V_0 \cos \theta$. Find J and E inside the sphere.

8-16 Assume that for an incandescent light bulb the tungsten filament is operated at a given temperature and that the lifetime of the filament is proportional to its diameter. Show that the lifetime is proportional to $I^{2/3}$, for current I, regardless of voltage rating or power. (Actually, lower-wattage bulbs are usually operated at lower temperature in order to increase their lifetime.)

8-17 Concentric cylindrical conductors of radius r_1 and r_2 are filled with material of uniform conductivity σ, Figure 8.13. Find the resistance of length l between the cylinders.

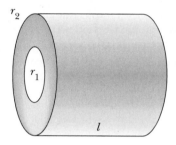

Figure 8.13

8-18 A cube of side s has conductivity increasing linearly with distance from 0 at side A to σ_0 at opposite side B. **(a)** Find the resistance between sides A and B. **(b)** Repeat for conductivity increasing as the square root of the distance.

8-19 Two charged conducting spheres of unequal size are separated by a distance large compared with either radius and are connected by a thin wire. Show that the configuration has minimum energy when the spheres are at equal potential.

8-20 A square loop of copper wire of side a and radius r falls out of a horizontal magnetic field B, Figure 8.14. Find the terminal velocity. Evaluate for $B = 1.2$ T, $a = 5$ cm, and $r = 1$ mm.

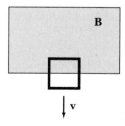

Figure 8.14

8-21 Repeat the analysis of Section 8.3 on the (simple but doubtful) assumption that the electrons start from rest after each collision. What's wrong with the result?

8-22 In a jumping-ring demonstration, an electromagnet provides a magnetic field of $B = 0.2 \cos 377t$ tesla over a circle of radius 1.2 cm. The magnetic flux passes through a copper ring of radius 1.5 cm and cross section 0.5 cm^2. Find an expression for the current in the ring.

8-23 An electric dipole $+p\hat{z}$ is situated in an otherwise uniform electric $E_o\,\hat{z}$; it is aligned parallel to the field. Find the spherical surface at which the field is perpendicular to the surface. Sketch the field lines and suggest a physical model that would produce the field outside of the sphere.

*8-24 A magnetic dipole $-m\hat{z}$ is situated in an otherwise uniform magnetic field $B_o\hat{z}$; it is aligned opposite to the field. Find the spherical surface that no field lines pass through. Sketch the field lines and suggest a physical model that would produce the field outside of the sphere.

8-25 For a thin conducting disk of charge Q and radius a, find the surface charge distribution σ as a function of radial distance ρ. (Consider starting with a spherical shell, Section 2.2.)

*8-26 For a thin conducting needle of charge Q and length $2a$, find the linear charge density λ. (Consider starting with a spherical shell, Section 2.2.)

C H A P T E R

9

DIELECTRICS

ielectric materials consist, effectively, of large numbers of electric dipoles; however, as we shall see, we can simplify the description of their behavior by introducing new kinds of fields that incorporate the effects of the dipoles in an efficient and natural way. Along the way we will learn how to calculate an important practical quantity called capacitance.

9.1. Fields in Dielectric Materials, D and P

If we insert a thick plastic dielectric between insulated charged plates, the voltage goes down. (A dielectric is a material that is not a conductor.) And when we remove it, V goes back up again. This is easily demonstrated using parallel plates and an electrometer. The charge on the plates is unaffected; what the dielectric does is to reduce the E field and thereby reduce V (remember that $V = -\int \mathbf{E} \cdot dl$ in the static case).

Figure 9.1a shows what happens to the field. In this case, $2/3$ of the field lines that start on the positive plate are swallowed by negative charges in the dielectric; they reappear at the other side. The field induces polarization in the dielectric: The negative charges move a little to the left under the influence of the applied field, and the positives move to the right. So, we have some negative charge sticking out on the left, and some positive on the right. These charges haven't gone far; in fact, they are still attached to their home molecules, because the dielectric is an insulator and has no unattached free charges. Each molecule has become a tiny electric dipole, Figure 9.2. The charge σ_b that is caused by polarization of the dielectric represents **bound charge** (or polarization charge), which means charge that can't leave its home atoms. The charge σ_f on the conducting plates is free to come and go as voltage considerations dictate; this is **free charge.**

The effect of the dielectric is entirely due to the bound charge. If we were somehow able to remove the dielectric material and leave the bound charge in place, the potential V would be exactly the same. To illustrate this, we will smooth over the internal structure of the material and assign it an average dipole moment

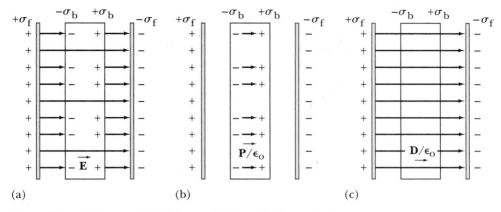

Figure 9.1   (a) Lines of **E**. (b) Lines of $\mathbf{P}/\epsilon_0$. (c) Lines of $\mathbf{D}/\epsilon_0$.

p per unit volume v,

$$\boxed{\frac{d\mathbf{p}}{dv} = \mathbf{P}} \tag{9.1}$$

called the **polarization,** units C/m^2. For one dipole **p**, oriented in the average direction (Section 7.1), the potential V at arbitrary radius **r** is

$$V = \frac{\mathbf{p} \cdot \hat{\mathbf{r}}}{4\pi\epsilon_0 r^2}$$

For a volume distribution of dipoles, we integrate over **P**:

$$V = \frac{1}{4\pi\epsilon_0} \int \frac{\mathbf{P} \cdot \hat{\mathbf{r}}}{r^2}\, dv = \frac{1}{4\pi\epsilon_0} \int \mathbf{P} \cdot \nabla\left(\frac{1}{r}\right) dv \tag{9.2}$$

The minus sign is missing because we are differentiating with respect to source variables (Sections 3.7, 6.3, 7.3). Now we apply a vector identity (inside front cover):

$$V = \frac{1}{4\pi\epsilon_0} \int \nabla \cdot \left(\frac{\mathbf{P}}{r}\right) dv - \frac{1}{4\pi\epsilon_0} \int \frac{\nabla \cdot \mathbf{P}}{r}\, dv$$

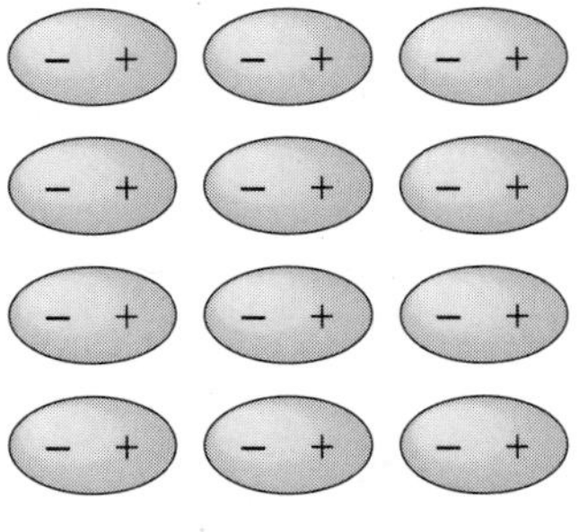

Figure 9.2   Dipole molecules.

The divergence theorem transforms the first integral:

$$V = \frac{1}{4\pi\epsilon_0} \int \frac{\mathbf{P}\cdot\hat{\mathbf{n}}}{r}\, da - \frac{1}{4\pi\epsilon_0} \int \frac{\nabla\cdot\mathbf{P}}{r}\, dv \tag{9.3}$$

Here $\hat{\mathbf{n}}\, da = d\mathbf{a}$; the unit vector in the $\mathbf{a}$ direction might better be called $\hat{\mathbf{a}}$, but $\hat{\mathbf{n}}$ for "normal" is customary. Now for free charge we have, from Section 6.3,

$$V = \frac{1}{4\pi\epsilon_0} \int \frac{\sigma}{r}\, da \qquad \text{and} \qquad \frac{1}{4\pi\epsilon_0} \int \frac{\rho}{r}\, dv$$

Comparing these integrals with Equation 9.3, we see that the role of the surface charge density is played by

$$\boxed{\sigma_{\mathrm{b}} = \mathbf{P}\cdot\hat{\mathbf{n}}} \tag{9.4}$$

and the volume charge density is

$$\boxed{\rho_{\mathrm{b}} = -\nabla\cdot\mathbf{P}} \tag{9.5}$$

We haven't yet specified exactly what closed surface we are integrating over. If we place the surface just barely inside of the polarized medium, like the dashed line in Figure 9.3, then we lose an infinitesimal contribution to the potential V, but the surface and volume integrals are both significant.

(If the surface surrounded all of the polarized material, then $\mathbf{P}$ would be zero on the surface, and the surface charge would be included in the overall volume integral. Placing the surface slightly inside of the dielectric material represents a neat way of separating out the surface charge automatically.)

Now, we have just found $-\nabla\cdot\mathbf{P} = \rho_{\mathrm{b}}$, so Gauss's law says (Section 5.1)

$$\epsilon_0 \nabla\cdot\mathbf{E} = \rho_{\mathrm{t}} = \rho_{\mathrm{f}} + \rho_{\mathrm{b}} = \rho_{\mathrm{f}} - \nabla\cdot\mathbf{P}$$

This equation clearly is asking for another divergence term for ρ_{f}, of the form

$$\boxed{\nabla\cdot\mathbf{D} = \rho_{\mathrm{f}}} \tag{9.6}$$

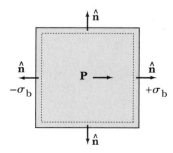

Figure 9.3   Bound surface charge.

so that the equation will read

$$\nabla \cdot \mathbf{D} = \epsilon_0 \nabla \cdot \mathbf{E} + \nabla \cdot \mathbf{P}$$

And this suggests that we define the **displacement D** as follows:

$$\boxed{\mathbf{D} = \epsilon_0 \mathbf{E} + \mathbf{P}} \tag{9.7}$$

(units C/m^2, like **P**). But what is the physical significance of **D**? **E** tells us the force on a test charge, and **P** is the smeared-out average of the molecular dipole moment per unit volume. Unfortunately it is necessary to admit that in itself **D** is somewhat lacking in clear-cut physical significance. The reason for this is that the distinction between ρ_f and ρ_b is artificial: Both represent real charge, and in fact bound charge can turn into free charge if the **E** field gets high enough to break down the dielectric.

However, **D** does turn out to be convenient in discussing current of various sorts. When **P** changes, charges move within molecules, and this represents current density, in particular in the sense that it makes magnetic field **B**. We have $\nabla \cdot \mathbf{P} = -\rho_b$ and so the equation of bound-charge continuity (conservation; see Section 5.1) reads

$$\frac{\partial \rho_b}{\partial t} = -\nabla \cdot \left(\frac{\partial \mathbf{P}}{\partial t} \right) = -\nabla \cdot \mathbf{J}_{pol}$$

Thus, $\partial \mathbf{P} / \partial t$ plays the role of **polarization current density,**

$$\mathbf{J}_{pol} = \frac{\partial \mathbf{P}}{\partial t} \tag{9.8}$$

Also, in Section 3.5 we remarked in connection with Ampère's law that $\epsilon_0 \partial \mathbf{E}/\partial t$ represents displacement current density. The sum

$$\frac{\partial \mathbf{D}}{\partial t} = \epsilon_0 \frac{\partial \mathbf{E}}{\partial t} + \frac{\partial \mathbf{P}}{\partial t} \tag{9.9}$$

may be regarded as a convenient generalization and will henceforth be called the **displacement current density.**

The vectors **E** and **P** need not be parallel: If the molecules in a crystal stretch more easily in one direction than another, then **P** may point in a direction somewhat different from **E**. Indeed, there are even materials, which are called **electrets** by analogy with magnets, which have their own permanent polarization; for these, **E** and **P** may be antiparallel. For example, in the electret in Figure 9.4, **E** is due to the bound surface charge and it points from right to left, inside the dielectric, whereas **P** and **D** point from left to right.

We discuss the physics of electrets a little further in Section 9.3. However, for now they are going to provide an instructive exercise in curls and divergences, as well as illustrating the physical meanings of the various definitions given above. Figure 9.4 represents a uniformly polarized electret in the shape of a cube; inside the cube, **P** is constant and points as shown. We can have curls and divergences of

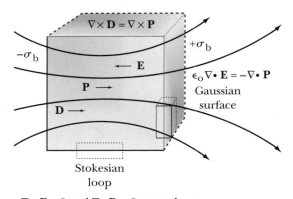

Figure 9.4 Electret, with **D** field lines.

$\nabla \times \mathbf{E} = 0$ and $\nabla \cdot \mathbf{D} = 0$ everywhere

E, **P**, and **D**, a total of $2 \times 3 = $ six quantities in all. All of them are zero over most of space.

QUESTION: Where are the curls and divergences of **E**, **P**, and **D** nonzero in Figure 9.4?

ANSWER: The six quantities are numbered below as they are presented. First, there is no free charge here; the electret is polarized all by itself. So,

(1) $\nabla \cdot \mathbf{D} = \rho_f = 0$

everywhere. **D** lines just keep going, like lines of magnetic field **B**. And there is no changing magnetic field, so

(2) $\nabla \times \mathbf{E} = -\partial \mathbf{B}/\partial t = 0$

everywhere. On the left and right, **P** encounters the surface normally. We have

(3) $\nabla \cdot \mathbf{P} = -\rho_b$

there, in the form of surface charge ($\sigma_b = \pm P$). This may be illustrated using the Gaussian surface provided. Flux of **P** comes in from the left, but none leaves on the right because **P** = 0 outside the dielectric. So there is a net flux of **P** into the Gaussian surface, and consequently there is a divergence on the surface of the dielectric, by the divergence theorem. We know

$$\mathbf{D} = \epsilon_0 \mathbf{E} + \mathbf{P}, \quad \text{so}$$

$$\nabla \cdot \mathbf{D} = \epsilon_0 \nabla \cdot \mathbf{E} + \nabla \cdot \mathbf{P} = 0 \quad \text{and}$$

(4) $\nabla \cdot \mathbf{E} = -\nabla \cdot \mathbf{P}/\epsilon_0 = \rho_b/\epsilon_0$

This means that **E** flux starts and ends on bound charge on the left and right surface.

On the top and bottom surfaces, **P** changes abruptly. In the Stokesian loop shown, there is a contribution to the line integral of **P** inside of the dielectric but not outside, signaling the

(5) presence of a nonzero curl $\nabla \times \mathbf{P}$ on the top and bottom surfaces.

Finally, starting once again with $\mathbf{D} = \epsilon_0 \mathbf{E} + \mathbf{P}$, we find

(6) $\nabla \times \mathbf{D} = \epsilon_0 \nabla \times \mathbf{E} + \nabla \times \mathbf{P} = \nabla \times \mathbf{P}$

providing a curl on the top and bottom surfaces. So that is where $\mathbf{D}$ comes from. The flux of $\mathbf{D}$ is similar to the magnetic flux of a solenoid, Figure 3.13; $\mathbf{D}$ has no divergence here, so the field lines don't end, but instead they make closed loops.

To summarize: In the electret shown, we have electric field $\mathbf{E}$ everywhere, provided only by $\nabla \cdot \mathbf{E}$, since $\nabla \times \mathbf{E} = 0$. We have $\mathbf{D}$ everywhere, provided only by $\nabla \times \mathbf{D}$, since $\nabla \cdot \mathbf{D} = 0$. We have $\mathbf{P}$ only inside the dielectric, due to cancellation of the effects of $\nabla \cdot \mathbf{P}$ and $\nabla \times \mathbf{P}$ outside.

Note that all four of the nonzero curls and divergences here are delta functions (Sections 2.4, 3.6, and 8.2), that is, they are infinite for infinitesimal distances. This is due to the fact that the polarization $\mathbf{P}$ changes suddenly at the surfaces: P is a step-function θ (Problem 2-37). Details are left for Problem 9-34.

These conclusions are good for the cube-shaped dielectric of Figure 9.4 but not in general. However, they do illustrate a point we have made before (Sections 1.2, 5.1): The divergence of a vector field tells us something about how the field changes in the direction the field points, and the curl tells us about how it changes in a direction transverse to that field. Together with the boundary conditions, the curl and divergence tell us all there is to know about the geometrical aspects of a field (Section 1.4).

The following example will illustrate two things: We will find the fields by replacing the polarized material with the bound charge at the surface, and we will simplify the geometry by employing the superposition of two simple solutions, namely the uniformly charged sphere with plus charge and with minus charge.

◀ **EXAMPLE 9-1**

Find the potential V inside and outside of a spherical dielectric of radius R and of uniform polarization $\mathbf{P}$.

ANSWER Figure 9.5a shows the potential inside. In a polarized dielectric, we have a positive spherical charge distribution, and a negative one, displaced by a distance we shall call s. This s is small, less than one atomic diameter, because the charges involved are bound and thus cannot leave their home molecules. As shown, some of the positive charge distribution sticks out above the sphere, and some of the negative charge is seen at the bottom. In the bulk of the dielectric, there is no significant change. The electrostatic effects observed are due to the bound charge, shown as $+$ and $-$ charges, on the surface of the dielectric. Inside of the uniform positive spherical charge distribution by itself, from Problem 6-5,

$$V_+ = \frac{Q}{4\pi\epsilon_0 R}\left(\frac{3}{2} - \frac{r^2}{2R^2}\right) \tag{9.10}$$

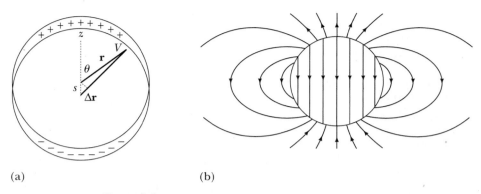

Figure 9.5   (a) Dipole potential. (b) Electret **E** field.

For the negative charge distribution the potential is analogous. Our next calculation is similar to that of Figure 7.1, Section 7.1. For the pair of equal + and −
spheres, we have

$$V_i = V_+ + V_- \implies -\frac{\partial V_+}{\partial r} \Delta r$$

for the potential inside the dielectric sphere. Differentiating, with $\Delta r = s \cos \theta$, we
obtain

$$V_i = \frac{Q}{4\pi\epsilon_0 R^3} \, r s \cos \theta \tag{9.11}$$

Now $Qs = p$, and $p = P\left(\frac{4}{3}\pi R^3\right)$, and $r \cos \theta = z$, so

$$V_i = \frac{Pz}{3\epsilon_0} \tag{9.12}$$

and

$$-\nabla V_i = \mathbf{E}_i = -\frac{P}{3\epsilon_0} \, \hat{\mathbf{z}} \qquad \text{inside}$$

So the **E** field is uniform inside the dielectric, Figure 9.5b.

Outside of each uniform spherical charge distribution, the field is the same as
if all of the charge were concentrated at the center. So we can use the derivation in
Section 7.1 for a dipole; the result is

$$V_0 = \frac{P \cos \theta}{3\epsilon_0} \frac{R^3}{r^2} \tag{9.13}$$

and

$$\mathbf{E}_0 = \frac{2P \cos \theta}{3\epsilon_0} \frac{R^3}{r^3} \, \hat{\mathbf{r}} + \frac{P \sin \theta}{3\epsilon_0} \frac{R^3}{r^3} \, \hat{\boldsymbol{\theta}}$$

To summarize: Inside, the **E** field is uniform; outside, the field is that of an
ideal dipole.

9.2. Ideal Dielectrics

The preceding is rather general. In what follows, we are going to concentrate on simple and practical dielectrics that will be approximately ideal:

Homogeneous: Their properties don't change with position in the medium.
Isotropic: Their properties don't depend on direction.
Linear: P is proportional to **E**.
Stationary.

Figure 9.1b shows lines of $\mathbf{P}/\epsilon_0$ for the ideal case. (We use $\mathbf{P}/\epsilon_0$ instead of $\mathbf{P}$ so that all the field lines represent commensurable quantities in Figure 9.1a–9.1c.) **P** fills in where **E** leaves off, because $\nabla \cdot \mathbf{P} = -\rho_b = -\epsilon_0 \nabla \cdot \mathbf{E}$: The source of one is the sink of the other. Because of the minus sign in its divergence equation, $\nabla \cdot \mathbf{P} = -\rho_b$, **P** points from $-$ to $+$, and so it points in the same direction as **E**.

Lines of $\mathbf{D}/\epsilon_0$, Figure 9.1c, are the sum of the other two. This means that the **D** lines plow straight through the dielectric, ignoring the bound charge because $\nabla \cdot \mathbf{D}$ depends only on free charge ρ_f. By the way, this doesn't mean that **D** itself depends only on free charge. Taking the curl of **D**, Equation 9.7,

$$\nabla \times \mathbf{D} = \epsilon_0 \nabla \times \mathbf{E} + \nabla \times \mathbf{P} = \nabla \times \mathbf{P} \tag{9.14}$$

since $\nabla \times \mathbf{E} = 0$ for this static case ($\partial \mathbf{B}/\partial t = 0$). At the sides of the dielectric, the tangential component $P_{\tan}$ changes abruptly; by Stokes' theorem that implies a curl of **P**, and so of **D**. This means that in the absence of free charge, lines of **D** don't begin or end, but they do form closed loops, like lines of magnetic field **B**.

The conditions of linearity and isotropy imposed above imply that we can write

$$\boxed{\mathbf{P} = \chi_e \epsilon_0 \mathbf{E}} \tag{9.15}$$

where the dimensionless constant of proportionality χ_e is called the **electric susceptibility** (Greek letter χ, "chi"). (Anisotropic crystals may polarize more easily in one direction than another, so **P** is not necessarily parallel to **E**, and χ_e is a tensor.) In Figure 9.1, P/ϵ_0 is twice as big as E inside the dielectric, so $\chi_e = 2$.

Since **P** is parallel to **E**, the equation $\mathbf{D} = \epsilon_0 \mathbf{E} + \mathbf{P}$ implies that **D** is also parallel to **E**. In fact, this equation becomes

$$D = \epsilon_0 E + \chi_e \epsilon_0 E$$

$$= (1 + \chi_e) \epsilon_0 E$$

$$\boxed{D = \epsilon_r \epsilon_0 E} \tag{9.16}$$

where

$$\boxed{\epsilon_r = (1 + \chi_e)}$$

Dielectric Constants ϵ_r	
Air	1.00059
Argon	1.00055
Helium	1.00007
Hydrogen chloride	1.0046
Steam (110°C)	1.013
Benzene C_6H_6	2.3
Ethanol C_2H_5OH	28
Fused silica SiO_2	3.8
Glass	~ 5
Mica	5.4
Mylar (polyester)	3.2
Polyethylene	2.3
Salt NaCl	6.1
Sulfur	4.0
Water (Fig. 9.10)	80
Barium titanate	~ 1000

is called the relative permittivity, or the **dielectric constant.** This may also be written simply

$$D = \epsilon E$$

where $\epsilon = \epsilon_r \epsilon_0$ is the **permittivity,** and ϵ_0 is the permittivity of free space. Since $\nabla \cdot \mathbf{D} = \rho_f$ (Section 8.1), this means

$$\nabla \cdot \epsilon \mathbf{E} = \rho_f$$

so

$$\nabla \cdot \mathbf{E} = \frac{\rho_f}{\epsilon} \qquad \textit{if } \epsilon \textit{ is constant} \qquad \textbf{(9.17)}$$

In Figure 9.1a, as mentioned, $\chi_e = 2$, and consequently the dielectric constant ϵ_r is $1 + \chi_e = 3$. D is the same inside and outside of the dielectric in this case (because its divergence is zero), so, in view of $D = \epsilon_r \epsilon_0 E$, it must be that E inside the dielectric is only 1/3 of its value outside. E is reduced by a factor of ϵ_r inside the dielectric, and 2/3 of the E field lines are swallowed by the bound surface charge of the dielectric. Note that if the plastic were a metal instead, then *all* of the field lines would disappear and E would be zero inside; in some ways, a metal is like a dielectric of infinite dielectric constant.

▼ EXAMPLE 9-2

For Figure 9.1, if $\sigma_f = 1 \ \mu C/m^2$ and $\epsilon_r = 3$, and the planes are approximately infinite in extent, find D, E, and P inside and outside of the dielectric (and between the charged plates), and find the bound charge.

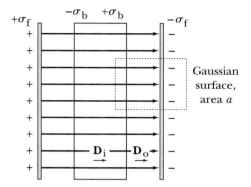

Figure 9.6   **D** in ideal dielectric.

ANSWER   We can evaluate D_o outside and D_i inside, Figure 9.6, just as we did E for two parallel plates, Figure 2.31, starting with $\nabla \cdot \mathbf{D} = \rho_f$. Applying the divergence theorem for the Gaussian surface shown,

$$\int \nabla \cdot \mathbf{D} \, dv = \int \rho_f \, dv$$

$$\int \mathbf{D} \cdot d\mathbf{a} = \int \sigma_f \, da$$

$$D_i a = \sigma_f a$$

(since $\mathbf{D} = 0$ on the right side of the Gaussian surface)

$$D_o = D_i = \sigma_f = 1 \ \mu\text{C/m}^2$$

$$E_o = \sigma_f / \epsilon_0$$

$$= 10^{-6}/8.85 \times 10^{-12} = 1.13 \times 10^5 \, \text{V/m}$$

$$E_i = E_o / \epsilon_r$$

$$= 1.13 \times 10^5 / 3 = 3.77 \times 10^4 \, \text{V/m}$$

$$P_o = 0$$

$$P_i = D_i - \epsilon_0 E_i \quad \text{(or } \chi_e \epsilon_o E_i)$$

$$= 10^6 - 8.85 \times 10^{-12} \times 3.77 \times 10^4 = 0.67 \ \mu\text{C/m}^2$$

$$\sigma_b = \mathbf{P}_i \cdot \hat{\mathbf{n}} = 0.67 \ \mu\text{C/m}^2$$

and

$$\rho_b = -\nabla \cdot \mathbf{P} = 0$$

since $\mathbf{P}$ is constant inside the dielectric. All of these are modest numbers for ordinary plastic.

Bound charge is usually confined to the dielectric surface, as the following argument illustrates. We have

$$\rho_b = -\nabla \cdot \mathbf{P} = -\nabla \cdot (\chi_e \epsilon_0 \mathbf{E})$$

$$= -\epsilon_0 \mathbf{E} \cdot \nabla \chi_e - \chi_e \nabla \cdot (\epsilon_0 \mathbf{E}) = -\epsilon_0 \mathbf{E} \cdot \nabla \chi_e - \chi_e(\rho_f + \rho_b)$$

using a vector identity and Gauss's law, so

$$\rho_b = -\frac{\epsilon_0 \mathbf{E} \cdot \nabla \chi_e}{1 + \chi_e} - \frac{\chi_e \rho_f}{1 + \chi_e} \tag{9.18}$$

Now if χ_e is constant then the first term is zero, and if there is no free charge ρ_f then the second term vanishes. So we find bound charge at the surface because ϵ_r suddenly changes there, but we find bound charge in the volume of the ideal dielectric only next to any free charge there.

Figure 9.7 shows the $\mathbf{E}$ field lines for a case in which there is free charge in a cavity inside a dielectric; it corresponds to Figure 8.2 for a conductor. (A metal is somewhat like a dielectric of $\epsilon_r \to \infty$.) Half of the field lines are absorbed by the dielectric, so $\epsilon_r = 2$. The dielectric remains uncharged. When there is free charge inside of a dielectric, then bound charge arises as shown and it tends to partly cancel out the $\mathbf{E}$ field due to the free charge.

$\mathbf{D}$ helps in calculating the refraction of a line of $\mathbf{E}$ at a dielectric interface, Figure 9.8. Since there is no free charge, $\nabla \cdot \mathbf{D} = 0$, and then the divergence theorem implies that the normal component of $\mathbf{D}$ is continuous across the interface, Figure 9.8c:

$$\int \mathbf{D} \cdot d\mathbf{a} = D_1 \cos \theta_1 \, a - D_2 \cos \theta_2 \, a = 0$$

$$D_1 \cos \theta_1 = D_2 \cos \theta_2$$

so

$$E_1 \cos \theta_1 = \epsilon_r E_2 \cos \theta_2 \tag{9.19}$$

And since there is no changing $\mathbf{B}$ field, $\nabla \times \mathbf{E} = 0$, and then Stokes' theorem

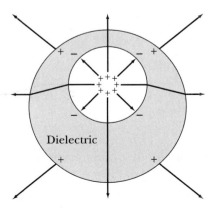

Dielectric

Figure 9.7   Charge in cavity in dielectric.

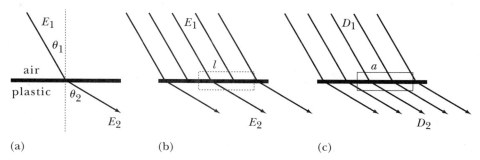

Figure 9.8 (a) Bend in line of **E**. (b) Stokesian loop. (c) Gaussian surface.

implies that the tangential component of **E** is continuous across the interface, Figure 9.8b:

$$\oint \mathbf{E} \cdot d\mathbf{l} = E_1 \sin \theta_1 \, l - E_2 \sin \theta_2 \, l = 0$$

(9.20)

$$E_1 \sin \theta_1 = E_2 \sin \theta_2$$

Dividing Equation 9.20 by Equation 9.19, we obtain

$$\tan \theta_1 = \frac{1}{\epsilon_r} \tan \theta_2$$

(9.21)

Unlike a ray of light, the line of **E** bends away from the normal in passing from air into a denser medium. (And there is no reflected ray.)

9.3. Inside Dielectrics

Materials have widely differing dielectric constants due to their varying internal structure. Here we shall peek inside three important types of dielectric.

A. Polar Dielectrics

Water is a typical polar liquid: Its constituent molecules have a permanent electric dipole moment (Section 7.1). Knowing that the water molecule is a dipole, Figure 9.9a, and that water has a large dielectric constant, we might guess that the dielectric constant is just a matter of lining up dipole molecules, Figure 9.9b. And in some sense that's almost true, but it leaves out the thermal motion. For example, we can now calculate the field of a bunch of aligned water molecules.

 **EXAMPLE 9-3**

The dielectric constant of water is 80. If its average molecular dipole moment is the dipole moment of the water molecule, 6.2×10^{-30} C · m, find the electric field.

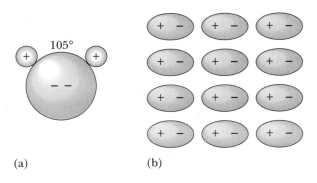

Figure 9.9 (a) Dipole: H_2O.
(b) Aligned dipoles.

(a)   (b)

ANSWER $P = Np$, where N is the number of molecules per unit volume, P is the polarization, and p is the dipole moment:

$$P = \frac{6.02 \times 10^{23} \text{ molecules}}{\text{mole}} \times \frac{1 \text{ mole}}{18 \text{ grams}} \times \frac{10^6 \text{ grams}}{1 \text{ m}^3} \times \frac{6.2 \times 10^{-30} \text{ C} \cdot \text{m}}{\text{molecule}}$$

$$= 0.21 \text{ C/m}^2$$

So in the water

$$E = P/\chi_e \epsilon_0$$

$$= 0.21/(80 - 1) \times 8.85 \times 10^{-12} = 3.0 \times 10^8 \text{ V/m}$$

This means that the water molecules are not all lined up until the field reaches three million V/cm, which is enough to break down the water and make an arc. At ordinary temperatures and fields, the average dipole moment is greatly reduced by thermal agitation: The molecules point every way, with only a slight tendency to line up with the field. Thermodynamical analysis yields the **Langevin formula;** in the low-field approximation,

$$\frac{P}{N} = p \frac{pE}{3kT} \qquad (9.22)$$

instead of just $P/N = p$; k is the Boltzmann constant, and T is absolute temperature, so kT is thermal energy, and pE is the energy of the dipole molecule when aligned with the effective field **E** (Section 7.1). The polarization is proportional to the applied E field, and inversely proportional to the absolute temperature. At small E field, P/N is much less than p, because of thermal motion. On the other hand, near absolute zero, if the water were still fluid, a small internal **E** field would line up almost all the molecules, so the effective dielectric constant would be enormous at first; but then adding more **E** field would not line up any more molecules because they would already be all lined up, so the effective dielectric constant for additional field would become rather small: The dielectric would be "saturated."

It is also true that it takes the water molecules a while, 10^{-10} seconds or so, to respond to the applied field, because of their moment of inertia. For molecules

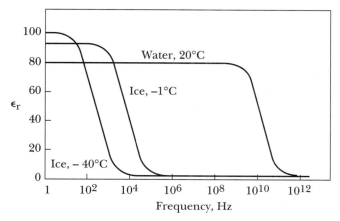

Figure 9.10 Dielectric constant of water.

locked in an ice crystal the time is even longer, 10^{-4} s, but eventually they find a way. Effects of temperature and frequency in the case of water are nicely displayed in Figure 9.10 (from Purcell).

B. Ferroelectrics

Since opposite charges attract, would it be possible for polar molecules to align themselves spontaneously as shown in Figure 9.9b? Yes, under peculiar circumstances. As we have seen for water, thermal motion usually keeps them pretty well shaken up. But there are materials called "ferroelectrics" in which the dipoles do align spontaneously at room temperature, Figure 9.11a. And, depending on crystal structure, they might decide to line up alternately instead, as in Figure 9.11b, because that arrangement also places plus charges near minus ones; this is called "antiferroelectricity."

For example, the barium titanate crystal structure is a cage with an ion rattling around inside, and it polarizes itself spontaneously over large areas called domains. Ferroelectricity is analogous to ferromagnetism, Chapter 10, and it even in-

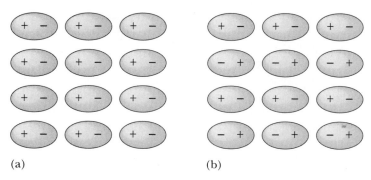

(a)   (b)

Figure 9.11 **(a)** Ferroelectric. **(b)** Antiferroelectric.

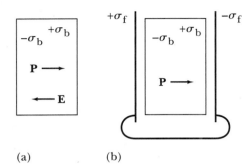

(a)   (b)   **Figure 9.12**   (a) Electret. (b) With keeper.

cludes hysteresis and a "Curie temperature" above which the effect vanishes ($120°C$ for $BaTiO_3$). Like ferromagnets, such as iron, ferroelectric materials are useful in practical applications because of their large susceptibility; but also like iron they are rather nonlinear (so non–ideal), which can be a nuisance.

Materials like $BaTiO_3$ can be used to illustrate the concept of the electret (Section 9.1 and Figure 9.12a). A magnet provides a convenient and portable source of **B** field, and an electret does the same for **E**, because it retains some polarization after an externally imposed **E** field is removed. But once its surfaces are covered with air-borne ions, its own **E** field is gone. Consequently, practical electrets are dynamic systems which recover after exposure to ions. For example, if you solidify carnauba wax under a field of 2×10^6 V/m, its long polar molecules tend to remain aligned with the field, and, more important, some free charge is injected into the wax and remains there, trapped in the wax. Over the course of years, the free charge diffuses to the surface of the electret. This causes the surface **E** field to build up fairly quickly to the point where air breaks down, around 10^6 V/m.

Electrets are less useful than magnets because, as mentioned, their surfaces tend to get coated with free ions from the air. Magnets don't face that problem because there are no magnetic monopoles. One can protect the electret surfaces with an "electret keeper," Figure 9.12b, which is a pair of plates, electrically connected, placed on the surfaces of the electret. The electret keeper blocks stray ions, and it also helps by reducing the "depolarizing" field which points opposite to the polarization inside the electret; analogously, a permanent magnet needs a magnet keeper, Section 10.9.

But of course to use an electret or a magnet, you have to remove the keeper.

To make an electret microphone, place a conducting foil near the face of the electret. As the foil moves back and forth, its voltage changes.

Polarized ferroelectrics display **piezoelectricity:** That is, under applied stress their polarization **P** changes and they produce surface charge; and conversely, applied voltage produces strain. Other kinds of crystals may also be piezoelectric. For quartz, which is not ferroelectric, the piezoelectric constant is 2.2×10^{-12} coulombs per newton, or meters per volt. So if you apply a newton of force perpendicular to the axis of a crystal of quartz, you get a charge of 2.2 pC (picocoulomb); if you apply a volt, the length changes by 2.2×10^{-12} m. The piezoelectric constants for barium titanate and Rochelle salt (potassium sodium tartrate, ferroelectric) may be a hundred times greater but are quite variable. These materials are widely used in transducers such as high-frequency electroacoustical devices (microphones, speakers). The potential involved may be a volt or more, and the charge is typically less than a microcoulomb. If you apply voltage to a quartz crystal at a frequency corresponding to a mode of mechanical oscillation, the resulting resonance is stable and precise; such oscillators have become commonplace, and in fact you are probably wearing one on your wrist right now.

C. Nonpolar Dielectrics

Molecules without any intrinsic dipole moment will acquire an induced dipole moment in an electric field, and so materials consisting of such molecules have a dielectric constant. The electrons shift position slightly inside their molecules; and they do so very quickly, around 10^{-15} second, so that temperature and frequency have little effect. We now undertake the calculation of dielectric constants for nonpolar materials.

First: As would be expected for a given molecular polarizability, the electric susceptibility χ_e depends on density; so, for example, if we double the atmospheric pressure, the dielectric constant ϵ_r of air increases approximately from 1.00059 to 1.00118. This kind of calculation is quite accurately verified by experiment.

And now we introduce our simple physical model of a nonpolar atom, Figure 9.13. The atom is spherical, of radius R and of uniform charge density ρ, which is certainly an idealization; real atoms are kind of fuzzy and tend to have greater electron density near the nucleus. But this may give us an order-of-magnitude result for nonpolar molecules and particularly for elements.

The atom is placed in a field $\mathbf{E}_a$, the "microscopic" field. The atom becomes distorted as shown. We regard the atom as a cloud of charge $-Q$, surrounding a nucleus of charge $+Q$. The nucleus moves until the $\mathbf{E}$ field due to the electron cloud equals that imposed from outside. The electron cloud is fixed by the surrounding material structure, and the nucleus moves by a distance z. The resulting induced dipole moment of the atom is

$$Qz = p$$

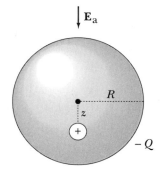

Figure 9.13   Polarized atom.

The field at z inside a uniformly charged sphere is, from Section 2.3 (with $Q = \frac{4}{3}\pi R^3 \rho$),

$$E_a = \frac{\rho r}{3\epsilon_0} = \frac{Qz}{4\pi\epsilon_0 R^3} \tag{9.23}$$

so

$$E_a = \frac{p}{4\pi\epsilon_0 R^3}$$

We define the **atomic polarizability** α for the atom:

$$p = \alpha E_a = \alpha \frac{p}{4\pi\epsilon_0 R^3}$$

so

$$\alpha = 4\pi\epsilon_0 R^3 \tag{9.24}$$

And now, what $\mathbf{E}_a$ field does the atom face? The $\mathbf{E}_d$ field inside a dielectric is actually a wildly fluctuating thing, as one measures it near a nucleus or outside of an electron cloud. But the field we employ in the preceding section is a sensible average. For example, if we drill a fine needle hole through the dielectric, parallel to $\mathbf{E}$, Figure 9.14a, then the $\mathbf{E}_a$ field in the hole is the same as $\mathbf{E}_d$ inside the dielectric; it is like the tangential $\mathbf{E}$ of Figure 9.8b, and it is continuous across the boundary. If somehow the average $\mathbf{E}_d$ inside the dielectric were actually larger than $\mathbf{E}_a$, then we could in principle violate conservation of energy by carrying a charge around a closed Stokesian loop, up through the needle hole and then down through the dielectric.

If instead we make a flat pancake-shaped cavity, Figure 9.14b, the field $\mathbf{E}_a$ is larger; the Gaussian box shown has some negative bound charge in it. From the preceding section, $\mathbf{E}_a = \epsilon_r \mathbf{E}_d$. (But you still can't gain energy by carrying a test charge around a loop here.)

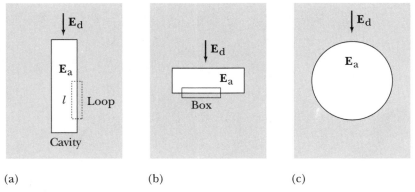

(a)   (b)   (c)

Figure 9.14   **(a)** Needle cavity. $\mathbf{E}_d = \mathbf{E}_a$ **(b)** Pancake cavity. $\mathbf{E}_a = \epsilon_r \mathbf{E}_d$ **(c)** Sphere cavity.

$$\mathbf{E}_a = \mathbf{E}_d + \frac{\mathbf{P}}{3\,\epsilon_0}$$

Our model atom is going to be in a jagged hole, but on the average the hole should be about the size of the atom itself, and round, Figure 9.14c. The polarized medium will then yield a field like that of Figure 9.5b, of magnitude $P/3\epsilon_0$. This is added to $\mathbf{E}_d$, giving

$$E_a = E_d + \frac{P}{3\epsilon_0} \tag{9.25}$$

which is an intermediate value between those of Figures 9.14a and 9.14b. Now, from above, $p = \alpha \mathbf{E}_a$, and $Np = P$, so

$$Np = N\alpha E_a$$

$$P = N\alpha \left(E_d + \frac{P}{3\epsilon_0} \right)$$

We set $P = \chi_e \epsilon_0 E_d = \epsilon_0(\epsilon_r - 1)E_d$, from Section 9.2, and solve for α:

$$\boxed{\alpha = \frac{3\epsilon_0}{N} \frac{\epsilon_r - 1}{\epsilon_r + 2}} \tag{9.26}$$

This is the **Clausius-Mossotti** equation. We can equate this to the expression for α found above, Equation 9.24, $\alpha = 4\pi\epsilon_0 R^3$, which yields

$$4\pi\epsilon_0 R^3 = \frac{3\epsilon_0}{N} \frac{\epsilon_r - 1}{\epsilon_r + 2} \qquad \text{so} \qquad \tfrac{4}{3}\pi R^3 N = \frac{\epsilon_r - 1}{\epsilon_r + 2} \tag{9.27}$$

First let's try it for argon, $\epsilon_r = 1.00055$. At STP, a mole is 22.4 liters, so Loschmidt's number is

$$N = \frac{6.02 \times 10^{23} \text{ molecules/mole}}{0.0224 \text{ m}^3/\text{mole}} = 2.69 \times 10^{25} \text{ molecules/m}^3$$

Then

$$R^3 = \frac{3}{4\pi N} \frac{\epsilon_r - 1}{\epsilon_r + 2} = \frac{3 \times 0.00055}{4\pi \, 2.69 \times 10^{25} \times 3.00055}$$

so

$$R = 1.2 \times 10^{-10} \text{ m}$$

a reasonable value.

And now for solids: The spherical atoms fill about half of the volume, depending on the crystal structure. Figure 9.15 shows a sphere inscribed in a cube, corresponding to a simple cubic structure. This crystal structure is rare among elements, and it does not represent the closest possible packing of spheres, but it will do for our approximate calculations. The volume of the cube is $\frac{1}{N} = 8R^3$, which is the volume per molecule; and the volume of the sphere is $\frac{4}{3}\pi R^3$. The ratio of these is 0.52, which is close enough to a half. If we set $\frac{4}{3}\pi R^3 N = 0.5$ in the expression in

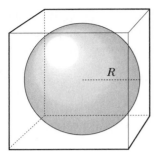

Figure 9.15 Sphere in cube.

Equation 9.27,

$$\tfrac{4}{3}\pi R^3\,N = \frac{\epsilon_r - 1}{\epsilon_r + 2}$$

then we find

$$\epsilon_r = 4$$

Thus, the theory predicts that the dielectric constants of nonconducting solid elements (which are all nonpolar) should be around 4. Examining the table in Section 9.2, we find the solid element sulfur has dielectric constant 4.0. Should we try another? Maybe not.

Clearly the dielectric constant is sensitive to the interatomic spacing, which depends on N. For example: Red phosphorus has specific gravity 2.20 and $\epsilon_r = 4.1$; for yellow phosphorus, which is another allotrope of the same element, the numbers are 1.82 and 3.6. We would expect $\chi_e(=\epsilon_r - 1)$ to be approximately proportional to the density, so:

$$\frac{2.20}{1.82} \approx \frac{4.1 - 1}{3.6 - 1}$$

$$1.21 \approx 1.19$$

And so that works pretty well.

To summarize: A look at the microscopic structure of nonpolar dielectrics yields insight into the physical basis of their relative permittivities; but great accuracy is not to be expected for our simple atomic model.

9.4. Capacitance, C

Capacitors store charge. If we charge up a capacitor of several hundred grams, even of low voltage rating, and then we use a wire to discharge it, we can get a jolly fat spark in a darkened room. One difference between a capacitor and a rechargeable battery is, the battery stores its charge mostly at a constant voltage, whereas in

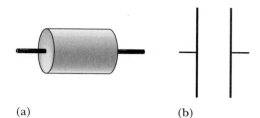

(a)   (b)   **Figure 9.16**   Two views of a capacitor.

the capacitor the charge is proportional to the voltage:

$$\boxed{Q = CV}$$ **(9.28)**

where C is the **capacitance** in C/V, or farads (F, named after Faraday).

Figure 9.16a shows the physical appearance of a typical capacitor (or "condenser"); it is a roll of foil. Figure 9.16b shows the schematic diagram of the same capacitor.

Capacitors also store energy. If one adds a little charge dQ to a capacitor at voltage V, the energy required is

$$dW = V\,dQ = \frac{Q}{C}\,dQ$$

Integrating, we obtain

$$W = \tfrac{1}{2}QV$$ **(9.29)**

which is a general result obtained also in Section 6.5. If almost all the charge were added at the same voltage, as in a storage battery, then the energy would be about QV. (Other battery characteristics: larger stored W, slower charging rate, limited number of recharges.) But for the capacitor the voltage starts out at zero, and the average is only $V/2$. Using $Q = CV$ we can write the energy in several useful ways:

$$\tfrac{1}{2}QV = \tfrac{1}{2}Q^2/C = \boxed{\tfrac{1}{2}CV^2 = W}$$ **(9.30)**

It is an indication of the fundamental nature of energy that the size of a capacitor depends much more on how much energy it can store than how much charge, as the following example will illustrate.

▼ EXAMPLE 9-4

For an electrolytic filter capacitor, about 3 cm by 5 cm, the rating is 4000 μF at 25 V (called WVDC). Find the maximum charge and energy it can store. Repeat for a ceramic "doorknob" television capacitor of about the same size, rated at 3000 pF at 30 kV.

ANSWER For the electrolytic, $Q = CV = 4000 \times 10^{-6} \times 25 = 0.10$ C, and

$$W = \tfrac{1}{2}CV^2$$

$$= 0.5 \times 4000 \times 10^{-6}\,\text{F} \times 25^2\,\text{V}^2 = 1.25\ \text{J}$$

For the doorknob,

$$Q = CV$$

$$= 3000 \times 10^{-12} \times 25 \times 10^3 = 75\ \mu\text{C}$$

and

$$W = \tfrac{1}{2}CV^2$$

$$= 0.5 \times 3000 \times 10^{-12} \times (30 \times 10^3)^2 = 1.35\,\text{J}$$

These two capacitors differ in capacitance by a factor of a million, and they differ in voltage rating by a factor of a thousand. And their construction is entirely different: The electrolytic capacitor consists largely of many meters of fine aluminum foil in a roll, and the ceramic capacitor is a pair of parallel plates with an inch-thick slab of barium titanate sandwiched between. Yet their energy storage is nearly the same because they are nearly the same size.

Now we shall undertake the practical task of calculating capacitances for various conducting objects. In calculating a capacitance for a particular pair of conductors, we begin by transferring charge Q from one conductor to the other; if there is only one conductor present, we suppose the other is at infinity. We calculate the voltage V, usually using $-\int E\,dl$. Then when we substitute V into the equation $C = Q/V$, the Q's in numerator and denominator cancel out. The units of ϵ_0 are farads per meter, so a calculation of a capacitance will usually result in an ϵ_0 in the numerator.

 EXAMPLE 9-5

A capacitor consists of a rolled-up pair of strips of aluminum foil of size 1.4 m $\times$ 2.5 cm, separated by 0.02 mm mylar of dielectric constant 3.2. Find its capacitance.

ANSWER It is approximately a parallel-plate capacitor, Figure 9.17; so $E = \sigma/\epsilon_0\epsilon_r$, since the field is reduced by the dielectric constant 3.2, and

$V = Es = \sigma s/\epsilon_0\epsilon_r$ where s is the separation 0.02 mm

$\quad = Qs/\epsilon_0\epsilon_r\,a$,   where a is the area $1.4 \times 0.025 \times 2$: both sides of the foil count

so

$Q/V = \boxed{C = \epsilon_0\epsilon_r\,a/s}$ (9.31)

$\quad = 8.85 \times 10^{-12}\,\text{F/m} \times 3.2 \times 1.4\,\text{m} \times 0.025\,\text{m} \times 2/0.02 \times 10^{-3}\,\text{m} = 0.10\ \mu\text{F}$

which is a reasonable value. The farad is a rather large unit.

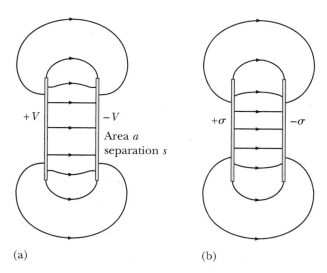

Figure 9.17   **(a)** Constant V. Conducting plates. **(b)** Constant σ.

In the preceding example, we proceeded as if constant V and constant σ on the plates were given. And that is usually a good approximation. However, Figure 9.17 is designed to show how for small plates or large separation, they are not the same. If V is constant on the plates, Figure 9.17a, then the charges must to some extent accumulate at the edges; this may be ascribed to the mutual repulsion of the excess charges. Whereas if σ is constant, Figure 9.17b, then the E field will be greater in the center of the space separating the plates, and so V is greater at the center than near the edges.

In a parallel-plate capacitor, the curved "fringing field" illustrated in Figure 9.17a reduces the E field slightly from its Gauss's-law value. (Compare with Fig. 5.2.) Note that the charge tends to accumulate at the ends of the conducting plates, so the density of field lines is greater there. Also, the field lines are constrained to be perpendicular to the metal surface, as mentioned in Section 8.2.

Figure 9.17b shows how, for constant σ, the same number of field lines leave each unit area, counting lines on both sides. Capacitors are made of conductors, of course; if the charge isn't free to come and go, the capacitor isn't much use. That means that Figure 9.17a is more practical and 9.17b is of merely academic interest.

◣ EXAMPLE 9-6

Find the capacitance of the earth, regarded as a conducting sphere of radius R, Figure 9.18.

ANSWER $V = Q/4\pi\epsilon_0 R$ for a sphere (as if all the charge were at the center).

$$Q/V = \boxed{C = 4\pi\epsilon_0 R}$$

(9.32)

$$= 6.4 \times 10^6/9 \times 10^9 = 710 \ \mu F$$

Figure 9.18 Conducting sphere.

which is surprisingly small, but then, the voltage rating is very high. (This is a "capacitance to ground"; the other electrode may be regarded as being at infinity.)

In two-dimensional geometry, one usually needs the capacitance per unit length, $C' = C/l$.

 **EXAMPLE 9-7**

Find the capacitance per unit length C' of a coaxial pair of cylinders of radii 0.2 cm and 0.7 cm, half-filled with liquid dielectric of relative permittivity 2.3, Figure 9.19a.

ANSWER First we find the capacitance of the upper half, C'_1. Symmetry dictates that the field lines are radial, so we can employ a half-cylinder gaussian surface, radius r, Figure 9.19a; no electric flux passes through the flat surface.

$$\int E\, da = Q/\epsilon_0$$

$$E\, 2\pi rl/2 = \lambda l/\epsilon_0$$

$$E = \lambda/\pi\epsilon_0 r$$

$$V = \int E\, dr = \int_{r_1}^{r_2} \frac{\lambda}{\pi\epsilon_0 r}\, dr = \frac{\lambda}{\pi\epsilon_0} \ln(r_2/r_1)$$

$$C = Q/V$$

$$C'_1 = C/l = \frac{Q/l}{V} = \frac{\lambda}{\dfrac{\lambda}{\pi\epsilon_0} \ln(r_2/r_1)} = \frac{\pi\epsilon_0}{\ln(r_2/r_1)}$$

Then for the lower half,

$$C'_2 = \frac{\pi\epsilon_0\epsilon_r}{\ln(r_2/r_1)}$$

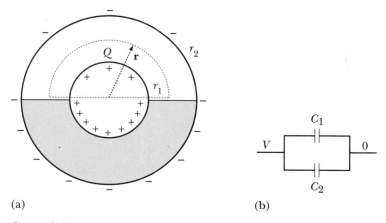

(a)   (b)

Figure 9.19   (a) Half-filled coaxial cylinders. (b) Capacitors in parallel.

Capacitances in parallel simply add, Figure 9.19b:

$$Q_1 = C_1 V \quad \text{and} \quad Q_2 = C_2 V$$

so

$$Q_1 + Q_2 = (C_1 + C_2) V$$

which means

$$\boxed{C = C_1 + C_2} \tag{9.33}$$

(In series, capacitors add as reciprocals; see Problem 9-31.) Consequently, for the case of Figure 9.19a,

$$C' = C'_1 + C'_2 = \frac{\pi \epsilon_0}{\ln(r_2/r_1)} (1 + \epsilon_{\mathrm{r}})$$

$$= \frac{\pi \epsilon_0}{\ln(0.7/0.2)} (3.3) = 7.3 \times 10^{-11} \text{ F/m} = 73 \text{ pF/m}$$

As would be expected, the capacitance per unit length goes to zero for large r_2 and for small r_1, and it goes to infinity for $r_2 \to r_1$.

 EXAMPLE 9-8

Find the capacitance per unit length of an infinite cylinder of radius a and linear charge density λ, Figure 9.20.

ANSWER   From Section 2.3, $E = \lambda/2\pi\epsilon_0 r$. So the potential is

$$V = \int_a^\infty \frac{\lambda \, dr}{2\pi\epsilon_0 r} = \frac{\lambda}{2\pi\epsilon_0} \ln r \bigg]_a^\infty \implies \infty$$

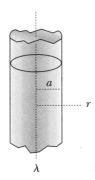

Figure 9.20   Infinite conducting cylinder.

and so

$$C' = \lambda / V = 0$$

If any finite line charge is placed on the cylinder, the voltage goes to infinity (as mentioned in Section 6.3). So the capacitance per unit length C' is zero. The actual situation is not so grave, however, because in practice there are usually plenty of other conductors around to hold the voltage down. A simple case is given in the preceding example, the coaxial infinite cylinders. Another easy one is parallel wires, which we save for Problem 9-17 at the end of the chapter. For the infinite conducting plane there is a similar situation, with infinite V and zero C' (capacitance per unit area); see Problem 9-30. (We will find an analogous case in the next chapter, Section 10.4: for an infinitely long straight wire carrying a current, the vector potential is infinite and the inductance per unit length L' is infinite.)

An isolated cylinder of finite size has a finite capacitance, of course. Calculating it exactly is no trivial problem, because the charge is not uniformly distributed. But we can get a useful approximation as follows. The divergence (infinite value) obtained above is due to the fact that the field we use falls off only as $1/r$, all the way to infinity. But for a cylinder of length l, once we get farther away than l, clearly E must fall off more rapidly, more like $1/r^2$ (this follows from Coulomb's law), causing the integral to converge. So we can get an approximate value for the capacitance by cutting off the above integration at l:

$$V \approx \frac{\lambda}{2\pi\epsilon_0} \ln r \Big]_a^l = \frac{\lambda}{2\pi\epsilon_0} \ln \frac{l}{a}$$

and so

$$C' = \frac{\lambda}{V} \approx \frac{2\pi\epsilon_0}{\ln \dfrac{l}{a}} \tag{9.34}$$

(It happens to be the same as one would obtain for constant λ; see Problem 9-29.)

 **EXAMPLE 9-9**

Find the capacitance of a wire of length 10 cm and radius 0.1 cm.

ANSWER

$$C = l\,C' = \frac{2\pi\epsilon_0 l}{\ln\dfrac{l}{a}}$$

$$= 2\pi\epsilon_0 0.1/\ln\,(0.1/0.001) = 1.2\ \mathrm{pf}$$

In a circuit, this constitutes capacitance to ground, and at TV frequencies, around 100 MHz, it can produce a significant loss of signal. To reduce stray capacitance, one reduces the size of components and increases their distance from other conductors.

9.5.   Time Constant *RC*

As we have remarked, capacitors store charge. If we put a resistor R across a capacitor C, which is originally charged to V_0, Figure 9.21a, the capacitor discharges more or less gradually, Figure 9.21b. To find the time constant, we begin with

$$Q = CV \qquad \text{and} \qquad V = IR$$

so

$$Q = CIR$$

Differentiating,

$$\frac{dQ}{dt} = CR\frac{dI}{dt} \qquad\qquad (9.35)$$

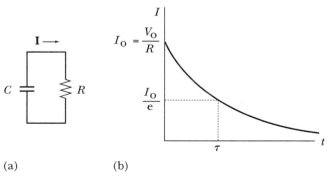

(a)   (b)

Figure 9.21   **(a)** Capacitor discharge. **(b)** Time constant.

that is,

$$-I = CR\frac{dI}{dt}$$

The solution of this differential equation is

$$I = \frac{V_0}{R}e^{-t/\tau}$$

as you can verify by direct substitution. The **time constant**

$$\boxed{\tau = RC}$$ (9.36)

is the time for the current I to drop to $1/e = 0.368$ of its original value, $I_0 = V_0/R$. ("τ" is the lowercase greek letter "tau.")

 EXAMPLE 9-10

For a 1-microfarad capacitor charged to 10 volts, what resistance is required for a 1-second time constant?

ANSWER

$$R = \tau/C$$

$$= \frac{1 \text{ s}}{10^{-6} \text{ C/V}} = 10^6 \frac{\text{V}}{\text{A}} = 1 \text{ M}\Omega \qquad \text{(one megohm, } 10^6 \text{ ohm)}$$

This can easily be demonstrated using a typical solid-state voltmeter, whose input impedance is likely to be 10 MΩ. An ordinary galvanometer-based voltmeter would probably be rated at 20,000 ohms per volt (full-scale reading), so at 10 volts its input impedance would be only 200 kΩ and it would tend to hide the effect of the 1-MΩ resistor. In this case a 100-μF capacitor and 10-K resistor would show the effect clearly.

We will return to time constants under the heading of Transients in Section 13.1.

9.6. Energy and Force

When a linear dielectric is placed in a given electrostatic field as in Figure 9.1, the field energy goes down. Consequently, the dielectric will tend to be drawn into an electric field. The energy density in the field itself is (from Section 6.1)

$$\tfrac{1}{2}\epsilon_0 E^2 = \tfrac{1}{2}\epsilon_0 \mathbf{E} \cdot \mathbf{E}$$

where now the **E** field includes the averaged fields of the dipoles. There is also mechanical energy involved in inducing the polarization. For each induced dipole, the mechanical energy is $\frac{1}{2}\mathbf{p}\cdot\mathbf{E}$ (Section 7.1), so the mechanical energy density will be

$$\tfrac{1}{2}\mathbf{P}\cdot\mathbf{E} \tag{9.37}$$

The same is true for molecules that are permanent dipoles, using an effective **p** in this case; because of thermal agitation, in either case the polarization is proportional to the **E** field. The total energy density is then

$$\frac{dW}{dv} = \tfrac{1}{2}(\epsilon_0\mathbf{E}\cdot\mathbf{E} + \mathbf{P}\cdot\mathbf{E})$$

so

$$\boxed{\frac{dW}{dv} = \tfrac{1}{2}\mathbf{D}\cdot\mathbf{E}} \tag{9.38}$$

We will apply these concepts in the case of a pair of parallel plates separated by a linear isotropic dielectric, Figure 9.22 (so **D** is parallel to **E**). We will disassemble it and find out how much energy we have. The charges $\pm\sigma$ on the plates are fixed. The initial field is E, and the final one, after all the dielectric molecules have been removed, is $E_\mathrm{f} = D/\epsilon_0 = \sigma/\epsilon_0$. We will not simply remove the molecular dipoles one by one as they stand, because as their number goes down the field goes up, and so their mechanical energy goes up, producing a rather complicated problem. So first we latch, or lock, or freeze, the mechanical energy, that is, the imaginary little springs inside the dipoles (Section 7.1). Then we have an assemblage of fixed dipoles of average dipole moment **p** and of mechanical energy $\frac{1}{2}pE$. Removing the first one takes energy pE. Removing the last one takes pD/ϵ_0; the field has increased. The average dipole takes energy $p(E + D/\epsilon_0)/2$ to remove. So the energy expended per unit volume in removing the dipoles is

$$-P(E + D/\epsilon_0)/2$$

Now we unlock the dipoles and cash in the mechanical energy, getting back an energy per unit volume of

$$+\tfrac{1}{2}PE$$

At the end we have a field in vacuum whose energy density is $\frac{1}{2}\epsilon_0 E_\mathrm{f}^2$, or

$$+\tfrac{1}{2}D^2/\epsilon_0$$

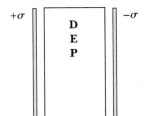

$+\sigma$ **D** $-\sigma$
       E
       P

Figure 9.22   Energy in dielectric.

The three terms add up to $\frac{1}{2}DE$, as expected:

$$-P(E + D/\epsilon_0)/2 + \tfrac{1}{2}PE + \tfrac{1}{2}D^2/\epsilon_0 = \tfrac{1}{2}DE \qquad (9.39)$$

 EXAMPLE 9-11

Two square parallel plates of width $w = 10$ cm, separated by $s = 1$ cm, are maintained at a potential difference of $V = 10$ kV. Find the field energy before and after the space between them is filled with dielectric of $\epsilon_r = 2$, Figure 9.22.

ANSWER In air,

$$\frac{W}{v} = \tfrac{1}{2}\epsilon_0 E^2$$

$$W = \tfrac{1}{2}\epsilon_0 E^2 \, v$$

$$= \tfrac{1}{2} \times 8.85 \times 10^{-12}\,\text{F/m} \times (10^4\,\text{V}/0.01\,\text{m})^2 \times (0.1 \times 0.1 \times 0.001\,\text{m}^3)$$

$$= 4.4 \times 10^{-4}\,\text{J}$$

In dielectric, at the same potential V, E is the same, and the charge and D are increased.

$$W = \tfrac{1}{2}DE\,v = \tfrac{1}{2}\epsilon_0\epsilon_r E^2 \, v$$

$$= 2 \times 4.4 \times 10^{-4} = 8.8 \times 10^{-4}\,\text{J}$$

The extra energy comes from the voltage source.

 EXAMPLE 9-12

The same plates are insulated at 10 kV in air, and then the space is filled with the dielectric. Find the field energy.

ANSWER In a dielectric, for the same charge, D is the same, and V and E are reduced.

$$W = \tfrac{1}{2}DE\,v = \tfrac{1}{2}D^2 \, v/\epsilon_0\epsilon_r$$

$$= 4.4 \times 10^{-4}/2 = 2.2 \times 10^{-4}\,\text{J}$$

The lost energy goes into the mechanism that introduces the dielectric.

The **force** on the dielectric is the negative gradient of the energy—a general rule: for example, water flows downhill. So,

$$\mathbf{F} = -\nabla W \qquad (9.40)$$

In dealing with this concept, it is necessary to include all of the energy involved. For example, in the second of the two examples immediately above, the force works in such a direction as to decrease the field energy, as would be naively expected. But in the first example, the energy goes up, and it is necessary to include the energy contribution from the voltage source in order to complete the calculation of the force. We do this in the following example.

EXAMPLE 9-13

For constant V, find the force on a square dielectric sheet of width w and thickness s that is introduced a distance x into the space between parallel plates, Figure 9.23.

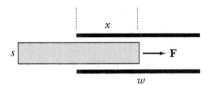

Figure 9.23 Force on dielectric.

ANSWER For constant V, it is convenient to differentiate the capacitor energy in the form (Section 9.4)

$$W = \tfrac{1}{2}CV^2$$

Then we have

$$F = -\frac{dW}{dx} = -\frac{d}{dx}\left(\tfrac{1}{2}CV^2\right) = -\tfrac{1}{2}V^2\frac{d}{dx}(C)$$

The capacitance C is the sum of the parallel capacitances with and without dielectric, so:

$$\frac{dW}{dx} = \frac{V^2}{2}\frac{d}{dx}\left(\epsilon_0\epsilon_r\frac{wx}{s} + \epsilon_0\frac{w-x}{s}w\right)$$

$$= \frac{V^2 w\epsilon_0}{2s}(\epsilon_r - 1)$$

As x increases, the energy in the system increases, which implies a repulsion. So the expression for force points in the wrong direction, so far. But we haven't put in the battery yet. For the battery, the energy input is negative (it puts out charge dQ at potential V):

$$dW = -V\,dQ = -V(V\,dC) = -V^2 d\left(\epsilon_0\epsilon_r\frac{wx}{s} + \epsilon_0\frac{w-x}{s}w\right)$$

so

$$\frac{dW}{dx} = -\frac{V^2 w\epsilon_0}{s}(\epsilon_r - 1)$$

which is twice the above energy, and in the other direction. As x increases, the battery supplies the increase in field energy, and it also supplies the energy absorbed by our hand as we slowly let the dielectric sheet into the gap. For the force, involving the total energy balance, we get

$$F = -\frac{dW}{dx} = \frac{V^2 w \epsilon_0}{2s} (\epsilon_r - 1) \qquad (9.41)$$

For the conditions of Example 9-11, this is

$$F = \frac{(10^4)^2 \, 0.10 \times 8.85 \times 10^{-12}}{2 \times 0.01} (2 - 1) = 4.4 \times 10^{-3} \, \text{N}$$

which is only a small force. The force is constant. Actually, fundamentally, the force is an edge effect, due to dipoles being drawn into an **E** field, as mentioned in Section 7.1. The energy calculation enables us to deal with the integral of this dipole force without actually having to evaluate it in detail.

The other case, involving constant charge Q, will be dealt with in Problem 9-23. In that case, one differentiates $W = Q^2/2C$, and the force is not constant.

9.7 SUMMARY

In a dielectric material, the electric dipoles may be represented using bound charges that respond to, and contribute to, the electric field. We introduce the P field and the D field to facilitate our mathematical treatment. We also introduce the practical quantity known as capacitance, C, and we find the energy density and the force.

In a dielectric material, we distinguish bound and free charge:

$$\nabla \cdot \mathbf{D} = \rho_f \quad \text{and} \quad \nabla \cdot \mathbf{P} = -\rho_b \quad (\text{and } \sigma_b = \mathbf{P} \cdot \hat{\mathbf{n}}) \qquad (9.4, 9.5, 9.6)$$

so

$$\mathbf{D} = \epsilon_0 \mathbf{E} + \mathbf{P} \quad \text{in general} \qquad (9.7)$$

When polarization changes, current flows:

$$\mathbf{J}_{\text{pol}} = \frac{\partial \mathbf{P}}{\partial t} \qquad (9.8)$$

For homogeneous, isotropic, linear dielectrics,

$$\mathbf{D} = \epsilon_r \epsilon_0 \mathbf{E} \quad \text{and} \quad \mathbf{P} = \chi_e \epsilon_0 \mathbf{E} \qquad (9.16, 9.15)$$

and

$$\epsilon = \epsilon_r \epsilon_0 \quad \text{and} \quad \epsilon_r = \chi_e + 1$$

We distinguish polar and nonpolar dielectrics, and ferroelectrics. Electrets are nonideal dielectrics having a permanent polarization.

The capacitance of a capacitor is given by

$$Q = CV \tag{9.28}$$

and its energy is

$$W = \tfrac{1}{2}CV^2 \tag{9.30}$$

With a resistor its time constant is

$$\tau = RC \tag{9.36}$$

The energy density is

$$\frac{dW}{dv} = \tfrac{1}{2}\,\mathbf{D}\cdot\mathbf{E} \tag{9.38}$$

and the force on a dielectric is

$$\mathbf{F} = -\nabla W \tag{9.40}$$

where W involves all of the energy participating in the interaction.

◢ PROBLEMS

9-1 A slice of polyethylene is placed across a field of 10^6 V/m, Figure 9.24a. At the center, ignoring edge effects, find **E**, **D**, **P**, and ρ_b inside, and σ_b on the surface.

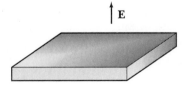

Figure 9.24a

9-2 A rod of polyethylene is placed parallel to a field of 10^6 V/m, Figure 9.24 b. At the center, ignoring end effects, find **E**, **D**, **P**, and ρ_b inside.

Figure 9.24b

9-3 Sulfur of specific gravity 2.1 is subjected to an internal field of 10^7 V/m. Find the dipole moment per atom. If all of the electrons in an atom are displaced by the same distance s relative to the nucleus, find s.

9-4 A sphere of $BaTiO_3$ of radius R has uniform polarization $\mathbf{P}$. Find its dipole moment $\mathbf{p}$.

9-5 A thin electret disk of radius a and thickness s has uniform polarization $\mathbf{P}$ parallel to its axis. Find E and V at distance $z > s$ on its axis. Check for large z ($z \gg a$).

9-6 A high-voltage parallel-plate capacitor is filled with a 2-cm slab of $BaTiO_3$ of radius 2 cm. Find the capacitance.

9-7 If the capacitor of the preceding problem is attached to a resistor of 700 kΩ, how long does it take to discharge to $1/e$ of its original charge?

9-8 If the capacitor of the preceding problem is charged to 30 kV, find the charge and the energy stored.

9-9 Find the capacitance per unit length C' of a pair of coaxial cylinders of radii r_1 and r_2, Figure 9.25, filled with material of dielectric constant ϵ_r. Evaluate for polyethylene with $r_2/r_1 = 2$.

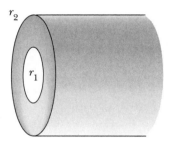

r_2

r_1

Figure 9.25

9-10 Find the capacitance of a pair of concentric spheres of radii a and b, filled with material of dielectric constant ϵ_r. Evaluate for polyethylene with $a = 1$ cm and $b = 2$ cm.

9-11 A dielectric sphere of constant ϵ_r and radius R has uniform free charge density ρ. Find the potential V at the center. Sketch E and V versus r for $\epsilon_r = 2$.

9-12 The permittivity of free space ϵ_0 appears in Coulomb's law as well as in capacitances. Show that its units, F/m, are the same as $C^2/N\ m^2$.

9-13 A spherical charge Q of radius R is imbedded in an infinite dielectric of ϵ_r. Find E, D, and P at $2R$ from the center of the charge.

9-14 In a parallel-plate capacitor of area A and separation s, the plates are separated by a material whose dielectric constant varies linearly from ϵ_1 to ϵ_2. Find the capacitance.

9-15 A mica capacitor has plates of area 3 cm^2 separated by 0.02 cm; find the capacitance.

9-16 Find the capacitance per unit length of a pair of concentric cylinders of radii 0.2 and 0.7 cm, half-filled with dielectric of relative permittivity 2.3, Figure 9.26.

9-17 Find the capacitance per unit length of a pair of parallel wires ("twinlead") of radius a and separation $s \gg a$ imbedded in material of dielectric constant ϵ_r, Figure 9.27. (See Problem 6-28.) Evaluate for $a = 1$ mm, $s = 5$ mm, $\epsilon_r = 2.3$.

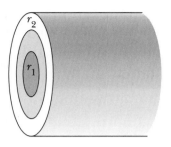

Figure 9.26

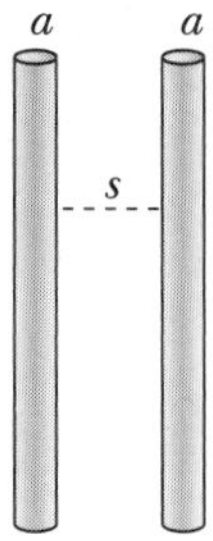

Figure 9.27

9-18 For Figure 9.5b, find the ratio E_i/E_o at $r = R$ and $\theta = 0°$. Repeat for $\theta = 90°$. Which of these two ratios is obvious by inspection?

9-19 A parallel-plate capacitor in vacuum has voltage V, separation s, area A, and mass m of each plate. The plates are insulated and released; find the time for them to come together.

*9-20 A parallel-plate capacitor in vacuum has voltage V, separation s, area A, and mass m of each plate. The plates are released while connected to V; find the time for them to come together.

9-21 A parallel-plate capacitor of area a and separation s is half-filled with dielectric of ϵ_r as shown in Figure 9.28a. Find the capacitance. Find the limits as $\epsilon_r \to 1$ and as $\epsilon_r \to \infty$.

Figure 9.28a

9-22 A parallel-plate capacitor of area a and separation s is half-filled with dielectric of ϵ_r as shown in Figure 9.28b. Find the capacitance. Find the limits as $\epsilon_r \to 1$ and as $\epsilon_r \to \infty$. Compare with the preceding problem.

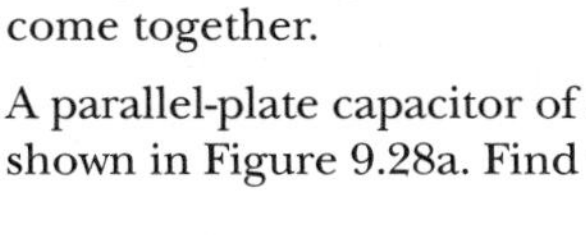

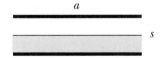

Figure 9.28b

9-23 Square parallel plates of width w and separation s are charged to potential difference V_0 and then isolated. A dielectric sheet of ϵ_r, width w, and thickness s is inserted to distance x (Fig. 9.23). Find the force on the dielectric. Evaluate for $\epsilon_r = 2$, $s = 0.01$ m, $V_0 = 10^4$ V, $w = 0.10$ m, $x = 5$ cm.

9-24 Vertical parallel plates of separation $s = 1$ cm and potential difference 10^4 V are inserted in a liquid dielectric of $\epsilon_r = 2.2$ and density $r = 1.2 \times 10^3$ kg/m³. How high does the liquid rise between the plates?

9-25 A spherical conductor of radius R is surrounded by dielectric of ϵ_r to a depth equal to R, Figure 9.29a. Find its capacitance. Find the limits as $\epsilon_r \rightarrow 1$ and as $\epsilon_r \rightarrow \infty$.

Figure 9.29a

9-26 A conducting sphere of radius R is half-submerged in dielectric of ϵ_r, as shown in Figure 9.29b. Find its capacitance.

Figure 9.29b

9-27 A dielectric disk of radius a and dielectric constant ϵ_r spins with angular velocity ω about its axis in a uniform axial field **B**, Figure 9.30. Calculate the polarization P and bound charge density ρ_b at radius r from the axis, and the bound surface charge density σ_b (where?), and the total bound charge.

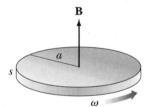

Figure 9.30

9-28 Show that the total bound charge on a dielectric object is zero. (Use the divergence theorem.)

9-29 Find the capacitance of an isolated straight wire of length l and radius a under the assumption that the linear charge density λ is constant (compare with Problem 8-26). Evaluate for $l = 10$ m and diameter 0.1 cm.

9-30 Show that an infinite conducting plane has zero capacitance per unit area.

9-31 Show that capacitors in series add as reciprocals, that is, $1/C = 1/C_1 + 1/C_2$.

9-32 Following Section 9.3, find the dipole moment of the HCl molecule.

9-33 A round parallel-plate capacitor of small separation contains dielectric material of small conductivity. The plates carry charge density σ and they gradually discharge. Find B inside the capacitor.

9-34 Evaluate the four nonzero curls and divergences discussed in connection with Figure 9.4, in terms of delta functions and polarization P.

*9-35 Find C for a thin conducting disk of radius a. (See Problem 8-25.)

C H A P T E R

10

MAGNETIC MATERIALS

agnetic materials consist, in effect, of large numbers of magnetic dipoles; but we shall see that we can simplify their description by introducing new kinds of fields that incorporate the effects of the dipoles in a natural and efficient way. In doing so we will encounter inductance and, most importantly, the transformer.

10.1. Fields in Magnetic Materials, H and M

When we place a dielectric between charged parallel plates, the voltage is reduced because the **E** field is *smaller* inside the dielectric. But when we place an iron rod in a solenoid, the vector potential is increased and so is the magnetic flux. The steady-state current in the wires is unaffected, but the **B** field inside the iron is *increased*.

Figure 10.1a is a review of the electrostatic situation illustrated in Figure 9.1, with arrows instead of field lines. The **E** field polarizes the dielectric: Each mole-

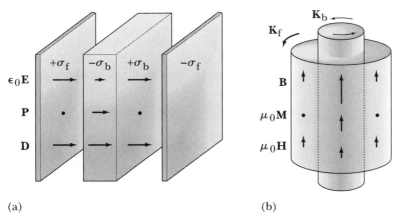

(a)   (b)

Figure 10.1   (a) Dielectric in **E** field. (b) Magnetic material in **B** field.

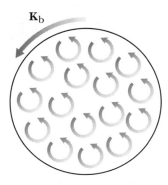

$\mathbf{K}_b$

Figure 10.2   Aligned spins.

cule becomes a tiny electric dipole. Negative charges move a little to the left, and positives to the right, resulting in the bound surface charge densities $\pm \sigma_b$. In this case the **E** field is *reduced* by a factor of 3.

Figure 10.1b shows the magnetic situation with an iron rod in a solenoid. The current in the wires, not shown, is **I**, and their number is N' turns per meter; so the surface current density due to current in the wires is $\mathbf{K}_f = N'\mathbf{I}$ amperes per meter. This is called **free current.** The directions are connected by the right-hand rule: Point your thumb in the direction of the **B** field, and your fingers curl in the direction of the current.

A spinning electron represents a tiny magnetic dipole. The **B** field tends to align these dipoles in the material. If we think of the electrons as classical current loops, Figure 10.2, the corresponding currents inside uniformly magnetized iron will tend to cancel each other except on the side surface. (We have mentioned nonclassical aspects of electrons in previous chapters. In modern quantum mechanics, electrons behave as point charges. However, classically, it is satisfactory to pretend they are little current loops.) These current loops cannot leave their home atoms. The surface current density $\mathbf{K}_b$ on the side, which is due to one side of the spinning aligned electrons, is called **bound current** (or effective or equivalent or magnetization current). It points in the same direction as the free current, and in this case it *increases* **B** by a factor of 3.

The effect of the magnetic material (iron) is entirely due to the bound current. If you were somehow able to remove the iron and leave the bound current in place, the vector potential **A** would be exactly the same. To illustrate this, we will smooth over the internal structure of the material and assign it an average magnetic dipole moment per unit volume

Dielectric analog:

$$\boxed{\frac{d\mathbf{m}}{dv} = \mathbf{M}} \qquad\qquad \textbf{(10.1)} \qquad \frac{d\mathbf{p}}{dv} = \mathbf{P}$$

called the **magnetization,** with units A/m. For one dipole, oriented in the average direction (Section 7.2),

$$\mathbf{A} = \frac{\mu_0 \mathbf{m} \times \hat{\mathbf{r}}}{4\pi\, r^2}$$

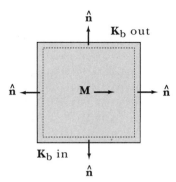

Figure 10.3   Bound surface current.

For a volume distribution of dipoles, as in Section 9.1,

$$\mathbf{A} = \frac{\mu_0}{4\pi} \int \frac{\mathbf{M} \times \hat{\mathbf{r}}}{r^2} \, dv = \frac{\mu_0}{4\pi} \int \mathbf{M} \times \nabla(1/r) \, dv \tag{10.2}$$

The minus sign is missing because we are differentiating with respect to source variables (Sections 3.7, 6.3, 7.3). Now we apply a vector identity (front cover):

$$\mathbf{A} = -\frac{\mu_0}{4\pi} \int \nabla \times \frac{\mathbf{M}}{r} \, dv + \frac{\mu_0}{4\pi} \int \frac{\nabla \times \mathbf{M}}{r} \, dv$$

Another identity transforms the first integral:

$$\mathbf{A} = \frac{\mu_0}{4\pi} \oint \frac{\mathbf{M} \times \hat{\mathbf{n}}}{r} \, da + \frac{\mu_0}{4\pi} \int \frac{\nabla \times \mathbf{M}}{r} \, dv$$

As in Section 9.1, the volume of integration is just inside of the magnetized material, Figure 10.3; and $\hat{\mathbf{n}} \, da = d\mathbf{a}$. Now for free current density $\mathbf{J}$ or surface current density $\mathbf{K}$ we have, from Section 6.3,

$$\mathbf{A} = \frac{\mu_0}{4\pi} \int \frac{\mathbf{K}}{r} \, da \quad \text{and} \quad \frac{\mu_0}{4\pi} \int \frac{\mathbf{J}}{r} \, dv \tag{10.3}$$

So the part of the bound surface current density is being played by

$$\boxed{\mathbf{K}_b = \mathbf{M} \times \hat{\mathbf{n}}} \tag{10.4} \qquad \sigma_b = \mathbf{P} \cdot \hat{\mathbf{n}}$$

and the bound volume current density is

$$\boxed{\mathbf{J}_b = \nabla \times \mathbf{M}} \tag{10.5} \qquad \rho_b = -\nabla \cdot \mathbf{P}$$

We have now encountered five **current densities:**

- $\mathbf{J}_t$: Total charge flow density, $= \mathbf{J}_f + \mathbf{J}_b + \mathbf{J}_{pol}$ (not including $\epsilon_0 \partial \mathbf{E}/\partial t$).
- $\mathbf{J}_f$: Free current density, in wires and so on.
- $\mathbf{J}_b$: Bound current density, due to spinning electrons that can't leave their molecules.

- $\mathbf{J}_{pol} = \partial\mathbf{P}/\partial t$: Changing $\mathbf{P}$ in a dielectric (Section 9.1).
- $\epsilon_0\partial\mathbf{E}/\partial t$: Called displacement current density although no charge flows (Section 3.5).

The latter two may be grouped in one term that is also referred to as displacement current density:

$$\frac{\partial\mathbf{D}}{\partial t} = \frac{\partial\mathbf{P}}{\partial t} + \epsilon_0\frac{\partial\mathbf{E}}{\partial t} \tag{10.6}$$

Now, we have just found $\nabla \times \mathbf{M} = \mathbf{J}_b$, so Ampère's law says (Section 3.6)

$$\nabla \times \mathbf{B} = \mu_0\mathbf{J}_t + \mu_0\epsilon_0\frac{\partial\mathbf{E}}{\partial t} = \mu_0\mathbf{J}_f + \mu_0\mathbf{J}_b + \mu_0\frac{\partial\mathbf{P}}{\partial t} + \mu_0\epsilon_0\frac{\partial\mathbf{E}}{\partial t}$$
$$= \mu_0\mathbf{J}_f + \mu_0\nabla \times \mathbf{M} + \mu_0\frac{\partial\mathbf{D}}{\partial t} \tag{10.7}$$

This equation could be regarded as wanting another curl term of the form

$$\boxed{\nabla \times \mathbf{H} = \mathbf{J}_f + \frac{\partial\mathbf{D}}{\partial t}} \tag{10.8}$$

so that the equation reads

$$\nabla \times \mathbf{B} = \mu_0\nabla \times \mathbf{H} + \mu_0\nabla \times \mathbf{M}$$

And this suggests that we define the **field H** as follows:

Dielectric analog:

$$\boxed{\mathbf{B} = \mu_0\mathbf{H} + \mu_0\mathbf{M}} \tag{10.9} \qquad \mathbf{D} = \epsilon_0\mathbf{E} + \mathbf{P}$$

Units of $\mathbf{H}$ are A/m (like $\mathbf{M}$). $\mathbf{H}$ is sometimes called the magnetic field, or the magnetic intensity; here we shall simply call it $\mathbf{H}$. In this connection, $\mathbf{B}$ is sometimes called the magnetic flux density, or the magnetic induction. We shall continue to call $\mathbf{B}$ the magnetic field, because that is the quantity which is physically significant. For example, in Faraday's law, it is $\mathbf{B}$, and not $\mathbf{H}$, which determines the emf. And in the Lorentz force, a fast-moving charge is found to respond to $\mathbf{B}$ and not $\mathbf{H}$. The $\mathbf{H}$ field draws an artificial distinction between free and bound currents, but they actually have the same physical effects on rapidly moving charges. (Compare Feynman, II-36–12; Griffiths, p. 260; Purcell, p. 392.)

The relationship between $\mathbf{B}$ and $\mathbf{H}$ is similar to that between $\mathbf{E}$ and $\mathbf{D}$: that is, $\mathbf{D}$ is an artificial quantity, and it is $\mathbf{E}$ that figures in the Lorentz force and in Faraday's law.

The vectors $\mathbf{B}$ and $\mathbf{M}$ need not be parallel, and indeed in a permanent *magnet* they may even point in opposite directions, as we shall see. We discuss magnets further in Section 10.9; however, for now they are going to provide an instructive exercise in curls and divergences, as well as illustrating the physical meanings of the various definitions given above. Figure 10.4 represents a uniformly magnetized

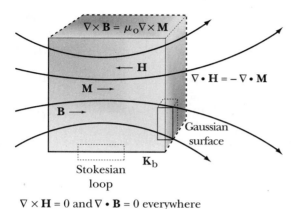

Figure 10.4   Permanent magnet, with **B** flux.

magnet in the shape of a cube (compare with Fig. 9.4). We can have curls and divergences of **B**, **H**, and **M**, a total of $2 \times 3 = $ six quantities in all.

QUESTION: Where are the curls and divergences of **B**, **H**, and **M** nonzero in Figure 10.4?

ANSWER: The six quantities are numbered below as they are presented. First, there is no free current here; the permanent magnet is magnetized all by itself. So,

(1) $\nabla \times \mathbf{H} = \mathbf{J}_f = 0$

everywhere. Also, there being no magnetic monopoles,

(2) $\nabla \cdot \mathbf{B} = 0$

everywhere; lines of **B** just keep on going.

On the left and right surfaces, the normal component of **M** suddenly changes. A net flux of **M** enters the gaussian surface shown, indicating the
(3) presence of a nonzero divergence $\nabla \cdot \mathbf{M}$, on the left and right surfaces. We know that

$$\mathbf{B} = \mu_0 \mathbf{M} + \mu_0 \mathbf{H} \quad \text{and} \quad \nabla \cdot \mathbf{B} = 0, \quad \text{so}$$

(4) $\nabla \cdot \mathbf{H} = -\nabla \cdot \mathbf{M}$ on the sides

On the top and bottom surfaces, **M** changes abruptly. In the Stokesian loop shown, there is a contribution to the integral of **M** inside of the magnet but not outside, signaling the presence of a nonzero curl:

(5) $\nabla \times \mathbf{M} = \mathbf{J}_b$

We still have $\mathbf{B} = \mu_0 \mathbf{M} + \mu_0 \mathbf{H}$, and $\nabla \times \mathbf{H} = 0$, so

(6) $\nabla \times \mathbf{B} = \mu_0 \nabla \times \mathbf{M}$

on the top and bottom surfaces. The curl of **B** is due to the bound surface current $\mathbf{K}_b$.

To summarize: In the magnet shown, we have magnetic field **B** everywhere, provided only by $\nabla \times \mathbf{B}$, since $\nabla \cdot \mathbf{B} = 0$. We have the field **H** everywhere, provided only by $\nabla \cdot \mathbf{H}$, since $\nabla \times \mathbf{H} = 0$ here. We have **M** only inside the magnet, because of the cancellation of the effects of $\nabla \cdot \mathbf{M}$ and $\nabla \times \mathbf{M}$ outside.

Note that all four of the nonzero curls and divergences here are delta functions (Sections 2.4, 3.6, 8.2, and 9.1), that is, they are infinite for infinitesimal distances. This is due to the fact that the magnetization **M** changes suddenly at the surfaces: M is a step-function θ (Problem 2-37). (The above conclusions are valid for the cube-shaped magnet of Figure 10.4 but not necessarily in general.)

10.2. Ideal Magnetic Materials

The preceding is rather general. In what follows, we are going to concentrate on ideal magnetic materials that are **homogeneous, isotropic,** and **linear** (not to mention stationary).

For the ideal case, Figure 10.1b and Figure 10.5 show the **M** and **H** corresponding to **B**. (We use μ_0 so that all of the arrows represent commensurable quantities in Fig. 10.1b.) On the sides, **B** inside the iron is increased over **B** outside by the addition of $\mu_0\mathbf{M}$. The bound surface current $\mathbf{K}_b$ produces a curl of **M** at the sides, and that is the source of **M** and of the additional **B**. The **B** field doesn't change at the top, of course (no magnetic monopoles).

By contrast, on the sides **H** is the same size inside as outside. The bound surface current does not affect **H** directly, since $\nabla \times \mathbf{H}$ excludes bound current. This doesn't mean that **H** itself depends only on free current. Taking the divergence of **B**, Equation 10.9,

$$\nabla \cdot \mathbf{B} = \nabla \cdot \mu_0\mathbf{H} + \nabla \cdot \mu_0\mathbf{M} = 0$$

always (no magnetic monopoles). So,

$$\nabla \cdot \mathbf{H} = -\nabla \cdot \mathbf{M} \tag{10.10}$$

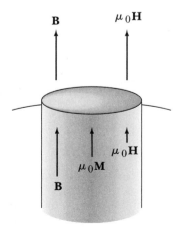

Figure 10.5 End of iron rod.

At the top of the iron rod, the lines of **M** suddenly end, implying a large negative divergence, Figure 10.5. This means that there is a large positive divergence of **H** at the top; lines of **H** spring up as if there were **H** monopoles there.

Figures 10.1 and 10.5 illustrate once again the significance of the curl and the divergence. Intuitively, the divergence tells you how the field changes as you go in the direction of the field, and the curl tells you how it changes as you go across it. Because there is a divergence of **M** on the end of the iron rod, and **M** is normal to the end, **M** changes there, and the same is true for **H**. And because there is a curl of **B** on the sides of the iron rod, and **B** is tangential, **B** changes there. Because of their complementary nature, the divergence and curl together tell you everything there is to know about how a field changes spatially.

For the ideal case, the fields are proportional to each other. In particular,

Dielectric analogs:

$$\boxed{\mathbf{M} = \chi_m \mathbf{H}} \qquad\qquad (10.11) \qquad \mathbf{P} = \chi_e \epsilon_0 \mathbf{E}$$

where the dimensionless constant of proportionality χ_m is called the **magnetic susceptibility.** (We could instead have chosen $\mathbf{M} = \chi_m \mathbf{B}$, with $\chi_m < 1$ always.) The **B**, **H**, and **M** fields are all parallel, and the general equation $\mathbf{B} = \mu_0 \mathbf{H} + \mu_0 \mathbf{M}$ becomes

$$B = \mu_0 H + \mu_0 \chi_m H$$

$$= (1 + \chi_m)\mu_0 H = \mu_0 \mu_r H = \boxed{\mu H = B} \qquad (10.12) \qquad D = \epsilon E$$

where

$$\boxed{\mu_r = (1 + \chi_m)}$$

is called the **relative permeability,** and $\mu = \mu_0 \mu_r$ is simply the permeability, and μ_0 is the permeability of free space. Since $\nabla \times \mathbf{H} = \mathbf{J}_f + \partial \mathbf{D}/\partial t$ (Section 10.1), this means

$$\nabla \times (\mathbf{B}/\mu) = \mathbf{J}_f + \frac{\partial \mathbf{D}}{\partial t}$$

so

$$\nabla \times \mathbf{B} = \mu \mathbf{J}_f + \epsilon\mu \frac{\partial \mathbf{E}}{\partial t} \qquad\qquad (10.13)$$

if ϵ and μ are constant (compare with Section 9.2).

 **EXAMPLE 10-1**

If the relative permeability of iron is 100, and if its average atomic dipole moment is that of two electrons, $2\mu_B$, find H and B.

ANSWER $M = Nm$, where N is the number of atoms per unit volume and m is the average dipole moment:

$$M = \frac{6.02 \times 10^{23} \text{ atoms}}{\text{mole}} \times \frac{1 \text{ mole}}{56 \text{ grams}} \times \frac{7.8 \times 10^6 \text{ grams}}{1 \text{ m}^3}$$
$$\times \frac{2 \times 9.3 \times 10^{-24} \text{ A} \cdot \text{m}^2}{\text{atom}}$$
$$= 1.56 \times 10^6 \text{ A/m}$$

So

$$H = M/\chi_{\mathrm{m}}$$
$$= 1.56 \times 10^6 \text{ A/m}/99 = 1.6 \times 10^4 \text{ A/m}$$

and

$$B = \mu H = \mu_0 \mu_{\mathrm{r}} H$$
$$= 4\pi \times 10^{-7} \text{ T} \cdot \text{m/A} \times 100 \times 1.56 \times 10^6 \text{ A/m}/99 = 2.0 \text{ T}$$

This value hardly depends on μ_{r} as long as it is fairly large; H is practically negligible here, and $B \approx \mu_0 M$. A field of 2 tesla (20,000 gauss) is attainable in an electromagnet, but it represents a kind of upper limit in that the iron is **saturated:** The two electrons per atom that provide the magnetization are all lined up, so the iron can provide no assistance in reaching higher fields.

Magnetic materials generally fall into one of three categories: ferromagnetic, paramagnetic, and diamagnetic.

In **ferromagnetic** materials such as iron (and cobalt and nickel), large numbers of electrons line up spontaneously in "domains," for quantum-mechanical reasons. They remain aligned up to the "Curie temperature" (around 770°C for iron). Such materials typically have large relative permeabilities, 100 to 1000 or more; and they are more or less nonlinear, so non–ideal. When you apply a **B** field, domains that are already pointing in the right direction grow at the expense of the others. Ferromagnetism is analogous to ferroelectricity, Section 9.3. There also exist antiferromagnetic materials in which atomic dipole moments spontaneously align alternately up and down and which have no net magnetic moment. Ferrites (ferrimagnets) are similar but have a net magnetization; an example is Fe_3O_4, magnetite (lodestone). As these materials have low conductivity, they can be useful in high-frequency applications because eddy-current losses are small.

In **paramagnetic** materials, each molecule has a permanent magnetic dipole moment (due to unpaired electrons) but the molecules do not align spontaneously; their susceptibilities are small at room temperature and are inversely proportional to absolute temperature because thermal motion knocks them out of line. They are analogous to polar dielectrics in the electrostatic case, and the Langevin analysis applies as in Section 9.3. In the low-field approximation,

Dielectric analog:

$$\frac{M}{N} = m\frac{mB}{3kT} \qquad \textbf{(10.14)} \qquad \frac{P}{N} = p\frac{pE}{3kT}$$

instead of just $M/N = m$; k is the Boltzmann constant, and T is absolute temperature, so kT is thermal energy, and mB is the energy of the magnetic dipole when aligned with the effective field B (Section 7.2). At low T, paramagnetic effects are increased; for example, oxygen is significantly paramagnetic, and liquid oxygen will cling to the poles of a magnet.

Diamagnetic materials are analogous to nonpolar dielectrics: they have no permanent dipole moment of their own, and their susceptibility depends little on temperature. However, unlike the electrostatic case, diamagnetic materials are *repelled* by a region of magnetic field: Their susceptibilities are negative. By a phenomenon analogous to Lenz's law, the induced magnetic dipoles are in such a direction as to oppose the field that produces them. The effect is typically masked in paramagnetic materials by the larger susceptibility which is due to the permanent dipoles.

Magnetic Susceptibilities χ_m	
Air	40×10^{-8}
Argon	-1.0×10^{-8}
Helium	-0.10×10^{-8}
Hydrogen	-0.22×10^{-8}
Nitrogen	-0.67×10^{-8}
Oxygen gas	200×10^{-8}
Aluminum	2.1×10^{-5}
Copper	-1.0×10^{-5}
Graphite	-14×10^{-5}
Lead	-1.6×10^{-5}
Oxygen liq	400×10^{-5}
Polyethylene	-0.2×10^{-5}
Salt NaCl	-1.3×10^{-5}
Silver	-2.4×10^{-5}
Sodium	0.85×10^{-5}
Sulfur	-1.2×10^{-5}
Water	-0.90×10^{-5}
Ferrite	~ 100
Iron	~ 1000

As is suggested by the table, the susceptibilities of ferromagnetic materials (iron, ferrite) are large but not very well-defined. Such materials are required for most practical applications of magnetism, because the susceptibilities of diamagnetic and paramagnetic materials are very small. This situation contrasts with the electrostatic one in that ferroelectric materials are available but most capacitors are made of other materials.

Unlike dielectric constants, magnetic susceptibilities may be expressed in various different units. For example: in the 63rd edition of *Handbook of Chemistry and Physics* we find that the magnetic susceptibility of "one gram formula weight" of aluminum is $+16.5 \times 10^{-6}$ "cgs units." To convert to SI we must always multiply by 4π (see Appendix D). Here we need also to multiply by the density, 2.70, and divide by the atomic weight, 27.0; the result is dimensionless:

$$\chi_m = 4\pi \, 16.5 \times 10^{-6} \frac{2.70}{27.0} = 2.1 \times 10^{-5}$$

In Figure 10.1b, we find that the **B** field increases by a factor of 3 when we enter the iron, implying that $\mu_r = 3$ in this case. That, however, is an unusual value; for ferromagnetic materials a relative permeability of several hundred would be more typical. Unfortunately, such materials are **nonlinear,** so μ_r is not constant, but an effective value of μ_r is useful.

▼ EXAMPLE 10-2

For Figure 10.1b, assuming the electromagnet is very long, if $\mu_r = 200$, $I = 3$ A, and $N' = 700$ turns per meter, find B, H, and M inside and outside of the iron, and find the bound current.

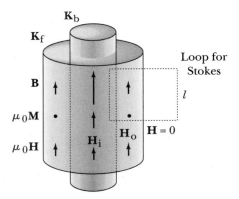

Figure 10.6   Solenoid with iron.

ANSWER   We find H just as we did B for an infinite solenoid, using Stokes' theorem, Figure 10.6.

$$\nabla \times \mathbf{H} = \mathbf{J}_f \Longrightarrow \oint \mathbf{H} \cdot d\mathbf{l} = N' l I$$

$$Hl = N' l I$$

$$H_o = H_i = N' I \quad (= K_f I)$$

$$= 700 \times 3 = 2100 \text{ A/m}$$

(H_o is outside the iron but inside the solenoid.)

$$B_{\rm o} = \mu_0 H$$
$$= 4\pi \times 10^{-7} \times 2100 = 0.0026 \text{ T}$$
$$B_{\rm i} = \mu_{\rm r} B_{\rm o}$$
$$= 200 \times 0.0026 = 0.53 \text{ T}$$
$$M_{\rm o} = 0$$
$$M_{\rm i} = (\mu_{\rm r} - 1) H$$
$$= 199 \times 2100 = 4.2 \times 10^5 \text{ A/m}$$
$$K_{\rm b} = M_{\rm i} = 4.2 \times 10^5 \text{ A/m} \qquad \text{and} \qquad J_{\rm b} = 0$$

All of these are reasonable numbers for an electromagnet. The bound surface current is surprisingly large: In a 10-cm segment of such a magnet it would amount to 40,000 amperes. The iron contributes 200 times as much current as the current-carrying wire. This is why iron has been so important to development of a technology based on electricity. Iron is also extraordinarily strong, malleable, and cheap. Its commonness is due largely to the fact that ^{56}Fe is one of the most stable nuclei and thus is the endpoint for much stellar evolution. Its tragic flaw is: It rusts.

We have suggested that the bound currents are usually found on the surface of an ideal magnetized material, not in the volume. For ideal materials:

$$\mathbf{J_b} = \nabla \times \mathbf{M} = \nabla \times (\chi_{\rm m} \mathbf{H})$$
$$= \chi_{\rm m} \nabla \times \mathbf{H} - \mathbf{H} \times \nabla \chi_{\rm m} \qquad \textbf{(10.15)}$$
$$= \chi_{\rm m} (\mathbf{J_f} + \partial \mathbf{D}/\partial t) - \mathbf{H} \times \nabla \chi_{\rm m}$$

If there is no free current and we are quasi-static, then the first term is zero, and if the permeability doesn't change then the second term is zero. So we find bound current at the surface because $\mu_{\rm r}$ changes suddenly there, but we find it in the volume only if there is free current flowing there. Analogously, we found in Section 9.2 that bound charge tends to be at the surface of dielectrics.

For magnetic materials, in analogy with materials consisting of electric dipoles, Section 9.6, if the magnetization is proportional to the magnetic field, then the energy density is

Dielectric

$$\boxed{\frac{dW}{dv} = \tfrac{1}{2} \mathbf{B} \cdot \mathbf{H}} \qquad \textbf{(10.16)} \qquad \frac{dW}{dv} = \tfrac{1}{2} \mathbf{D} \cdot \mathbf{E}$$

10.3. Inside Diamagnets

Ferromagnets are similar to ferroelectrics, and paramagnets are similar to polar dielectrics; we have mentioned them in the preceding section, and we have nothing to add here. On the other hand, diamagnetism is something that can be treated classically in an interesting fashion, and that is significantly different from the dielectric case.

Now, the magnetism of materials is intimately tied up in quantum mechanics, so our results will be of only limited accuracy and validity. However, as in the case of the almost-classical Bohr atom, there are some features that correspond fairly well to experiment.

When we studied dielectrics, we had to recognize that the field inside the dielectric was significantly different from that outside, Section 9.3. But here we are going to find that the effect of the diamagnetic material is quite small (see the table, preceding section), so we will be able to assume that the **B** field is approximately the same everywhere.

In diamagnetism, the material repels the magnet, due to the operation of Faraday's law plus Lenz's law. We imagine that each electron, of mass m_e and charge $-e$, circulates about the positive nucleus at speed v in a circular orbit of radius r, Figure 10.7. Then when **B** increases, if r is constant,

$$\text{emf} = -\pi r^2 \frac{dB}{dt} = 2\pi r E$$

$$\pi r^2 B = -2\pi r \, Et = 2\pi r \, \Delta p / e \qquad (10.17)$$

$$\Delta p = m_e \, \Delta v = m_e r \, \Delta \omega = \frac{erB}{2}$$

So

$$\Delta \omega = \frac{eB}{2m_e} \qquad (10.18)$$

(This is sometimes called the Larmor frequency.) But is it in fact true that the radius is unchanged? We have for the centripetal force F, with v $= \omega r$,

$$F = \frac{m_e v^2}{r} = m_e \omega^2 r$$

Then we set the Lorentz force evB equal to the change in centripetal force, at constant r:

$$evB = \Delta F = 2m_e \omega r \, \Delta \omega = 2m_e \omega r \, \frac{eB}{2m_e} = e\omega rB = evB$$

It balances, so the radius doesn't change.

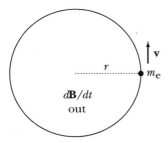

Figure 10.7   Electron orbit in increasing **B**.

The magnetic moment m in general is current I times area πr^2:

$$m = I \, \pi r^2 = -\pi r^2 \frac{e}{t} = -\pi r^2 \frac{e}{2\pi r/\mathrm{v}} = -\frac{er^2 \omega}{2}$$

Then in particular if this ω is the $\Delta\omega$ found above, then the change in magnetic moment is

$$m = -\frac{er^2}{2} \frac{eB}{2m_e} = -\frac{e^2 r^2 B}{4m_e} \qquad (10.19)$$

This is the excess magnetic moment of one electron due to the imposed field **B**. By Lenz's law it opposes the imposed field, and so we have diamagnetism.

And now we will look at the total m for a material that has Z electrons per atom. We'll be employing plausible approximations, because a complete quantum-mechanical treatment would be beyond us here, and anyway it would constitute overkill: We're mainly interested in seeing whether a simple model predicts reasonable values.

We begin by assuming that $1/3$ of the electron orbits point in each of the x, y, and z directions, so only $Z/3$ electron orbits are perpendicular to the **B** field. Then we imagine that they are in a flat uniform disk, Figure 10.8, with constant surface density σ; in actuality, the density is somewhat greater near the nucleus, and electrons closer to the nucleus make a smaller contribution, so our prediction will be too large. For a disk of radius a,

$$\sigma = -\frac{Ze}{3\pi a^2} \qquad (10.20)$$

$$m = -\frac{e^2 r^2 B}{4m_e} \implies \int \frac{e}{m_e} \frac{r^2 B}{4} \sigma \, 2\pi r \, dr \qquad (e/m_e \text{ is constant})$$

$$= -\frac{eB \, Ze}{m_e \, 6a^2} \int_0^a r^3 \, dr = -\frac{e^2 BZ}{24 \, m_e} a^2 \qquad (10.21)$$

Then (from Section 10.2)

$$\chi_m = \frac{M}{H} = \frac{Nm}{B/\mu_0}$$

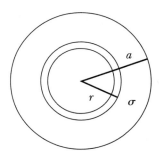

Figure 10.8   Integrating over an atom.

so we predict

$$\chi_m = -\frac{\mu_0 N Z e^2 a^2}{24 \, m_e} = -1.5 \times 10^{-15} \, N Z a^2 \tag{10.22}$$

(Quantum mechanically we get $\chi_m = -\dfrac{\mu_0 N Z e^2}{6 \, m_e} <r^2>$, where $<r^2>$ is average r^2, which is less than a^2; so we're not far off.)

Let's try it out. For argon gas at STP, $N = 2.69 \times 10^{25}$ atoms/m^3 (Loschmidt's number), $Z = 18$, and $\chi_m = -1.0 \times 10^{-8}$; then

$$a^2 = \frac{-\chi_m}{1.5 \times 10^{-15} \, N Z} = \frac{1.0 \times 10^{-8}}{1.5 \times 10^{-15} \times 2.69 \times 10^{25} \times 18}$$

so we get

$$a = 1.2 \times 10^{-10} \text{ m} \qquad \text{a reasonable value}$$

And now we try sulfur, $Z = 16$. First we need:

$$N = 2.0 \, \frac{\text{g}}{\text{cm}^3} \times \frac{1}{32} \, \frac{\text{mole}}{\text{g}} \times 6.02 \times 10^{23} \, \frac{\text{atoms}}{\text{mole}} \times 10^6 \, \frac{\text{cm}^3}{\text{m}^3}$$

$$= 3.8 \times 10^{28} \text{ atoms/m}^3$$

and (Section 9.3)

$$a \approx \frac{1}{2 \sqrt[3]{N}} = 1.5 \times 10^{-10} \text{ m}$$

So

$$\chi_m = -1.5 \times 10^{-15} \, N Z a^2$$

$$= -1.5 \times 10^{-15} \times 3.8 \times 10^{28} \times 16 \times (1.5 \times 10^{-10})^2 = -2.0 \times 10^{-5}$$

The experimental value is -1.2×10^{-5}; so we're a little high, but at least our model seems to be in qualitative agreement with experiment.

10.4. Inductance L

Like capacitors, inductors store energy, but they do so when current passes through them. If we pass a current through an inductor of a kilogram or two, and we suddenly break electrical contact, a spark may be seen, in which the stored energy is dissipated.

Typically an inductor (or "choke") consists of a coil of wire about a magnetic core, Figure 10.9, but some have air cores, and even a straight bare wire has inductance. The magnetic flux Φ is proportional to the current I; for N turns:

Analog:

$$\boxed{N \Phi = L I} \tag{10.23}$$ $Q = CV$

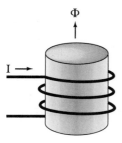

Figure 10.9   Inductor.

The proportionality constant L is the **inductance,** or self-inductance, in Wb/A, or V · s/A, or henries (H, named after Joseph Henry, American physicist). It contains the permeability and geometric factors, and it is always positive. Differentiating with respect to time, we obtain

$$N\frac{d\Phi}{dt} = L\frac{dI}{dt}$$

so

$$-\,\mathrm{emf} = L\frac{dI}{dt}$$

in view of Faraday's law of induction, Section 4.1. If we change the current, thus changing the flux, a **back emf** is produced *by* the inductor which, because of Lenz's law, tends to oppose the change we impose, so there is a negative sign. The expression is typically written

$$\boxed{V = L\frac{dI}{dt}} \tag{10.24}$$

where now V is the voltage imposed *on* the inductor, so the sign is positive. Analogously, when we push on a cart, the cart pushes back with equal force, and yet the cart may accelerate. In this and other ways, inductance is analogous to mass (Section 13.1).

◢ EXAMPLE 10-3

We apply a voltage of 15 V to an inductance of 8 H and 17 Ω; what is the initial rate of increase of current?

ANSWER

$$dI/dt = V/L$$

$$= \frac{15\ \mathrm{V}}{8\ \mathrm{H}} = 1.9\,\frac{\mathrm{V}}{\mathrm{V\cdot s/A}} = 1.9\ \mathrm{A/s}$$

In practice, the internal resistance of the inductor will usually limit the current in a second or less.

Power is VI, volts times amps, so the energy stored is

$$dW = VI\, dt$$

	Dielectric analog:

$$\int LI \frac{dI}{dt}\, dt = \boxed{\tfrac{1}{2} LI^2 = W} \qquad \textbf{(10.25)} \qquad W = \tfrac{1}{2} CV^2$$

Inductors typically store less energy than capacitors, by a factor of ten per unit mass: Capacitors store around a joule per kilogram, and inductors store only 0.1 joule per kilogram. In a charged capacitor the dielectric is stressed to near the breaking point, whereas in iron the energy is stored merely by flipping a couple of electrons per atom.

We shall now calculate some inductances in more or less practical cases. In calculating inductance, we typically begin by introducing a current I. We calculate the B field and the flux. Then when we substitute Φ in the equation $L = N\Phi/I$, the I's cancel out. The units of μ_0 are henries per meter, so a calculation of an inductance will usually result in a μ_0 in the numerator.

 EXAMPLE 10-4

An iron ring of $\mu_r = 200$ and of major and minor radii 7 cm and 0.8 cm respectively is uniformly wrapped with 300 turns of wire carrying current 4 A. Find the magnetic field, the inductance, and the energy stored. (Assume constant B inside the iron.)

ANSWER   See Figure 10.10. As usual, we find B using H.

$$\nabla \times \mathbf{H} = \mathbf{J}_f$$

$$2\pi R H = NI$$

$$B = \mu H = \mu_0 \mu_r NI / 2\pi R$$

$$= 4\pi \times 10^{-7} \times 200 \times 300 \times 4/2\pi\, 0.07 = 0.69\ \text{T}$$

$$\Phi = \pi r^2 B$$

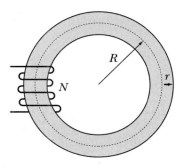

Figure 10.10   Iron ring.

Then,

$$L = N\Phi/I$$

$$= N\pi r^2\, \mu_0\mu_\mathrm{r}NI/2\pi RI = N^2 r^2\, \mu_0\mu_\mathrm{r}/2R$$

$$= 300^2 \times (0.008\ \mathrm{m})^2 \times 4\pi \times 10^{-7}\,\mathrm{H/m} \times 200/2 \times 0.07\ \mathrm{m}$$

$$= 10\ \mathrm{mH}$$

The inductance L usually involves N^2.

$$W = \tfrac{1}{2}\, LI^2$$

$$= \tfrac{1}{2} \times 0.01\ \mathrm{H} \times (4\ \mathrm{A})^2 = 8 \times 10^{-2}\,\mathrm{J}$$

which is about all you would expect for such a small ring.

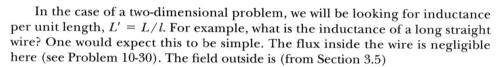

In the case of a two-dimensional problem, we will be looking for inductance per unit length, $L' = L/l$. For example, what is the inductance of a long straight wire? One would expect this to be simple. The flux inside the wire is negligible here (see Problem 10-30). The field outside is (from Section 3.5)

$$B = \frac{\mu_0 I}{2\pi r}$$

We integrate the flux for length l over the area, Figure 10.11. We imagine, if you like, that the current return path is at infinity. Thus,

$$\Phi = \int B\, l\, dr = \int_a^\infty \frac{\mu_0 I}{2\pi r}\, l\, dr \longrightarrow \infty \qquad (10.26)$$

The flux goes logarithmically to infinity, and so the inductance per unit length is infinite:

$$L' \longrightarrow \infty \qquad (10.27)$$

It would be impossible to change the current I in such a wire. The situation here is similar to that involving the capacitance of a long cylinder, Section 9.4. There, the

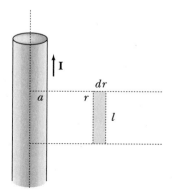

Figure 10.11 Infinite wire.

scalar potential V was infinite; here, the vector potential $\mathbf{A}$ is infinite, and this entails an infinite flux.

For some other infinite geometries the result is finite and the calculation is easy, for example the coaxial cable, and "twinlead" (two parallel wires). We shall save these for problems.

Now a finite straight wire has a finite inductance, and we could calculate it as accurately as we like, in principle: First we find the field everywhere using the Biot-Savart law (Section 3.2), and then we integrate the result numerically to find the flux. However, in this case we typically only need an approximate value for L, especially since a wire is never truly isolated, and the "return" current path may be via capacitance involving neighboring conductors. The divergence (infinite value) obtained above is due to the fact that the field only falls off as $1/r$, all the way to infinity. But for a wire of length l, once we get farther away than l, clearly B must fall off more rapidly, more like $1/r^2$ (this follows from the Biot-Savart law), causing the integral to converge. So we can obtain an approximation by cutting off the integration at l (compare with Section 9.4 for capacitance):

$$\Phi \approx \int_a^l \frac{\mu_0 I}{2\pi r} \, l \, dr = \frac{\mu_0 I}{2\pi} \, l \ln \frac{l}{a}$$

and so

$$L = \frac{\Phi}{I} \approx \frac{\mu_0 l}{2\pi} \ln \frac{l}{a} \tag{10.28}$$

◤ EXAMPLE 10-5

Estimate the inductance of a straight wire of length $l = 10$ cm and radius $a = 0.1$ cm, Figure 10.12.

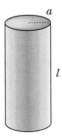

Figure 10.12   Short wire.

ANSWER

$$L \approx \frac{\mu_0 l}{2\pi} \ln \frac{l}{a}$$

$$= \frac{\mu_0}{2\pi} \, 0.1 \ln (0.1/0.001) = 9.2 \times 10^{-8} \text{ H}$$

Even such a small inductance is often significant at television frequencies, 100 MHz or so. It can behave somewhat like a resistor in impeding the progress of a high-frequency signal, and it can change the frequency of a resonant circuit (Chapter 13). In circuit construction it may be desirable to reduce the lengths of isolated wires, and to increase their diameters (or using conducting ribbon), in order to reduce inductance. If capacitance is also a problem (Section 9.4), then the best solution to both problems may simply be to reduce the length of the wire as much as possible. Mutual inductance (cross talk; see Section 10.6) between wires can also cause trouble; it is reduced by "dressing" leads away from each other.

10.5.   Time Constant *L/R*

Inductors oppose changes in current. If we have a current flowing in an inductor connected to a resistor, Figure 10.13a, the current gradually lessens, Figure 10.13b, as energy stored in the inductor gets dissipated in the resistor. To find the time constant, we begin with V due to the inductor,

$$V = -L\frac{dI}{dt} \qquad \text{and} \qquad V = IR$$

so

$$-L\frac{dI}{dt} = IR \qquad \qquad (10.29)$$

The solution of the differential equation is

$$I = I_0 e^{-t/\tau}$$

The time constant

$$\boxed{\tau = L/R} \qquad \qquad (10.30)$$

is the time for the current I to drop to $1/e = 0.368$ of its original value, I_0. (For capacitors, the analogous time constant is $\tau = RC$, Section 9.5.)

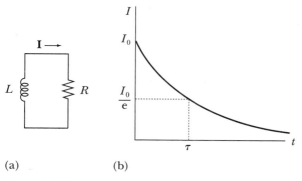

(a)   (b)

Figure 10.13   **(a)** Inductor current. **(b)** Time constant.

 **EXAMPLE 10-6**

What is the time constant for a 1-henry inductor, short-circuited, with 30 ohms internal resistance?

ANSWER

$$\tau = L/R$$

$$= \frac{1\ \text{H}}{30\ \Omega} = \frac{1\ \text{V}\cdot\text{s/A}}{30\ \text{V/A}} = 0.033\ \text{second}$$

Such values are reasonable for ordinary components. An ordinary voltmeter won't show it, but you can verify it with an oscilloscope and an external resistor.

We return to time constants under the heading of Transients in Section 13.1.

10.6.   Mutual Inductance M

Mutual inductors are pairs of circuits which may be electrically separate but which are linked by magnetic flux. In practical applications they are commonly referred to as "transformers," and we shall direct our attention to ideal transformers in the next section. Here we examine mutual inductance in general.

The magnetic flux Φ linking two circuits, Figure 10.14, is proportional to the current I_1 in the primary circuit; for N_2 turns in the secondary,

$$\boxed{N_2\Phi = MI_1}$$ (10.31)

Analogs:

$$Q = CV$$

$$N\Phi = LI$$

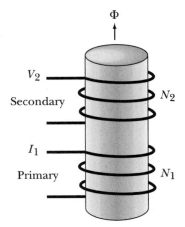

Figure 10.14   Transformer.

The proportionality constant M is the **mutual inductance** in henries (H). It contains the permeability and geometric factors, including N_1, the number of turns in the primary. Differentiating with respect to time, we obtain

$$N_2 \frac{d\Phi}{dt} = M \frac{dI_1}{dt}$$

so

$$-\,\mathrm{emf}_2 = M \frac{dI_1}{dt} \tag{10.32}$$

by Faraday's law. We may write this as $V_2 = M\, dI_1/dt$. (M may be positive or negative.)

In calculating a mutual inductance, we typically begin by introducing a current I_1 in one circuit. We calculate the B field and the flux Φ linking the two circuits. Then when we substitute Φ in the equation $M = N_2\Phi/I_1$, the I_1's cancel out. There should remain a μ_0 in the numerator in order to provide the units, henries.

▼ EXAMPLE 10-7

Calculate M for a pair of coaxial solenoids, Figure 10.15.

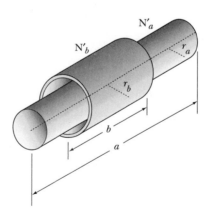

Figure 10.15   Mutual inductor.

ANSWER We assume that the length a is enough larger than the radius r_a that the B field is approximately that of an infinitely long solenoid. Solenoid b has N_b turns and N'_b turns per meter. We introduce current I_a in solenoid a. Then,

$$B_a = \mu_0 N'_a I_a$$

All of the flux in solenoid a passes through solenoid b, so

$$\Phi = B_a \pi r_a^2 = \mu_0 N'_a I_a \pi r_a^2$$

Then

$$M = N_b \Phi / I_a$$

$$M = \frac{N_b \mu_0 N'_a I_a \, \pi \, r_a^2}{I_a} = \frac{\mu_0 N_a N_b \, \pi \, r_a^2}{a}$$

Mutual inductance usually involves the product $N_a N_b$.

Let us put in reasonable numbers: $N_a = N_b = 1000$ turns, $r_a = 1$ cm, $a = 10$ cm. Then we find

$$M = \frac{\mu_0 N_a N_b \, \pi \, r_a^2}{a}$$

$$= \frac{4\pi \times 10^{-7} \, \text{H/m} \, 1000 \times 1000 \, \pi \, (0.01 \, \text{m})^2}{0.1 \, \text{m}} = 0.0039 \, \text{H}$$

So if we increase the current in the primary at a rate of 250 A/s, we get only 1 V in the secondary. But it is not a very well-designed transformer.

Now we will redo the calculation starting with I_b in solenoid b:

$$B_b = \mu_0 N'_b I_b$$

$$\Phi = \mu_0 N'_b I_b \, \pi \, r_a^2$$

since not all of the flux in b passes through a.

Solenoid a is longer than b, and only a fraction b/a of its turns are in the flux from b:

$$M = \frac{N_a \left(\dfrac{b}{a} \right) \mu_0 N'_b I_b \, \pi \, r_a^2}{I_b} = \frac{\mu_0 N_a N_b \, \pi \, r_a^2}{a}$$

as before. We get the same mutual inductance either way.

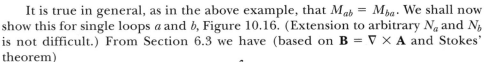

It is true in general, as in the above example, that $M_{ab} = M_{ba}$. We shall now show this for single loops a and b, Figure 10.16. (Extension to arbitrary N_a and N_b is not difficult.) From Section 6.3 we have (based on $\mathbf{B} = \nabla \times \mathbf{A}$ and Stokes' theorem)

$$\Phi_b = \oint \mathbf{A}_a \cdot dl_b \qquad (10.33)$$

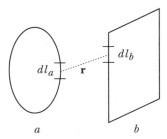

Figure 10.16   Arbitrary loops.

We insert the **A** integral from Section 6.3, Equation 6.22:

$$\Phi_b = \oint \left(\frac{\mu_0 I_a}{4\pi} \oint \frac{dl_a}{r} \right) \cdot dl_b = \frac{\mu_0 I_a}{4\pi} \oint \oint \frac{dl_a \cdot dl_b}{r}$$

We define $\Phi_b = M_{ba} I_a$, so

$$M_{ba} = \frac{\mu_0}{4\pi} \oint \oint \frac{dl_a \cdot dl_b}{r}$$

which is **Neumann's formula.** It is symmetric in a and b, which obviously implies that

$$\boxed{M_{ab} = M_{ba}} \tag{10.34}$$

as stated previously.

The self and mutual inductances are related via the "coupling coefficient" $\sqrt{k_a k_b}$, where for example k_a is the fraction of the flux from loop a that penetrates loop b. Using $M_{ab} = M_{ba}$ it is not hard to show (see Problem 10-29) that

$$M = \pm \sqrt{k_a k_b} \sqrt{L_a L_b} \tag{10.35}$$

For an ideal transformer, $\sqrt{k_a k_b} = 1$.

For long conductors we might be interested in the mutual inductance per unit length M', as in the following examples.

Figure 10.17a shows a pair of parallel wires ("twinlead"), labeled "1" and "2." Their separation is s and their radii are a. In Figure 10.17b they are connected in a single circuit, for calculation of self-inductance. We imagine that they are very

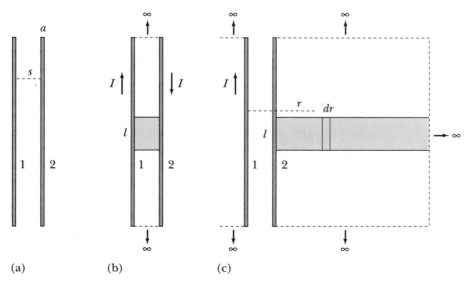

Figure 10.17   **(a)** Parallel wires. **(b)** Self-inductance. **(c)** Mutual inductance.

long, so that the connections between them recede to infinity, as indicated. In order to calculate L, we introduce current I in both wires, as shown. The flux through the loop, for length l, occupies the shaded region. The calculation is left as Problem 10-7 at the end of the chapter, and the answer is given in the back of the book: $L' = \dfrac{\mu_0}{\pi} \ln \dfrac{s}{a}$.

But if we try to calculate the mutual inductance of these wires, using Figure 10.17c, we encounter the kind of infinity in M' that we have already become accustomed to, for the infinite cylinder, in L', Section 10.4, and for C', Section 9.4. For mutual inductance, wires 1 and 2 are in separate circuits whose outer elements recede to infinity, as before. In order to calculate M', we introduce I only in wire 1. The flux due to 1, linking 2, for length l, is shaded. There is obviously a lot more flux than in Figure 10.17b, and that is what causes the problem. The field due to 1 is

$$B_1 = \frac{\mu_0 I_1}{2\pi r}$$

Then,

$$\Phi_2 = \frac{\mu_0 I_1}{2\pi} \int_s^\infty \frac{l\, dr}{r} \longrightarrow \infty$$

and so

$$M' = M/l = \Phi_2/l I_1 \longrightarrow \infty \qquad (10.36)$$

So any change in current in wire 1 produces an infinite emf in wire 2. Following Section 10.4 for L, we can obtain a suitable approximation for wires of finite length l by cutting off the integral:

$$\Phi_2 \approx \frac{\mu_0 I_1}{2\pi} \int_s^l \frac{l\, dr}{r}$$

so that

$$M \approx \frac{\mu_0 l}{2\pi} \ln \frac{l}{s} \qquad (10.37)$$

This effect, along with capacitance, results in "cross talk" between neighboring wires carrying high-frequency signals unless they are properly shielded.

10.7. Transformers

When Edison began marketing electricity to American homes, it was direct current, DC. So why did Westinghouse and Tesla win out with their alternating current, AC? DC will do almost everything that AC will do: light lamps, run motors, and so on. And DC is somewhat safer at a given voltage; indeed, Edison proposed that the verb "westinghouse" be used in place of "electrocute."

The answer is: the transformer. A transformer can change voltage levels without dissipating excessive power, but it only works on AC. And transportation of power over long distances is dependent on the transformer.

For example, if we wish to provide a megawatt to a town 100 miles away, we can supply it as 10,000 amperes at 100 volts, or as 10 amperes at 100,000 volts. But the latter involves only 1/1000 of the current, with a corresponding reduction in the cost for copper or aluminum wire. Insulation of the 100,000 volts is only a small problem, because air is an excellent dielectric and it is free.

But we cannot supply 100,000 volts to outlets inside people's homes; 120 V is hazardous enough. So we transform the voltage up at the generating station, transport it to the town via high-voltage wires, and then transform it back down.

In the United States, AC usually means 120 V rms (see Section 13.2) at 60 Hz. In Europe, typically it is 220 V and 50 Hz. Much of Europe was originally wired for 110 V; but as electricity use grew after World War II, it was found sensible to double the voltage to 220 V, rather than to rewire, because the insulator is already in place: air.

Figure 10.18 shows a transformer from three different viewpoints: its actual appearance, an idealized sketch, and a schematic diagram.

In Figure 10.18a you can see that the primary and secondary coils are wrapped on top of each other, and that the iron core is laminated to reduce eddy currents. Voltage and current are introduced into the primary, and the transformer returns voltage and current out of the secondary. The iron outside of the coils provides a return path for the magnetic flux, and flux leakage is small. In Figure 10.18c the parallel lines signify that the core of the transformer is a magnetic material; their absence would indicate an air core.

In Figure 10.18b, the dots mark the leads that are in phase, that is, they are both positive at the same time. It is paradoxical that the currents flow in opposite directions in the primary and secondary coils, in obedience to Lenz's law, and yet the voltages are in the same phase.

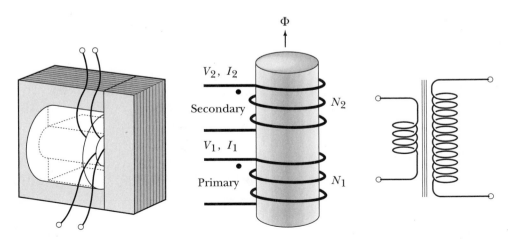

Figure 10.18   Three views of a transformer.

Each coil in an ideal transformer surrounds the same flux Φ, and so the emf in each coil is the same, $- d\Phi/dt$; in the primary this is the back emf, and in the secondary it is the induced emf. Thus, the voltage is proportional to the number of turns of wire:

$$\frac{V_1}{N_1} = \frac{V_2}{N_2}$$

(10.38)

Transformers are usually fairly efficient: The input and output power are nearly the same, so

$$V_1 I_1 = V_2 I_2$$

(10.39)

The currents in the primary and secondary are inversely proportional to the number of turns, and they flow in opposite directions; this implies that the fluxes due to these currents almost cancel out. The magnetic flux inside of a working transformer is much less than it would be if either the primary or secondary current were somehow eliminated.

Impedance Z is somewhat like resistance R: it is voltage divided by current (Section 13.4). A transformer transforms impedance levels as well as voltages and currents.

 **EXAMPLE 10-8**

120 V AC is fed to the primary of a $9:1$ ($= N_1/N_2$) step-down transformer that has a load of 8 ohms in its secondary. Find the voltage and current in the primary and secondary, the power, and the impedance in the primary.

ANSWER

$$V_1 = 120 \text{ V}$$

$$V_2 = 120 \text{ V}/9 = 13.3 \text{ V}$$

$$I_2 = 13.3 \text{ V}/8 \, \Omega = 1.67 \text{ A}$$

$$I_1 = 1.67 \text{ A}/9 = 0.186 \text{ A}$$

$$P = VI = 120 \times 0.186 = 22 \text{ W}$$

$$Z_1 = 120 \text{ V}/0.186 \text{ A} = 650 \, \Omega$$

So the impedance is transformed as the square of the turns ratio: $8 \times 9^2 = 648 \, \Omega$.

The mutual inductance M between a pair of circuits is a quantity fundamental to the operation of transformers. However, it is little used in practice, because, as

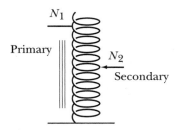

N_1

Primary

N_2

Secondary

Figure 10.19   Autotransformer.

we have just seen, the salient features of transformers are derived without it. All that is usually required is that M be sufficiently large that the transformer may be treated as ideal. This approximation breaks down at low frequency.

Here are a couple of practical transformers.

The variable autotransformer (Variac), Figure 10.19, uses the same coil for primary and secondary. It is efficient but a little lacking in safety: Ordinary transformers isolate the house current from the secondary, but the autotransformer may connect them directly together. The output voltage may be greater than the input.

In Figure 10.20a we see a spark plug analog demonstration. The neon light or spark plug flashes when the switch is opened, but not when it is closed. Why?

The transformer began life as a 28-V filament transformer, so a $4:1$ step-down from house current of 120 VAC; but here it is being operated backwards as a $1:4$ step-up transformer. It weighs half a kilogram—it has to store some energy.

The neon bulb does not conduct until the voltage reaches $V_2 = 60$ V. When the switch is closed, the 1.5-V battery is applied to the primary as V_1. The $1:4$ transformer immediately supplies secondary voltage $V_2 = 6$ V, but that isn't enough to light the neon bulb, by a factor of ten. Then the current in the primary I_1 increases toward V_1/R_1, Figure 10.20b, with the time constant L_1/R_1, where L_1 and R_1 are the inductance and resistance of the primary; this time constant is at

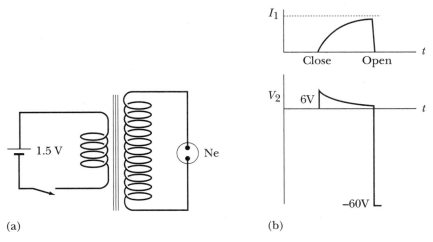

(a)

1.5 V

Ne

(b)

I_1

Close   Open t

V_2 6V t

−60V

Figure 10.20   **(a)** Spark plug analog. **(b)** I and V for spark plug analog.

least 0.1 s, because the transformer must be able to operate efficiently at 60 Hz. Meanwhile V_2 decreases toward zero, and no current flows in the secondary. But . . .

When the switch is opened, the primary current I_1 is suddenly required to be zero. The rate of change of current dI_1/dt is arbitrarily large, and so the induced voltage is large enough to operate the neon bulb. The primary voltage V_1 is not attached to the battery any more, and it promptly goes to -15 V, because at that point the secondary voltage is -60 V, Figure 10.20b. Then current flows in the secondary, through the Ne bulb, and the stored energy is dissipated.

In this case, the applied voltage is being increased by a factor of ten, by cutting off the current suddenly. The same principle is often involved in generation of high voltages in various contexts. In particular, the spark in the spark plug of an engine comes when the primary current is stopped, not when it is started.

10.8.   Hysteresis

Ferromagnetic materials are very nice, but they are not linear. When you demagnetize a piece of iron, it does not follow the same curve as when you magnetize it. So if you start with an ordinary unmagnetized nail, you can hold it to a permanent magnet, and it will pick up small paper clips and nails because of **magnetic induction.** But when you remove the permanent magnet, the nail retains a little residual magnetism and it will still pick up a paper clip or two.

To measure the magnetic properties of a given material, we can build it into a Rowland ring, which is a toroidal transformer, Figure 10.21a. The current in the primary provides H, because of the fundamental equation $\nabla \times \mathbf{H} = \mathbf{J}_f$ (for no $\partial\mathbf{D}/\partial t$); and B is found by monitoring the secondary, using Faraday's law, $\nabla \times \mathbf{E} = -\partial\mathbf{B}/\partial t$. For an ideal magnetic material, a plot of B versus H would be a straight line, Figure 10.21b: $B = \mu_0\mu_r H = \mu H$.

But useful magnetic materials are generally not ideal. Real results are shown in Figure 10.22. All such materials tend to level off above 1 tesla due to saturation: Most of the available electrons are already lined up (Section 10.2). In Figure

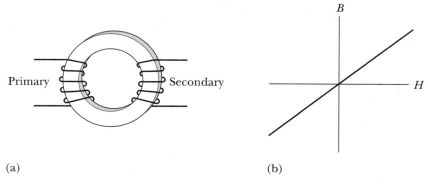

(a)   (b)

Figure 10.21   **(a)** Rowland ring. **(b)** Ideal: $B = \mu H$.

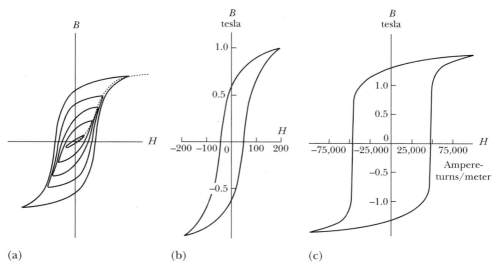

Figure 10.22 Hysteresis curves.

10.22a we see many hysteresis loops for one material under different cycles of H. The dashed line is the "magnetization curve." When the current returns to zero, then H is zero, but the magnetic field B is not yet zero because of residual magnetization. (The Greek root of "hysteresis" implies a lag.) It is necessary to run current in the primary in the opposite direction in order to return B to zero. The path on a B–H plot depends not only on present position but also on history.

Figure 10.22b shows a hysteresis curve for a transformer iron, and Figure 10.22c is for Alnico V, a permanent magnet material. Note the difference in horizontal scales: The H field required to saturate transformer iron is much smaller.

As we have mentioned (Section 10.2), the field energy density in a magnetic material is $\frac{1}{2}\mathbf{B}\cdot\mathbf{H}$. Correspondingly, the energy involved in going around a closed loop on a B–H plot is proportional to the area of the loop. This energy is lost as heat. So for transformer iron, Figure 10.22b, which goes around the loop 60 times a second, one wants a skinny hysteresis loop that wastes little energy.

For a permanent magnet, on the other hand, one wants a material that stays magnetized when the current goes off, Figure 10.22c. For the Rowland ring, when H is reduced to zero, the residual B field in the Alnico V is still 1.3 T, a rather large value.

Unfortunately, the permanent magnet material usually has to have a gap in it. Figure 10.23, in order to get at the field. This reduces the residual B field significantly, for the following reason: When the current I is turned off, then from Ampère's law $\oint \mathbf{H}\cdot d\mathbf{l} = NI = 0$. If H is positive in the air gap, then it must be negative in the iron, in order that the integral around the entire loop be zero. But B just keeps going; it is positive all around the loop, because it has no divergence, $\nabla\cdot\mathbf{B} = 0$. So in the hysteresis loop, Figure 10.22c, the permanent magnet is usually found not where the curve crosses the B axis but to the left, in a region of posi-

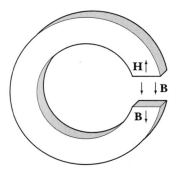

Figure 10.23   Permanent magnet.

tive B and negative H; the precise location depends on the geometry of the permanent magnet. (We look at some magnets in the next section.) As is seen from Figure 10.22c, the B field is less than it would be if we could stop at $H = 0$.

Magnetic material which retains little magnetic field, like that of Figure 10.22b, is often called magnetically **soft.** Such materials are also typically relatively soft mechanically. Ordinary iron nails fit this description fairly well: They are soft magnetically, in that they do not make good permanent magnets, and they are in some sense soft mechanically, as iron goes; at least they are malleable, they bend fairly easily. Permanent magnet materials tend to be hard mechanically; for example, there are Alnico bar magnets that shatter like glass when dropped on a hard floor. More exotic materials can't even be formed by casting; instead, they are ground up and sintered together.

For soft materials, the relative permeability μ_r is not altogether useless, although it is not constant, Figure 10.22b. For permanent magnets μ_r is simply meaningless; there is no $B = \mu H$ and in fact **B** and **H** can point in opposite directions. The superposition principle does not hold, in that B is not proportional to H or the free current.

10.9. Magnets; mmf

Magnet design is a vast and practical subject involving as much art as science. Here we will introduce a couple of the more elementary concepts.

In Figure 10.24a we see a "C"-shaped iron electromagnet, approximately square with side a, with a small gap s. We seek the magnetic field in the gap. We assume that there is no stray magnetic flux, such as that shown in the fringing field in Figure 10.24b; we also assume that the field is uniform across the iron, and that the iron is ideal, with a well-defined μ_r. These assumptions are likely to be significantly wrong, 20% or more, but they lead to reliable order-of-magnitude results. In building a real magnet, first you do some calculations, like these but more sophisticated, and then you build a model and see how it came out; great accuracy is not to be expected.

We have in general

$$\nabla \times \mathbf{H} = \mathbf{J}_f + \frac{\partial \mathbf{D}}{\partial t}$$

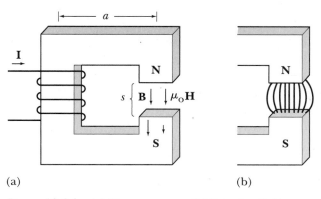

Figure 10.24   **(a)** Electromagnet. **(b)** Fringing field.

so for the quasi-static case, Ampère's law for **H** is

$$\oint \mathbf{H} \cdot d\boldsymbol{l} = \int \mathbf{J}_f \cdot d\mathbf{a} = \text{total current through coil} = NI \qquad \textbf{(10.40)}$$

The quantity $\oint \mathbf{H} \cdot d\boldsymbol{l}$ is called the **mmf,** "magnetomotive force," or "magneto-motance," by analogy with emf $= \oint \mathbf{E} \cdot d\boldsymbol{l}$ (Chapter 4); needless to say, it is not really a force. Continuing, the distance the magnetic flux travels in iron is $4a - s$, and the distance in air is s, to complete one complete path. Ampère's law becomes

$$(4a - s)\, H_i + s\, H_o = NI$$

where H_i is inside the iron, H_o is in the gap. Now $B = \mu_0 H_0 = \mu H_i$, since B is the same inside and out (no magnetic monopoles), so

$$(4a - s)\, B/\mu_r + sB = \mu_0\, NI \qquad \textbf{(10.41)}$$

Here we will simplify by neglecting s compared with $4a$. Solving for B,

$$B = \frac{\mu_0\, NI}{(4a/\mu_r) + s} \qquad \textbf{(10.42)}$$

◤ **EXAMPLE 10-9**

For $N = 100$ turns, $I = 2$A, $a = 4$ cm, $s = 0.1$ cm, $\mu_r = 100$, we find

$$B = \frac{4\pi \times 10^{-7}\,\text{T} \cdot \text{m/A} \times 100 \times 2\,\text{A}}{(4 \times 0.04\,\text{m}/100) + 0.001\,\text{m}} = 0.097\,\text{T}$$

If the gap were zero instead of 0.1 cm, we would find $B = 0.16$ T. The gap of only 1 millimeter of air, in a total path of about 16 cm in iron, reduces the field by one third. This illustrates the talent of iron in guiding and encouraging magnetic field lines. It takes a lot more mmf to make magnetic field lines go through air than through iron.

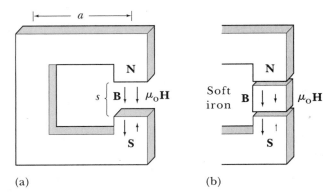

Figure 10.25 **(a)** Permanent magnet. **(b)** With keeper.

Figure 10.25a shows a permanent magnet in the same geometry as the preceding electromagnet, and with the same field in the gap. Note that here H_i points in the direction opposite to **B**. In this case, as in the preceding one, the magnitude of **H** is larger in the air than in the iron; but in general their relative size depends on the particular geometry of the magnet. A key to permanent magnets is (from Ampère's law again):

$$\oint \mathbf{H} \cdot d\mathbf{l} = 0 \qquad (10.43)$$

There is no free current and no mmf. So if you have a positive **H** in the air, you must have a corresponding negative contribution in the iron, so that the total line integral can be zero.

In the case of Figure 10.25a, for the same **B**, we have $NI = 0$, so

$$(4a - s) H_i + s H_o = 0$$

We still have $B = \mu_0 H_o$, of course. Solving for H_i (again we neglect an s):

$$H_i = -\frac{s}{4a} H_o = -\frac{s}{4\mu_0 a} B \qquad (10.44)$$

So in the permanent magnet the H_i field is of different size and points in the opposite direction, compared with that in the electromagnet, even though we have set the parameters so that **B** and H_o are the same in the two cases.

In both cases, H_i is small, and B is approximately equal to $\mu_0 M$. This is no general rule, however; it depends on the geometry. In the bar magnet below, $H_i > H_o$ and so B is smaller than $\mu_0 M$.

For the electromagnet, **H** points in the same direction as **M**, so **H** helps make a large **B**; whereas for the permanent magnet, **H** is "demagnetizing": It points opposite to **M**. This means, roughly speaking, one can get a large magnetic field more easily with an electromagnet than a permanent magnet.

In linear magnetic materials, **H** produces **M** (Section 10.2: $\mathbf{M} = \chi_m \mathbf{H}$), and they point in the same direction. It is characteristic of permanent-magnet material that its magnetization stubbornly persists despite the opposing, demagnetizing **H** field. In most permanent magnets, the demagnetizing **H** field gradually whittles away at **M**, reducing the useful life of the magnet. But you can reduce **H** with a soft iron "magnet keeper," Figure 10.25b, of relative permeability μ_r. As before, we have $\oint \mathbf{H} \cdot d\mathbf{l} = 0$. We will replace $\mathbf{H}_o$ with $\mathbf{H}_g$ in the gap, which is much smaller; and now $\mathbf{B} \approx \mu_0 \mu_r \mathbf{H}_g$. The preceding expression for H_i becomes

$$H_i = -\frac{s}{4\mu_0\mu_r a} B \qquad (10.45)$$

The demagnetizing field H_i is reduced by a factor of μ_r, which will be several hundred, so H becomes negligible everywhere. To explain it in another way, there is no air gap, so there is no place where we need a large H to push flux through air.

So much for the "C" magnet. Now we go over to the bar magnet. Figure 10.26a shows the **B** field lines for a solenoid, and the solid lines in Figure 10.26b show the **B** lines for a bar magnet; the **B** fields are similar because in either case the field is due to $\nabla \times \mathbf{B}$ on the sides. For a long solenoid, the axial field at the end is only half of the value well within the solenoid. This is proven by superposition: If we bring in another identical solenoid and attach it end-to-end, the axial field at the interface is doubled. The magnitude of B increases near the edge of the solenoid, or the corner of the magnet, due to an increasing radial component of B. In Figure 10.26b, lines of **H** are shown as dashed lines. Outside of the iron,

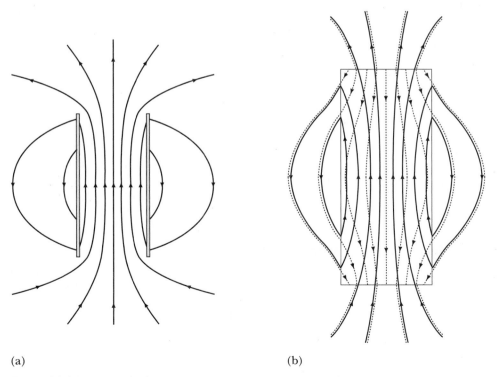

(a)   (b)

Figure 10.26   (a) Solenoid. Solid lines: **B**. (b) Permanent bar magnet. Dashed lines: **H**.

$\mathbf{B} = \mu_0\mathbf{H}$, so lines of $\mathbf{H}$ follow $\mathbf{B}$, but inside of a permanent magnet they are completely different. In fact, the lines of $\mathbf{H}$ are like those of $\mathbf{E}$ for a parallel-plate capacitor: In both cases there is no curl, and the divergence is at the ends; see Figure 10.4. The $\mathbf{H}$ field just inside this magnet at the ends is larger than that just outside, as well as opposite in direction. (The situation is different for a bar electromagnet; see Problem 10-24.)

10.10 SUMMARY

In a magnetic material, the magnetic dipoles may be represented as bound currents that respond to, and contribute to, the magnetic field. We introduce the $\mathbf{M}$ field and the $\mathbf{H}$ field to facilitate calculations. We encounter the practical quantity known as inductance, and we introduce that pillar of the electric age, the transformer.

In a magnetic material, we distinguish bound and free current:

$$\nabla \times \mathbf{H} = \mathbf{J}_f + \frac{\partial \mathbf{D}}{\partial t} \quad \text{and} \quad \nabla \times \mathbf{M} = \mathbf{J}_b \quad \text{and} \quad \mathbf{K}_b = \mathbf{M} \times \hat{\mathbf{n}}$$

$$(10.4, 10.5, 10.8)$$

So

$$\mathbf{B} = \mu_0\mathbf{H} + \mu_0\mathbf{M} \quad \text{in general} \tag{10.9}$$

For homogeneous, isotropic, linear magnetic materials,

$$\mathbf{B} = \mu_0\mu_r\mathbf{H} \quad \text{and} \quad \mathbf{M} = \chi_m\mathbf{H} \tag{10.11, 10.12}$$

and

$$\mu = \mu_r\mu_0 \quad \text{and} \quad \mu_r = \chi_m + 1$$

The energy density is

$$\frac{dW}{dv} = \tfrac{1}{2}\mathbf{B} \cdot \mathbf{H} \tag{10.16}$$

The inductance of an inductor is given by

$$N\Phi = LI \tag{10.23}$$

with

$$V = L\frac{dI}{dt} \tag{10.24}$$

and energy

$$W = \tfrac{1}{2}LI^2 \tag{10.25}$$

In a circuit its time constant is

$$\tau = \frac{L}{R} \qquad (10.30)$$

For an ideal transformer,

$$\frac{V_1}{N_1} = \frac{V_2}{N_2} \qquad \text{and} \qquad V_1 I_1 = V_2 I_2 \qquad (10.38, 10.39)$$

The mutual inductance of a transformer is given by

$$N_2 \Phi = M I_1 \qquad (10.31)$$

For magnetic calculations, the magnetomotive force may be useful:

$$\text{mmf} = \oint \mathbf{H} \cdot d\mathbf{l} \qquad (10.40)$$

 **PROBLEMS**

10-1 An iron slab, $\mu_r = 200$, is placed across a field of 10^{-3} T, Figure 10.27a. At the center, ignoring edge effects, find B, H, M, and J_b inside.

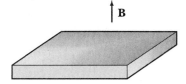

Figure 10.27a

10-2 A rod of iron, $\mu_r = 200$, is placed parallel to a field of 10^{-3} T, Figure 10.27b. At the center, ignoring end effects, find B, H, M, J_b, and K_b.

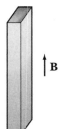

Figure 10.27b

10-3 Repeat the preceding problem for a field of 1 T. Assume that the iron saturates completely at $M = 10^6$ A/m.

10-4 A magnetic sphere of radius R has uniform magnetization M. What is its dipole moment?

10-5 A thin magnetic disk of radius a and thickness s has uniform magnetization **M** parallel to its axis. Find **A** and **B** at distance $z > s$ on its axis. Check for large z ($z \gg a$).

10-6 A long straight wire carrying current I lies on the axis of an iron cylindrical shell of relative permeability μ_r and of inner and outer radii r_1 and r_2. Find **A** and **B** at $r < r_1$, at $r_1 < r < r_2$, and at $r > r_2$. Find the bound surface current density on both surfaces of the iron, and verify that the total is zero.

10-7 Find the inductance per unit length L' of a pair of parallel wires ("twinlead") of radius a and separation s, Figures 9.27 and 10.17b; neglect field inside metal. Evaluate for $a = 1$ mm, $s = 5$ mm.

10-8 Find the inductance per unit length L' of a pair of concentric conductors (coaxial cable) of radii r_1 and r_2 ($> r_1$), Figure 10.28; neglect field inside metal.

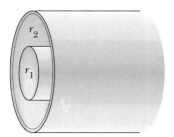

Figure 10.28

10-9 An emf of 10 volts is applied to an inductor of 0.2 H and internal resistance 40 Ω. How long does it take for the current to reach 95% of its final value?

10-10 A long straight wire lies on the axis of a toroid of N turns and of major and minor radii R and r, respectively. Find the mutual inductance assuming constant B inside the toroid.

10-11 Find the inductance of a square iron toroid of μ_r, wrapped with N turns, of side s, at distance a from the axis, Figure 10.29. Evaluate for $N = 200$, $\mu_r = 300$, $a = 1$ cm, $s = 3$ cm.

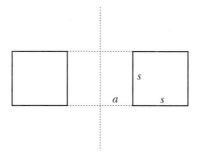

Figure 10.29

*10-12 Find the inductance of an iron toroid of μ_r, wrapped with N turns, of major and minor radii R and r respectively, Figure 10.30. Then find L under the assumption that B is uniform inside the torus.

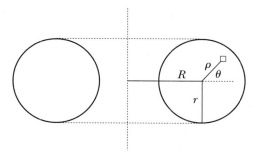

Figure 10.30

10-13 A loop of radius r lies in the plane of a long wire, with its center at $R > r$ from the wire. Find the mutual inductance.

10-14 A square loop of side s lies in the plane of a long wire, at distance a from it. Find the mutual inductance. Evaluate for $a = 1$ cm, $s = 3$ cm.

10-15 A neon-sign transformer provides 450 VA (volt-amps) at 15,000 V. The primary is provided with "house current," 120 V AC at 60 Hz. Find the primary and secondary currents and the turns ratio.

10-16 A soldering gun is a step-down transformer with a one-turn secondary whose resistance is mainly in a piece of iron of length 1.0 cm and cross section 1.3 mm × 3.0 mm. It dissipates 100 W. Find the secondary voltage and current, the primary current, and the number of turns in the primary. (For iron use conductivity 1.0×10^7 S/m.)

10-17 Find the inductance of the example electromagnet of Section 10.9, with and without gap, assuming a cross-sectional area of 1 cm².

10-18 Find B in the gap of a toroidal iron ring of major and minor radii 8.1 and 0.72 cm respectively, Figure 10.31, gap $s = 0.15$ cm, $N = 100$ turns, $\mu_r = 150$, $I = 0.7$ A.

10-19 Find B in the gap of a toroidal permanent magnet of magnetization 10^5 A/m, of major and minor radii 8.1 and 0.72 cm respectively, Figure 10.31, gap 0.7 cm (neglecting stray field).

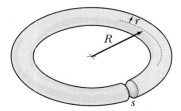

Figure 10.31

10-20 Repeat the preceding problem in the case in which 1/10 of the circumference is permanent magnet of 10^5 A/m and the rest (except for the gap) is soft iron of $\mu_r = 100$.

10-21 An iron toroid, Figure 10.21, of $\mu_r = 200$ and of major radius 7.3 cm and minor radius 0.82 cm is wound with a primary coil of 37 turns and a secondary of 137 turns. Find the mutual inductance. (Assume constant B.)

10-22 Show that a magnetic field line encountering a magnetic material of μ_r bends in accordance with $\tan \theta_1 = \tan \theta_2 / \mu_r$ (compare with Fig. 9.8).

10-23 A long iron rod of $\mu_r = 300$ is uniformly wrapped with 10^3 turns per meter of current 1.7 A. Find the field in the center of the iron, and approximate the field at the end of the iron rod.

10-24 Sketch **B** and **H** for a bar electromagnet shaped like that of Figure 10.26b.

10-25 An iron wire of susceptibility χ_m and radius a carries current density J. Find B, H, and J_b inside the wire and K_b at its surface. Compare the total free current, bound volume current, and bound surface current for $\chi_m = 200$. Verify that the total bound current is zero.

10-26 Find B at position z on the axis of a cylindrical permanent magnet of length L, radius a, and axial magnetization M. (See Problem 3-18. It is proportional to the solid angle subtended by the surface current.)

10-27 Estimate the inductance of an isolated wire of length 10 m and radius 0.1 cm.

10-28 A contour, lying entirely in air, surrounds a magnetic object. Show that the total bound current is zero through any surface bounded by the contour (compare with Problem 9-28).

10-29 Show that $M = \sqrt{k_a k_b}\sqrt{L_a L_b}$ (Eq. 10.35).

10-30 For a long straight wire of radius a carrying current I, find the distance r, from the axis, at which the flux between r and the surface of the wire equals the flux inside the wire. (This tells us something about what we're missing when we neglect flux inside conductors.)

10-31 In our technological society, we often encounter a need for temporary storage of energy, especially in connection with private and public transportation, and finding a better way to store energy is a high priority. Compare the energy stored in joules per kilogram in the following cases:

(a) Capacitor, $E = 10^8$ V/m, $\epsilon_r = 2.3$, specific gravity 1.0 (that is, density $\rho = 1000$ kg/m^3).
(b) Iron inductor, $B = 2$ T, $\mu_r = 200$, sp. gr. 7.8.
(c) Superconducting inductor, $B = 10$ T; it contains no ferromagnetic material, but it needs steel containment, so try sp. gr. 7.8 again.
(d) Car battery, 12 V, 50 ampere-hours, 20 kg.
(e) Flywheel, mass 1 kg, disk of radius 10 cm, 7200 rpm (revolutions per minute).
(f) Gasoline, 5×10^7 J/kg, 10% efficiency.

The values in Parts a–e may increase by an order of magnitude using the most modern, and most expensive, technologies.

10-32 We have found the energy in a long but finite solenoid, Figure 10.32, by integrating $\mathbf{A} \cdot \mathbf{J}/2$ in Section 6.5, and by integrating $B^2/2\mu_0$ in Problem 6-1. Now find it by calculating the inductance L and then finding $LI^2/2$. Evaluate for $l = 10$ cm, $R = 1$ cm, $N' = 10^4$ turns/meter, and $I = 2$A.

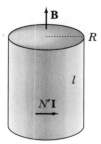

Figure 10.32

10-33 A long cylindrical neodymium magnet of diameter 7/8 inch clings to a thick iron plate with a force of 28 pounds. Estimate the normal B_0 field of the magnet by itself.

*10-34 Figure 10.33 (from Fig. 4.11) shows a dynamo that generates current and magnetic field, assuming correct direction of relative rotation. The mutual inductance between loop and disk is $M = \Phi/I$, and the emf is $\Phi\omega/2\pi$, ω being the angular velocity. Find the differential equation governing the current. Then find the minimum ω required, and find the condition for stable operation. (Note that the current supplies a retarding torque that can reduce ω.) What direction does **B** point?

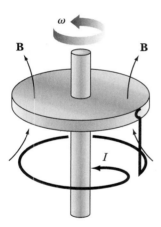

Figure 10.33

10-35 A magnetic monopole passes through a superconducting loop of inductance L. Find the resulting current in the loop.

C H A P T E R

11

MAXWELL'S EQUATIONS
FOR MATERIALS

Since we last paused to display and admire Maxwell's equations, in Chapter 5, we have introduced various new fields designed to facilitate analysis of materials. This would be a good time to gather these results together and to have another look at Maxwell with the broader understanding that the new developments have brought.

11.1. Forms of Maxwell's Equations

In Chapter 5 we encountered Maxwell's equations in all their glory and generality:

$$\nabla \cdot \mathbf{E} = \frac{\rho_t}{\epsilon_0}$$

$$\nabla \cdot \mathbf{B} = 0$$

$$\nabla \times \mathbf{E} = -\frac{\partial \mathbf{B}}{\partial t} \qquad\qquad (11.1)$$

$$\nabla \times \mathbf{B} = \mu_0 \mathbf{J}_t + \mu_0 \epsilon_0 \frac{\partial \mathbf{E}}{\partial t}$$

In these equations, the t subscript implies that *all* electrical charge and charge flow is included in ρ and $\mathbf{J}$, including any material effects. These equations are always true, in all materials, at all points, at all speeds, within the confines of classical electrodynamics.

However, in the presence of materials, it is often convenient to write Maxwell's equations in the following form:

$$\nabla \cdot \mathbf{D} = \rho_f \qquad \text{(Section 9.1)}$$

$$\nabla \cdot \mathbf{B} = 0$$

$$\nabla \times \mathbf{E} + \frac{\partial \mathbf{B}}{\partial t} = 0 \qquad\qquad (11.2)$$

$$\nabla \times \mathbf{H} - \frac{\partial \mathbf{D}}{\partial t} = \mathbf{J}_f \qquad \text{(Section 10.1)}$$

We have written the fields on the left and the sources on the right. The sources are the free charge and current, and the subscript f is often omitted.

Advantages The ϵ_0's and μ_0's are gone, and the characteristics of materials are easier to include. As sources, the free charge and free current are more or less under our control. The analogies between electricity and magnetism are displayed: **D** is analogous to **B**, and **E** to **H**.

Disadvantages The fields **D** and **H** are artificial, and it is **E** and **B** that actually determine the forces on charges. These equations simply ignore bound charge and bound current; they imply conservation of free charge, but bound charge is left to fend for itself. These equations are no longer complete; we need constitutive relations, the first of which are (Sections 9.1 and 10.1)

$$\mathbf{D} = \epsilon_0\mathbf{E} + \mathbf{P} \qquad \mathbf{B} = \mu_0\mathbf{H} + \mu_0\mathbf{M}$$
$$\nabla \cdot \mathbf{P} = -\rho_b \qquad \nabla \times \mathbf{M} = \mathbf{J}_b \tag{11.3}$$

There is another form of Maxwell's equations that is true only for ideal media but that includes materials explicitly:

$$\nabla \cdot \mathbf{E} = \frac{\rho_f}{\epsilon} \qquad \text{(Equation 9.17)}$$

$$\nabla \cdot \mathbf{B} = 0$$

$$\nabla \times \mathbf{E} = -\frac{\partial \mathbf{B}}{\partial t} \tag{11.4}$$

$$\nabla \times \mathbf{B} = \mu\mathbf{J}_f + \mu\epsilon\frac{\partial \mathbf{E}}{\partial t} \qquad \text{(Equation 10.13)}$$

For these equations we have used $\epsilon_0\epsilon_r = \epsilon$ and $\mu_0\mu_r = \mu$. Their main disadvantage is that they don't work where μ or ϵ change: At the end of a polarized dielectric, they suggest that $\nabla \cdot \mathbf{E} = 0$, and on the side of a magnetized material they suggest that $\nabla \times \mathbf{B} = 0$, neither of which is true. Here are some more constitutive relations for this case (Sections 9.2 and 10.2):

$$\mathbf{D} = \epsilon\mathbf{E} \qquad \mathbf{B} = \mu\mathbf{H}$$
$$\mathbf{P} = \chi_e\epsilon_0\mathbf{E} \qquad \mathbf{M} = \chi_m\mathbf{H}$$
$$\epsilon_r = 1 + \chi_e \qquad \mu_r = 1 + \chi_m \tag{11.5}$$

We note once again the analogies which permeate the formalism:

$$\boxed{\mathbf{D} \Longleftrightarrow \mathbf{B} \qquad \epsilon_0\mathbf{E} \Longleftrightarrow \mu_0\mathbf{H} \qquad \mathbf{P} \Longleftrightarrow \mu_0\mathbf{M}} \tag{11.6}$$

 **EXAMPLE 11-1**

Derive $\nabla \cdot \mathbf{E} = \rho_f/\epsilon$ from $\nabla \cdot \mathbf{D} = \rho_f$.

ANSWER

$$\nabla \cdot \mathbf{D} = \rho_f$$

$$\mathbf{D} = \epsilon_0 \epsilon_r \mathbf{E} \qquad \text{a constitutive equation is required}$$

$$\nabla \cdot \epsilon_0 \epsilon_r \mathbf{E} = \rho_f$$

$$\nabla \cdot \mathbf{E} = \rho_f / \epsilon_0 \epsilon_r = \rho_f / \epsilon$$

(The derived form assumes constant ϵ_r.)

 EXAMPLE 11-2

What is the difference between $\nabla \times \mathbf{H} - \partial \mathbf{D}/\partial t = \mathbf{J}_f$ and $\nabla \times \mathbf{B} = \mu_0 \mathbf{J}_f + \mu_0 \epsilon_0 \partial \mathbf{E}/\partial t$?

ANSWER $\nabla \times \mathbf{M} = \mathbf{J}_b$. That is, we add $\nabla \times \mathbf{M} = \mathbf{J}_b$ to $\nabla \times \mathbf{H} - \partial \mathbf{D}/\partial t = \mathbf{J}_f$, obtaining

$$\nabla \times \mathbf{H} + \nabla \times \mathbf{M} = \mathbf{J}_f + \mathbf{J}_b + \partial \mathbf{D}/\partial t$$

so

$$\nabla \times \mathbf{B} = \mu_0(\mathbf{J}_f + \mathbf{J}_b + \mathbf{J}_{pol} + \epsilon_0 \partial \mathbf{E}/\partial t) = \mu_0 \mathbf{J}_t + \mu_0 \epsilon_0 \partial \mathbf{E}/\partial t$$

11.2.   The Use of Analogies

In the preceding section we mentioned the analogies between various quantities, and these can be quite useful in simplifying problems: When we already have the solution to an easy problem, we may be able to use it to solve a more difficult one by analogy. A fine example is the sphere of radius R that we provided with a uniform charge ρ in Section 2.3 and in Problem 6-5, and a uniform polarization $\mathbf{P}$ in Section 9.1 (Fig. 11.1).

To find the field for uniform ρ, we simply applied Gauss's law inside and out to find $\mathbf{E}$, and then we integrated $\mathbf{E}$ to find V; the results were, using $Q = \frac{4}{3} \pi R^3 \rho$:

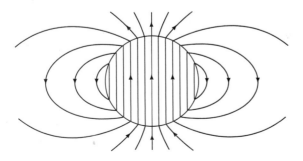

Figure 11.1   B of uniformly magnetized sphere.

Uniform ρ

$$\text{Inside:} \quad \mathbf{E} = \frac{\rho r}{3\epsilon_0}\,\hat{\mathbf{r}} \qquad \text{and} \qquad V = \frac{\rho R^2}{3\epsilon_0}\left(\frac{3}{2} - \frac{r^2}{2R^2}\right)$$

$$\text{Outside:} \quad \mathbf{E} = \frac{\rho R^3}{3\epsilon_0 r^2}\,\hat{\mathbf{r}} \qquad \text{and} \qquad V = \frac{\rho R^3}{3\epsilon_0 r} \qquad \textbf{(11.7)}$$

For uniform $\mathbf{P}$, Section 9.1, we superposed two uniform spheres, of opposite charge, separated by small distance, s, obtaining the dipole field outside, with $Qs = p = P(\frac{4}{3}\pi R^3)$, and a uniform $\mathbf{E}$ field inside:

Uniform P

$$\text{Inside:} \quad \mathbf{E} = -\frac{P}{3\epsilon_0}\,\hat{\mathbf{z}} \qquad\qquad\qquad \text{and} \qquad V = \frac{Pr\cos\theta}{3\epsilon_0}$$

$$\text{Outside:} \quad \mathbf{E} = \frac{2PR^3\cos\theta}{3\epsilon_0 r^3}\,\hat{\mathbf{r}} + \frac{PR^3\sin\theta}{3\epsilon_0 r^3}\,\hat{\boldsymbol{\theta}} \quad \text{and} \quad V = \frac{PR^3\cos\theta}{3\epsilon_0 r^2} \qquad \textbf{(11.8)}$$

which is not so hard; at least we didn't have to integrate anything to get it. We have used superposition, but still no analogies.

And now how do we cope with uniform $\mathbf{M}$? Well, the straightforward way is to integrate the surface bound–current density $\mathbf{K}_b$ (Section 10.1); but there are easier things to do. A half-century ago we might have followed the uniform $\mathbf{P}$ form above, but using magnetic pole density in analogy to electric charge density. We would have imagined two spheres of uniform North and South pole density, slightly displaced from each other. But poles have pretty much gone out of fashion, largely because of the evident nonexistence of magnetic monopoles, and so we will not be discussing that kind of analogy. However, the related analogy between $\mathbf{E}$ and $\mathbf{H}$ provides the kind of bridge we want from the solved electrostatic problem of uniform $\mathbf{P}$ to the about-to-be-solved one of uniform $\mathbf{M}$. There is no free charge here, and no free current, so the equations are

Electrostatic	**Magnetostatic**	
$\nabla \times \mathbf{E} = 0$ and $\nabla \cdot \mathbf{D} = 0$	$\nabla \times \mathbf{H} = 0$ and $\nabla \cdot \mathbf{B} = 0$	**(11.9)**

so

$$\nabla \cdot \mathbf{E} = -\nabla \cdot \mathbf{P}/\epsilon_0 \qquad \text{and} \qquad \nabla \cdot \mathbf{H} = -\nabla \cdot \mathbf{M}$$

These differential equations for $\mathbf{E}$ and $\mathbf{H}$ (both curl and divergence) are of the same form, and $\mathbf{P}$ and $\mathbf{M}$ are functions of the same form. So we can substitute $\mathbf{H}$ for $\mathbf{E}$ and $\epsilon_0\mathbf{M}$ for $\mathbf{P}$ in solutions for uniform polarization, Equation 11.8. Then, to find $\mathbf{B}$, we can use $\mathbf{B} = \mu_0\mathbf{H} + \mu_0\mathbf{M}$. Note that $\mathbf{B}$ is continuous across the surface at $\theta = 0$. The vector potential $\mathbf{A}$ will take a little longer, but we'll display it here for convenience.

Uniform M

Inside: $\mathbf{B} = \dfrac{2\mu_0 M}{3}\,\hat{\mathbf{z}}$ and $\mathbf{A} = \dfrac{\mu_0 M\, r\sin\theta}{3}\,\hat{\boldsymbol{\phi}}$

$$(11.10)$$

Outside: $\mathbf{B} = \dfrac{2\mu_0 M\, R^3 \cos\theta}{3\, r^3}\,\hat{\mathbf{r}} + \dfrac{\mu_0 M\, R^3 \sin\theta}{3\, r^3}\,\hat{\boldsymbol{\theta}}$ and $\mathbf{A} = \dfrac{\mu_0 M\, R^3 \sin\theta}{3\, r^2}\,\hat{\boldsymbol{\phi}}$

So we find **B** easily using analogies.

For V and **A** we can start with the following formulas (Griffiths, *Am. J. Phys.*):

(Section 2.2) $\mathbf{E} = \displaystyle\int \dfrac{\rho\,\hat{\mathbf{r}}}{4\pi\epsilon_0 r^2}\, dv = \dfrac{1}{\epsilon_0}\,\rho\left(\displaystyle\int \dfrac{\hat{\mathbf{r}}}{4\pi\, r^2}\, dv\right)$

(Section 9.1) $V = \displaystyle\int \dfrac{\mathbf{P}\cdot\hat{\mathbf{r}}}{4\pi\epsilon_0 r^2}\, dv = \dfrac{1}{\epsilon_0}\,\mathbf{P}\cdot\left(\displaystyle\int \dfrac{\hat{\mathbf{r}}}{4\pi\, r^2}\, dv\right)$ (11.11)

(Section 10.1) $\mathbf{A} = \displaystyle\int \dfrac{\mu_0\mathbf{M}\times\hat{\mathbf{r}}}{4\pi\, r^2}\, dv = \mu_0\mathbf{M}\times\left(\displaystyle\int \dfrac{\hat{\mathbf{r}}}{4\pi\, r^2}\, dv\right)$

The ρ, **P**, and **M** are constant, so we can take them out of the integral. Now the point is, we are left with the same integral in all three cases. So after we find the solution for **E**, it's easy to find the solutions for V and **A**:

$$V = \dfrac{1}{\rho}\,\mathbf{P}\cdot\mathbf{E} \quad\text{and}\quad \mathbf{A} = \dfrac{\mu_0\epsilon_0}{\rho}\,\mathbf{M}\times\mathbf{E} \qquad (11.12)$$

Starting with the **E** for the uniformly charged sphere, we immediately confirm the V and **A**, displayed above in Equation 11.10. As you see, we have obtained them quickly and easily using analogies.

So: Does Nature really love simplicity, or is she just a little lacking in imagination? Either way, we tend to find the same simple equations, with analogous solutions, popping up all over the field of physics. In the next chapter we will be dealing with one of Nature's particular favorites, called Laplace's equation.

◤ PROBLEMS

11-1   Derive $\nabla\times\mathbf{B} = \mu\mathbf{J}_f + \mu\epsilon\dfrac{\partial\mathbf{E}}{\partial t}$ from $\nabla\times\mathbf{B} = \mu_0\mathbf{J}_t + \mu_0\epsilon_0\dfrac{\partial\mathbf{E}}{\partial t}$.

11-2   Show that the corresponding analogy for **B** is $\mathbf{B} = \dfrac{\mu_0\epsilon_0}{\rho}\mathbf{J}\times\mathbf{E}$. Further, show that J and ρ needn't be constant, but need merely be mutually proportional.

CHAPTER

12

BOUNDARY-VALUE PROBLEMS

We already have at least eight ways to solve various elecromagnetic problems: See the summary in Section 6.3. Those procedures involve either integrating over charge distributions or taking advantage of symmetry to, in effect, do the integration for us. But there is yet another way to solve problems in electrostatics and magnetostatics, a powerful procedure based on the properties of differential equations and relying heavily on a uniqueness theorem.

12.1. Poisson's Equation, Laplace's Equation, and the Uniqueness Theorem

By way of introduction, we will give a rough summary of the remarkable kind of idea involved here: Suppose you are given the charge distribution inside a region and the potentials at the boundaries. And suppose that by hook or by crook you dream up a solution to an *easier* problem that involves the same boundary conditions and charges. Then your result is also the solution to the original problem, because of the **uniqueness theorem,** which follows.

Note that it isn't enough just to specify the charge distributions. In Figures 2.24c and 2.25 we saw two different fields for the same ρ. But if you also specify potentials on boundaries, then the result is unique. (For example, we might specify that the charge distribution of Figure 2.24c is located between parallel conducting plates, and that of Figure 2.25 is inside a conducting sphere.)

To help put the matter in perspective: As you know, the standards for proof in mathematics are more stringent than in a court of law. "Beyond a reasonable doubt" just isn't good enough. But in a boundary-value problem you need only find that the suspect *might* have done it! Uniqueness does the rest.

We will be needing the differential equation for *V*. We have Gauss's law, Section 2.4,

$$\nabla \cdot \mathbf{E} = \frac{\rho}{\epsilon_0}$$

and also the equation involving the potential, from Section 6.2,

$$\mathbf{E} = -\nabla V - \partial\mathbf{A}/\partial t \longrightarrow -\nabla V \qquad \text{for the static case}$$

so, putting them together,

$$-\nabla \cdot \mathbf{E} = \nabla \cdot (\nabla V) = \boxed{\nabla^2 V = -\frac{\rho}{\epsilon_0}} \qquad \text{(static)} \qquad \textbf{(12.1)}$$

which is **Poisson's equation.** This equation contains all electrostatics, in principle: if we solve it to find V, then we can find $\mathbf{E}$ using $\mathbf{E} = -\nabla V$.

When $\rho = 0$, Poisson's equation simplifies into **Laplace's equation:**

$$\boxed{\nabla^2 V = 0} \qquad \text{(static)} \qquad \textbf{(12.2)}$$

The expression $\nabla^2 V$ is called the **Laplacian** of V, Section 1.3; its components are displayed explicitly, for various coordinate systems, inside the front cover.

We can also use Laplace's equation in magnetostatics under appropriate circumstances, We define a **magnetic scalar potential** V_m:

$$\mathbf{H} = -\nabla V_m \qquad \textbf{(12.3)}$$

But this only works if $\nabla \times \mathbf{H} = 0$, because the curl of a gradient is always zero. So, since $\nabla \times \mathbf{H} = \mathbf{J}_f + \partial\mathbf{D}/\partial t$, the magnetic scalar potential is limited to magnetostatic situations (zero time derivatives) with no free current $\mathbf{J}_f$. Then since $\mathbf{B} = \mu_0\mathbf{H} + \mu_0\mathbf{M}$, we have

$$\nabla \cdot \mathbf{H} = -\nabla \cdot \mathbf{M}$$

so we get

$$\nabla \cdot (\nabla V_m) = \nabla^2 V_m = \nabla \cdot \mathbf{M} \qquad \textbf{(12.4)}$$

which is Poisson's equation in magnetostatics; it becomes Laplace's equation when $\nabla \cdot \mathbf{M} = 0$.

It is interesting and important that for Laplace's equation there is no minimum or maximum of V in any uncharged region, except at the boundary of that region. You can't trap a charged particle in a charge-free electrostatic field; there is no point of stable equilibrium. This is **Earnshaw's theorem,** mentioned in

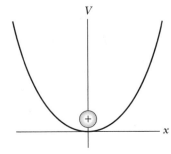

Figure 12.1   No potential trap.

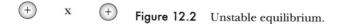

Figure 12.2   Unstable equilibrium.

Section 2.3. One simple argument goes as follows. Suppose that there were such a trap. At such a point, we must have $\partial V/\partial x = 0$, and $\partial^2 V/\partial x^2 > 0$, Figure 12.1, and the same for y and for z. You can imagine for example that the positive charge shown is a mass rolling about in a gravitational well. But Laplace's equation says that the three second derivatives in x, y, and z must add to zero:

$$\nabla^2 V = \frac{\partial^2 V}{\partial x^2} + \frac{\partial^2 V}{\partial y^2} + \frac{\partial^2 V}{\partial z^2} = 0 \tag{12.5}$$

so if one is positive, then at least one must be negative, providing an escape for the positive particle.

Of course, there are exceptions. If $\rho \neq 0$, an extremum is possible. Also, you can trap a charged particle, or even levitate it, using changing fields. And you can have *unstable* equilibrium in the static case, Figure 12.2. In this case a test charge would be in equilibrium at the point marked x, halfway between two equal charges: The **E** field is zero there, so the test charge would have no tendency to move. But if it were displaced infinitesimally up or down, it would enter a region where **E** pushed it farther away from the equilibrium position; that means the equilibrium is unstable.

Here is another somewhat intuitive but instructive proof of Earnshaw (Purcell). We begin by proving: *the average potential V_{av} over the surface of a charge-free sphere equals the potential V_c at the center of the sphere.* Observe that, in Figure 12.3a, the charge Q produces a potential V_c at the center of the sphere, and an average potential V_{av} over the surface. We intend to prove that $V_c = V_{av}$. The size of the sphere is irrelevant, but Q must be outside.

To facilitate determination of V_{av}, we spread a test charge q *uniformly* over the surface of the sphere, Figure 12.3b. Then we bring the two charges, q and Q, to separation r, as shown, in two different ways:

1. We place q, and then we bring Q in from infinity to r. The test charge q behaves as if it were all at the center of the sphere, for this purpose (this is a consequence of Gauss and so of Laplace); so the energy involved in bringing Q from infinity is $W = Qq/4\pi\epsilon_0 r$ (Section 6.5).

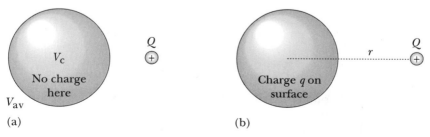

(a)   (b)

Figure 12.3   **(a)** Average V on sphere. **(b)** Test charge q.

2. We place Q, and then we bring the sphere with q from infinity. Each element of q arrives at a different potential, that is, a different distance from Q; the average potential on the surface of the sphere due to Q is V_{av}, and so the energy required is $W = qV_{av}$.

(We don't have to consider the self-energy of q on the sphere because it doesn't change.)

The energies in the two processes must be the same:

$$W = qV_{av} = \frac{Qq}{4\pi\epsilon_0 r} \quad \text{so} \quad V_{av} = \frac{Q}{4\pi\epsilon_0 r} = V_c \quad (12.6)$$

So we have proven our thesis for the single charge Q. If we have other charges outside the sphere, we just add up the potentials, relying on superposition. So $V_{av} = V_c$ in any charge-free sphere, in the static case. And this entails Earnshaw for the following reason: If we find a charge-free point of stable equilibrium for a positive charge, then we can draw a little sphere around it, and the potential will be higher everywhere on the sphere than at the center. But that's impossible, because the average V on the sphere must be the same as at the center.

And now we must tackle the **uniqueness theorem.** Now it may seem obvious to the casual observer that, yes, there is sure to be a solution for Poisson's equation, and it is going to be the only possible one. However, it so happens that for a $1/r^n$ force law, a uniqueness theorem holds only for $n = 2$ (Bartlett), which by good fortune is pertinent to the Coulomb's-law universe in which we live. So at least the point is worth looking into.

We may as well admit right up front: Our result will be of only limited validity. To complete the proof in all generality would take us far afield. But what we do will be conceptually simple and, it is to be hoped, illuminating.

We want to prove that, given the charge distribution inside a region and the potentials at the boundaries, there is only one possible potential distribution inside the region. We proceed as follows:

Suppose that there are two different possible potentials, V_1 and V_2, inside the region. They are equal at the boundaries, and they both obey Poisson's equation:

$$\nabla^2 V_1 = -\frac{\rho}{\epsilon_0} \quad \text{and} \quad \nabla^2 V_2 = -\frac{\rho}{\epsilon_0} \quad (12.7)$$

Set their difference equal to V_3:

$$V_3 = V_2 - V_1 \quad (12.8)$$

Then $V_3 = 0$ at all boundaries, and V_3 satisfies Laplace's equation:

$$\nabla^2 V_3 = \nabla^2 V_2 - \nabla^2 V_1 = \frac{\rho}{\epsilon_0} - \frac{\rho}{\epsilon_0} = 0 \quad (12.9)$$

Now suppose that $V_3 < 0$ somewhere inside the region. Then it must have a minimum value somewhere inside the region, because it has to get back to zero by the time it reaches the boundary. But as we have seen above, Laplace's equation permits no local minima or maxima. Intuitively: Starting at the supposed minimum value, V_3 cannot go down, whatever direction you move. This implies that

$$\frac{\partial^2 V_3}{\partial x^2} \geq 0 \quad \text{and} \quad \frac{\partial^2 V_3}{\partial y^2} \geq 0 \quad \text{and} \quad \frac{\partial^2 V_3}{\partial z^2} \geq 0$$

Since the three second derivatives must add to zero and none of them can be negative, all of them must be zero. The same idea holds of course for any supposed local maximum. All this can be true only if $V_3 = 0$ everywhere. (See Problem 12-1 for another approach.)

So since $V_3 = 0$, it must be that $V_1 = V_2$ everywhere, and the two supposedly different solutions are in fact the same. And that proves the uniqueness theorem for this case.

Unfortunately there are other cases. This proof works when there are conductors of known V inside the region. But suppose an isolated conductor has specified charge Q but unspecified potential V and unspecified distribution of its charge. Then the proof may be extended by drawing a Gaussian surface around that conductor. There can also be dielectrics (including permanently polarized dielectrics for which $\mathbf{D} \cdot \mathbf{E}$ is negative). Uniqueness theorems hold for all of these cases, but the proofs become tedious. The bottom line is, in these other cases we are simply going to accept uniqueness as a fortunate fact of life.

12.2.   Images

The most spectacular example of the use of the uniqueness theorem, and a very useful tool, is the image solution.

Suppose we place a positively charged pith ball, $+Q$, at distance D from a grounded (or just hand-held) conducting plate, Figure 12.4a. The pith ball is at-

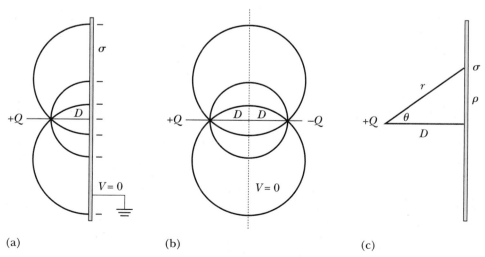

(a)   (b)   (c)

Figure 12.4   (a) Charge and conducting plane. (b) Charge and image. (c) Surface charge.

tracted to the plate. The field would seem to be a complicated thing to calculate in this case. But observe that the potential V is zero on the plane, because it's grounded. If we place a negative charge $-Q$ at the image position, at distance D on the other side of the plane, and we take away the conducting plane, Figure 12.4b, then by symmetry the potential is still zero at the former position of the plane. Then the uniqueness conditions apply on the left of the plane: V is still zero on the plane, as well as at infinity, and the $+Q$ is still in place. So the field must be exactly the same as in Figure 12.4a. Of course the field on the right is completely different, but we don't care about that side. (In fact, we already know $E = 0$ on the right, from Fig. 12.4a.)

 EXAMPLE 12-1

Find the force on a charge of 1 μC at a distance of 1 cm from a large grounded metal plane.

ANSWER $F = Q^2/4\pi\epsilon_0 r^2$ using Coulomb's law between the charge and its image,

$$F = 9 \times 10^9 \times (10^{-6})^2/0.02^2 = 22.5 \text{ N}$$

The total charge induced on the metal surface is equal to $-Q$, because all of the field lines that start out from $+Q$ eventually wend their way to the surface. **E** is perpendicular to the surface. The surface charge density σ can be found using $\sigma = \epsilon_0 E$. The field at the surface is twice the normal component due to the charge $+Q$. (The tangential components cancel out.) If Q is positive, then σ is negative. So (Fig. 12.4c):

$$\sigma = 2\,\epsilon_0 \frac{-Q}{4\pi\epsilon_0 r^2} \cos\theta$$

$$= \frac{-Q}{2\pi r^2} \cos\theta = \frac{-QD}{2\pi(D^2 + \rho^2)^{3/2}}$$

(12.10)

Its integral is of course $-Q$ (see Problem 12-5).

What if the conducting plane is at some specified potential, like 17 volts; then what is the force? Well, ideally, for an infinite plane, the **E** field is constant, and so it must be essentially zero, in order that its integral from infinity result in a finite potential. (So the corresponding surface charge density is zero.) That implies that it doesn't matter what the voltage on the plane is. But in practice there are no infinite planes, and so the force will depend somewhat on the surroundings.

We can extend the image concept to a right-angle corner (between semi-infinite metal planes), $\theta = 90°$, Figure 12.5. A charge $+Q$ and three images suffice to produce $V = 0$ at the conducting surfaces, by symmetry. The field due to the images is correct only in the quadrant where the real charge $+Q$ is located. If the real charge were in one of the other three quadrants, there would be no simple image solution, because you can't put an image inside a region where you are trying to find V. In fact, such image solutions exist only for $\theta = 180°/n$, where n is a positive integer.

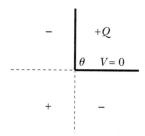

Figure 12.5 Three images in a corner.

There is also an image solution for a charge and a semi-infinite plane dielectric, Figure 12.6. We will now find the images. We begin in a straightforward fashion by calculating the bound surface charge σ_b. In magnitude it is equal to the normal component of the polarization P at the surface. The vector **P** points to the right, and $\hat{\mathbf{n}}$ points to the left. E_n is positive and σ_b is negative. Just inside the dielectric,

$$\sigma_b = \mathbf{P} \cdot \hat{\mathbf{n}} = -P_n = -\epsilon_0(\epsilon_r - 1) E_{ni} \tag{12.11}$$

where the subscript ni implies normal and just inside the dielectric. Conceptually we have now replaced the dielectric with its bound surface charge. The electric field is due both to σ_b and to $+Q$; its normal component is, just inside the dielec-

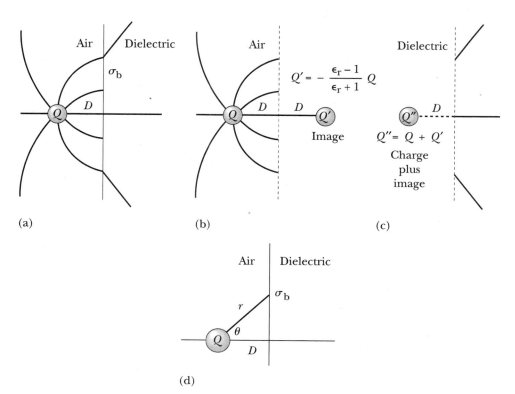

Figure 12.6 (a) Charge and dielectric. (b) For field in air. (c) For field in dielectric. (d) For bound surface charge.

tric (Fig. 12.6d):

$$E_{ni} = \frac{\sigma_b}{2\epsilon_0} + \frac{+Q}{4\pi\epsilon_0 r^2} \cos\theta \qquad (12.12)$$

Substituting the above value of E_{ni},

$$-\frac{\sigma_b}{\epsilon_0(\epsilon_r - 1)} = \frac{\sigma_b}{2\epsilon_0} + \frac{+Q}{4\pi\epsilon_0 r^2} \cos\theta$$

So, solving for σ_b,

$$\sigma_b = -\frac{(\epsilon_r - 1)\, Q \cos\theta}{2\pi(\epsilon_r + 1)r^2} \qquad (12.13)$$

For large ϵ_r this expression becomes the same as that for a metal plane, namely, $-Q \cos\theta/2\pi r^2$. As before (Section 9.2), we find that in some ways a metal is like a dielectric of large ϵ_r.

In principle, this problem is solved: We can find fields and potentials just by integrating σ_b and combining the result with $+Q$. However, we wish now to squeeze an image representation out of the result.

The normal component of **E** just inside the dielectric is a combination of the effects of $+Q$ and σ_b; as above,

$$E_{ni} = \frac{\sigma_b}{2\epsilon_0} + \frac{+Q}{4\pi\epsilon_0 r^2} \cos\theta \qquad (12.14)$$

When we substitute the value of σ_b, this becomes

$$E_{ni} = \left(1 - \frac{\epsilon_r - 1}{\epsilon_r + 1}\right) \frac{+Q \cos\theta}{4\pi\epsilon_0 r^2}$$

But this is simply the field of a charge Q'' consisting of the original charge Q plus an image charge Q' located at the same place, Figure 12.6c, of magnitude

$$\boxed{Q' = -\frac{\epsilon_r - 1}{\epsilon_r + 1} Q} \qquad (12.15)$$

For large ϵ_r, $\boldsymbol{E}_{ni}$ becomes zero, just as inside a metal.

Similarly, just outside of the dielectric, we find

$$E_{no} = \left(1 + \frac{\epsilon_r - 1}{\epsilon_r + 1}\right) \frac{+Q \cos\theta}{4\pi\epsilon_0 r^2}$$

which is the field of the original charge Q plus an image charge Q' located at D on the other side of the dielectric surface, Figure 12.6b. When ϵ_r is large, this is the field at a metal surface.

Another simple image situation is seen in Figure 12.7a. We place a charge Q at a distance D from the center of a grounded conducting sphere of radius $a < D$. Almost miraculously, if we replace the conductor with an image charge Q' at distance b from the center, as shown, then the surface of the sphere is still the $V = 0$

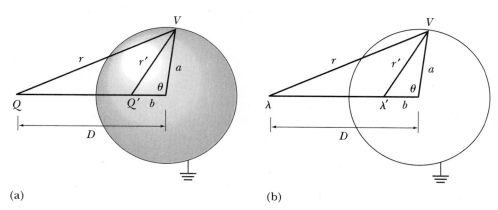

(a)   (b)

Figure 12.7   **(a)** Image in sphere. **(b)** Image in cylinder.

equipotential surface, provided that

$$Q' = -\frac{b}{a}\, Q = -\frac{a}{D}\, Q \qquad\qquad (12.16)$$

and

$$b = \frac{a^2}{D} \qquad\qquad (12.17)$$

This is easily seen using similar triangles. The triangles whose common angle is θ are similar, since

$$\frac{a}{D} = \frac{b}{a} \qquad \text{from the definition of } b$$

Then

$$\frac{a}{r} = \frac{b}{r'}$$

Also,

$$\frac{Q'}{b} = -\frac{Q}{a} \qquad \text{from the definition of } Q'$$

So

$$\frac{Q}{r} = -\frac{Q'}{r'}$$

and

$$\frac{Q}{4\pi\epsilon_0 r} + \frac{Q'}{4\pi\epsilon_0 r'} = 0$$

which means that the potential is zero on the surface, regardless of θ. (Only the zero equipotential is spherical.)

   If the real charge is *inside* a conducting spherical shell, then the image is outside, and the same derivation applies (see Problem 12-9).

▼ EXAMPLE 12-2

A charge of 1 nC is placed at 10 cm from the center of an *uncharged* isolated metal sphere of radius 5 cm. What is the potential of the sphere?

ANSWER The total charge on the sphere must be zero. After we place the usual image Q' at b, we must place a second image, of value $-Q'$, at the center of the sphere; then the sphere remains an equipotential, and its total charge is still zero. Q and Q' together contribute zero to the potential on the sphere, so we need only find the potential due to the image at the center, $-Q'$.

$$-Q' = aQ/D$$
$$= 5 \times 10^{-9}/10 = 5 \times 10^{-10} \text{ C}$$
$$V = -Q'/4\pi\epsilon_0 r$$
$$= 9 \times 10^9 \text{ N} \cdot \text{m}^2/\text{C}^2 \times 5 \times 10^{-10} \text{ C}/0.05 \text{ m} = 90 \text{ volts}$$

(If the real charge is *inside* a spherical shell, no second image charge is needed: The field inside a conducting sphere is independent of the charge on the sphere.)

The image solution for a line charge λ parallel to a grounded circular **cylinder** involves an equal opposite line charge image $\lambda' = -\lambda$, Figure 12.7b. Equation 12.17 still applies. The potential is $V = \lambda \ln(r'/r)/2\pi\epsilon_0$ (Problem 6-28); so for any equipotential (not just $V = 0$), $r'/r = $ constant, which is the case in both Figure 12.7a and Figure 12.7b. For the cylinder, the capacitance per unit length C' is zero (Section 9.4), so we can adjust its potential if necessary with a negligible linear charge density at the center.

When we place a charged conducting sphere, radius a, next to a conducting plane, we get an infinite series of images; Figure 12.8 shows several of these images. Fortunately the series converges quickly. The *total* charge on the sphere consists of the various charges Q, $-Q'$, Q'', and so forth. The charge Q at the center of the sphere produces an image $-Q$ in the metal; these two charges make the plane surface an equipotential, but the sphere is no longer an equipotential. The $-Q$ then produces an image $-Q'$ in the sphere, which makes the sphere an equipo-

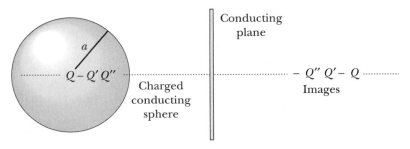

Figure 12.8   Series of images.

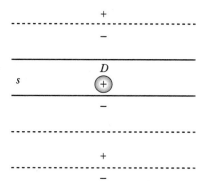

Figure 12.9   Series of images.

tential but messes up the plane. And so on. Each successive image pair produces a smaller effect, and in the limit both surfaces become equipotentials. If Q is positive, then the images in the sphere are all positive, because of the minus sign in the image formula; and the images on the other side of the conducting plane are all negative. The potential on the sphere is only due to Q; the other charges combine with their images to provide zero volts on the spherical surface.

So as we would expect, introducing the metal plane has reduced the voltage on the sphere, assuming that the total charge is unchanged. Without the plane, the potential is

$$V = (Q - Q' + Q'' + - \ldots)/4\pi\epsilon_0 a \tag{12.18}$$

all terms being positive if Q is positive; with the plane, it is only

$$V = Q/4\pi\epsilon_0 a$$

See Problem 12-13 for an example involving this geometry.

Figure 12.9 provides another example of an infinite series of images. Two parallel grounded conducting planes are separated by distance s, and a charge $+Q$ is located at distance $D < s$ from one plane. The force on the charge may be expressed in a convergent infinite series. The charges on the plates are $-QD/s$ and $-Q(s - D)/s$, but that is not so easy to show directly: The images don't get smaller, so the corresponding series is "conditionally convergent." One way to proceed is to take the limit of concentric spheres as the radii go to infinity (Dick). Another invokes superposition: see Problem 12-12.

12.3.   Solutions of Laplace's Equation

Most problems just don't have image solutions. In many cases we're forced to solve Laplace's equation, subject to boundary conditions, a process that often involves infinite series and other mathematical techniques. So, let us begin.

Laplace's equation, as we have said, is $\nabla^2 V = 0$. We shall first illustrate its use in *rectangular* coordinates:

$$\nabla^2 V = \frac{\partial^2 V}{\partial x^2} + \frac{\partial^2 V}{\partial y^2} + \frac{\partial^2 V}{\partial z^2} = 0$$

It can be solved by a process called "separation of variables." We write

$$V(x, y, z) = X(x)\ Y(y)\ Z(z) \tag{12.19}$$

in the fond hope that it may be possible to express V in this form. (In fact, we find that not only is this possible, but also there are other kinds of separations; for example, $V = X(x) + Y(y) + Z(z)$. But this will serve our purposes.) Substituting into Laplace's equation, we obtain

$$\frac{\partial^2 XYZ}{\partial x^2} + \frac{\partial^2 XYZ}{\partial y^2} + \frac{\partial^2 XYZ}{\partial z^2} = 0$$

For convenience we introduce

$$X'' = \frac{d^2 X}{dx^2} \qquad \text{and so on} \tag{12.20}$$

Then Laplace's equation becomes

$$X''YZ + XY''Z + XYZ'' = 0$$

Dividing through by XYZ, we get

$$\frac{X''}{X} + \frac{Y''}{Y} + \frac{Z''}{Z} = 0 \tag{12.21}$$

Now, the first term involves all of the x dependence. If we change x any way we want, the other terms aren't affected. So it must be that the first term is constant. The same goes for the second and third terms. Each term is constant. We have

$$\frac{X''}{X} = C_x \qquad \text{and} \qquad \frac{Y''}{Y} = C_y \qquad \text{and} \qquad \frac{Z''}{Z} = C_z$$

where the C's are constants such that

$$C_x + C_y + C_z = 0 \tag{12.22}$$

So through the magic of mathematics we have separated the variables: One differential equation in three variables has turned into three equations of one variable each.

The solutions of $\frac{X''}{X} = C_x$ are some trivial ones, $X = 1$ and $X = x$, and then some more useful ones. For positive C_x, set

$$X = Ae^{kx}$$

Substituting,

$$\frac{X''}{X} = \frac{k^2 Ae^{kx}}{Ae^{kx}} = k^2 = C_x$$

So Ae^{kx} is a solution, A is arbitrary, and k can be positive or negative. If C_x is negative, set $X = A \sin kx$. Substituting,

$$\frac{X''}{X} = \frac{-k^2 A \sin kx}{A \sin kx} = -k^2 = C_x$$

The same holds for cos kx. To summarize, permissible solutions are of the form:

Rectangular Coordinates   1 x

$\qquad\qquad\qquad\qquad$ sin kx cos kx $\qquad\qquad\qquad\qquad$ **(12.23)**

$\qquad\qquad\qquad\qquad$ e^{kx} e^{-kx} (or sinh kx and cosh kx)

You usually get a lot of solutions to Laplace's equation, and you can add them together as it suits your purpose: Obviously if $X_1 YZ$ satisfies Laplace, and $X_2 YZ$ does too, then so does $X_1 YZ + X_2 YZ$. It is the boundary conditions that determine which solutions we discard.

In *cylindrical* coordinates, one often encounters problems in which V shows no z dependence. Then Laplace's equation becomes

$$\nabla^2 V = \frac{1}{\rho}\frac{\partial}{\partial\rho}\left(\rho\frac{\partial V}{\partial\rho}\right) + \frac{1}{\rho^2}\frac{\partial^2 V}{\partial\phi^2} = 0$$

Similarly, in *spherical* coordinates we often have azimuthal symmetry, with no ϕ dependence, so

$$\nabla^2 V = \frac{1}{r^2}\frac{\partial}{\partial r}\left(r^2\frac{\partial V}{\partial r}\right) + \frac{1}{r^2\sin\theta}\frac{\partial}{\partial\theta}\left(\sin\theta\frac{\partial V}{\partial\theta}\right) = 0$$

The variables are easily separated; see Problem 12-19. Some solutions are given below; you may verify them by direct substitution.

Cylindrical Coordinates   1 $\ln\rho$

For V independent of z $\qquad$ $\rho^n\cos n\phi$ $\rho^{-n}\cos n\phi$ $\qquad\qquad$ **(12.24)**

$\qquad\qquad\qquad\qquad\qquad$ $\rho^n\sin n\phi$ $\rho^{-n}\sin n\phi$

Spherical Coordinates
For V independent of ϕ

1 $\qquad\qquad\qquad\qquad\qquad\qquad\qquad$ r^{-1} $\qquad\qquad\qquad\qquad\qquad$ **(12.25)**

$r\cos\theta$ $\qquad\qquad\qquad\qquad\qquad\qquad$ $r^{-2}\cos\theta$

$r^2(3\cos^2\theta - 1)/2$ $\qquad\qquad\qquad$ $r^{-3}(3\cos^2\theta - 1)/2$

$r^3(5\cos^3\theta - 3\cos\theta)/2$ $\qquad$ $r^{-4}(5\cos^3\theta - 3\cos\theta)/2$

$r^4(35\cos^4\theta - 30\cos^2\theta + 3)/8$ $\qquad$ $r^{-5}(35\cos^4\theta - 30\cos^2\theta + 3)/8$

$r^5(63\cos^5\theta - 70\cos^3\theta + 15\cos\theta)/8$ $\quad$ $r^{-6}(63\cos^5\theta - 70\cos^3\theta + 15\cos\theta)/8$

$\qquad$ . . . $\qquad\qquad\qquad\qquad\qquad\qquad$ . . .

$r^n P_n(\cos\theta)$ $\qquad\qquad\qquad\qquad\qquad$ $r^{-n-1}P_n(\cos\theta)$

where $P_n(\cos\theta)$ is a **Legendre polynomial**. These polynomials are obtainable using Rodrigues' formula:

$$P_n(x) = \frac{1}{2^n n!}\left(\frac{d}{dx}\right)^n (x^2 - 1)^n$$

The solutions of Laplace's equation are the "harmonic functions"; they are the approved bricks that satisfy the building code. You can construct whatever you want with them.

◥ EXAMPLE 12-3

There are two semi-infinite planes of $V = 0$ located at $y = 0$ and $y = b$, parallel to the xz plane, extending in the $+x$ direction, Figure 12.10a. On the yz plane the potential is V_0, a constant. Find the potential V in the region between the planes, for $x > 0$.

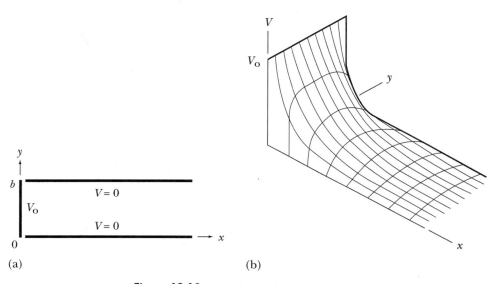

Figure 12.10   (a) Electrodes. (b) Potential.

ANSWER Rectangular coordinates will be appropriate here. As above, we begin with separation of variables: $V = XYZ$. Nothing depends on z, so we set

$$\boxed{Z = 1}$$ (12.26)

a constant, and so

$$C_z = \frac{Z''}{Z} = 0$$

Now we look back at Equation 12.23 for acceptable solutions. It appears that Y must consist of terms of the form

$$\boxed{Y = B_n \sin \frac{n\pi y}{b}}$$ (12.27)

where B_n is an arbitrary constant, and n is an integer; we omit $n = 0$ because it contributes zero to the potential, and we employ no cosines because V must be zero when $y = 0$ or $y = b$. Differentiating, we find

$$C_y = \frac{Y''}{Y} = -(n\pi/b)^2$$

Also, from Equation 12.22,

$$C_x + C_y + C_z = 0 \qquad \text{and} \qquad C_z = 0$$

so

$$C_x = +(n\pi/b)^2 = \frac{X''}{X}$$

which means X must consist of terms of the form

$$\boxed{X = A_n\, e^{-n\pi x/b}} \tag{12.28}$$

with no positive exponentials, because the potential must diminish toward zero at large x. A_n is another arbitrary constant. So V must be of the form

$$V = XYZ = C_n\, e^{-n\pi x/b} \sin\frac{n\pi y}{b} \tag{12.29}$$

where C_n is an as-yet arbitrary constant. This term satisfies most of the boundary conditions:

$$V = 0 \quad \text{at } y = 0, \quad \text{at } y = b, \quad \text{and at large } x$$

However, when $x = 0$ it is *not* V_0. In fact, no such simple term can possibly satisfy all of these conditions. We might at first fear that we are asking too much of Laplace's equation here. But surprisingly, there is a solution. We can find it by combining a number of such terms, with different n's. In particular: We can construct a square-wave function using a **Fourier series** composed of simple terms of the above form but whose *sum* will be V_0 at $x = 0$, Figure 12.11.

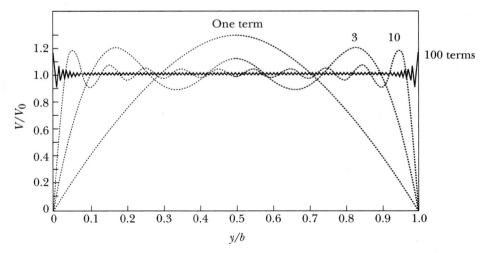

Figure 12.11   Fourier series for a square function.

We begin with a series of terms of the above form:

$$V(x, y) = \sum_{n=1}^{\infty} C_n \, e^{-n\pi x/b} \sin \frac{n\pi y}{b} \tag{12.30}$$

Each term of this infinite series is still 0 at $y = 0$ and $y = b$, and goes to zero at large x. So far so good. To get the total to be V_0 at $x = 0$ we will need to have

$$V(0, y) = V_0 = \sum_{n=1}^{\infty} C_n \sin \frac{n\pi y}{b}$$

And now we have to evaluate the coefficients C_n. There is a standard trick for doing this: We multiply both sides by $\sin \dfrac{m\pi y}{b}$ and integrate from 0 to b.

$$\int_0^b V_0 \sin \frac{m\pi y}{b} \, dy = \sum_{n=1}^{\infty} C_n \int_0^b \sin \frac{n\pi y}{b} \sin \frac{m\pi y}{b} \, dy \tag{12.31}$$

On the right side: It is a remarkable fact that only one term survives, due to the **orthogonality** of these functions:

$$\int_0^b \sin \frac{n\pi y}{b} \sin \frac{m\pi y}{b} \, dy = \frac{b}{2} \delta_{mn}$$

where the **Kronecker delta** is

$$\delta_{mn} = 0 \quad \text{unless} \quad m = n \quad \text{and} \quad \delta_{mn} = 1 \quad \text{if} \quad m = n \tag{12.32}$$

For example, for $m = n = 1$:

$$\int_0^\pi \sin y \sin y \, dy = \tfrac{1}{2} (y - \sin y \cos y) \Big]_0^\pi = \tfrac{\pi}{2} = \tfrac{\pi}{2} \delta_{11}$$

and for $m = 2$, $n = 1$:

$$\int_0^\pi \sin y \sin 2y \, dy = \int_0^\pi \sin y \, 2 \sin y \cos y \, dy = \tfrac{2}{3} \sin^3 y \Big]_0^\pi = 0 = \tfrac{\pi}{2} \delta_{21}$$

So all but one of the infinite number of terms is zero; the sole survivor is the $m = n$ term, $C_n \dfrac{b}{2}$. (One can do the same sort of thing with a cosine series; but here we need a sine series because we want it to be zero at the edges.)

Meanwhile, back on the left side of Equation 12.31, V_0 being constant, we find

$$V_0 \int_0^b \sin \frac{m\pi y}{b} \, dy = -\frac{b}{m\pi} V_0 \cos \frac{m\pi y}{b} \Big]_0^b$$

$$= \frac{2b}{m\pi} V_0 \quad \text{if } m \text{ is odd} \tag{12.33}$$

$$= 0 \quad \text{if } m \text{ is even}$$

Putting it all together,

$$\frac{2b}{n\pi} V_0 = C_n \frac{b}{2} \quad \text{if } n \text{ is odd}$$

$$= 0 \text{ if } n \text{ is even}$$

(12.34)

so

$$C_n = \frac{4}{n\pi} V_0 \quad (\text{odd } n)$$

and the Fourier series of Equation 12.30 becomes:

$$V(x, y) = \frac{4}{\pi} V_0 \sum_{\text{odd } n}^{\infty} \frac{1}{n} e^{-n\pi x/b} \sin \frac{n\pi y}{b}$$

(12.35)

This is plotted in Figure 12.10b. In the x direction, V just falls off exponentially. In the z direction, it is constant, so z is not shown at all. The y cross sections of V have more structure. At $x = 0$, the y cross section is a square function, zero at the ends and a constant V_0 in between. It is the short wavelengths, those of high n (which are high-frequency components in some applications), that produce the sharp corners; see Figure 12.11, which shows how the plot of the Fourier series becomes more angular as more terms are included. As x increases, the short wavelength sine waves are reduced because of the exponential function $e^{-n\pi x/b}$, and so the y cross sections in Figure 12.10b become smoother, approaching a half–sine wave.

For a computer approach, see Appendix C.

 EXAMPLE 12-4

Find the sum of the first three terms of Equation 12.35 for $V_0 = 1$, $x = 0$, $y = 0.5b$.

ANSWER $V = \dfrac{4}{\pi} \left(\dfrac{1}{1} - \dfrac{1}{3} + \dfrac{1}{5} \right) = 1.10$

which corresponds to the value shown on the graph, Figure 12.11, for three terms, at $y/b = 0.5$.

 EXAMPLE 12-5

Find V outside of a grounded spherical conductor of radius a inserted in a uniform electrostatic field E_0, Figure 12.12. Note that the field lines are perpendicular to the surface of the conductor, as in Figure 8.1.

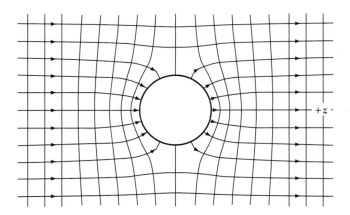

Figure 12.12   Metal sphere in uniform **E** field.

ANSWER Spherical coordinates will be appropriate here; and there is no ϕ dependence. We have

$$V = 0 \qquad \text{for} \qquad r = a$$

and for large r, placing the **E** field in the $+z$ direction,

$$V = -E_0 z = -E_0 r \cos \theta$$

By contrast with our usual practice, here we don't assume that V is zero at infinity.

In principle, the solution is a sum of the terms given in Equation 12.25, above, for Laplace's equation:

$$V(r, \theta) = \sum_{n=0}^{\infty} A_n r^n P_n(\cos \theta) + \sum_{n=0}^{\infty} B_n r^{-n-1} P_n(\cos \theta) \qquad \textbf{(12.36)}$$

Like the sines of the previous example, the Legendre polynomials are orthogonal:

$$\int_{-1}^{+1} P_m(\cos \theta)\, P_n(\cos \theta)\, d(\cos \theta) = \frac{2}{2n+1}\, \delta_{mn}$$

So we could integrate, as before, to find the coefficients (see the Problems).

But there is an easier way.

The solution for large r is $-E_0 r \cos \theta$. No larger power of r is permissible. As for smaller powers, we want one that will cancel out the $\cos \theta$, so that V can be zero on the spherical surface. Looking among the acceptable solutions, we find that the following one works:

$$V(r, \theta) = -E_0 r \cos \theta + E_0 \frac{a^3 \cos \theta}{r^2} \qquad \textbf{(12.37)}$$

This solution yields $V = 0$ at $r = a$. And for large r the second term disappears as required. So this is the required solution. We haven't proven directly that the other terms can't be there. But Uniqueness (the "by hook or by crook" algorithm) says we need look no farther. There is only one solution. If you have found a solution that works, then it's the only one there is.

The $V(r, \theta)$ consists of a superposition of two terms. The first involves a uniform field. The second you might recognize as the potential of a dipole (Section 7.1) of dipole moment $p = 4\pi\epsilon_0 E_0 a^3$. The external field of the sphere is that of an ideal dipole. In Problem 12-17 we will find this dipole using image techniques.

 EXAMPLE 12-6

Find V in the case of an uncharged ideal dielectric sphere of radius a inserted in a uniform field E_0, Figure 12.13. Note that these E field lines enter the dielectric at an oblique angle, as in Figure 9.8.

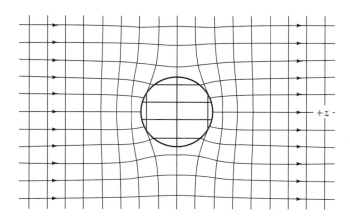

Figure 12.13   Dielectric sphere in uniform field.

ANSWER As above, **E** points in the $+z$ direction. We need a $V(r, \theta)$ such that:

(1) $V \to -E_0 r \cos \theta$ at large r.
(2) V is continuous across the surface of the dielectric.
(3) The normal component of **D** is continuous (Section 9.2), so $\partial V_o/\partial r = \epsilon_r \partial V_i/\partial r$ at $r = a$. (V_o outside, V_i inside.)

As in the previous example, it would be possible to set up an infinite series and evaluate the coefficients, but there is a better way. Encouraged by our success with the metal sphere, we will consider only terms in $\cos \theta$, Equation 12.25.

Outside the dielectric:

$$V_o(r, \theta) = -\left(1 + A\frac{a^3}{r^3}\right) E_0 r \cos \theta \qquad \text{(12.38)}$$

where A is a constant to be determined. No other positive r power is permissible because $V \to -E_0 r \cos \theta$ at large r.

Inside the dielectric:

$$V_i(r, \theta) = B E_0 \, r \cos \theta$$

B being another constant. No negative r power is permissible because the field can't diverge at $r = 0$.

Requirement (1) above is already met. For (2), we need $V_o(a, \theta) = V_i(a, \theta)$, so

$$- (1 + A) = B$$

For (3),

$$- (1 - 2A) = \epsilon_r B$$

Solving for A and B, we obtain:

$$A = - \frac{\epsilon_r - 1}{\epsilon_r + 2} \quad \text{and} \quad B = - \frac{3}{\epsilon_r + 2}$$

So the potentials are

$$V_i = \frac{-3}{\epsilon_r + 2} E_0 \, r \cos \theta$$

$$V_o = - \left(1 - \frac{\epsilon_r - 1}{\epsilon_r + 2} \frac{a^3}{r^3} \right) E_0 \, r \cos \theta \tag{12.39}$$

This is the unique solution. We didn't have to check negative powers of r outside the sphere, nor positive ones inside, because we already have an expression that works. Once more, we have used some intuition to save some mathematics.

As expected, the solution becomes that for a metal when $\epsilon_r \to \infty$.

12.4.   Examples of Poisson's Equation

To illustrate the use of Poisson's equation, $\nabla^2 V = -\rho/\epsilon_0$, we will begin by finding V for the simple case of a sphere of radius a and of uniform charge density ρ, Figure 12.14. We already did this in Problem 6-5, of course, but the approach here is quite different.

Figure 12.14   Charged sphere.

EXAMPLE 12-7

For the potential outside of the sphere, $\nabla^2 V_o = 0$, Laplace reigns, and we can just throw in harmonic functions as needed. Obviously there is no angular dependence, and the potential must go to zero as $r \to \infty$. So for V_o we have only one possible term:

$$V_o = A \frac{1}{r} \tag{12.40}$$

Inside, however, we can't use only the prefabricated solutions, because their Laplacians are zero rather than ρ. If necessary we can fall back on the general solution of $\nabla^2 V = -\rho/\epsilon_0$, which is (Section 6.3):

$$V = \int \frac{\rho \, dv}{4\pi\epsilon_0 r} \tag{12.41}$$

But we don't need to do that here. If there is no angular dependence then Poisson's equation becomes

$$\nabla^2 V_i \longrightarrow \frac{1}{r^2} \frac{\partial}{\partial r} \left(r^2 \frac{\partial V_i}{\partial r} \right) = -\rho/\epsilon_0 \tag{12.42}$$

It is easily integrated twice, giving

$$V_i = -\frac{\rho r^2}{6\epsilon_0} + B + \frac{C}{r}$$

with integration constants B and C. We can throw out C because the potential can't blow up at $r = 0$. We must have $V_o = V_i$ at the surface, $r = a$, so

$$A \frac{1}{a} = -\frac{\rho a^2}{6\epsilon_0} + B$$

Also, there being no surface charge, $E_o = E_i$, which means $\partial V_o/\partial r = \partial V_i/\partial r$, at the surface:

$$-\frac{A}{a^2} = -\frac{\rho a}{3\epsilon_0}$$

Solving for A and B, and substituting them above, we find

$$V_o = \frac{\rho a^3}{3\epsilon_0 r} \quad \text{and} \quad V_i = \frac{\rho}{6\epsilon_0}(3a^2 - r^2) \tag{12.43}$$

which are the same as the solutions found in Problem 6-5 (with $Q = \frac{4}{3}\pi R^3 \rho$).

 EXAMPLE 12-8

Find the potential between the plates of a vacuum diode, Figure 12.15a. The plate at $x = 0$ is held at $V = 0$ and constitutes an unlimited source of electrons at rest; in practice, it is a heated cathode. The other plate is at $x = s$ and is at potential $V = V_o$.

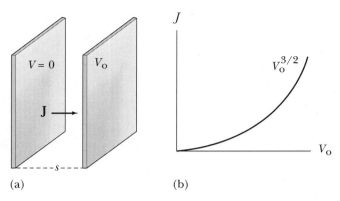

(a)   (b)

Figure 12.15   (a) Vacuum diode. (b) Child-Langmuir law.

ANSWER This is a one-dimensional problem in Cartesian coordinates. Poisson's equation becomes

$$\nabla^2 V \Longrightarrow \frac{d^2 V}{dx^2} = -\frac{\rho}{\epsilon_0} \tag{12.44}$$

The electrons move at speed v, and the current density is $J = \rho v$. By conservation of energy,

$$\tfrac{1}{2} mv^2 = eV$$

so

$$\frac{d^2 V}{dx^2} = \frac{J}{\epsilon_0} \sqrt{\frac{m}{2eV}}$$

Integrating once,

$$\left(\frac{dV}{dx}\right)^2 = 4 \frac{J}{\epsilon_0} \sqrt{\frac{mV}{2e}} + C \tag{12.45}$$

(you can differentiate it to check). The integration constant C is zero because: At the cathode, where $V = 0$, there are lots of electrons available. Electrons flow out freely, establishing a negative "space charge" in the vicinity of the cathode that tends to discourage more electrons from coming out. Equilibrium is established when $E = -dV/dx \approx 0$ at $V = 0$. Integrating again,

$$V^{3/2} = \frac{9J}{4\epsilon_0} \sqrt{\frac{m}{2e}} x^2$$

The integration constant is zero because $V = 0$ at $x = 0$. When we set $V = V_0$ at $x = s$, we find

$$J = \frac{4\epsilon_0}{9s^2} \sqrt{\frac{2e}{m}} V_0^{3/2} \tag{12.46}$$

and indeed it is found experimentally that the "Child-Langmuir law" holds under appropriate circumstances: The current is proportional to $V_0^{3/2}$, Figure 12.15b.

12.5 SUMMARY

We have developed several mathematical problem-solving techniques that are powerful but are limited to the static case. They rely on the uniqueness theorem, and they involve images, Laplace's equation, and Poisson's equation.

Poisson's equation for the electrostatic potential is

$$\nabla^2 V = -\frac{\rho}{\epsilon_0} \tag{12.1}$$

If $\rho = 0$, this becomes Laplace's equation,

$$\nabla^2 V = 0 \tag{12.2}$$

Laplace's equation admits no local maxima or minima of V.

The uniqueness theorem indicates that the V field in a region is uniquely determined by the internal charge distribution and the V at the boundaries.

There are image solutions for a charge near a conducting plane, a dielectric of plane surface, a conducting sphere, and other simple geometries. For the dielectric plane,

$$Q' = -\frac{\epsilon_r - 1}{\epsilon_r + 1} Q \tag{12.15}$$

For the conducting sphere,

$$Q' = -\frac{b}{a} Q \quad \text{and} \quad b = \frac{a^2}{D} \tag{12.16, 12.17}$$

In other cases, one may solve Laplace's equation formally. Typically we first perform a separation of variables; in rectangular coordinates,

$$V(x, y, z) = X(x) \, Y(y) \, Z(z) \tag{12.19}$$

Solutions may be built up of already known harmonic functions, for example via Fourier series.

◥ PROBLEMS

12-1 Another approach to the uniqueness theorem: Continuing from Section 12.1, show that

$$0 = \int V_3 \nabla \, V_3 \cdot d\mathbf{a}$$

over the surface of the region. Use this to show that $V_3 = 0$ inside the region.

12-2 Sketch the images for a charge in a conducting corner of angle $60°$. Then try $55°$. Finally, do a three-dimensional corner, comprising three mutually perpendicular conducting planes.

12-3 A charge Q is somewhere inside a sphere of radius R. Find the average potential V_{av} on the surface of the sphere. (Compare with Section 12.1.)

12-4 Object 1 comprises a charge distribution ρ_1 that produces potential $V(\mathbf{r}) = V_1$ everywhere. Similarly, a separate object 2 has ρ_2 and makes V_2. **(a)** Derive Green's reciprocity relationship:

$$\int \rho_1 V_2 \, d\tau = \int \rho_2 V_1 \, d\tau$$

the integrals extending over all space (use energy and superposition). **(b)** Then show that for two separate conductors, if Q on one conductor produces V on the second, then the same Q on the second produces the same V on the first. Demonstrate for a charge outside of an uncharged conducting sphere.

12-5 For a grounded conducting plane at distance D from a charge Q, Figure 12.4, find the total charge on the plane by integrating the surface charge density σ.

12-6 For a semi-infinite dielectric at distance D from a charge Q, Figure 12.6, find the total bound charge on the surface by integrating the bound charge density σ_b.

12-7 Find the force on a charge Q located at distance D from the center of an uncharged conducting sphere of radius $a < D$, Figure 12.7. Find the potential of the sphere.

12-8 Find the force on a 2-nC charge located at 20 cm from the center of a grounded conducting sphere of radius 7 cm. Find the potential of the sphere.

12-9 Repeat Problems 12-7 and 12-8 with a and D interchanged; that is, Q is *inside*, Figure 12.16.

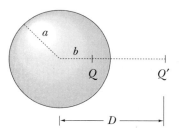

Figure 12.16

*12-10 Find the force between a charge q and a metal sphere of equal charge q and of radius a. At what distance is the force zero?

12-11 Two parallel grounded conducting planes are separated by distance s. A charge $+Q$ is located at distance $D < s$ from one plane. Sketch some of the images and develop an expression for the force on the charge (Fig. 12.9.)

12-12 Two infinite parallel grounded conducting planes are separated by distance s, and a charge $+Q$ is located between them at distance $D < s$ from one plane (Fig. 12.9). Find the image charge induced on each plane. (Consider superposition leading to a uniformly charged plane.)

12-13 A metal sphere of radius 10 cm and potential 100 V is centered 20 cm from an infinite metal plane, Figure 12.8. Estimate the charge on the sphere to within 1%.

12-14 A metal sphere of radius 2 cm holds charge 1 nC, while another of radius 4 cm is grounded. Their centers are separated by 10 cm. Sketch some of the images. Estimate the potential of both spheres to 1%.

12-15 A dipole **p** is parallel to a grounded conducting plane and situated at distance D from it. Find the force on the dipole. (See Problem 7-17.)

12-16 Find the energy W required to remove a charge Q from distance D from a grounded conducting plane to infinity.

12-17 A conducting sphere of radius a is placed between two opposite charges that produce two internal images, Figure 12.17. Increase the external charges and their separation in such a way that the **E** field remains constant. Show that the dipole moment of the sphere approaches that indicated in connection with Figure 12.12.

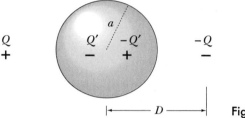

$$Q \atop +$$ $$Q' \quad -Q'$$ $$- \quad +$$ $$-Q \atop -$$

|←———— D ————→| **Figure 12.17**

*12-18 Show that the capacitance per unit length C' between two long cylinders of radius a, with axes separated by s, Figure 12.18, is $\pi\epsilon_0/\cosh^{-1}(s/2a)$; a is not necessarily small. Compare with Problem 9-17.

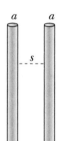

a a

s

Figure 12.18

12-19 Separate Laplace's equation in cylindrical coordinates, with no z dependence, and in spherical coordinates, with no ϕ dependence.

*12-20 In Example 12-3, show that the potential may be written as

$$V = \frac{2V_0}{\pi} \arctan \frac{\sin \pi y/b}{\sinh \pi x/b}$$

12-21 In Example 12-3, add the condition: the potential is 0 at $x = a$. Find V in the region between the plates.

*12-22 Find the potential $V(x, y, z)$ inside a cube that has constant potential V_0 on the plane at $x = 0$, and $V = 0$ at $x = b$, $y = 0$, $y = b$, $z = 0$, and $z = b$.

12-23 Find the potential everywhere inside and outside of a sphere of radius R having a specified potential $V = V(R, \theta)$ on its surface.

*12-24 Find the potential everywhere inside and outside of a sphere of radius R having $+ V_0$ on its upper hemisphere and $- V_0$ on the lower one, where V_0 is a constant. Evaluate for three terms.

12-25 A sphere of radius a carries surface charge density $\sigma = \sigma_0 \cos \theta$. Find V inside and out. (Compare with Problems 7-6 and 8-15, and Sections 9.1 and 11.2. Match $\partial V/\partial r$ as well as V at the surface; note that Gauss's law provides a condition on ΔE_r.)

12-26 Find the potential everywhere inside and outside of a sphere of radius R having a specified surface charge density $\sigma(\theta)$.

12-27 An uncharged dielectric cylinder of radius a is placed perpendicular to a previously uniform field E_0 in the x direction; find V inside and outside of the cylinder.

12-28 A grounded conducting cylinder of radius a is placed perpendicular to a previously uniform field E_0; find V inside and outside of the cylinder.

12-29 Find $\mathbf{B}_i$ and $\mathbf{B}_o$ inside and outside of a spherical permanent magnet of uniform magnetization $\mathbf{M}$ and radius R, Figure 12.19. (Compare with Sections 9.1 and 11.2.) Then find the ratio B_i/B_o at the surface at $\theta = 0°$ and at $\theta = 90°$. (Consider using V_m.) Which of these ratios is obvious by inspection?

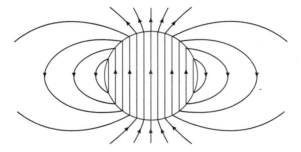

Figure 12.19

*12-30 A copper wire carries current I across a magnetic field B_0. Show that if the copper is replaced by iron, the internal field is increased by about a factor of two. (Neglect free current relative to bound current.) Does this imply greater force on the iron wire?

CHAPTER 13

CIRCUITS

T his chapter represents a kind of intermezzo involving simple calculations of practical importance in alternating-current circuitry. The techniques, which involve the use of complex numbers in describing solutions of differential equations, will also serve us well in the chapters to follow, in which we will encounter electromagnetic radiation in various of its manifestations.

13.1. Transients

At this point, in separate treatments, we have done resistors (Section 8.1), capacitors (9.4), and inductors (10.4). We found: for the resistor, $V = IR$ (Ohm's law); for the capacitor, $V = Q/C$; and for the inductor, $V = LdI/dt$. Now we can put these circuit elements together in elementary circuits and observe the resulting behavior. We divide the subject into two parts, corresponding to transient or steady-state behavior, and we begin with transients. We have already encountered transients when we found time constants in Sections 9.5 and 10.5; we shall start by reviewing and extending this work.

In the **RC circuit** of Figure 13.1a, the capacitor is initially uncharged. When we close the switch "s" at time $t = 0$, current $I(t)$ flows, and the capacitor voltage begins to go up. To simplify the notation, we shall omit the "(t)" suffix during the derivation and reattach it at the end. We have, for C and R,

$$Q = CV_C \qquad \text{and} \qquad V_R = IR$$

so the total voltage is

$$V_0 = \frac{Q}{C} + IR \qquad (13.1)$$

Differentiating,

$$0 = \frac{dQ}{dt} + RC\frac{dI}{dt}$$

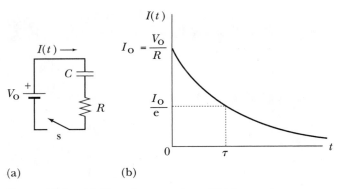

Figure 13.1   **(a)** Charging capacitor. **(b)** Time constant.

that is,

$$I = -RC\frac{dI}{dt} \tag{13.2}$$

The solution is

$$I(t) = \frac{V_0}{R}\,e^{-t/\tau} = I_0 e^{-t/\tau}$$

as you can verify by direct substitution. The **time constant**

$$\tau = \boxed{\tau_C = RC} \tag{13.3}$$

is the time for the current $I(t)$ to drop to $1/e = 0.368$ of its original value, $I_0 = V_0/R$. ("τ" is the lowercase greek letter "tau.") The current approaches zero asymptotically, Figure 13.1b.

◢ EXAMPLE 13-1

For a 1-microfarad capacitor charged by a 10-volt battery, what resistance is required for a 1-second time constant?

<u>ANSWER</u> The voltage is irrelevant. We have

$$R = \tau/C$$
$$= 1\,\text{s}/10^{-6}\,\text{C} = 1\,\text{M}\Omega \qquad (1 \text{ megohm, } 10^6 \text{ ohm}).$$

One can readily monitor this charging process using a typical solid-state voltmeter, whose input impedance is likely to be around 10 MΩ.

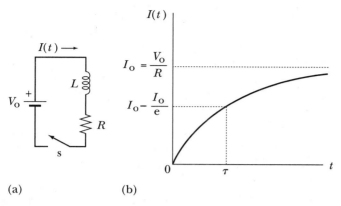

Figure 13.2 **(a)** Inductor. **(b)** Time constant.

Now we do an **RL circuit.** In Figure 13.2a the switch is closed at time $t = 0$, current $I(t)$ begins to flow, and voltage appears across the inductor (V_L) and resistor (V_R):

$$V_L = L\frac{dI}{dt} \qquad \text{and} \qquad V_R = IR$$

so the total voltage is

$$V_0 = L\frac{dI}{dt} + IR \tag{13.4}$$

The solution of this differential equation is

$$I(t) = I_0\,(1 - e^{-t/\tau})$$

as you may verify by direct substitution. The time constant

$$\tau = \boxed{\tau_L = \frac{L}{R}} \tag{13.5}$$

is the time for the current to rise to within I_0/e of its final value, Figure 13.2b.

◢ EXAMPLE 13-2

What is the time constant for a 1-henry inductor, with 30 ohms internal resistance, being charged by a 10-volt battery?

ANSWER The voltage is irrelevant. We have

$$\tau = L/R$$

$$= 1\,\text{H}/30\,\Omega = 0.033\,\text{s}$$

This would be hard to verify with a voltmeter; an oscilloscope would show it clearly, however. Since this R is internal to the inductor, you would also need a small external resistance in order to measure the current. Practical inductors typically have short time constants because of their internal resistance; whereas for capacitors this is usually no problem.

When a capacitor discharges through an inductor with no resistance, Figure 13.3a, oscillation ensues, Figure 13.3b. When the capacitor charge reaches zero, current is flowing in the inductor, which charges the capacitor up in the opposite direction; and this process repeats at a well-defined **resonant frequency.** For this **LC circuit** we have $V = 0$ around the closed circuit:

$$0 = \frac{Q}{C} + L\frac{dI}{dt}$$

Differentiating,

$$\frac{dQ}{dt} = I = -CL\frac{d^2I}{dt^2} \qquad (13.6)$$

A solution is

$$I(t) = I_0 \sin \omega t$$

where

$$\omega = \boxed{\omega_0 = \frac{1}{\sqrt{LC}}} \qquad (13.7)$$

Here the **angular frequency** (or circular frequency) ω is $\omega_0 = 2\pi f_0$ radians per second, where f_0 is the resonant frequency in Hz. $\sqrt{LC}$ is a kind of time constant (see Problem 13-4), with units of seconds.

The amplitude of the oscillation doesn't go down with time, so the energy in the circuit is clearly not getting dissipated. For ideal L, C, and R, it is only R that

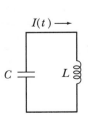

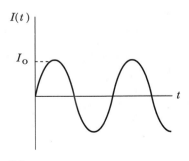

Figure 13.3   **(a)** LC circuit.
(b) Oscillation.

(a)   (b)

can dissipate energy into heat. In the case of the LC circuit, the energy appears alternately in the capacitor and the inductor. In more detail: When the current is zero, the capacitor is fully charged and the energy resides in the electric field of the capacitor; and when the capacitor is empty the current is flowing and the energy is in the magnetic field of the inductor.

 EXAMPLE 13-3

Find the resonant frequency of the combination of a 1-μF capacitor and a 10-mH inductor.

ANSWER

$$f = \omega/2\pi = 1/2\pi\sqrt{LC}$$

$$= 1/2\pi\sqrt{10^{-2}\,\text{H} \times 10^{-6}\,\text{C}} = 1600\ \text{Hz}$$

This frequency is in the audible range, 20 to 20,000 Hz. For AM broadcast, frequencies range from 540 to 1600 kHz, so LC values in such circuits may be smaller by a factor of a million.

The resonant circuit of Figure 13.3 is analogous to a simple harmonic oscillator consisting of a mass on a spring, Figure 13.4 We employ the usual notation: F is force, k is spring constant, m is mass, a is acceleration, and x is displacement from the equilibrium position. We start with

$$F = -kx \qquad \text{and} \qquad F = ma = m\frac{d^2x}{dt^2}$$

so

$$-kx = m\frac{d^2x}{dt^2}$$

A solution is

$$x = x_0 \sin \omega t$$

with

$$\omega = \sqrt{\frac{k}{m}}$$

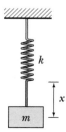

Figure 13.4 Mass on spring.

The corresponding formula for the LC oscillator (Equation 13.7) is

$$\omega = \frac{1}{\sqrt{LC}} = \sqrt{\frac{1/C}{L}}$$

In this situation the mass m is analogous to the inductance L: Each tends to slow down the oscillation by what amounts to a kind of inertia. And the spring constant k is analogous to the reciprocal of the capacitance, $1/C$: a larger k means a stiffer spring, but a larger C means a capacitor that is easier to stuff charge into. Force F corresponds to voltage V, and displacement x corresponds to charge Q; thus $F = -kx$ corresponds to $V = Q/C$. In analogy to the LC circuit, total energy resides alternately in the potential energy of the spring and the kinetic energy of the mass.

Electrical resistance corresponds to mechanical friction with force proportional to velocity, and it "damps out" oscillations by absorbing energy. Inductors made with copper wire always present substantial resistance, so we shall now consider the resulting damping of oscillations.

Above we have dealt with combinations of R, L, and C two at a time. Now we must tackle the more difficult case in which we have all three: R, L, and C. We set them in series, Figure 13.5a. The capacitor is initially at voltage V_0. The switch is closed at time $t = 0$, and voltage V_0 appears across L. The voltage across R begins at zero and builds up as current increases.

The voltage around the entire circuit remains at zero:

$$0 = IR + \frac{Q}{C} + L\frac{dI}{dt} \tag{13.8}$$

(As before, we omit the "(t)" during the derivation.) We differentiate with respect to time, knowing that $I = dQ/dt$; we obtain the second-order linear differential equation:

$$0 = \frac{I}{C} + R\frac{dI}{dt} + L\frac{d^2I}{dt^2}$$

A solution is

$$\boxed{I(t) = \frac{V_0}{\omega_n L}\, e^{-Rt/2L}\sin \omega_n t} \qquad \text{(underdamped)} \tag{13.9}$$

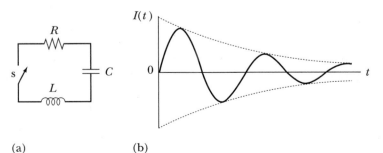

(a)   (b)

Figure 13.5   (a) Series RLC circuit. (b) Damped oscillation.

where

$$\omega_n = \sqrt{\frac{1}{LC} - \frac{R^2}{4L^2}}$$ (13.10)

This solution satisfies the condition that the current $I = 0$ at $t = 0$ (the inductance keeps it from jumping immediately to some finite value). The "natural" angular frequency ω_n is found by substituting the solution into the differential equation; and the leading factor, $V_0/\omega_n L$, is obtained by setting

$$V_0 = L\frac{dI}{dt} \qquad \text{at} \qquad t = 0$$

So if ω_n is real the result is damped sinusoidal oscillation that falls off exponentially, Figure 13.5b.

For small L/R or large LC, ω_n can become imaginary; then the sine turns into a hyperbolic sine, $\sinh x = (e^x - e^{-x})/2$, and there is no oscillation. This case is called "overdamped." The $\omega_n = 0$ case is called "critically damped" (see Fig. 13.8, below), and the other cases are "underdamped" (Fig. 13.5b, from Reitz) and "undamped" (Fig. 13.3b). These will be examined in the Problems. For the underdamped case the frequency ω_n is a little less than $\omega_0 = \sqrt{1/LC}$, but the difference is usually negligible in cases where the current oscillates many times before dying out.

Now, when you strike a bell, it rings for a while, while its energy goes down. The number of oscillations before it falls silent is a measure of the purity and duration of the tone: its "quality," so to speak. The corresponding **quality factor Q** is important in all matters involving sinusoidal oscillations. For the underdamped oscillator, roughly speaking, **Q** is the number of radians of oscillation occurring while the energy falls by a factor of e; and one complete oscillation corresponds to 2π radians. The energy stored in an inductor is proportional to the square of the current: $W = \frac{1}{2}LI^2$, Section 10.4. In this case all of the energy is stored in the inductor when $I = I_m$, the maximum current, or the amplitude of the oscillation, which is shown by the dashed line in Figure 13.5. The current I_m falls off by the exponential factor $e^{-Rt/2L}$ as time goes by; this is shown by the dashed line in Figure 13.6. So the energy falls off by the square of that factor, $(e^{-Rt/2L})^2 = e^{-Rt/L}$, as shown by the solid line in Figure 13.6, where $I_0 = V_0/\omega_n L$. (The line actually wiggles because energy is dissipated more rapidly at maximum current flow.) This means that the time for the energy to fall to $1/e$ of the original energy is L/R seconds. The angular frequency is approximately $\omega_0 = \sqrt{1/LC}$ rad/s; the distinc-

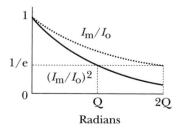

Figure 13.6   Energy falloff.

tion between ω_n and ω_0 is important only at low Q, as a few calculations will convince you. So the number Q of radians for the energy to fall to $1/e$ is L/R seconds times ω_0 radians per second:

$$Q = \frac{\omega_0 L}{R} \qquad \text{radians} \qquad \text{(13.11)}$$

Here Q means quality factor, *not* charge (this is standard notation). Dimensionally, Q is a pure number. And we might repeat, all of this is for the *under*damped oscillator.

▼ EXAMPLE 13-4

Find ω_n and Q for the circuit of Figure 13.7. In this case, the capacitor is initially uncharged and its energy builds up with time.

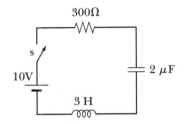

300Ω

10V

2 μF

3 H

s

Figure 13.7 Series *RLC* circuit.

ANSWER

$$\omega_n = \sqrt{\frac{1}{LC} - \frac{R^2}{4L^2}}$$

$$= \sqrt{\frac{1}{3 \times 2 \times 10^{-6}} - \frac{300^2}{4 \times 3^2}}$$

$$= \sqrt{166700 - 2500} = 405 \text{ rad/s}$$

Note that $\omega_0 = \sqrt{1/LC} = 408$ rad/s here, so the undamped frequency is nearly the same as the damped frequency, despite Q being fairly low. So

$$Q = \frac{\omega_0 L}{R}$$

$$= \frac{408 \times 3}{300} = 4.1 \text{ rad}$$

The exponential falloff shown in Figure 13.5b and Figure 13.6 corresponds roughly to this Q value. The Q of a typical bell, or a typical resonant circuit, is in the hundreds or even thousands.

▽

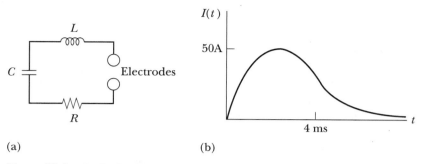

Figure 13.8   **(a)** Defibrillator circuit. **(b)** Defibrillating wave form.

An approximately critically–damped LCR circuit, Figure 13.8a, is used medically in many cases of severe **electric shock.** Typically the victim has been shocked by a current of between 100 and 200 ma (0.1 to 0.2 ampere), which has initiated "ventricular fibrillation": The normal heart rhythm is lost. An electrical "defibrillator" is employed to provide an even stronger shock, which clamps the heart muscle, after which it may resume normal beating. The defibrillating shock is provided by a charged capacitor C, and L is present in order to shape the pulse. The resistance R comes both from the circuit itself and from the resistance of the body being defibrillated; it absorbs enough energy to damp out unwanted oscillations. If you grasp the leads of an ohmmeter you will find your resistance appears to be more than 100 kΩ; but most of this resistance is on the surface, and once the electricity gets into the salt solution inside, it's clear sailing. So large electrodes are used, pressed against the skin, with conducting paste.

Figure 13.8b shows a typical wave form for the defibrillator. Approximately, peak current is 50 amperes, duration is 4 milliseconds, and the energy deposited is 200 joules (or around 2 J/kg of body weight).

13.2.   **Reactance X; Complex Notation**

Now we shall apply sinusoidal alternating voltage to individual circuit elements and see what steady-state AC current flows. In this context, "steady-state" means that the only time dependence remaining is in the applied single-frequency sinusoidal oscillation. Any transients damp out after a while, assuming there is resistance present to absorb the initial energy. This kind of calculation is of great practical importance because it is AC voltage that is supplied to us by power companies (Section 10.7), and also because sinusoidal waves of various frequencies are used in all kinds of communication from the wolf's howl to the television signal.

The AC wave is shown for the usual wall outlet in Figure 13.9a. The voltage involved could be expressed as $V(t) = V_0 \sin \omega t$, where the frequency is $f = \omega/2\pi$ and the amplitude is V_0, a constant. The frequency supplied is 60 Hz, and the peak voltage is $V_0 = 170$ V. However, in specifying voltages, it is customary to give the **root-mean-square** (rms) value V_{rms} rather than V_0; this is an appropriate and useful average value, as we shall now attempt to convince you. The average voltage

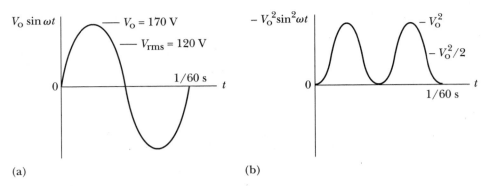

Figure 13.9 (a) Sine wave. (b) Sine square.

is just zero in Figure 13.9a, which doesn't help much in specifying the magnitude; we need a better kind of average. Power is generally proportional to voltage squared, Figure 13.9b; the average is $V_0^2/2$. The square root of this is

$$V_{\text{rms}} = V_0/\sqrt{2} = 120 \text{ volts} \tag{13.12}$$

Figure 13.9a. It is the square root of the mean (or average) of the square of the voltage: the root-mean-square, or rms, or "effective" voltage. Similarly, $I_{\text{rms}} = I_0/\sqrt{2}$. This provides an indication of the average power. In particular, the average power supplied to an ideal resistor is the same for 120 VAC, rms, as for 120 VDC, direct current (Section 8.1):

$$\begin{aligned} \text{For DC:} \quad & P = VI \\ \text{For AC:} \quad & P_{\text{av}} = V_{\text{rms}} I_{\text{rms}} \end{aligned} \tag{13.13}$$

As we have said, the voltage could be expressed as $V(t) = V_0 \sin \omega t$. Or it could be a cosine of ωt, or some combination of sine and cosine. In dealing with combinations of trigonometric functions, with various possible phase relationships, it is often easiest to express them in complex exponential form, using Euler's relationship

$$e^{i\theta} = \cos \theta + i \sin \theta$$

We use i for the **imaginary number** $\sqrt{-1}$. Leibnitz: "The imaginary numbers are a wonderful flight of God's spirit; they are almost an amphibian between being and not being." We can write

$$V(t) = V_0 e^{i\omega t} \tag{13.14}$$

with V_0 a real constant. This equation sets the phase of the oscillation, relative to which ϕ, the relative phase of I, will be expressed:

$$I(t) = I_0 e^{i(\omega t + \phi)} \tag{13.15}$$

where I_0 is also a real constant. The voltage $V(t)$ is **complex,** that is, it has both **real** and **imaginary** parts:

$$V(t) = V_r(t) + i V_i(t) = V_0 \cos \omega t + i V_0 \sin \omega t$$

where $V_r(t)$ and $V_i(t)$ are real. If we are interested in a real, instantaneous voltage, $V_r(t)$, we discard the "parasitic" imaginary term at the end of the calculation:

$$V_r(t) = \text{Re}(V_0 e^{i\omega t}) = V_0 \cos \omega t \tag{13.16}$$

("Re" means "real part of")and V_i doesn't matter. On the other hand, if we want to find the magnitude $|V(t)|$, or the amplitude V_0, we can employ the **complex conjugate** $V^*(t)$ (with $-i$ substituted for i):

$$|V(t)| = \sqrt{V(t) V^*(t)} = \sqrt{V_0 e^{i\omega t} V_0 e^{-i\omega t}} = V_0$$

In terms of V_r and V_i, this is (omitting the "(t)" for clarity here)

$$V_0 = \sqrt{(V_r + iV_i)(V_r - iV_i)} = \sqrt{V_r^2 + V_i^2}$$

Here $V_r(t)$ and $V_i(t)$ are of equal significance; also, they are functions of time, but V_0 is not.

So far we have had to write a lot of things as functions of t, but the t dependence is always the same: $e^{i\omega t}$ (once the transients have departed); so it would be convenient to cancel it out at the beginning of calculations. Also, we have employed V_0 and I_0, but it is V_{rms} and I_{rms} that the AC voltmeter and ammeter read. Finally, we would like to get rid of unnecessary subscripts if possible. We want to retain the simplicity of $V = IR$. Einstein: "Everything should be made as simple as possible, but not simpler." So we introduce a new, more convenient notation:

$$V = V_{rms} \qquad \text{and} \qquad I = I_{rms} e^{i\phi}$$

Sometimes we will face a phase angle ϕ_v other than zero for the voltage, in which case we can write: $V = V_{rms} e^{i\phi_v}$. V and I are possibly complex, but they are constant. To put it another way,

$$\boxed{V = \frac{V(t)}{\sqrt{2} e^{i\omega t}} \qquad \text{and} \qquad I = \frac{I(t)}{\sqrt{2} e^{i\omega t}}} \tag{13.17}$$

Then the actual voltage at a given instant is

$$V_r(t) = \text{Re}(\sqrt{2}\, V e^{i\omega t}) = \sqrt{2}\, V_{rms} \cos \omega t = V_0 \cos \omega t$$

and the current is

$$I_r(t) = \text{Re}(\sqrt{2}\, I e^{i\omega t}) = \sqrt{2}\, I_{rms} \cos(\omega t + \phi) = I_0 \cos(\omega t + \phi)$$

as we would wish.

And now we apply the above ideas. If we attach a voltage $V(t)$ to a **resistor** R, Figure 13.10, we find a current $I(t)$ given by Ohm's law, Section 8.1, at every instant, since ideal resistance has no frequency dependence:

$$V(t) = I(t)\, R$$

which means

$$V_0 e^{i\omega t} = I_0 e^{i(\omega t + \phi)}\, R$$

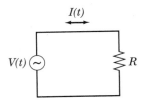

Figure 13.10   Resistance.

Canceling out the $e^{i\omega t}$ and the $\sqrt{2}$ as above, we get

$$V = IR \tag{13.18}$$

using the convenient notation we have just introduced. Since V is real and R is real, I must be real in this case, and ϕ is zero.

◤ EXAMPLE 13-5

We stick a $\frac{1}{2}$-watt resistor of 30 K (that is, 30 kΩ) into a wall outlet of 120 VAC. What current flows?

ANSWER This is too easy. $I = V/R$

$$I = 120/30,000 = 4.0 \text{ mA}$$

The V and I are rms values. And incidentally the power is

$$P_{av} = V_{rms} I_{rms}$$
$$= 120 \times 0.0040 = 0.48 \text{ W}$$

which is less than $\frac{1}{2}$ watt, so it's safe.

Now we apply an alternating voltage to a **capacitor** and see what current flows, Figure 13.11. At each instant, $Q(t) = CV(t)$, Section 9.4. Differentiating,

$$\frac{dQ}{dt} = C\frac{dV(t)}{dt} = C\frac{d}{dt} V_0 e^{i\omega t} \tag{13.19}$$

so, since dQ/dt is $I(t)$, the rate of charge flow,

$$I(t) = I_0 e^{i(\omega t + \phi)} = C i\omega V_0 e^{i\omega t}$$

and, canceling $e^{i\omega t}$,

$$I_0 e^{i\phi} = i\omega C V_0$$

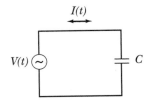

Figure 13.11   Capacitive reactance.

which means

$$e^{i\phi} = i \quad \text{so} \quad \phi = \frac{\pi}{2}$$

Dividing $I_0 e^{i\phi} = i\omega C\, V_0$ by $\sqrt{2}$, we get

$$I = i\omega C\, V \quad \text{so} \quad V = \frac{I}{i\omega C}$$

The **capacitive reactance** is customarily given as

$$\boxed{X_C = -\frac{1}{\omega C}} \tag{13.20}$$

In terms of X_C,

$$\frac{V}{I} = \frac{1}{i\omega C} = -\frac{X_C}{i} = iX_C$$

So for a capacitor we have a relationship analogous to Ohm's law:

Resistor: $V = IR$
Capacitor: $V = I\,iX_C$ $\tag{13.21}$

Also,

$$\frac{|V|}{|I|} = |i\,X_C| = \frac{1}{\omega C} = \frac{V_0}{I_0} = \frac{V_{rms}}{I_{rms}}$$

The reactance behaves in some ways like a resistance; major differences, however, include the following: It changes with frequency, it entails a phase change, and it dissipates no energy.

◤ EXAMPLE 13-6

Can we plug a 1-microfarad capacitor into a wall outlet?

ANSWER In general it is poor practice to plug miscellaneous things into wall outlets. In this case, we have $f = 60$ Hz so

$$\omega = 2\pi\, 60 = 377 \text{ rad/s}$$

Then,

$$\frac{V}{I} = iX_C = \frac{1}{i\omega C} = -i\,\frac{1}{377 \text{ rad/s} \times 1 \times 10^{-6}\,C} = -i\,2650\ \Omega$$

$$|I| = I_{rms} = V/X_C$$
$$= 120/2650 = 45 \text{ ma}$$

which is a small current; circuit breakers or fuses can typically take 15 A or more. The primary concern is with the voltage rating, which should be around 200 VDC, since the AC voltage swings from + 170 V to − 170 V. Also, for such a purpose you can't use an electrolytic capacitor that is designed to withstand voltage in only one direction.

As we have remarked above, for a capacitor $\phi = \pi/2$, which means that the current leads the voltage by 90°, Figure 13.12. This implies that no energy is dissipated. For half of the cycle, $V_r(t)$ and $I_r(t)$ have the same sign, so power $P = V_r I_r > 0$ is stored in the capacitor (we omit "(t)" for clarity). For the rest of the time, V_r and I_r have opposite sign, $P < 0$, and energy is fed from the capacitor back into the AC outlet. One such region is marked in Figure 13.12. So the average power is zero. And in fact, our ideal capacitor has no resistance, so there is no way electrical energy can be dissipated (turned into heat energy).

Capacitors serve somewhat like frequency-dependent resistors. Their reactance is $|X_C| = 1/\omega C$, so their effective resistance goes down as ω goes up. For direct current or low frequencies, no current flows after the capacitor has charged up. But at high frequency, the capacitor is being charged and discharged frequently, so there is substantial charge flow and substantial current. A capacitor passes high frequencies and blocks low frequencies.

And now for the **inductor**, Figure 13.13. From Section 10.4, $V(t) = L\, dI(t)/dt$, so

$$V_0 e^{i\omega t} = L\frac{dI(t)}{dt} \qquad (13.22)$$

Integrating,

$$\int V_0 e^{i\omega t}\, dt = \int L\, dI(t)$$

$$\frac{V_0 e^{i\omega t}}{i\omega} = L\, I(t) = L\, I_0 e^{i(\omega t + \phi)}$$

so, canceling $e^{i\omega t}$,

$$V_0 = I_0 e^{i\phi}\, i\omega L$$

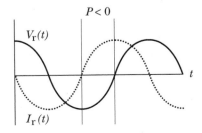

Figure 13.12   Current leads voltage.

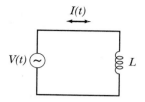

Figure 13.13   Inductive reactance.

so

$$e^{i\phi} \, i = 1 \quad \text{and} \quad \phi = -\frac{\pi}{2}$$

Dividing $V_0 = I_0 e^{i\phi} \, i\omega L$ by $\sqrt{2}$, we obtain

$$V = I i\omega L$$

The **inductive reactance** is customarily given as

$$\boxed{X_L = \omega L} \tag{13.23}$$

so

$$V = I \, i \, X_L$$

and

$$\frac{|V|}{|I|} = |i \, X_L| = \omega L = \frac{V_0}{I_0} = \frac{V_{rms}}{I_{rms}}$$

Again, the reactance behaves in some ways like a resistance.

◤ **EXAMPLE 13-7**

What happens when we plug a 1-henry inductor into a wall outlet?

ANSWER   We still have $\omega = 377$ rad/s, as above. Then,

$$I = V/i\omega L$$

$$|I| = 120/377 \times 1 = 0.32 \text{ A}$$

which won't blow a circuit breaker. But the inductor must be designed to carry 0.32 A without saturating the iron core (Section 10.8) or overheating. Real inductors typically have an internal resistance which is a substantial fraction of their reactance at design frequency. In this example, the inductor carries 120 V × 0.32 A = 38 volt-amperes (volt-amp is the unit of choice when the product does not represent watts of power dissipated); so the inductor should be massive enough to handle a few watts.

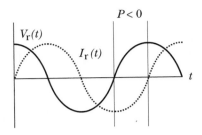

Figure 13.14   Voltage leads current.

As we have remarked previously, for an inductor $\phi = -\pi/2$, which means that the current lags the voltage by 90°, Figure 13.14. This implies that no energy is dissipated. For half of the cycle, $V_r(t)$ and $I_r(t)$ have the same sign, so power $P = V_r I_r > 0$ is stored in the capacitor (we omit "(t)" for clarity). For the rest of the time, V_r and I_r have opposite sign, $P < 0$, and energy is fed from the inductor back into the AC outlet. One such region is marked in Figure 13.14. So the average power is zero. And in fact our ideal inductor has no resistance (nor hysteresis, Section 10.8), so there is no way electrical energy can be dissipated (turned into heat energy).

Inductors, like capacitors, can serve as what might be called frequency-dependent resistors. Their reactance is $|X_L| = \omega L$, so their effective resistance goes up with ω. So inductors are complementary to capacitors in that they block high frequencies and pass low ones. This goes back to Faraday's law: The "back emf" is proportional to the rate of change of current.

In practical circuits one usually finds far more capacitors than inductors. Some reasons: The capacitor can store more than ten times more energy for a given weight or volume; the capacitor typically has low resistance and so dissipates less energy; and the capacitor usually has better high-frequency characteristics. (Ferrites work well at high frequency, but their effective μ_r is small.)

13.3. Power

Power dissipation for a resistor is $P(t) = V(t) I(t)$, Section 8.1, all real functions of time. For reactive elements, we need only the real parts of $V(t)$ and $I(t)$.

$$P = (\text{Re } V(t))(\text{Re } I(t)) \tag{13.24}$$

As we have seen, $V(t)$ and $I(t)$ are not necessarily in phase. For $V(t)$ we can write

$$V(t) = V_0 e^{i\omega t} = V_0(\cos \omega t + i \sin \omega t)$$

so

$$\text{Re } V(t) = V_0 \cos \omega t$$

Now if $I(t)$ is out of phase with $V(t)$ by the angle ϕ, then

$$I(t) = I_0 e^{i(\omega t + \phi)} = I_0 e^{i\omega t} e^{i\phi} = I_0(\cos \omega t + i \sin \omega t)(\cos \phi + i \sin \phi)$$

so

$$\text{Re } I(t) = I_0 (\cos \omega t \cos \phi - \sin \omega t \sin \phi)$$

and the power is

$$P = V_0 \, (\cos \omega t) \, I_0 \, (\cos \omega t \cos \phi - \sin \omega t \sin \phi)$$

$$= V_0 I_0 \cos^2 \omega t \cos \phi - V_0 I_0 \cos \omega t \sin \omega t \sin \phi$$

The average over a whole cycle, of time $2\pi/\omega$, is given by

$$P_{av} = \frac{\displaystyle\int_0^{2\pi/\omega} P \, dt}{\displaystyle\int_0^{2\pi/\omega} dt} = \tfrac{1}{2} \, V_0 \, I_0 \cos \phi$$

So

$$\boxed{P_{av} = V_{rms} \, I_{rms} \cos \phi} \qquad (13.25)$$

The cosine of the phase angle ϕ is called the "power factor." In the above examples involving a capacitor and an inductor, we have $\cos \phi = 0$, so no power dissipation.

There is another useful expression for power:

$$P_{av} = \tfrac{1}{2} \, V_0 I_0 \cos \phi \qquad \text{from above}$$

$$= \tfrac{1}{2} \, \text{Re} \, (V_0 e^{i\omega t} I_0 e^{-i\omega t - i\phi})$$

$$\boxed{P_{av} = \tfrac{1}{2} \, \text{Re} \, (V(t) \, I^*(t))} \qquad (13.26)$$

which lets us keep V and I in complex notation.

For maximum power transfer from a source to a user, **matching impedances** are needed. We will limit ourselves here to real V and I, because that is where the power is (ideal L and C absorb no average power). Suppose we have a power source containing an emf V and an internal resistance r, and we secure power from it with a load R, Figure 13.15. The current is

$$I = \frac{V}{r + R}$$

and the power obtained in R is

$$I^2 R = \frac{V^2 R}{(r + R)^2}$$

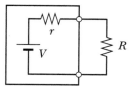

Source Load   **Figure 13.15**   Impedance matching.

When we set the derivative with respect to R equal to zero, we obtain the requirement that $r = R$ for maximum power.

13.4. Impedance Z

Now we can combine components of R, L, and C into circuits and see how they behave. Each element contributes a resistance and/or a reactance, as in the preceding section; and these combine to produce an **impedance** Z, which may be complex. (One may also employ the **admittance** $Y = 1/Z$.) Then Ohm's law $V = IR$ becomes, in terms of the V and I notation introduced in the preceding section,

$$\boxed{V = IZ} = I(R + iX) \tag{13.27}$$

For example, for the series circuit of Figure 13.16, we have

$$Z = R + iX_L + iX_C$$
$$= R + i\omega L + \frac{1}{i\omega C} \tag{13.28}$$

The quantities V, I, and Z are complex constants in general, although where it is convenient we adjust the phase of I in order to make V real. The value of the simple V and I notation introduced in the preceding section becomes apparent here: The magnitudes of this V and I are the rms voltage and current respectively: $|V| = V_{rms}$ and $|I| = I_{rms}$; and the various reactances behave mathematically just like resistances, except that they are complex.

For instantaneous values of current and voltage we have the assistance of **Kirchhoff's laws:**

1. Net current into any point is zero. This is illustrated in Figure 13.17a: 3 A flows into the branch or "node," and a total of $2 + 1 = 3$ A flows out. This implies that charge does not accumulate in conductors, which is a low-frequency approximation; at high frequencies the capacitance of circuit elements can become important. As we shall see, this Kirchhoff law does not apply to average or rms currents, but only to currents at a particular instant.

2. Net voltage change around a closed circuit is zero. In Figure 13.17b a 3-volt battery supplies voltages of 2 and 1 volts to two resistors in series. This implies no changing magnetic field, which can produce a net emf around a closed circuit. As with currents, this law does not apply to average voltages, but only to the voltage at any instant.

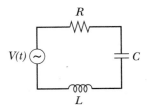

Figure 13.16   Impedance.

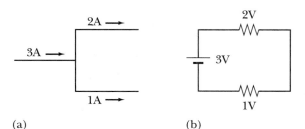

Figure 13.17   **(a)** First law.
(b) Second law.

(a)   (b)

That's all there is to it, in principle. The way to really understand how it all works together is to do a lot of problems, so let's start in.

If we put all three elements in series with the voltage source, as in Figure 13.16, then we can write for the circuit as a whole:

$$Z = R + i\omega L + \frac{1}{i\omega C} = R + i\left(\omega L - \frac{1}{\omega C}\right)$$

We can already see a **resonance** coming: If $\omega L = 1/\omega C$, then Z is minimum. Continuing,

$$I = \frac{V}{Z} = \frac{V}{R + i\left(\omega L - \frac{1}{\omega C}\right)} \tag{13.29}$$

To rationalize, we multiply numerator and denominator by the complex conjugate of the denominator:

$$I = \frac{V}{R + i\left(\omega L - \frac{1}{\omega C}\right)} \frac{R - i\left(\omega L - \frac{1}{\omega C}\right)}{R - i\left(\omega L - \frac{1}{\omega C}\right)}$$

$$I = \frac{VR - i\,V\left(\omega L - \frac{1}{\omega C}\right)}{R^2 + \left(\omega L - \frac{1}{\omega C}\right)^2}$$

Also,

$$|I| = \sqrt{II^*} = \frac{|V|}{\sqrt{R^2 + \left(\omega L - \frac{1}{\omega C}\right)^2}} = \frac{|V|}{|Z|} \tag{13.30}$$

The current magnitude $I_{\text{rms}} = |I|$ is plotted as a function of angular frequency ω in Figure 13.18a. It reaches a maximum of V/R at

$$\omega L = \frac{1}{\omega C} \qquad \text{so} \qquad \omega = \boxed{\omega_0 = \frac{1}{\sqrt{LC}}} \tag{13.31}$$

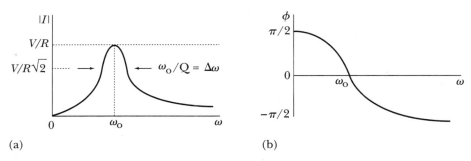

(a)   (b)

Figure 13.18   **(a)** Series resonance. **(b)** Current phase shift.

where $f_0 = \omega_0/2\pi$ is the resonant frequency; the phenomenon and formula were encountered in Section 13.1. This is **series resonance,** the capacitor and inductor being in series. The full width $\Delta\omega$ of the curve at $\sqrt{2}$ of the peak (which is half-power) is about

$$\Delta\omega \approx \omega_0/Q$$

where Q is the quality factor encountered in Section 13.1,

$$Q = \frac{\omega_0 L}{R} \tag{13.32}$$

(see Problem 13-16); so a larger Q gives a narrower, sharper resonance curve.

At low frequency, the reactance is capacitive (the capacitor is blocking the current), and so the current phase ϕ is positive: Current leads voltage, Figure 13.18b. The phase angle passes through zero at resonance.

EXAMPLE 13-8

For each component in the circuit of Figure 13.19, find the (rms) current through it and voltage across it.

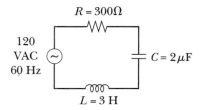

Figure 13.19   Series circuit.

ANSWER We'll work this very instructive example out in great detail. The expression given above makes considerable use of the combination $\omega L - 1/\omega C$, so let's get a number for that first:

$$\omega = 2\pi f$$

$$= 2\pi \, 60 = 377 \text{ rad/s}$$

$$\omega L - \frac{1}{\omega C} = 377 \times 3 - \frac{1}{377 \times 2 \times 10^{-6}}$$

$$= 1131 - 1326 = -195 \ \Omega$$

Since this is negative, the overall circuit shows capacitive reactance, and current will lead voltage. The inductive and capacitive reactances are nearly equal, so we are near resonance; in fact, the resonant frequency is $1/2\pi\sqrt{LC} = 65.0$ Hz.

Now we'll go ahead and find the current; from above:

$$I = \frac{VR - i\,V\left(\omega L - \dfrac{1}{\omega C}\right)}{R^2 + \left(\omega L - \dfrac{1}{\omega C}\right)^2} \tag{13.33}$$

$$= \frac{120 \times 300 + 120 \times 195\,i}{300^2 + 195^2} = 0.281 + 0.183\,i$$

The rms current is what we want:

$$I_{\text{rms}} = |I| = \sqrt{0.281^2 + 0.183^2} = 0.335 \text{ A}$$

and it leads the voltage by

$$\phi = \arctan\,(0.183/0.281) = 33.1°$$

The current is the same in all components.

The rms voltage across the resistor is

$$|V_{\text{R}}| = |I|\,R$$

$$= 0.335 \times 300 = 101 \text{ V}$$

For the capacitor,

$$|V_{\text{C}}| = |I|/\omega C$$

$$= 0.335/377 \times 2 \times 10^{-6} = 444 \text{ V}$$

For the inductor,

$$|V_{\text{L}}| = |I|\,\omega L$$

$$= 0.335 \times 377 \times 3 = 379 \text{ V}$$

Those are the rms voltages requested in the problem. They don't add up to 120 volts, because they are not in phase. In fact, at resonance, for $R = 0$, the voltages across the inductor and capacitor become infinite and opposite. But when phases are taken into account, the voltages do add up to 120 V. We have set the phase angle equal to zero for the applied voltage, Section 13.2. But here the voltages across the various components are not all in phase; each component has its own voltage phase angle. In principle, we begin with $V = IZ$, all complex, for each component:

$$V_R = |V_R| e^{i\phi_R} = IR$$

$$= (0.281 + 0.183i)\ 300 = 84 + 55\ i\ \text{volts}$$

$$V_C = |V_C| e^{i\phi_C} = I/i\omega C \tag{13.34}$$

$$= (0.281 + 0.183i)/i\ 377 \times 2 \times 10^{-6} = 243 - 373\ i$$

$$V_L = |V_L| e^{i\phi_L} = I i\omega L$$

$$= (0.281 + 0.183i)\ i\ 377 \times 3 = -207 + 318\ i$$

and so

$$V = (\text{sum of above}) = 120\ \text{volts, all real}$$

Note also that V_C and V_L are 180° out of phase with each other:

$$\arctan \frac{\text{Im}(V_C)}{\text{Re}(V_C)} = \arctan \frac{-373}{243} = -57°$$

$$\tag{13.35}$$

$$\arctan \frac{\text{Im}(V_L)}{\text{Re}(V_L)} = \arctan \frac{318}{-207} = 123°$$

The power is

$$V_{\text{rms}}\ I_{\text{rms}} \cos \phi = 120 \times 0.335 \times \cos 33.1° = 33.7\ \text{watts}$$

All of it is dissipated in the resistor:

$$|V_R||I| = 101 \times 0.335 = 33.8\ \text{W} \qquad \text{close enough}$$

As for Q, in this case

$$Q = \omega_0 L/R$$

$$= 2\pi 65 \times 3/300 = 4.1$$

(we found the 65 Hz above). This relatively small Q corresponds to Figure 13.18.

And now, a question. We are bathed, night and day, in man-made electromagnetic waves that are broadcast at all sorts of frequencies, attempting to bring us news of distant events or of headache remedies. Yet if we turn on the radio, it selects only one station from among the many signals present. How does it do it?

The answer is, it exploits the resonant frequency of the LC circuit, which we have just explored. We adjust L or C in the radio until the resonant frequency matches that of the broadcasting station. (In a heterodyne circuit, we also adjust an internal oscillator, but that gets beyond the scope of this book.) The effective Q

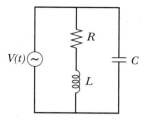

Figure 13.20   Parallel circuit.

should be high, to shut out unwanted frequencies, but not so high as to cut into the station's bandwidth. For example: AM stations broadcast at around 1000 kHz. Neighboring AM stations are separated by 10 kHz, so the bandwidth is around 10 kHz (this enables them to broadcast audio signals up to around 10 kHz). Then Q needs to be no larger than around

$$Q = \frac{\omega_0}{\Delta\omega} = \frac{1000 \text{ kHz}}{10 \text{ kHz}} = 100 \tag{13.36}$$

And now let's try **parallel resonance,** Figure 13.20. The parallel impedances add as reciprocals:

$$\frac{1}{Z} = \frac{1}{R + i\omega L} + \frac{1}{1/i\omega C} = \frac{1}{R + i\omega L} + i\omega C \tag{13.37}$$

$$= \frac{R}{R^2 + \omega^2 L^2} + i\left(\omega C - \frac{\omega L}{R^2 + \omega^2 L^2}\right) \tag{13.38}$$

To find $|I|$ we first multiply Equation 13.37 by its complex conjugate:

$$\frac{1}{|Z|^2} = \left(\frac{1}{R + i\omega L} + i\omega C\right)\left(\frac{1}{R - i\omega L} - i\omega C\right)$$

which finally yields

$$|I| = \frac{|V|}{|Z|} = |V|\sqrt{\frac{\omega^2 R^2 C^2 + (1 - \omega^2 L C)^2}{R^2 + \omega^2 L^2}} \tag{13.39}$$

A resonance seems dimly perceptible here. It appears that $|I|$ will be a minimum at or near $\omega^2 L C = 1$, corresponding to the same frequency that produced maximum current in series resonance. However, further analysis reveals that the minimum current actually occurs at a slightly lower frequency. In fact, there turn out to be three differently defined resonant frequencies: **(1)** the frequency for which $|I|$ is a minimum; **(2)** the frequency for which the current and voltage are in phase (the imaginary part vanishes in Eq. 13.38); **(3)** the frequency for which $\omega = 1/\sqrt{LC}$ (as in Eq. 13.31). These frequencies are all the same for series resonance, and they become the same in parallel resonance when Q is large. We'll look into the matter further in the problems.

Note that at resonance: The series LC has minimum Z, and the parallel LC has maximum Z (for frequency (1) above, anyway).

13.5 SUMMARY

Our modern technology depends on AC circuits, and here we have learned how to do some of the relevant calculations for current and power.

The time constant for a capacitor is $\tau_C = RC.$ (13.3)

and that for an inductor is $\tau_L = \dfrac{L}{R}.$ (13.5)

For an LC circuit, resonance occurs at $\omega = \dfrac{1}{\sqrt{LC}}.$ (13.7)

AC voltage and current are expressible in the form

$$V(t) = V_0 e^{i\omega t} \quad \text{and} \quad I(t) = I_0 e^{i(\omega t + \phi)} \qquad \text{(13.14, 13.15)}$$

The root-mean-square value is a suitable average value:

$$V_{\text{rms}} = V_0/\sqrt{2} \qquad \text{(13.12)}$$

So we introduce the convenient averages

$$V = \dfrac{V(t)}{\sqrt{2}e^{i\omega t}} \quad \text{and} \quad I = \dfrac{I(t)}{\sqrt{2}e^{i\omega t}} \qquad \text{(13.17)}$$

The reactance of a capacitor is

$$X_C = -\dfrac{1}{\omega C} \qquad \text{(13.20)}$$

and that of an inductor is

$$X_L = \omega L \qquad \text{(13.23)}$$

Ohms law is generalized to include impedance:

$$V = IZ \qquad \text{(13.27)}$$

with

$$Z = R + iX$$

Average power dissipated is

$$P_{\text{av}} = V_{\text{rms}} I_{\text{rms}} \cos \phi \qquad \text{(13.25)}$$

where $\cos \phi$ is the power factor.

The quality factor Q is a measure of the width of a resonance:

$$Q = \dfrac{\omega_0 L}{R} \qquad \text{(13.11, 13.32)}$$

 **PROBLEMS**

13-1 A 20-μF capacitor is charged to 10 V and attached to a 1-M resistor. How long does it take to drop to 1 volt? 0 volt?

13-2 A 2-H inductor, of internal resistance 400 Ω, is attached to a source of 10 VDC. How long does it take for the current to reach 20 ma? 30 ma?

13-3 A 1000-pF capacitor resonates at 500 kHz with an ideal inductor. Find the inductance.

13-4 For Figure 13.3, find the time for the first radian of oscillation, in terms of L and C, and in seconds, using 1 μF and 10 mH.

13-5 Find $I(t)$ in the case of an overdamped circuit, Figure 13.5a. Use $\omega' = \sqrt{\dfrac{R^2}{4L^2} - \dfrac{1}{LC}}$.

Sketch $I(t)$ versus t.

13-6 Repeat the preceding problem for a critically damped circuit.

13-7 An infinite plane array of 1-K resistors is shown, Figure 13.21, connected in a square grid. Find the resistance across one resistor, for example between points a and b.

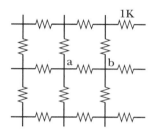

Figure 13.21

13-8 Twelve 1-K resistors are connected as the edges of a cube, Figure 13.22; find the resistance between points **(a)** a and d **(b)** b and d **(c)** c and d.

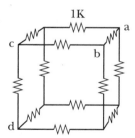

Figure 13.22

13-9 In Figure 13.23, $R = 500 \Omega$, $C = 0.03 \mu$F, and $L = 27$ mH. Find the frequency of oscillation and the time for the energy to drop to 10%, for large t.

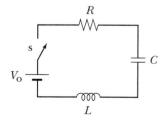

Figure 13.23

13-10 Repeat the preceding problem for $R = 5$ K $(= 5$ k$\Omega)$.

13-11 For the preceding problem, with $C = 0.03$ μF and $L = 27$ mH, find R for critical damping.

13-12 For the defibrillator of Figure 13.8, estimate: V, R, C, and L.

13-13 Find the reactance of a 100-pF capacitor (a picofarad is 10^{-12} F) at a frequency of 1.6 MHz.

13-14 Find the frequency at which the reactance of a 100-μH inductor is equal to the negative of that found in the preceding problem.

13-15 Find the resonant frequency of a 4000-pF capacitor and a 300-μH inductor, connected in series. Repeat for parallel.

13-16 Show that in the underdamped case, $Q \approx \omega_0 / \Delta\omega$, where $\Delta\omega$ is the full width of the resonance curve of Figure 13.18a.

13-17 Evaluate I and I_{rms} in Figure 13.24 using the parameters provided (as in Fig. 13.19).

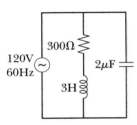

120V
60Hz
300Ω
3H
2μF

Figure 13.24   Parallel circuit.

13-18 For the preceding problem, find the power dissipated in the resistor and in the circuit as a whole.

13-19 Find the three resonant frequencies in the preceding problem, and find the limit for small resistance.

13-20 For the preceding problem, find Q at resonance. Plot I_{rms} versus ω and compare with Figure 13.18a.

13-21 For Figure 13.20, find Z.

13-22 For the series-parallel circuit of Figure 13.25, find the impedance.

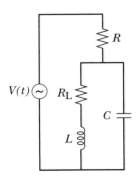

$V(t)$ R_L R C L

Figure 13.25   Series-parallel circuit.

13-23 Find Z at resonance for series and parallel resonance in the case of large Q.

13-24 An American light bulb dissipates 100 W at 120 VAC and 60 Hz. Add an inductor so that the bulb will function properly in a French circuit of 220 VAC, 50 Hz. (Neglect the resistance of the inductor and the inductance of the light bulb.) Find the power dissipated.

13-25 Under what condition does Figure 13.26 represent a differentiating circuit:

$$V_0 \approx RC\frac{dV_i}{dt}$$

13-26 Draw a circuit similar to that of Figure 13.26 that can serve as an integrating circuit, and provide the corresponding equation and condition.

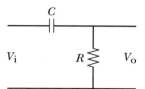

Figure 13.26

13-27 For the circuit of Figure 13.27, show that the impedance between the ends (marked "a") is independent of frequency if $RC = L/R$.

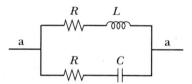

Figure 13.27

*13-28 The Maxwell bridge, Figure 13.28, is used for measuring the inductance and resistance of an inductor. When $V = 0$ in the detector, find L and R_L in terms of the other parameters. Qualitatively, what determines the values of ω and C?

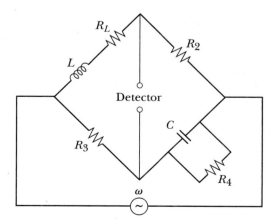

Figure 13.28

C H A P T E R

14

RADIATION

We now embark upon a study, in four chapters, of electromagnetic radiation. This first chapter deals in general terms with the transport of energy, the wave equations and their solutions, and the actual process of radiation from a simple charge. The discussion will be based largely on Maxwell's equations as displayed in Chapter 11.

14.1. Poynting's Vector S. Energy Flow and Momentum

From our standpoint as humans, the primary function of the electromagnetic field is transportation of energy, whether in radiation (solar energy is essential to life) or in voltages and currents on copper wires. So far we have studied various aspects of the electromagnetic field, including its energy density, but we have yet to deal with the question of how exactly energy gets from one place to another. The results may surprise you.

The units can give us a clue. E is measured in volts per meter, and H comes in amperes per meter. If we take the product, we get watts per square meter, which is exactly what we need for a measure of energy flux. We would prefer the cross product, because we want a vector that will tell us what direction the energy is flowing. This suggests that we try $\mathbf{E} \times \mathbf{H}$. We will remain in the realm of ideal materials and so we can use the third form of Maxwell's equations from Chapter 11, with $B = \mu H$ and $D = \epsilon E$. So we shall now investigate the quantity

$$\mathbf{S} = \mathbf{E} \times \mathbf{B}/\mu \tag{14.1}$$

which is called **Poynting's vector** after an Englishman who discovered it in 1884.

If $\mathbf{S}$ actually represents flow of energy, then its divergence should be expressible in terms of changes in electromagnetic energy. That is because, referring to Figure 14.1, if there is a flux of $\mathbf{S}$ through the surface $\mathbf{a}$ of the volume v shown, then by the divergence theorem there must be a positive divergence of $\mathbf{S}$ inside:

$$\oint \mathbf{S} \cdot d\mathbf{a} = \int \nabla \cdot \mathbf{S} \, dv$$

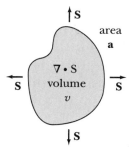

Figure 14.1   Poynting's vector.

So let's take a look at $\nabla \cdot \mathbf{S}$.

$$\nabla \cdot \mathbf{S} = \nabla \cdot (\mathbf{E} \times \mathbf{B}/\mu)$$
$$= \mathbf{B} \cdot \nabla \times \mathbf{E}/\mu - \mathbf{E} \cdot \nabla \times \mathbf{B}/\mu \qquad \text{(A vector identity)}$$
$$= -\mathbf{B} \cdot \partial\mathbf{B}/\mu\partial t - \mathbf{E} \cdot \left(\mathbf{J} + \epsilon \frac{\partial \mathbf{E}}{\partial t} \right) \qquad \text{(Maxwell equations)}$$

$\qquad\qquad\qquad\qquad\qquad\qquad\qquad\qquad\qquad\qquad\qquad\qquad$ **(14.2)**

$$\nabla \cdot \mathbf{S} = -\frac{\partial}{\partial t} (B^2/2\mu + \epsilon E^2/2) - \mathbf{E} \cdot \mathbf{J} \qquad \text{(Rearranging)} \qquad \textbf{(14.3)}$$

So $\nabla \cdot \mathbf{S}$ does indeed represent loss of electromagnetic energy in the volume. The first term on the right involves electromagnetic field energy (Section 6.1); if it decreases, then it tends to make the divergence of $\mathbf{S}$ positive, meaning that the energy is leaving the volume. The second term, $\mathbf{E} \cdot \mathbf{J}$, can be due to Joule heating (Section 8.1), which converts electrical energy into heat; this reduces the energy available for $\mathbf{S}$.

That's all very well, but it is incomplete: We have tied $\nabla \cdot \mathbf{S}$ to energy conservation, but $\nabla \times \mathbf{S}$ is not yet nailed down. In other words, we could write $\mathbf{S} = \mathbf{E} \times \mathbf{B}/\mu_0 + \nabla \times \mathbf{U}$, where $\mathbf{U}$ is an arbitrary vector, and the above derivation would still hold. So we shall finish, as is customary and consistent with other results, simply by defining Poynting's vector as $\mathbf{S} = \mathbf{E} \times \mathbf{B}/\mu_0$, and interpreting it as energy flow per unit area.

This is an astonishing concept! It suggests that if we find a place where $\mathbf{E}$ is perpendicular to $\mathbf{B}$, then there is energy flowing there. It also says that when energy flows, as in radiation, the flow is transverse to both $\mathbf{E}$ and $\mathbf{B}$. Since energy has mass, and thus carries momentum, it also brings us news concerning radiation pressure and momentum transfer. It is really rather unbelievable at first. So we will try to convince you with a few examples.

In Figure 14.2 we have a battery V that provides power to a resistor R. Since $\mathbf{E}$ and $\mathbf{J}$ are in opposite directions in the battery, $\mathbf{E} \cdot \mathbf{J}$ is negative there; so instead of removing power as Joule heat, the battery is providing electromagnetic energy to the system. $\mathbf{B}$ circulates about the battery; so we see that Poynting's vector $\mathbf{S}$ points out into the air, and not along the wires.

And when we examine $\mathbf{E}$ and $\mathbf{B}$ in the vicinity of the resistor, we find that $\mathbf{S}$ points into the resistor from the air, and again *not* along the wires.

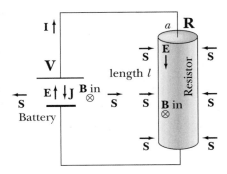

Figure 14.2   Resistor with battery.

EXAMPLE 14-1

Integrate **S** over the surface of the resistor, Figure 14.2; show that the result equals *VI*, which is the electrical power dissipated in the resistor.

ANSWER

The electric field is $E = V/l$.
The magnetic field is $B = \mu I / 2\pi a$ (at the surface).
So, Poynting's vector is $S = EB/\mu = VI/2\pi al$.

Poynting's vector **S** is constant over the surface, so the total power is the area times S:

$$P = 2\pi alS = VI \tag{14.4}$$

which demonstrates quantitatively that the power which heats the resistor enters through the side of the cylinder and not through the wires. So the energy is in the fields; the conductors merely provide boundary conditions and guide the fields. The electromagnetic energy goes out of the battery into the air, and then goes into the resistor from the air.

EXAMPLE 14-2

Find S for a half-watt resistor (Fig. 14.2, $l = 1$ cm, $a = 2$ mm) that is in fact dissipating 0.5 W.

ANSWER

$$S = P/2\pi al$$

$$= 0.5 \text{ W}/2\pi\, 0.01 \text{ m} \times 0.002 \text{ m} = 4000 \text{ W/m}^2$$

which is nearly three times the solar constant (1400 W/m^2 at the top of the earth's atmosphere).

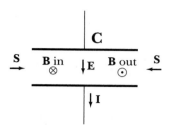

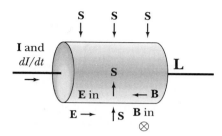

Figure 14.3   Capacitor.

Figure 14.4   Inductor.

In Figure 14.3 we have a capacitor C that is being charged. On the left, at the point marked $\otimes$, **B** points into the paper, and so it points out on the right. The **E** field points down, as shown. So the energy flux $\mathbf{S} = \mathbf{E} \times \mathbf{B}/\mu$ is entering through the air at the edges of the capacitor C. The energy W is being stored in the volume v of the **E** field; in terms of the energy density, $dW/dv = \frac{1}{2}\,\epsilon\,E^2$.

In Figure 14.4 the current in the inductor L, a solenoid, is increasing. Outside of the inductor, the situation is like that for the resistor, Figure 14.2: **E** points to the right, and **B** is in. Inside, the effective directions of **E** and **B** are rotated by 90°, and this **E** is due to Faraday's law. In each case, Poynting's vector brings energy inwards, where it is stored in $\frac{1}{2}\,\mathbf{B}^2/\mu$. We will save the details for Problem 14-4.

In Figure 14.5, a charge moves to the right at velocity **v**, and it has an **E** field and a **B** field, each of which contains energy. As we have remarked before, fields don't move. So how does the field energy get out to where the charge is going?

The answer is, Poynting's vector picks up energy from behind the charge, totes it along, and drops it in front of the charge. Use your right hand to verify the direction of **S** all around the moving charge. I have always felt that this is a little like workers moving a block of stone toward a pyramid, rolling it on wooden rollers; they pick up the rollers from behind the block and place them in front. However, if you ride on the block, the workers and rollers are still there; but if you move along with the charge, all the complicated machinery disappears: there is no **B**, no **S**, and no transport of energy. In the chapter on Relativity we will discuss how fields change as you change reference frames.

The flow of energy is not always so obvious, especially if there is an exchange between electrical energy and mechanical energy. In Figure 14.6 there is a

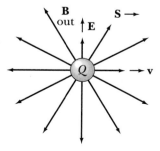

Figure 14.5   Moving charge.

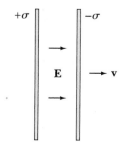

Figure 14.6   Moving capacitor.

+σ   **K →**
                                    → v
▔▔▔▔▔▔▔▔▔▔▔▔▔
↓**E**   ⊗ **B** in   **→ S**

▔▔▔▔▔▔▔▔▔▔▔▔▔   **→ v**
−σ   **K ←**   **Figure 14.7**   Moving capacitor.

charged capacitor moving in a direction perpendicular to its parallel plates. The energy in the field **E** is moving to the right, but there is no Poynting's vector. In this case, it is necessary to supply a force to move the negative plate to the right against the attraction of the positive plate. This force, applied to the moving negative plate, provides the power required to create the electric field. An opposite force on the positive plate absorbs the energy. The power is transferred between the plates mechanically, through the supports that keep the plates apart.

Figure 14.7 shows a charged capacitor moving (slowly) parallel to its plates. The moving charged plates provide surface current density $\mathbf{K} = \sigma\mathbf{v}$, and so a magnetic field $B = \mu_0 K$ is present. But the resulting S turns out to be twice as great as needed to carry the field energy to the right. The excess is transferred to the plates by the fringing fields (Herrmann).

Energy has mass, and so Poynting's vector carries momentum as well as energy. For light, the energy is moving at speed c, the speed of light. Figure 14.8 shows a block of light (you can think of it as a group of photons), of total energy W, which will arrive at the area a in time t. Poynting's vector is energy per unit area and per unit time:

$$S = \frac{W}{at} \tag{14.5}$$

The mass of the energy is $m = W/c^2$, and its momentum is $p = mc$, even relativistically, when m includes the mass of the energy involved ("relativistic mass"). So we get

$$S = \frac{pc}{at}$$

The volume is act, so the **momentum density** is

$$\frac{p}{act} = \boxed{g = \frac{S}{c^2}} \tag{14.6}$$

This formula turns out to be general, although we have derived it only for light.

The force F is p/t, so the **radiation pressure** P on the area a is given by

$$\frac{F}{a} = \frac{p}{at} = \boxed{P = \frac{S}{c}} \tag{14.7}$$

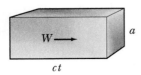

W → a

ct **Figure 14.8**   Light momentum.

if a absorbs the radiation. For a reflective surface, the change in momentum is twice as great, so the pressure is doubled.

 EXAMPLE 14-3

Find the pressure of sunlight at the earth's surface, using intensity $S = 700$ W/m^2. (At the top of the earth's atmosphere, $S = 1400$ W/m^2.)

ANSWER

$P = S/c$

$P = 700/3 \times 10^8 = 2.3 \times 10^{-6}$ pascals (Pa)   which is 2.3×10^{-11} atmospheres

Radiation pressures are typically small even for high light power levels, because photons move so fast. Analogously, a bullet and the gun that fires it acquire momenta of equal magnitude, but the bullet has much more kinetic energy; otherwise you might as well keep the bullet and send the gun.

Electromagnetic field momentum explains the presence of angular momentum in the interaction of a stationary positron ($+e$) and magnetic monopole (Q_m), Figures 5.3, 14.9, and 14.11. In this configuration, **S** points out of the paper on the right, and in on the left. Since **S** carries momentum, angular momentum is present that here, using the right-hand rule, points in the direction from $+e$ to Q_m. The electromagnetic field swirls invisibly about this apparently inert system; see Problem 14-15.

We have used relativistic concepts ($W = mc^2$) in finding the radiation pressure P, but it was known classically before relativity. The following simplified discussion will indicate how. We need to import two results for light from Section 14.5: **E** is perpendicular to **B**, and $E = cB$; we jump the gun a little because we want to present the two approaches near each other, for comparison. Now: **S** impinges normally on area a of a low-density medium (nonreflecting) obeying Ohm's law, $J = \sigma E$. The energy dissipated in thickness Δz, Figure 14.10, is

$$\Delta W = -a\,\Delta S = -EJ a\,\Delta z \tag{14.8}$$

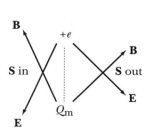

Figure 14.9 Monopole.

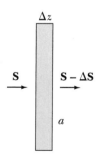

Figure 14.10 Radiation pressure.

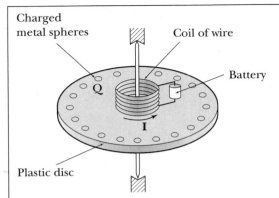

Charged metal spheres — Coil of wire — Battery

Q

I

Plastic disc

Figure 17-5. Will the disc rotate if the current I is stopped?

Imagine that we construct a device like that shown in Fig. 17–5. There is a thin, circular plastic disc supported on a concentric shaft with excellent bearings, so that it is quite free to rotate. On the disc is a coil of wire in the form of a short solenoid concentric with the axis of rotation. This solenoid carries a steady current I provided by a small battery, also mounted on the disc. Near the edge of the disc and spaced uniformly around its circumference are a number of small metal spheres insulated from each other and from the solenoid by the plastic material of the disc. Each of these small conducting spheres is charged with the same electrostatic charge Q. Everything is quite stationary, and the disc is at rest. Suppose now that by some accident—or by prearrangement—the current in the solenoid is interrupted, without, however, any intervention from the outside. So long as the current continued, there was a magnetic flux through the solenoid more or less parallel to the axis of the disc. When the current is interrupted, this flux must go to zero. There will, therefore, be an electric field induced which will circulate around in circles centered at the axis. The charged spheres on the perimeter of the disc will all experience an electric field tangential to the perimeter of the disc. This electric force is in the same sense for all the charges and so will result in a net torque on the disc. From these arguments we would expect that as the current in the solenoid disappears, the disc would begin to rotate. If we knew the moment of inertia of the disc, the current in the solenoid, and the charges on the small spheres, we could compute the resulting angular velocity.

But we could also make a different argument. Using the principle of the conservation of angular momentum, we could say that the angular momentum of the disc with all its equipment is initially zero, and so the angular momentum of the assembly should remain zero. There should be no rotation when the current is stopped. Which argument is correct? Will the disc rotate or will it not?

The answer is that if you have a magnetic field and some charges, there will be some angular momentum in the field. It must have been put there when the field was built up. When the field is turned off, the angular momentum is given back. So the disc in the paradox *would* start rotating. This mystic circulating flow of energy, which at first seemed so ridiculous, is absolutely necessary, There is really a momentum flow. It is needed to maintain the conservation of angular momentum in the whole world.

Figure 14.11 Excerpts from *The Feynman Lectures on Physics* are displayed here: pages II-17–6 and II-27-11. This paradox vividly demonstrates the need to include electromagnetic field momentum in momentum conservation considerations. (From THE FEYNMAN LECTURES ON PHYSICS Vol. II, by Richard Feynman, Robert Leighton, and Sands; Copyright (c) 1964 by the California Institute of Technology. Reprinted by permission of Addison Wesley Longman, Inc.)

so

$$J\Delta z = \frac{\Delta S}{E}$$

When the current J flows, B gets its hands on it, with $\mathbf{F} = Q\mathbf{v} \times \mathbf{B}$ becoming BIl in the direction of $\mathbf{S}$:

$$\Delta F = BJa\,\Delta z$$

so, using $J\Delta z$ from above,

$$\frac{\Delta F}{a} = \Delta P = BJ\Delta z = -\frac{B}{E}\Delta S = -\frac{\Delta S}{c} \tag{14.9}$$

Now when we integrate down from initial S to zero, and up from zero pressure to P, we get $P = S/c$ as above.

14.2.   Maxwell Stress Tensor

We have already found the pressure exerted by static fields in certain situations, Section 6.1; and we have calculated the flow of energy and of momentum, Section 14.1. We shall now carry through the more general task of finding the force on an arbitrary charge distribution strictly in terms of the **E** and **B** fields.

To begin with, if we have a charge density ρ and a current density **J**, in a a region where there are **E** and **B** fields, then the Lorentz force **F** per unit volume v is

$$\frac{d\mathbf{F}}{dv} = \rho\mathbf{E} + \mathbf{J} \times \mathbf{B} \tag{14.10}$$

We use Maxwell's equations (Section 11.1) to eliminate ρ and **J**.

$$\frac{d\mathbf{F}}{dv} = (\epsilon\nabla\cdot\mathbf{E})\,\mathbf{E} + \left(\nabla \times \mathbf{B}/\mu - \epsilon\frac{\partial\mathbf{E}}{\partial t}\right) \times \mathbf{B}$$

Now

$$\frac{\partial\mathbf{E}}{\partial t} \times \mathbf{B} = \frac{\partial}{\partial t}\,(\mathbf{E} \times \mathbf{B}) - \mathbf{E} \times \frac{\partial\mathbf{B}}{\partial t} = \mu\frac{\partial\mathbf{S}}{\partial t} + \mathbf{E} \times (\nabla \times \mathbf{E})$$

so

$$\frac{d\mathbf{F}}{dv} = \epsilon[(\nabla\cdot\mathbf{E})\,\mathbf{E} - \mathbf{E} \times (\nabla \times \mathbf{E})] - \frac{1}{\mu}\mathbf{B} \times (\nabla \times \mathbf{B}) - \epsilon\mu\frac{\partial\mathbf{S}}{\partial t} \tag{14.11}$$

Using a vector identity (for $\nabla(\mathbf{A} \cdot \mathbf{B})$, inside front cover), we find that

$$\mathbf{E} \times (\nabla \times \mathbf{E}) = \tfrac{1}{2}\nabla(E^2) - (\mathbf{E}\cdot\nabla)\,\mathbf{E}$$

and **B** is similar. Also, we add the zero term $(\nabla \cdot \mathbf{B})\,\mathbf{B}$ in order to make the expression symmetrical in **E** and **B**. We arrive at

$$\frac{d\mathbf{F}}{dv} = \epsilon[(\nabla\cdot\mathbf{E})\,\mathbf{E} + (\mathbf{E}\cdot\nabla)\,\mathbf{E}] + \frac{1}{\mu}[(\nabla\cdot\mathbf{B})\,\mathbf{B} + (\mathbf{B}\cdot\nabla)\,\mathbf{B}]$$
$$- \tfrac{1}{2}\nabla(\epsilon\,E^2 + B^2/\mu) - \epsilon\,\mu\frac{\partial\mathbf{S}}{\partial t} \tag{14.12}$$

At this point we introduce the Maxwell stress tensor **T** (Section 1.6), with components T_{ij}:

$$T_{ij} = \epsilon\,(E_iE_j - \tfrac{1}{2}\delta_{ij}\,E^2) + \frac{1}{\mu}\,(B_iB_j - \tfrac{1}{2}\delta_{ij}\,B^2) \tag{14.13}$$

where the Kronecker delta δ_{ij} is 1 if i = j, and is zero otherwise. As before, we have numbered the coordinates: $x \to x_1$, $y \to x_2$, and $z \to x_3$; thus, i and j can take the values 1, 2, or 3. We can take the divergence of this tensor; the jth component is:

$$(\nabla\cdot\mathbf{T})_j = \frac{\partial T_{ij}}{\partial x_i}$$

where summation is implied over the repeated index, i. (This divergence is still a vector.) When we carry this through, we find that the result fits in very neatly with the force density already given in Equation 14.12, which becomes:

344 Radiation

$$\frac{d\mathbf{F}}{dv} = \nabla \cdot \mathbf{T} - \epsilon\mu \frac{\partial \mathbf{S}}{\partial t} \tag{14.14}$$

(As in the case of Poynting's vector, Section 14.1, the definition of **T** is incomplete in that any divergence-free tensor may be added to **T** without affecting Eq. 14.14.)

Now we integrate over the volume of interest; we can use the divergence theorem on **T**.

$$\mathbf{F} = \oint \mathbf{T} \cdot d\mathbf{a} - \frac{\partial}{\partial t} \int \frac{\mathbf{S}}{c^2} \, dv \tag{14.15}$$

Physical Significance The second term on the right involves the volume integral of the momentum density $g = S/c^2$, Section 14.1. So that integral is the momentum of the electromagnetic field. If it changes with time, there must be an associated force, in accordance with Newton's second law, $\mathbf{F} = d\mathbf{p}/dt$. This verifies that our rather limited deduction of $g = S/c^2$ in Section 14.1 actually produced a general result.

The first term on the right of Equation 14.15 involves electromagnetic field pressure. In the static case, that's all there is. (Note that, despite appearances, $\mathbf{T} \cdot d\mathbf{a}$ is a vector, because **T** is a tensor.) The tensor **T** is the stress, that is, the force per unit area, and when it is integrated over the area the result is force. This is a little like Gauss's law: Instead of integrating over charges in a volume, we can integrate over the surface boundary of the volume.

In tensor notation, for $\mathbf{S} = 0$, Equation 14.15 becomes

$$F_i = \int T_{ij} \, da_j$$

and **T** is:

j = 1	2	3
$\frac{\epsilon}{2}(E_x^2 - E_y^2 - E_z^2) + \frac{1}{2\mu}(B_x^2 - B_y^2 - B_z^2)$	$\epsilon E_x E_y + B_x B_y/\mu$	$\epsilon E_x E_z + B_x B_z/\mu$
$\epsilon E_y E_x + B_y B_x/\mu$	$\frac{\epsilon}{2}(E_y^2 - E_x^2 - E_z^2) + \frac{1}{2\mu}(B_y^2 - B_x^2 - B_z^2)$	$\epsilon E_y E_z + B_y B_z/\mu$
$\epsilon E_z E_x + B_z B_x/\mu$	$\epsilon E_z E_y + B_z B_y/\mu$	$\frac{\epsilon}{2}(E_z^2 - E_x^2 - E_y^2) + \frac{1}{2\mu}(B_z^2 - B_x^2 - B_y^2)$

Now if we have only E_x and da_x, then the only applicable term is at the upper left of the matrix. T_{xx} applies a pressure $\frac{\epsilon}{2} E_x^2$ in the x direction—the rubber bands pull along their length. If we have only E_y and da_x, then the pressure T_{xx} is

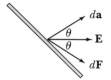

Figure 14.12   **E** bisects angle.

$-\dfrac{\epsilon}{2} E_y^2$ in the other direction—the rubber bands repel each other sideways. The magnitude of the pressure is equal to the energy density, as we saw in Section 6.1.

And if we have both E_x and E_y, there is a sideways pressure T_{xy}—a shear. The field **E** bisects the angle between the force $d\mathbf{F}$ and the normal to the surface, $d\mathbf{a}$, Figure 14.12 (see Problem 14-11).

◢ EXAMPLE 14-4

(a) Using the stress tensor, find the force between charges of $+Q$ and $-Q$, separated by distance D (in free space). **(b)** Remove the $-Q$ charge and find the force on the region of space that it formerly occupied.

ANSWER (a) The integral of **T** over an arbitrary surface will provide the force on the charge inside the surface. Figure 14.13 shows the two charges on the z axis, separated by distance D. We choose a hemispherical surface of radius R, shown in dotted outline in Figure 14.13a, with its flat surface on the xy plane (which bisects D), and we permit R to increase without limit. The E field at distance R becomes a dipole field and so E^2 is proportional to $1/R^6$, whereas the area of the hemispherical surface is proportional only to R^2; so the contribution of the spherical part of the surface is negligible at large R. On the flat surface, $E = E_z$ by symmetry, and the surface normal is entirely in the $-z$ direction, $da = da_z$. So only one term in the **T** matrix is involved, namely the $\dfrac{\epsilon_0}{2} E_z^2$ in the lower right-hand element. The

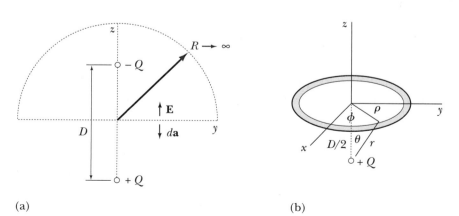

(a)   (b)

Figure 14.13   **(a)** Hemisphere. **(b)** Circle in xy plane.

force is

$$F_z = \int \frac{\epsilon_0}{2} E_z^2 \, da_z$$

integrated over the xy plane. We use circular elements of area $da_z = 2\pi r \, dr$, Figure 14.13b (we can use $\rho \, d\rho = r \, dr$ since $(D/2)^2 + \rho^2 = r^2$). So

$$F_z = \pi\epsilon_0 \int \left(\frac{2Q}{4\pi\epsilon_0 r^2} \cos \theta \right)^2 r \, dr$$

We substitute $D/2 = z$, $r = z/\cos \theta$, and $dr = z \sin \theta \, d\theta/\cos^2\theta$, obtaining

$$F_z = \pi\epsilon_0 \left(\frac{2Q}{4\pi\epsilon_0 z^2} \right)^2 \int_0^{\pi/2} \cos^3\theta \sin \theta \, d\theta$$

$$= \frac{Q^2}{4\pi\epsilon_0 (2z)^2} = \frac{Q^2}{4\pi\epsilon_0 D^2}$$

this corresponds to Coulomb's law, of course, but here we have obtained it using the stress tensor.

(b) Now we remove the $-Q$ charge and repeat, using the same hemispherical surface. The preceding example exercised only one term of **T**, but this one will bring much more of the machinery into play despite the basic simplicity of the problem. We have x, y, and z field components, so the entire right-hand column of the matrix contributes:

$$F_x = \int \epsilon_0 E_x E_z \, da_z$$

$$F_y = \int \epsilon_0 E_y E_z \, da_z$$

$$F_z = \int \frac{\epsilon_0}{2} (E_z^2 - E_x^2 - E_y^2) \, da_z$$

Our hope, of course, is to find that each of these components is zero, since there is no charge inside the surface.

First we try F_x:

$$F_x = \int_0^\infty \int_0^{2\pi} \epsilon_0 E_x E_z \rho \, d\phi \, d\rho$$

The E component in the xy plane is $E \sin \theta$; and the x component of this is $E_x = E \sin \theta \cos \phi$, which gives zero when integrated over ϕ (since neither E nor E_z depends on ϕ). So

$$\boxed{F_x = 0}$$

as desired.

The situation for F_y is similar, with $E_y = E \sin \theta \sin \phi$, so:

$$\boxed{F_y = 0}$$

And now for F_z. We have

$$E_z^2 - E_x^2 - E_y^2 = E^2(\cos^2\theta - \sin^2\theta\cos^2\phi - \sin^2\theta\sin^2\phi)$$

$$= E^2(\cos^2\theta - \sin^2\theta)$$

So

$$F_z = \frac{\epsilon_0}{2}\int\left(\frac{Q}{4\pi\epsilon_0 r^2}\right)^2 (\cos^2\theta - \sin^2\theta)^2\, 2\pi r\, dr$$

Again we substitute $D/2 = z$, $r = z/\cos\theta$, and $dr = z\sin\theta\, d\theta/\cos^2\theta$, obtaining

$$F_z = \pi\epsilon_0\left(\frac{Q}{4\pi\epsilon_0 z^2}\right)^2\left(\int_0^{\pi/2}\cos^3\theta\sin\theta\, d\theta - \int_0^{\pi/2}\sin^3\theta\cos\theta\, d\theta\right)$$

so

$$\boxed{F_z = 0}$$

So the force does turn out to be zero. Still, it's interesting how much seems to be going on in this region of empty space.

14.3. Wave Equations. Lorentz Condition

When we tweak the electromagnetic field, a disturbance tends to travel out as a wave at the speed of light. And in fact, at a fundamental level that is the story of electromagnetic radiation, including light, x-rays, radio waves, and all that. We now develop the wave equations governing this radiation, still for ideal media.

We begin with Faraday's law:

$$\nabla \times \mathbf{E} = -\frac{\partial \mathbf{B}}{\partial t}$$

Taking the curl,

$$\nabla \times (\nabla \times \mathbf{E}) = -\frac{\partial(\nabla \times \mathbf{B})}{\partial t} \tag{14.16}$$

Then we invoke a vector identity: $\nabla \times (\nabla \times \mathbf{E}) = \nabla(\nabla \cdot \mathbf{E}) - \nabla^2\mathbf{E}$; and we employ a couple of Maxwell equations. We find

$$\nabla^2\mathbf{E} - \nabla(\nabla \cdot \mathbf{E}) = \mu\frac{\partial}{\partial t}\left(\mathbf{J} + \epsilon\frac{\partial \mathbf{E}}{\partial t}\right)$$

$$\boxed{\nabla^2\mathbf{E} - \mu\epsilon\frac{\partial^2\mathbf{E}}{\partial t^2} = \mu\frac{\partial \mathbf{J}}{\partial t} + \nabla\frac{\rho}{\epsilon}} \tag{14.17}$$

Similarly, when we take the curl of Ampère's law, we arrive at (see Problem 14-13)

$$\boxed{\nabla^2\mathbf{B} - \mu\epsilon\frac{\partial^2\mathbf{B}}{\partial t^2} = -\mu\nabla \times \mathbf{J}} \tag{14.18}$$

These are the **nonhomogeneous wave equations** for **E** and **B**. It is encouraging to note that, in the absence of free charge and current, **E** and **B** have the same wave equation so they travel along together at speed $v = 1/\sqrt{\mu\epsilon}$; we will need both fields in order to provide the **E** × **B** for Poynting's vector.

Observe that both of these wave equations require the $\partial\mathbf{E}/\partial t$ in Ampère's law. That's one reason Maxwell added this term.

We would hope that **A** and V would be fellow travelers too. We want the radiation to be able to travel along in more or less stable configuration for millions of miles. We begin with

$$\mathbf{E} = -\nabla V - \frac{\partial\mathbf{A}}{\partial t}$$

If we take the curl, we arrive immediately at Faraday's law. So we take the divergence:

$$\nabla \cdot \mathbf{E} = \frac{\rho}{\epsilon} = -\nabla^2 V - \frac{\partial\nabla \cdot \mathbf{A}}{\partial t} \tag{14.19}$$

So far, no wave equation. However, up to this point we have had no particular need to specify the divergence of **A**. Only its curl has been specified: $\mathbf{B} = \nabla \times \mathbf{A}$ (Section 6.2). We can define $\nabla \cdot \mathbf{A}$ so as to produce the desired wave equation by introducing the **Lorentz condition:**

$$\boxed{\nabla \cdot \mathbf{A} = -\mu\epsilon\frac{\partial V}{\partial t}} \tag{14.20}$$

With the Lorentz condition, Equation 14.9 becomes the desired nonhomogeneous wave equation for V:

$$\boxed{\nabla^2 V - \mu\epsilon\frac{\partial^2 V}{\partial t^2} = -\frac{\rho}{\epsilon}} \tag{14.21}$$

(Use of the Lorentz condition implies what may be called the Lorentz gauge, or the radiation gauge. There are other gauges; in particular, for the Coulomb gauge we set $\nabla \cdot \mathbf{A} = 0$. This gauge is interesting in the static case because Ampère's law becomes

$$\nabla \times \mathbf{B} = \mu_0\mathbf{J} + \mu_0\epsilon_0\partial\mathbf{E}/\partial t$$

$$\nabla \times (\nabla \times \mathbf{A}) = \nabla(\nabla \cdot \mathbf{A}) - \nabla^2\mathbf{A} = \mu_0\mathbf{J} \tag{14.22}$$

$$\nabla^2\mathbf{A} = -\mu_0\mathbf{J}$$

like Poisson's equation, which simplifies some magnetostatic calculations. The Lorentz gauge becomes the Coulomb gauge in the static case.)

Similarly, starting with $\mathbf{B} = \nabla \times \mathbf{A}$, if we take the divergence, it all goes away. So we take the curl, and again we employ the Lorentz condition; we find (see

Problem 14-14)

$$\boxed{\nabla^2 \mathbf{A} - \mu\epsilon \frac{\partial^2 \mathbf{A}}{\partial t^2} = -\mu\mathbf{J}}$$
(14.23)

So as expected, in the absence of free charge and current all four wave equations are of the same form, and $\mathbf{E}$, $\mathbf{B}$, V, and $\mathbf{A}$ march through space in lockstep.

The wave equations for $\mathbf{A}$ and V, with the Lorentz condition, are equivalent to Maxwell's equations in the following sense: If we can find solutions for $\mathbf{A}$ and V, then we can find $\mathbf{E}$ and $\mathbf{B}$ using $\mathbf{E} = -\nabla V - \partial \mathbf{A}/\partial t$ and $\mathbf{B} = \nabla \times \mathbf{A}$. They are more economical in that they contain only four independent components (one for V and three for $\mathbf{A}$) instead of six.

The corresponding waves contain information of all kinds; indeed, one may say, the history of the universe is written, on the universe itself, in the form of electromagnetic waves. In Section 14.4 we take a look at general solutions of the wave equations, which are fundamentally important but not very manageable on a practical level; and in Section 14.5 we shall simplify down to plane waves, which are usually all we really need.

14.4.   Solutions; Retarded Potentials

Solutions of the wave equations for V and $\mathbf{A}$ may be shown to be the **retarded potentials**

$$\boxed{\begin{aligned} V(t) &= \int \frac{\rho(t - r/\mathrm{v})}{4\pi\epsilon\, r}\, dv = \int \frac{[\rho]}{4\pi\epsilon\, r}\, dv \\[2mm] \mathbf{A}(t) &= \int \frac{\mu\mathbf{J}(t - r/\mathrm{v})}{4\pi\, r}\, dv = \int \frac{\mu[\mathbf{J}]}{4\pi\, r}\, dv \end{aligned}}$$
(14.24)

where the square brackets indicate **retarded** quantities, that is, the quantities evaluated at an earlier time $t - r/\mathrm{v}$, v being the speed of propagation of the corresponding waves. Conceptually these solutions are simple: The potential at a given point at time t depends on the position of neighboring charges at an earlier time, because it takes time r/v for information regarding these charges to reach the point where the potential is to be calculated. *Notation:* Here we use V for scalar potential, v for velocity, and v for volume.

So now, for those of you who are keeping score, we have written Maxwell's equations; we have rewritten them in terms of V and $\mathbf{A}$; and we have solved them, in general, in terms of the retarded potentials.

We shall now demonstrate by direct substitution that the V solution is consistent with its wave equation. The procedure for $\mathbf{A}$ is similar.

First we take the gradient of the given retarded V:

$$4\pi\epsilon\, \nabla V = \int \nabla\, \frac{\rho(t - r''/\mathrm{v})}{r''}\, dv$$
(14.25)

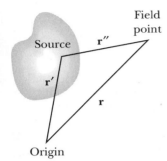

Field
point

Source $\mathbf{r}''$

$\mathbf{r}'$

$\mathbf{r}$

Origin

Figure 14.14   Potential V at a field point.

We will not bother with brackets on ρ, to simplify the notation. We have exchanged the order of differentiation and integration because different variables are involved, Figure 14.14, as in Sections 3.7 and 7.3; and as before we are going to call $\mathbf{r}'' = \mathbf{r} - \mathbf{r}'$ the vector from source to field point. So when we differentiate with respect to $\mathbf{r}$, the numerator and denominator both contribute. We get

$$4\pi\epsilon\,\nabla V = \int\left(\frac{1}{r''}\,\nabla\rho(t - r''/v) + \rho\,\nabla\frac{1}{r''}\right)dv$$

Now

$$\frac{\partial\rho}{\partial r} = \frac{d\rho}{d[t]}\,\frac{\partial[t]}{\partial r''} = -\dot\rho\,\frac{1}{v}$$

so

$$4\pi\epsilon\,\nabla V = \int\left(-\frac{\dot\rho}{v}\,\frac{\hat{\mathbf{r}}''}{r''} - \rho\,\frac{\hat{\mathbf{r}}''}{r''^2}\right)dv$$

Now we take the divergence, in order to finish up with $\nabla^2 V$:

$$4\pi\epsilon\,\nabla^2 V = \int\left(-\frac{\dot\rho}{v}\,\nabla\cdot\frac{\hat{\mathbf{r}}''}{r''} - \frac{\hat{\mathbf{r}}''}{r''}\cdot\nabla\frac{\dot\rho}{v} - \rho\nabla\cdot\frac{\hat{\mathbf{r}}''}{r''^2} - \frac{\hat{\mathbf{r}}''}{r''^2}\cdot\nabla\rho\right)dv \quad \textbf{(14.26)}$$

$$= \int\left(-\frac{\dot\rho}{v}\,\frac{1}{r''^2} + \frac{1}{v^2 r''}\,\frac{d^2\rho}{d[t]^2} - \rho 4\pi\delta\,(r'') + \frac{\dot\rho}{v}\,\frac{1}{r''^2}\right)dv$$

The delta function was introduced in Section 2.4. Now we divide by $4\pi\epsilon$:

$$\nabla^2 V = \frac{1}{v^2}\int\frac{d^2\rho}{d[t]^2}\,\frac{dv}{4\pi\epsilon r''} - \int\frac{\rho}{\epsilon}\,\delta(r'')\,dv$$

The ρ's are still retarded, but the integral V is not.

$$\nabla^2 V = \frac{1}{v^2}\,\frac{\partial^2 V}{\partial t^2} - \frac{\rho(r'')}{\epsilon} \quad \textbf{(14.27)}$$

And the remaining ρ is not retarded; its value is significant only at $r'' = 0$, the field point, because the delta function is zero everywhere else. This equation is in fact the wave equation for V, with $1/v^2 = \epsilon\mu$, showing that the retarded potential is consistent with the wave equation.

The following example will illustrate results of this and the preceding section.

▼ EXAMPLE 14-5

At time $t = 0$, uniform surface current density $\mathbf{K}$ begins to flow in the $+x$ direction in the xy plane, Figure 14.15, with constant rate of change $\dot{K}$, so that $\mathbf{K} = t\,\dot{K}\,\hat{\mathbf{x}}$. Find and sketch the fields $\mathbf{A}$, V, $\mathbf{E}$, and $\mathbf{B}$ in charge-free, current-free space on the $+z$ axis.

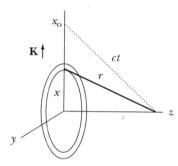

Figure 14.15   Retarded $\mathbf{A}$.

ANSWER Starting at time $t = 0$, a disturbance in the electromagnetic field propagates outward from the xy plane in both the $+z$ and $-z$ directions. For the scalar potential, from Equation 14.24,

$$V(t) = \int \frac{\rho(t - r/\mathrm{v})}{4\pi\epsilon\,r}\,dv = 0$$

everywhere, since there is no ρ anywhere. The vector potential is

$$\mathbf{A}(t) = \int \frac{\mu \mathbf{J}(t - r/\mathrm{v})}{4\pi\,r}\,dv = \int \frac{\mu_0 \mathbf{K}(t - r/\mathrm{v})}{4\pi\,r}\,da$$

We must find $\mathbf{A}$ by integrating the current contributions of $\mathbf{K}$ in the xy plane at the earlier, retarded time. ($\mathbf{K}$ starts to flow everywhere in the xy plane at $t = 0$, but its effect is felt at z at a later time.) The circular area element shown in the xy plane is $da = 2\pi x\,dx$, which is the same as $2\pi r\,dr$ (since $x^2 + z^2 = r^2$). Retardation is constant around this circle. The retarded time is zero at x_0; only points closer than x_0 to the origin contribute to $\mathbf{A}$. With retardation, setting $\mathrm{v} = c$, the time is $(t - r/c)$, so retarded $\mathbf{K}$ is

$$[\mathbf{K}] = (t - r/c)\,\dot{K}\hat{\mathbf{x}}$$

and $\mathbf{A}$ becomes

$$\mathbf{A} = \mu_0\dot{K}\,\hat{\mathbf{x}}\int_z^{ct} \frac{(t - r/c)}{4\pi\,r}\,2\pi r\,dr$$

so

$$\mathbf{A} = \frac{\mu_0\dot{K}}{4c}\,(ct - z)^2\,\hat{\mathbf{x}}$$

Then

$$\mathbf{E} = -\nabla V - \frac{\partial \mathbf{A}}{\partial t}$$

$$= 0 - \frac{\mu_0 \dot{K}}{2}(ct - z)\,\hat{\mathbf{x}}$$

and

$$\mathbf{B} = \nabla \times \mathbf{A} = \frac{\partial A_x}{\partial z}\,\hat{\mathbf{y}}$$

$$= -\frac{\mu_0 \dot{K}}{2c}(ct - z)\,\hat{\mathbf{y}}$$

The fields are sketched in Figure 14.16. As time proceeds, the field disturbance moves farther from the xy plane. Note that Poynting's vector $\mathbf{E} \times \mathbf{B}/\mu_0$ points in the $+z$ direction; it carries the energy that the expanding fields will need. To check: The wave equations for V, $\mathbf{E}$, and $\mathbf{B}$, Section 14.3, are satisfied trivially: There is no $\mathbf{J}$ or ρ, and all second derivatives are zero. And in the wave equation for $\mathbf{A}$, Equation 14.23, we find that $\nabla^2 \mathbf{A} = \mu_0 \epsilon_0 \dfrac{\partial^2 \mathbf{A}}{\partial t^2}$ when we set $\epsilon_0 \mu_0 = 1/c^2$. And of course $\mathbf{E}$ and $\mathbf{B}$ satisfy Maxwell's equations.

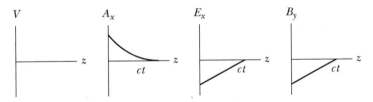

Figure 14.16   Fields.

Unfortunately, the solutions to the $\mathbf{E}$ and $\mathbf{B}$ wave equations are *not* just the retarded $\mathbf{E}$ and $\mathbf{B}$ fields; they are much more complicated than that. Usually they may simply be given in terms of the retarded potentials, as in the preceding example:

$$\mathbf{E} = -\nabla V - \frac{\partial \mathbf{A}}{\partial t} \qquad \text{and} \qquad \mathbf{B} = \nabla \times \mathbf{A}$$

Feynman in 1950 gave a particularly perspicuous explanation of the fields for a simple charge Q in arbitrary motion in free space (Janah; Heaviside found them in 1902):

$$\mathbf{E} = \frac{Q}{4\pi\epsilon_0}\left[\frac{\hat{\mathbf{r}}}{r^2} + \frac{r}{c}\frac{d}{dt}\left(\frac{\hat{\mathbf{r}}}{r^2}\right) + \frac{d^2\hat{\mathbf{r}}}{c^2 dt^2}\right] \qquad (14.28)$$

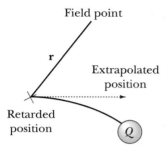

Field point

r

Extrapolated
position

Retarded
position

Q

Present position   **Figure 14.17**   Accelerated charge.

and

$$cB = [\hat{r}] \times E \tag{14.29}$$

In these equations, *all* of the r's and $\hat{r}$'s are retarded; we omit some brackets for clarity. (c is speed of light.)

The *first term* of **E** in Feynman's formula is the retarded Coulomb's-law field; it is similar to the retarded-potential formulas above. The field due to this term points away from the retarded position, not the present position. *But* there are two more terms.

The *second term* is of the form $\dfrac{df}{dt} \Delta t$, which is just a linear extrapolation of f; the time for information to travel the distance r is $\Delta t = r/c$, Figure 14.17. What's inside the parentheses in the second term is just the retarded Coulomb's law again. So the second term extrapolates the retarded field right up to the current time. That is why, when a charge moves uniformly, the **E** field keeps right up with it, rather than pointing behind the charge to its retarded position; the linear extrapolation in the second term tries to keep it up to date. But of course if the charge accelerates, then the extrapolation doesn't come out quite right.

The *third term* involves the rate of change of the unit vector $\hat{r}$. Since it has constant magnitude of 1, it can only change direction, and this means that its derivative is transverse, perpendicular to $\hat{r}$. If the charge accelerates, this term tends to generate a transverse **E** field; so this term is the one that is responsible for radiation. We can illustrate its effect for a simple case. In Figure 14.18 we see a charge Q undergoing constant acceleration a for time t, starting from rest. It travels a distance $\frac{1}{2} at^2$ at angle θ to the radial distance r; as before, all of these r's are retarded.

Meanwhile the tip of the unit vector $\hat{r}$ has moved by $\frac{1}{2} \dfrac{d^2 \hat{r}}{dt^2} t^2$. Again, this is just uniform acceleration starting from rest. For $\frac{1}{2} at^2 \ll r$ the similar triangles provide the equation

$$\frac{\frac{1}{2} \left| d^2 \hat{r} / dt^2 \right| t^2}{|\hat{r}|} = \frac{\frac{1}{2} at^2 \sin \theta}{r}$$

$$\left| \frac{d^2 \hat{r}}{c^2 \, dt^2} \right| = \frac{a \sin \theta}{c^2 r}$$

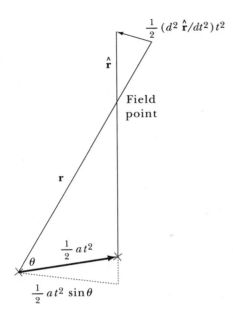

$\frac{1}{2}(d^2\,\hat{\mathbf{r}}/dt^2)\,t^2$

$\hat{\mathbf{r}}$

Field
point

$\mathbf{r}$

$\frac{1}{2}\,a t^2$

θ

$\frac{1}{2}\,a t^2\,\sin\theta$

Figure 14.18   For the third term.

with $|\hat{\mathbf{r}}| = 1$. So the third term gives us a transverse radiated **E** field of magnitude

$$E = \frac{Qa\sin\theta}{4\pi\epsilon_0 c^2 r} \tag{14.30}$$

The variables are of course retarded. The equation for **B** provides the corresponding transverse radiated **B** field:

$$B = \frac{Qa\sin\theta}{4\pi\epsilon_0 c^3 r} \tag{14.31}$$

We will encounter these fields again in Sections 14.6 and 15.1.

14.5. Plane Waves; Complex Notation

The wave equations (Section 14.3) are very accepting and nonjudgmental. Solutions can be three-dimensional sine waves, square waves, single pulses, and so forth (but not sound waves). Radar, visible light, and x-rays are all represented; see the accompanying table. To limit the discussion, we shall idealize down to a plane wave of **E** field moving in the absence of free charge and current in the $\hat{\mathbf{z}}$ direction:

$$\frac{\partial^2 \mathbf{E}}{\partial z^2} = \mu\epsilon\,\frac{\partial^2 \mathbf{E}}{\partial t^2} \tag{14.32}$$

Electromagnetic Spectrum

Frequency	Wavelength	Name
100 Hz	3000 km	elf, extremely low-frequency*
540–1610 kHz	556–186 m	AM broadcast
27 MHz	11 m	CB, citizens band
47 MHz	6.4 m	Cordless phone
54–60 MHz	5.6–5.0 m	TV channel 2
82–88 MHz	3.7–3.4 m	TV channel 6
88–108 MHz	3.4–2.8 m	FM broadcast
174–180 MHz	1.72–1.67 m	TV channel 7
800–900 MHz	37–34 cm	Cellular phone
1–300 GHz	30–0.1 cm	Radar, microwave
300 GHz	0.1 cm	Cosmic 3° background radiation†
10^{11}–4.3×10^{14} Hz	3000–0.70 μm	Infrared
4.3×10^{14} Hz	0.70 μm	Red light
7.5×10^{14} Hz	0.40 μm	Blue light
7.5×10^{14}–10^{18} Hz	400–0.30 nm	Ultraviolet
10^{17}–10^{21} Hz	3.0 nm–3×10^{-13} m	X-rays
10^{19} Hz–∞	3.0×10^{-11} m–0	Gamma rays

The transition from one sort of radiation to another is gradual and not necessarily well-defined. For example, x-rays and gamma rays overlap over a large region of wavelength. The distinction is: x-rays are usually regarded as coming from electronic processes and transitions, whereas gamma radiation comes from nuclear transitions and from elementary particles.

* "elf" radiation has been used in communicating with submarines under water (Section 16.3); the land-based antennas must be the size of a state for effective radiation.

† Cosmic "3°" background is blackbody radiation presumably left over from the Big Bang. At $T = 2.7$ K, Wien's displacement law, $\lambda_m T = .0029$ m · K, yields $f = 280$ GHz, and Stefan-Boltzmann gives $S = \sigma T^4 = 5.7 \times 10^{-8} \times 2.7^4 = 3 \ \mu\text{W/m}^2$.

There are of course miscellaneous solutions of the form $E = z$ and $E = t$, but for waves we try the solution:

$$\mathbf{E} = \mathbf{E}_0 \sin(kz - \omega t) \qquad (14.33)$$

This is a traveling sinusoidal wave (Fig. 14.19). $\mathbf{E}_0$ is an arbitrary constant vector. The wave has no x or y dependence; it is the same, in the xy plane, all the way across space. (But $\mathbf{E}_0$ may have x and y components.) In the z direction, it repeats

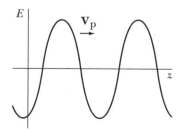

Figure 14.19   Traveling wave.

itself when $\omega t = 2\pi, 4\pi$, and so on, so the angular frequency is

$$\omega = 2\pi f$$

and it repeats when $kz = 2\pi$, and so on, so the **propagation constant** (wave number) is

$$k = \frac{2\pi}{\lambda} \quad \left(\text{sometimes written } \frac{1}{\lambdabar}\right)$$

where λ is wavelength.

The wave is **polarized** with its **E** vector in the $\mathbf{E}_0$ direction. It moves in the $+z$ direction. For constant phase, we have

$$kz - \omega t = \text{constant}$$

Differentiating,

$$k\,dz - \omega\,dt = 0$$

so

$$\frac{dz}{dt} = \boxed{v_p = \frac{\omega}{k}} = \lambda f, \quad \text{the \textbf{phase velocity}} \quad (14.34)$$

When we substitute the solution into the differential equation, we find, after carrying through the appropriate partial differentiations,

$$-k^2 \mathbf{E}_0 \sin(kz - \omega t) = -\mu\epsilon\omega^2 \mathbf{E}_0 \sin(kz - \omega t)$$

so

$$\mu\epsilon \frac{\omega^2}{k^2} = \mu\epsilon v_p^{\,2} = 1$$

In empty space,

$$\boxed{\mu_0\epsilon_0 c^2 = 1} \quad (14.35)$$

where c is the speed of light in vacuum. Let's check it:

$$\mu_0\epsilon_0 c^2 = 4\pi \times 10^{-7} \times 8.85 \times 10^{-12} \times (3 \times 10^8)^2 = 1.001, \quad \text{close enough.}$$

It is a spectacular triumph of theoretical physics to be able to predict the speed c of rapidly moving, rapidly fluctuating light from two *static* electric and magnetic measurements involving, in principle, ϵ_0 and μ_0. Maxwell understood this very well when he introduced his displacement current, $\epsilon_0 \partial \mathbf{E}/\partial t$, upon which electromagnetic radiation depends.

All of the above also holds for **B**, V, and **A**, of course. A disturbance in a plane electromagnetic wave travels at speed c in free space.

We needn't use a sine wave; the wave equation is satisfied by a cosine, or a square wave, or in fact *any* (differentiable) vector of the form $\mathbf{f}(kz - \omega t)$. If we want to find the relationship between **E** and **B**, with Poynting's vector in mind, we will find it convenient to use **complex notation** with solutions of the form

$$\mathbf{E} = \mathbf{E}_0 e^{i(kz-\omega t)} \qquad \text{and} \qquad \mathbf{B} = \mathbf{B}_0 e^{i(kz-\omega t)} \tag{14.36}$$

We use i for the **imaginary number** $\sqrt{-1}$. We employ Euler's relationship

$$e^{i\theta} = \cos\theta + i\sin\theta$$

this often facilitates calculations involving trigonometric functions. We are only interested in real fields, so we discard the "parasitic" imaginary term at the end of the calculation. (We did the same when dealing with AC circuits in Section 13.2.)

To relate **E** to **B**, we use Faraday's law:

$$\nabla \times \mathbf{E} = -\frac{\partial \mathbf{B}}{\partial t}$$

For our solutions,

$$\nabla \times (\mathbf{E}_0 e^{i(kz-\omega t)}) = -\frac{\partial}{\partial t}(\mathbf{B}_0 e^{i(kz-\omega t)}) \tag{14.37}$$

We introduce a vector identity on the left: $\nabla \times (f\mathbf{A}) = f\nabla \times \mathbf{A} - \mathbf{A} \times \nabla f$

$$e^{i(kz-\omega t)}\nabla \times \mathbf{E}_0 - \mathbf{E}_0 \times \nabla e^{i(kz-\omega t)} = i\omega\,\mathbf{B}_0 e^{i(kz-\omega t)}$$

$\mathbf{E}_0$ is a constant vector so $\nabla \times \mathbf{E}_0$ is zero. The gradient $\nabla e^{i(kz-\omega t)}$ involves only the z component. So,

$$-\mathbf{E}_0 \times (ik\,e^{i(kz-\omega t)}\hat{\mathbf{z}}) = i\omega\,\mathbf{B}_0 e^{i(kz-\omega t)}$$

and finally

$$\boxed{\hat{\mathbf{z}} \times \mathbf{E} = \frac{\omega}{k}\,\mathbf{B}} \tag{14.38}$$

This means that **B** is perpendicular to $\hat{\mathbf{z}}$ and to **E**, but we also want to show that **E** is perpendicular to $\hat{\mathbf{z}}$. To get this done, we start with Maxwell's term in Ampère's law:

$$\nabla \times \mathbf{B} = \mu\epsilon\frac{\partial \mathbf{E}}{\partial t}$$

Proceeding as before, we arrive at

$$\mathbf{B} \times \hat{\mathbf{z}} = \frac{k}{\omega}\,\mathbf{E}$$

The relationships among **E**, **B**, and Poynting's vector $\mathbf{S} = \mathbf{E} \times \mathbf{B}/\mu_0$ are summarized in Figure 14.20. They are all mutually perpendicular, and **S** points in the direction of $\hat{\mathbf{z}}$. The ratio of E to B is the phase velocity ω/k derived above. In free space it is c, so

$$\boxed{E = cB} = \mu_0 cH = 377\,H \tag{14.39}$$

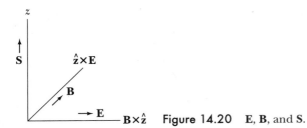

Figure 14.20   E, B, and S.

where 377 Ω is "the impedance of free space." Consequently, for the radiated fields in free space, the electric and magnetic energy densities are equal:

$$\tfrac{1}{2}\,\epsilon_0 E^2 = \tfrac{1}{2}\,B^2/\mu_0$$

and Poynting's vector is

$$\boxed{S = \epsilon_0 c E^2} = E^2/377 \qquad\qquad (14.40)$$

So S is field energy density times speed, which is intuitively reasonable.

EXAMPLE 14-6

For sunlight, $S_{av} = 1400$ W/m^2 at the top of the atmosphere; find E_0 and B_0. (Treat it as if $E(t)$ were sinusoidal.)

ANSWER $S = 1400$W/m^2 is the average value of $\epsilon_0 c E^2$. The average value of $\sin\theta$ is zero, but the average of $\sin^2\theta$ is $1/2$. So the maximum value is

$$S_{max} = \epsilon_0 c E_0^2 = 2800 \text{ W/m}^2$$

Then

$$E_0 = \sqrt{S/\epsilon_0 c}$$
$$= \sqrt{2800 \times 377} = 1030 \text{ V/m}$$

and

$$B_0 = E_0/c = 3.4 \times 10^{-6} \text{ T}$$

These are rather small fields. The magnetic field is less than $1/20$ of the earth's magnetic field. One could increase the intensity by a factor of a million without breaking down the air. (And in fact there are lasers that *can* break down the air.)

The **root-mean-square** value of E is the square root of the mean of the square of E (Section 13.2). It is useful because when you employ rms fields you get average values of power. In the above example (Fig. 14.21):

$$E_{rms} = E_0/\sqrt{2} = 730 \text{ V/m}$$
$$B_{rms} = B_0/\sqrt{2} = 2.4 \times 10^{-6} \text{ T} \qquad\qquad (14.41)$$

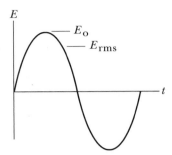

Figure 14.21 Root-mean-square **E**.

and so

$$S_{\text{av}} = E_0 B_0 / 2\mu_0 = E_{\text{rms}} B_{\text{rms}} / \mu_0$$

$$= 730 \times 2.4 \times 10^{-6} / 4\pi 10^{-7} = 1400 \text{ W/m}^2$$

which is the correct average value, as given.

For Poynting's vector in complex notation, we can't just multiply E and B together; first we have to take the real parts:

$$\mathbf{S} = \text{Re } \mathbf{E} \times \text{Re } \mathbf{B} / \mu \tag{14.42}$$

In magnitude,

$$\frac{\mu S}{\sin \theta} = \text{Re } E \text{ Re } B$$

Now if E and B are similar waves with phase difference ϕ, a simplification is possible (we omit z for simplicity):

$$\frac{\mu S}{\sin \theta} = \text{Re } E_0 e^{-i\omega t} \text{ Re } B_0 e^{-i\omega t} e^{i\phi}$$

$$= E_0 B_0 (\cos^2(\omega t) \cos \phi + \cos(\omega t) \sin(\omega t) \sin \phi)$$

The time average of $\cos^2(\omega t)$ is $1/2$, and the average of $\cos(\omega t) \sin(\omega t)$ is zero. So,

$$\frac{\mu S_{\text{av}}}{\sin \theta} = \tfrac{1}{2} E_0 B_0 \cos \phi \tag{14.43}$$

And now consider the expression $\text{Re } \frac{1}{2} E B^*$, where B^* is the complex conjugate of B, that is, the same function with i replaced by $-$i. We find

$$\text{Re } \tfrac{1}{2} E B^* = \tfrac{1}{2} \text{Re}(E_0 e^{-i\omega t} B_0 e^{+i\omega t} e^{-i\phi})$$

$$= \tfrac{1}{2} \text{Re}(E_0 B_0 e^{-i\phi}) = \tfrac{1}{2} E_0 B_0 \cos \phi$$

as before, Equation 14.43. So we may find it convenient to keep E and B in complex notation when calculating S (as in Section 13.3):

$$\boxed{\mathbf{S}_{\text{av}} = \tfrac{1}{2} \text{Re } (\mathbf{E} \times \mathbf{B}^* / \mu)} \tag{14.44}$$

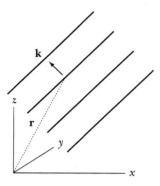

Figure 14.22 Plane waves.

For example, for the fields we have been using,

$$\mathbf{S}_{av} = \tfrac{1}{2} \operatorname{Re} (\mathbf{E}_0 e^{i(kz-\omega t)} \times \mathbf{B}_0 e^{-i(kz-\omega t)}/\mu)$$
$$= \tfrac{1}{2} \mathbf{E}_0 \times \mathbf{B}_0/\mu = \tfrac{1}{2} (E_0 B_0/\mu) \,\hat{\mathbf{z}} = (E_{rms} B_{rms}/\mu) \,\hat{\mathbf{z}}$$

as expected. The same trick works for energy density; for example,

$$\tfrac{1}{2} \epsilon E_{av}{}^2 = \tfrac{1}{4} \epsilon \mathbf{E} \cdot \mathbf{E}^*$$

(which is always real).

What if the plane wave is not traveling in the z direction? Then we can turn k into the **propagation vector k**, Figure 14.22. It is perpendicular to the plane of the waves, and so it is perpendicular to **E** and **B**. Then we have

$$\mathbf{E} = \mathbf{E}_0 e^{i(\mathbf{k}\cdot\mathbf{r} - \omega t)}$$

If we have a plane wave in the z direction, as in the preceding case, then $\mathbf{k} = k_z\hat{\mathbf{z}}$, and

$$\mathbf{k} \cdot \mathbf{r} = (k_z\hat{\mathbf{z}}) \cdot (x\hat{\mathbf{x}} + y\hat{\mathbf{y}} + z\hat{\mathbf{z}}) \tag{14.45}$$

so the expression reverts to $\mathbf{E} = \mathbf{E}_0 e^{i(kz-\omega t)}$

We can construct a **circularly polarized wave** out of two linearly polarized ones. For example,

$$\mathbf{E} = \mathbf{E}_0 e^{i(kz-\omega t)} (\hat{\mathbf{x}} - i\,\hat{\mathbf{y}}) \tag{14.46}$$

is a right-hand circularly polarized plane wave traveling in the $+z$ direction. Because of the factor of $-i = e^{-i\,\pi/2}$, the y direction lags the x direction by 90°. Thus as we follow the wave in the $+z$ direction with our right thumb, at a given time, the **E** vector rotates, pointing first to x and then to y (Fig. 14.23), and this rotation is in the direction our fingers are pointing. (At a given z, as time advances, the vector goes around the other way.)

We should also have a look at the **potentials** for the infinite plane wave, although they will turn out to be not very helpful at this point. In particular, with potentials there is often some leeway, as we have seen before (Section 6.2), and correspondingly we find here that V and **A** cannot be completely determined without more information.

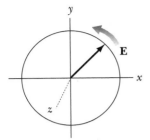

Figure 14.23   Circular polarization.

As with the fields, we begin by assuming plane waves moving in the positive z direction:

$$V = V_0 e^{i(kz-\omega t)} \quad \text{and} \quad \mathbf{A} = \mathbf{A}_0 e^{i(kz-\omega t)} \tag{14.47}$$

where $\mathbf{V}_0$ and $\mathbf{A}_0$ are constants. Then for $\mathbf{B}$ we have

$$\mathbf{B} = \nabla \times \mathbf{A}$$
$$= \nabla \times \mathbf{A}_0 e^{i(kz-\omega t)} = -\mathbf{A}_0 \times (ik\, e^{i(kz-\omega t)}\, \hat{\mathbf{z}})$$

Once again, we have used $\nabla \times (f\mathbf{A}) = f\nabla \times \mathbf{A} - \mathbf{A} \times \nabla f$ with $\nabla \times \mathbf{A}_0 = 0$. So,

$$\mathbf{B} = -ik\, \mathbf{A} \times \hat{\mathbf{z}} = ik\,(A_x\hat{\mathbf{y}} - A_y\hat{\mathbf{x}})$$

Setting the components equal, we find that

$$A_x = -\frac{iB_y}{k} \quad \text{and} \quad A_y = \frac{iB_x}{k}$$

and that $\mathbf{B}$ doesn't depend on A_z.

And now we'll try $\mathbf{E}$:

$$\mathbf{E} = -\nabla V - \partial \mathbf{A}/\partial t$$
$$\mathbf{E} = -V_0\, ik\, e^{i(kz-\omega t)}\, \hat{\mathbf{z}} + i\omega\, \mathbf{A}_0 e^{i(kz-\omega t)} = -ikV\hat{\mathbf{z}} + i\omega \mathbf{A}$$

Remembering that $\mathbf{E}$ is transverse, so $E_z = 0$, we immediately find both

$$\mathbf{E} = i\omega\,(A_x\hat{\mathbf{x}} + A_y\hat{\mathbf{y}})$$

and

$$\boxed{c\, A_z = V} \tag{14.48}$$

a cute relationship for plane waves. Unlike $\mathbf{E}$ and $\mathbf{B}$, $\mathbf{A}$ need not be transverse.

That's as far as we can go without further information. We can calculate A_x and A_y, but A_z and V are indeterminate. That's okay, because later we will find that V may or may not be present, depending on the circumstances: In Section 15.1, A_z and V are both present, and they cancel as far as the radiation is concerned; and in Section 15.2, V and A_z aren't present at all. In practice we can usually assume that $\boxed{V = 0}$ for infinite plane waves of indeterminate source, because the V and A_z involved have little physical significance: they contribute nothing to $\mathbf{E}$ or $\mathbf{B}$, which depend only on A_x and A_y, as seen directly above. (But in a wave guide we must instead set $A_z \neq 0$; see Section 17.4.)

 EXAMPLE 14-7

For sunlight, with $S_{av} = 1400$ W/m^2, and average wavelength $\lambda = 0.55$ μm, find A_{rms} with $V = 0$.

ANSWER $cA_z = V = 0$, so $A_z = 0$.

$$\mathbf{S}_{av} = \tfrac{1}{2} \, \text{Re}(\mathbf{E} \times \mathbf{B}^*/\mu)$$
$$= \tfrac{1}{2} \, \text{Re}(i\omega \, (A_x\hat{\mathbf{x}} + A_y\hat{\mathbf{y}}) \times (ik \, (A_x\hat{\mathbf{y}} - A_y\hat{\mathbf{x}}))^*/\mu)$$
$$= \frac{\omega k}{2\mu} \, (A_{0x}{}^2 + A_{0y}{}^2) \, \hat{\mathbf{z}} = \frac{\omega k}{\mu} \, A_{rms}{}^2 \, \hat{\mathbf{z}}$$

$$A_{rms} = \sqrt{\mu S_{av}/\omega k} = \sqrt{\mu S_{av} \, \lambda^2/c(2\pi)^2}$$
$$= \sqrt{4\pi \times 10^{-7} \times 1400 \times (0.55 \times 10^{-6})^2/3 \times 10^8 \times 6.28^2}$$
$$= 2.12 \times 10^{-13} \, \text{T·m}$$

A is numerically small but yet vital to the radiation.

14.6.   Simple Charge; Larmor Formula

Now that we know something quantitative about electromagnetic radiation, we can push a charge and see what happens to the field.

In Figure 14.24a a single charge Q sits unsuspectingly. At this instant we kick it, applying a sudden acceleration a, resulting in a velocity v that is near $c/2$ in this case. The charge then coasts for time t, and we arrive at Figure 14.24b. Far from the charge, the field lines haven't yet "gotten the word" that the charge has moved, and they continue to point back at where the charge was. Closer in, they point back to the present position (Feynman, Section 14.4). In obedience to Gauss's law, the field lines must be continuous. The kinks in the field lines travel outward at the speed of light. In fact, these transverse field components represent electromagnetic radiation. That is how a charge radiates.

The transverse field, represented by the kinks, is in the plane of the paper, rather than perpendicular to it. Correspondingly, the radiation is polarized with its **E** vector in the plane defined by the acceleration vector and the viewing direction. The direction of the field in the kink is generally antiparallel to the acceleration, corresponding to the fact that when we increase current in a wire, neighboring wires experience an emf in the opposite direction. This explains the emf found in Figure 4.2c, Faraday's law due to charge acceleration.

(The same arguments apply to the gravitational field g of an accelerated mass m. However, for a given ma, there is always an equal and opposite ma somewhere,

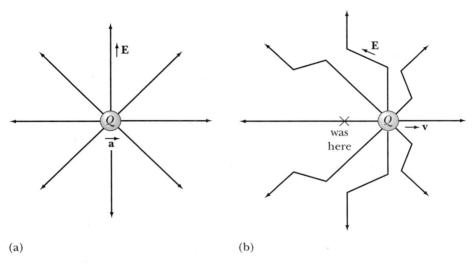

(a)   (b)

Figure 14.24   **(a)** Kick. **(b)** Coast.

so the radiation tends to get canceled out; the best you can do is quadrupole radiation, not treated here.)

Once we have the picture, the rabbit is in the hat. Various assumptions have
gone into the picture, of course: The electromagnetic disturbance travels at speed
c, the field lines of the moving charge point directly away from the charge, and so
forth. But all of these assumptions appear plausible and in fact most of them follow from Feynman's formula, Section 14.4. (The idea goes back to J.J. Thomson,
1904. Appendix C deals further with this model.) And now we can proceed to pull
the rabbit out of the hat.

A charge Q receives an acceleration for small time t_a, Figure 14.25, and
thereby acquires (small) speed v; its acceleration is therefore $a = v/t_a$. The charge
then coasts for time t, traversing a distance vt. A single field line is displayed.
Following the arguments given in connection with Figure 14.24b, the field line as-

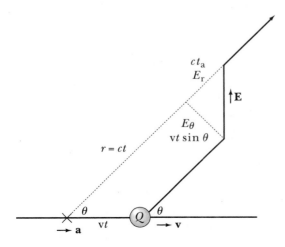

Figure 14.25   An **E** field line.

sumes the form shown. For the field line segment labeled **E**, the ratio of transverse to longitudinal field can be read from the picture:

$$\frac{E_\theta}{E_r} = \frac{vt\sin\theta}{ct_a} \tag{14.49}$$

The radial field comes from Coulomb's law:

$$E_r = \frac{Q}{4\pi\epsilon_0 r^2}$$

Consequently, the transverse field is

$$E_\theta = \frac{Qvt\sin\theta}{4\pi\epsilon_0 r^2\, ct_a}$$

so, with $r = ct$ and $a = v/t_a$,

$$\boxed{E_\theta = \frac{Qa\sin\theta}{4\pi\epsilon_0 c^2 r}} = \frac{\mu_0 Qa\sin\theta}{4\pi r} \tag{14.50}$$

This is the radiated E field. It is transverse, as promised by Poynting's vector and by the wave equations. It falls off as $1/r$, rather than $1/r^2$, because in order to conserve energy it is S that must fall off as $1/r^2$ (see below). Because of the $\sin\theta$ factor, the radiation goes out mostly sideways from the direction of acceleration, as suggested by Figure 14.24b.

The corresponding **B** field comes from $E = cB$ (Sections 14.4 and 14.5):

$$\boxed{B_{\text{rad}} = \frac{\mu_0 Qa\sin\theta}{4\pi rc}} \tag{14.51}$$

Of course all of B_{rad} is B_ϕ.

These fields are compatible with $\nabla \times \mathbf{E} = -\dfrac{\partial \mathbf{B}}{\partial t}$ and $\nabla \times \mathbf{B} = \mu_0\epsilon_0 \dfrac{\partial \mathbf{E}}{\partial t}$ (assuming no free current), and also with $\mathbf{E} = -\nabla V - \partial \mathbf{A}/\partial t$, using the transverse component of $\mathbf{A} = \mu_0 Q\mathbf{v}/4\pi r$.

If we accelerate the charge again, we get another kink in the field lines. For continuous acceleration the **E** field is smooth, of course, and it assumes the form shown in Figure 14.26b, as you may verify by doing some sketching. For uniform motion, not too fast, the form is that of Figure 14.26a (in agreement with Section 14.4).

Now we have enough to find Poynting's vector, $\mathbf{E} \times \mathbf{B}/\mu_0$, for a simple charge:

$$\boxed{\mathbf{S} = \frac{Q^2 a^2 \sin^2\theta\, \hat{\mathbf{r}}}{16\pi^2\epsilon_0 c^3 r^2}} \tag{14.52}$$

Figure 14.27 is a polar plot of Poynting's vector for a slowly moving charge: The distance r is proportional to S at the angle θ. Three-dimensionally this is a donut, isotropic in angle ϕ. The pattern doesn't depend on whether $\mathbf{v}$ is to right or left, nor on $\mathbf{a}$ in the same sense. The radiation is polarized with $\mathbf{E}$ in the plane of the paper.

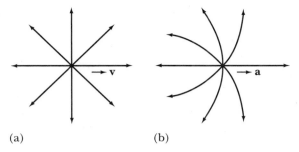

(a)   (b)

Figure 14.26   **(a)** Uniform motion. **(b)** Uniform acceleration.

(In Section 18.6 we will deal with a rapidly moving charge using relativistic techniques.)

Poynting's vector must be inverse-square for conservation of energy. As the radiation moves farther from the source, it passes through concentric spheres of increasing radius. The area of these spheres increases as r^2, so the intensity must decrease as $1/r^2$ in order that the total power not change. Correspondingly, the radiated **E** field must be $1/r$ instead of $1/r^2$, as we have mentioned, and so must the radiated **B**.

To find the total power emitted by an accelerated charge, we integrate **S** over a sphere, Figure 14.28. We have (with da as an area element)

$$\text{Power} = \frac{dW}{dt} = \int S \, da$$

$$= \int_0^\pi \frac{Q^2 a^2 \sin^2\theta}{16\pi^2\epsilon_0 c^3 r^2} \, 2\pi r \sin\theta \, r \, d\theta$$

$$\frac{dW}{dt} = \boxed{P = \frac{2}{3} \frac{Q^2 a^2}{4\pi\epsilon_0 c^3}} \tag{14.53}$$

This is **Larmor's formula**.

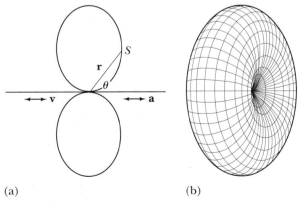

(a)   (b)

Figure 14.27   **(a)** Radiation pattern. **(b)** 3-D view.

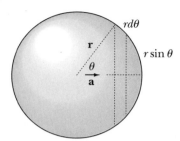

Figure 14.28   For integrating Poynting's vector.

Larmor's formula is independent of θ because we integrated over it; and it is independent of r in order to conserve energy, as mentioned above in connection with **S**. Consequently, it is rather simple as well as powerful.

▼ EXAMPLE 14-8

A charge of 1 C undergoes an acceleration of 1 g ($= 9.8$ m/s^2). Find the power radiated.

ANSWER

$$P = \frac{2}{3} \frac{Q^2 a^2}{4\pi\epsilon_0 c^3}$$

$$= 9 \times 10^9 \times \frac{2}{3} \times 1^2 \times 9.8^2/(3 \times 10^8)^3 = 2.1 \times 10^{-14} \text{ W}$$

The point of this is, to get much radiation we will need lots of coulombs and lots of g's.

Would you believe you can now estimate the cross section for photon scattering by electrons? Surprisingly, the calculation is easy and the result is not bad.

In Figure 14.29 a plane wave strikes an electron; its **E** field accelerates the electron, which then reradiates some of its energy. For the plane wave,

$$S = \epsilon_0 c E^2 \qquad \text{in watts per meter}^2$$

and for the radiation from the electron,

$$P = \frac{2}{3} \frac{e^2 a^2}{4\pi\epsilon_0 c^3} \qquad \text{in watts}$$

The ratio of these is a cross sectional area σ:

$$\sigma = \frac{P}{S} = \frac{2}{3} \frac{e^2 a^2}{4\pi\epsilon_0^2 c^4 E^2} \qquad \text{in meters}^2 \qquad \textbf{(14.54)}$$

For example, if we shine light over an area of 1 m^2, and 1% of the light is scattered, it implies that the area of the scatterer is 0.01 m^2.

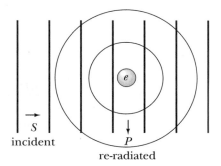

incident

re-radiated

Figure 14.29   Scattering of radiation.

The acceleration is given by Newton's second law:

$$F = ma = eE$$

Substituting E in P/S, Equation 14.54, we find that the a^2 cancels out, and we obtain

$$\frac{P}{S} = \frac{8\pi}{3}\left(\frac{e^2}{4\pi\epsilon_0 mc^2}\right)^2$$

So

$$\frac{P}{S} = \boxed{\sigma_T = \frac{8\pi}{3}r_e^2} \qquad (14.55)$$

$$= 0.67 \times 10^{-28}\ \text{m}^2 = 0.67\ \text{barn}$$

This is called the **Thomson cross section;** it is about the size of a nucleus. Here r_e is the classical electron radius, 2.8×10^{-15} m; we encountered it in Section 6.5 and we remarked that the electron is actually much smaller. But despite the fact that r_e is somewhat lacking in physical significance, σ_T turns out to be fairly close to the measured value for Compton effect photons in the 50 to 100 keV range. Below this, the photoelectric effect takes over; and at higher energy, the cross section goes down because the photon mass becomes comparable to the rest mass of the electron, 511 keV in energy units (with $W = mc^2$).

14.7 SUMMARY

We have introduced many of the basic concepts of radiation. We began by examining the flow of energy in radiation. Then we presented the electromagnetic wave equations and their solutions. We studied the plane-wave solutions in some detail. Finally we developed Larmor's formula for the power radiated by an accelerated charge.

Energy flux is given by Poynting's vector:

$$\mathbf{S} = \mathbf{E} \times \mathbf{B}/\mu \qquad (14.1)$$

Radiation pressure is $\mathbf{S}/c$.

The nonhomogeneous wave equations are

$$\nabla^2 \mathbf{E} - \mu\epsilon \frac{\partial^2 \mathbf{E}}{\partial t^2} = \mu \frac{\partial \mathbf{J}}{\partial t} + \nabla \frac{\rho}{\epsilon} \tag{14.17}$$

$$\nabla^2 \mathbf{B} - \mu\epsilon \frac{\partial^2 \mathbf{B}}{\partial t^2} = -\mu \nabla \times \mathbf{J} \tag{14.18}$$

$$\nabla^2 V - \mu\epsilon \frac{\partial^2 V}{\partial t^2} = -\frac{\rho}{\epsilon} \tag{14.21}$$

$$\nabla^2 \mathbf{A} - \mu\epsilon \frac{\partial^2 \mathbf{A}}{\partial t^2} = -\mu \mathbf{J} \tag{14.23}$$

where we employ the Lorentz condition:

$$\nabla \cdot \mathbf{A} = -\mu\epsilon \frac{\partial V}{\partial t} \tag{14.20}$$

Solutions include the retarded potentials:

$$V(t) = \int \frac{\rho(t - r/v)}{4\pi\epsilon\, r}\, dv \quad \text{and} \quad \mathbf{A}(t) = \int \frac{\mu \mathbf{J}(t - r/v)}{4\pi\, r}\, dv \tag{14.24}$$

For plane waves traveling in the $+z$ direction,

$$\mathbf{E} = \mathbf{E}_0 e^{i(kz - \omega t)} \quad \text{and} \quad v_p = \frac{\omega}{k} \tag{14.36, 14.34}$$

In vacuum,

$$\mu_0 \epsilon_0 c^2 = 1 \quad \text{and} \quad E = cB \tag{14.35, 14.39}$$

In complex notation, Poynting's vector may be expressed as

$$\mathbf{S}_{\text{av}} = \tfrac{1}{2} \operatorname{Re}(\mathbf{E} \times \mathbf{B}^*/\mu) \tag{14.44}$$

For an accelerated simple charge,

$$\mathbf{S} = \frac{Q^2 a^2 \sin^2\theta\, \hat{\mathbf{r}}}{16\pi^2 \epsilon_0 c^3 r^2} \tag{14.52}$$

and the Larmor formula for radiated power is

$$P = \frac{2}{3} \frac{Q^2 a^2}{4\pi\epsilon_0 c^3} \tag{14.53}$$

◤ PROBLEMS

14-1 An electric field of 1.7×10^6 V/m, pointing North-East, coexists with a magnetic field of 0.73 T pointing South. Find Poynting's vector there.

14-2 A radio station broadcasts 50 kW (effective) at 810 kHz. Find rms E and B at a distance of 1 km. (The given effective power assumes spherical symmetry; in practice the power is less because they send it out horizontally.)

14-3 Find Poynting's vector inside a parallel plate capacitor, of radius a and separation s in air, which is being charged by current $\mathbf{I}$, Figure 14.30. Show that the integral of $\mathbf{S}$ over a cylindrical surface of radius r equals the rate of energy deposition in the field inside the surface.

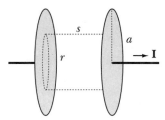

Figure 14.30

14-4 Compare Poynting's vector $\mathbf{S}$ just inside and just outside of a long air-core solenoid, Figure 14.4, of radius ρ, length l, and turns N, with increasing current $dI/dt = \dot{\mathbf{I}}$. Show that the integral of $\mathbf{S}$ over the cylindrical surface equals the rate of energy deposition in the field.

14-5 A coaxial cable of inner and outer radii 0.1 cm and 0.5 cm carries $I = 300$ ma at $V = 25$ V on the inner conductor, and the outer conductor is grounded. Integrate Poynting's vector over the space between the conductors and compare with the power VI.

14-6 In Figure 14.7, show that S is twice as large as necessary to transport field energy.

14-7 A Venusian tentacle-held ray gun delivers 10^6 joules in 0.1 second over an area of 17 sq cm. Find the recoil force.

14-8 A solar sailboat has an aluminized-mylar sail of area 1 square kilometer. What force is obtainable at right angles to the sun's gravitational field, at the earth's orbit (solar constant 1400 W/m^2), Figure 14.31? How long would it take a 1-ton mass to travel $x = 100$ million miles, in a straight line, starting from rest, under the influence of this force?

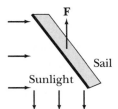

Figure 14.31

14-9 Find the radius r of a black spherical dust particle, of specific gravity 5.0, for which the sun's gravitational force and radiation pressure force are equal. (Smaller particles are blown out of the solar system.)

*14-10 Repeat Problem 14-9 for a shiny sphere.

14-11 Show that the $\mathbf{E}$ field at a surface bisects the angle between the normal $d\mathbf{a}$ and the corresponding Maxwell stress force $d\mathbf{F}$, Figure 14.12.

*14-12 Find the force of repulsion between the two hemispheres of a spherical charge distribution of uniform charge density ρ and radius R, using the Maxwell stress tensor (compare with Problem 2-33).

14-13 Derive the wave equation for $\mathbf{B}$ starting with Ampère's law.

14-14 Derive the wave equation for $\mathbf{A}$ starting with $\mathbf{B} = \nabla \times \mathbf{A}$.

*14-15 (a) Find the field angular momentum L of a conducting magnetized sphere of radius a, charge Q, and magnetization M. (b) Find the angular velocity ω of an iron sphere having the same mechanical angular momentum; the sphere has radius $a = 2$ cm and it is magnetically saturated ($\mu_0 M = 2$ tesla) and charged to 20 kV.

14-16 For sunlight, starting with $A_{rms} = 2.12 \times 10^{-13}$ T · m at $\lambda = 0.55\ \mu$m, find E_{rms}.

14-17 A spherical electrode of radius 2 cm is connected to 120 VAC; use the Lorentz condition to find A_{rms} at the surface.

14-18 A conducting sphere of charge Q and radius R is expanding slowly at the rate of $dR/dt = \dot{R}$. Find A and B inside and out.

14-19 A plane wave travels in the z direction and is polarized with its $\mathbf{E}$ vector in the x direction. Its average energy flux is $7\ \text{mW/m}^2$ and its frequency is 100 MHz. Find the rms emf induced in a loop of radius 10 cm located in the xz plane.

14-20 Repeat the preceding problem for a 10-cm rod oriented along the x axis.

14-21 Show how a linearly polarized wave may be constructed out of left- and right-hand circularly polarized waves.

14-22 A plane wave $\mathbf{E} = \mathbf{E}_0 e^{i(\mathbf{k}\,\cdot\,\mathbf{r}-\omega t)}$ propagates in free space ($\rho = 0, J = 0$). Show that

$$\mathbf{k} \cdot \mathbf{E} = 0$$
$$\mathbf{k} \cdot \mathbf{B} = 0$$
$$\mathbf{k} \times \mathbf{E} = \omega \mathbf{B}$$
$$\mathbf{k} \times \mathbf{B} = -\omega \mathbf{E}/c^2$$

14-23 For a charge accelerated at $9.8\ \text{m/s}^2$, find the distance at which the Coulomb field equals the transverse radiated E field at 90°.

14-24 Find an expression for the radiated gravitational field g_0 due to an accelerated mass m.

14-25 A nonrelativistic electron moves across (perpendicular to) a magnetic field of 1.5 T. How long does it take to lose 10% of its energy through radiation?

*14-26 A nonrelativistic proton moves across a magnetic field of 1.5 T. How long does it take to lose 10% of its energy through radiation?

14-27 A 10-keV electron comes to rest under uniform acceleration in a distance of $s = 10^{-10}$ m (which is about one atomic diameter). What fraction of its kinetic energy W is radiated away?

14-28 In the Bohr model, Figure 14.32, an electron circles a proton (assume infinite mass). Its centripetal force is provided by the Coulomb's-law attraction, and its angular momentum is $\hbar$ in the ground state. Find the power radiated. (In 1913 Bohr simply postulated that it wouldn't radiate at all, for reasons unknown at that time.)

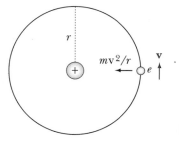

Figure 14.32

C H A P T E R

15

RADIATING SYSTEMS

Building on results of the preceding chapter, we will look in some detail at systems that produce electromagnetic radiation. By virtue of a reciprocity theorem, many of our conclusions will be found to hold also for antennae that receive such radiation.

15.1. Radiating Electric Dipole; Radiation Resistance

We begin with the radiating electric dipole. We shall employ the model dipole of Figure 15.1; this may be regarded as a metal dumbbell with current going back and forth along the bar between the spheres. We shall find the retarded potentials V and $\mathbf{A}$, and then we shall differentiate them to find the fields $\mathbf{E}$ and $\mathbf{B}$. As in Section 7.1, the dipole approximation requires $r \gg s$; but it is not necessary that $r > \lambda$, the radiated wavelength.

We suppose that charge moves in such a way that the charge on the sphere marked $+Q$ varies sinusoidally: $Q = Q_0 e^{-i\omega t} = Q_0[\cos(-\omega t) + i\sin(-\omega t)]$. (The complex notation was introduced earlier, Sections 13.2 and 14.5. At any instant the charge is $Q_0\cos(-\omega t)$; the imaginary term may be discarded.) So in Figure 15.1 we

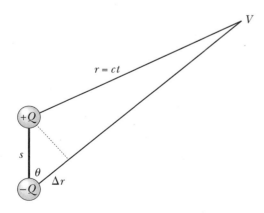

Figure 15.1 Radiating dipole.

have a dipole whose moment is $p = Q_0 e^{-i\omega t}s$. It takes time for the charge situation at the dipole to make itself felt at the field point at V; so the potential at that point is the "retarded" V corresponding to the dipole moment at the retarded time $t - r/c$, because information concerning the charges travels the distance r at speed c. Following the discussion in Sections 7.1 and 14.4, and with brackets representing retarded quantities, the potential due to the positive charge is

$$V_+(r) = \frac{[Q]}{4\pi\epsilon_0 r}$$
$$= \frac{Q_0 e^{-i\omega(t-r/c)}}{4\pi\epsilon_0 r} = \frac{Q_0 e^{i(kr-\omega t)}}{4\pi\epsilon_0 r} \tag{15.1}$$

with $\omega/c = k$. The corresponding potential due to the negative charge almost cancels it; it is

$$V_-(r) = -V_+(r + \Delta r)$$

The total potential due to both $+$ and $-$ charges is

$$V = V_+ + V_- \tag{15.2}$$
$$= V_+(r) - V_+(r + \Delta r) \approx -\frac{\partial V_+}{\partial r}\Delta r$$
$$= -\frac{\partial}{\partial r}\left(\frac{Q_0 e^{i(kr-\omega t)}}{4\pi\epsilon_0 r}\right)(s\cos\theta) \tag{15.3}$$

so, taking the partial derivative, the retarded scalar potential is

$$V = \frac{[p]k\cos\theta}{4\pi\epsilon_0 r}\left(\frac{1}{kr} - i\right) \tag{15.4}$$

where $[p]$ means $Q_0 e^{-i\omega(t-r/c)}s$, the retarded dipole moment.

And now for the vector potential $\mathbf{A}$, Figure 15.2, which we didn't have to worry about in the static case. It's easier. We have from Section 14.4:

$$\mathbf{A} = \frac{\mu_0}{4\pi}\int \frac{[\mathbf{J}]}{r}dv \tag{15.5}$$

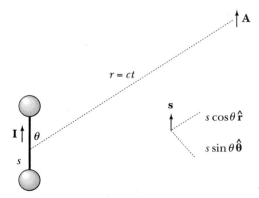

Figure 15.2 Radiating dipole.

so for the dipole with retardation and $\mu_0 = 1/\epsilon_0 c^2$, we find

$$\mathbf{A} = \frac{\mu_0}{4\pi r} I_0 e^{-i\omega(t-r/c)} \mathbf{s}$$

$$= \frac{1}{4\pi\epsilon_0 rc^2} [I]\, s\, (\cos\theta\, \hat{\mathbf{r}} - \sin\theta\, \hat{\boldsymbol{\theta}}) \tag{15.6}$$

in spherical coordinates, since $\mathbf{s}$ is parallel to the z axis, Figure 15.2. In the scalar potential V, the positive and negative ends of the dipole tend to cancel each other out, so we take the difference in the V's due to the two ends; but for the vector potential $\mathbf{A}$ all of the current in the bar connecting the charges is in the same phase, so it all just adds up.

Now V involves p, and $\mathbf{A}$ involves I. We would like to trade in the p for an I somehow, to make it easier to put V and $\mathbf{A}$ into the same equation. Since $Q = Q_0 e^{-i\omega t}$ and $I = dQ/dt$, we find that $Qs = ps$ becomes

$$Is = -i\omega p \tag{15.7}$$

The factor of i $(= e^{i\pi/2})$ represents a phase change of $\pi/2$ radians, or $90°$; for example, the current flow is maximum when the dipole moment is zero, and vice versa. We also have $k = \omega/c$. Thus, V can be written as

$$V = \frac{[I]\, s\cos\theta}{4\pi\epsilon_0 cr}\left(1 + \frac{i}{kr}\right) \tag{15.8}$$

For large kr, A and V are in phase. However, for small kr, V lags by $90°$ because of the i in the equation. This is reasonable because: the current I is maximum when the charge Q is zero, so you would expect A and V to be out of phase at first. As the waves move out, the **phase velocity** of V exceeds c in such a way that ultimately it makes up its $90°$ phase lag. These matters are explored further in the Problems.

Now we can calculate $\mathbf{E}$, in spherical coordinates, using the general formula $\mathbf{E} = -\nabla V - \partial\mathbf{A}/\partial t$. We find first that

$$(\nabla V)_r = \frac{\partial V}{\partial r} = \frac{k\,[I]\, s\cos\theta}{4\pi\epsilon_0 cr}\left(-2\frac{1}{kr} + i - \frac{2i}{(kr)^2}\right)$$

$$(\nabla V)_\theta = \frac{\partial V}{r\partial\theta} = \frac{k\,[I]\, s\sin\theta}{4\pi\epsilon_0 cr}\left(-\frac{1}{kr} - \frac{i}{(kr)^2}\right) \tag{15.9}$$

$$(\partial\mathbf{A}/\partial t)_r = -\frac{ik\,[I]\, s}{4\pi\epsilon_0 cr}\cos\theta$$

$$(\partial\mathbf{A}/\partial t)_\theta = \frac{ik\,[I]\, s}{4\pi\epsilon_0 cr}\sin\theta$$

Then,

$$E_r = \frac{k\,[I]\, s\cos\theta}{4\pi\epsilon_0 cr}\left(\frac{2}{kr} + \frac{2i}{(kr)^2}\right)$$

$$E_\theta = \frac{k\,[I]\, s\sin\theta}{4\pi\epsilon_0 cr}\left(\frac{1}{kr} + \frac{i}{(kr)^2} - i\right) \tag{15.10}$$

The radiated field is proportional to $1/r$, so all of it is contained in the one "$-i$" term in parentheses in E_θ; and that term comes entirely from the transverse component of $-\partial A/\partial t$. The static field is proportional to $1/r^3$, like that given in Section 7.1.

In the $\cos\theta$ term, there is a cancellation between $1/r$ terms from V and A_r. This was anticipated in Section 14.5 in Equation 14.48, $cA_z = V$. There may be a longitudinal A component, but it contributes nothing to the radiation because V cancels it out. The **B** field is easier because $B_r = B_\theta = 0$:

$$B_\phi = (\nabla \times \mathbf{A})_\phi = \frac{\partial}{r\partial r} rA_\theta - \frac{\partial A_r}{r\partial\theta} = \frac{k[I] s\sin\theta}{4\pi\epsilon_0 c^2 r}\left(\frac{1}{kr} - i\right) \tag{15.11}$$

The real parts of the fields are:

$$\text{Re } E_r = \frac{I_0 ks\cos\theta}{4\pi\epsilon_0 cr}\left(\frac{2}{kr}\cos(kr - \omega t) - \frac{2}{(kr)^2}\sin(kr - \omega t)\right)$$

$$\text{Re } E_\theta = \frac{I_0 ks\sin\theta}{4\pi\epsilon_0 cr}\left(\frac{1}{kr}\cos(kr - \omega t) + \left(1 - \frac{1}{(kr)^2}\right)\sin(kr - \omega t)\right) \tag{15.12}$$

$$\text{Re } B_\phi = \frac{I_0 ks\sin\theta}{4\pi\epsilon_0 c^2 r}\left(\frac{1}{kr}\cos(kr - \omega t) - \sin(kr - \omega t)\right)$$

The **E** field lines are displayed in Figure 15.4.

All of the radiation is contained in the $1/r$ terms; the others have effect only at short distances. As we remarked in Section 14.6, for energy conservation the radiated fields must fall off as $1/r$. Since our primary interest lies in the radiated energy that escapes from the dipole, we shall now require that $r \gg \lambda$, or since $k = 2\pi/\lambda$,

$$kr \gg 1 \qquad \text{now}$$

The $1/r$ radiated **E** and **B** fields are the same as those obtained in Sections 14.4 and 14.6 if we achieve the same p with a different model: constant charges $\pm Q$ oscillating sinusoidally with amplitude $s/2$. Differentiating Qs, we find that we can substitute Qa for $-i\omega Is$. Thus, for the radiated E field:

$$E_\theta = -i\frac{k[I] s\sin\theta}{4\pi\epsilon_0 cr} = \frac{[Qa]\sin\theta}{4\pi\epsilon_0 c^2 r} \tag{15.13}$$

as in Equation 14.50, using $k = \omega/c$. And correspondingly

$$B_\phi = -i\frac{k[I] s\sin\theta}{4\pi\epsilon_0 c^2 r} = E_\theta/c \tag{15.14}$$

(so $E = cB$ for the radiated fields, Section 14.5). So, the radiated E and B fields are the same as if two opposite charges were physically oscillating back and forth, emitting dipole radiation in phase.

This is the third time we have come upon these radiated E and B fields. The first time, we extracted them directly from Feynman's formulas for the fields of a simple charge in arbitrary motion, Section 14.4. The second time, we employed an intuitively appealing picture with bent E lines, Section 14.6. And this time we started with the retarded potentials.

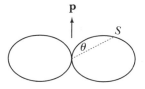

Figure 15.3   Poynting's vector.

We can immediately state the average Poynting's vector (Section 14.5 and Fig. 14.27), Figure 15.3, in terms of $I_{rms} = I$:

$$\mathbf{S}_{av} = \tfrac{1}{2} \operatorname{Re} (\mathbf{E} \times \mathbf{B}^*/\mu_0) = \frac{1}{16\pi^2\epsilon_0 c}\left(\frac{ks\, I \sin\theta}{r}\right)^2 \hat{\mathbf{r}}$$

Then the average radiated power is

$$P = \frac{dW}{dt} = \int \mathbf{S}\cdot d\mathbf{a} = \int_0^\pi \frac{(ksI)^2 \sin^2\theta}{16\pi^2\epsilon_0 cr^2}\, 2\pi r \sin\theta\, r\, d\theta$$

$$= \frac{2}{3}\frac{1}{4\pi\epsilon_0 c}(ks\, I)^2 \tag{15.15}$$

$$= \frac{2}{3}\frac{\omega^4 p^2}{4\pi\epsilon_0 c^3}\qquad (\text{for } p_{rms})$$

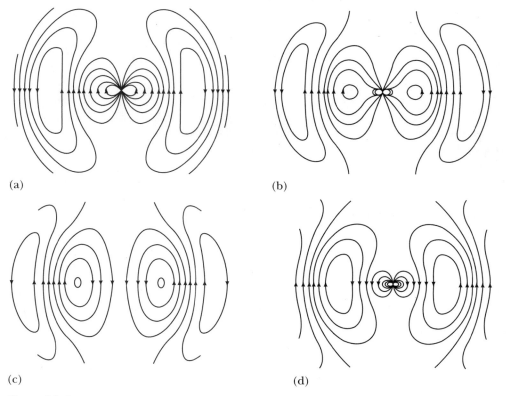

(a)

(b)

(c)

(d)

Figure 15.4

Now for a resistor the power dissipated into heat is $P = I^2 R$, and here we also have a power P that is proportional to I^2; so it is convenient to characterize the dipole by a **radiation resistance;** from Equation 15.15,

$$P = I^2 R_{\text{rad}} \tag{15.16}$$

with

$$R_{\text{rad}} = \frac{2}{3} \frac{1}{4\pi\epsilon_0 c} (ks)^2$$

Numerically,

$$R_{\text{rad}} = 20.0 \ (ks)^2 \tag{15.17}$$

ks is dimensionless, and the 20.0 has units of ohms. When current flows in the antenna, energy is dissipated, not into Joule heating, but into radiation.

The **E** field lines of a radiating dipole are plotted for various times in a half-cycle in Figure 15.4. The **B** field lines are circles perpendicular to the paper. In Figure 15.4a the dipole is fully developed. In Figure 15.4b, loops are starting to break off, at a certain distance from the dipole. Close to the dipole, the loops retreat into the dipole; farther off, they can't get back by closing time, so they strike out into the world. In Figure 15.4c the dipole moment is zero, and in Figure 15.4d it is starting to build up in the opposite direction. (For computer animation, see Appendix C.)

▼ EXAMPLE 15-1

A 1-meter antenna carries $I_{\text{rms}} = 1$ A at 1 MHz; find the average power radiated.

ANSWER First we establish that the wavelength λ is much longer than the antenna:

$$\lambda = c/f = 3 \times 10^8/10^6 = 300 \text{ m}$$

so the dipole approximation is okay.
    Now $k = \omega/c = 2\pi/\lambda = 0.0209 \text{ m}^{-1}$

$$R_{\text{rad}} = 20.0 \ (0.0209 \times 1)^2 = 0.0088 \ \Omega$$

$$P = I^2 R_{\text{rad}}$$

$$= 1^2 \times 0.0088 = 0.0088 \text{ W}$$

To get more power we would need an antenna whose length is closer to $\lambda/2$.

15.2. Radiating Magnetic Dipole

How will the radiating magnetic dipole compare with the electric one?

In Figure 15.5 we have a magnetic dipole whose moment is $m = s^2 I_0 e^{-i\omega t}$. It takes time for the effect of the current to make itself felt at the field point at **A**; the potential at that point is the "retarded" **A** corresponding to the dipole moment at the retarded time $t - r/c$, because information concerning the current travels the distance r at speed c. Following the discussion in Sections 7.2 and 14.4, we have

$$A_1 = \frac{\mu_0[I]s}{4\pi r} = \frac{\mu_0 I_0 e^{-i\omega(t-r/c)}s}{4\pi r} = \frac{\mu_0 I_0 e^{i(kr-\omega t)}s}{4\pi r} \tag{15.18}$$

Then,

$$A = \frac{\partial A_1}{\partial r}\Delta r = \frac{\partial}{\partial r}\left(\frac{\mu_0 I_0 e^{i(kr-\omega t)}s}{4\pi r}\right)s\sin\theta$$

$$\mathbf{A} = \frac{\mu_0[m]k\sin\theta}{4\pi r}\left(-\frac{1}{kr}+i\right)\hat{\boldsymbol{\phi}}$$

The current loop is of very low impedance, so we assume that

$$V = 0 \qquad \text{everywhere} \tag{15.19}$$

(This is connected to the observation in Section 14.5 that $cA_z = V$. There is no longitudinal A here, so no V, as far as radiation is concerned.)

Now we shall find **E**, in spherical coordinates, using the general formula $\mathbf{E} = -\nabla V - \partial\mathbf{A}/\partial t$. With $V = 0$,

$$E_\phi = -\frac{\partial A}{\partial t} = \frac{\mu_0[m]k^2 c}{4\pi r}\sin\theta\left(1 + \frac{i}{kr}\right) \tag{15.20}$$

The **B** field is given by

$$\mathbf{B} = \nabla \times \mathbf{A} \tag{15.21}$$

so

$$B_r = \frac{\mu_0[m]k^2}{4\pi r}\left(\frac{2}{k^2 r^2} - \frac{2i}{kr}\right)\cos\theta$$

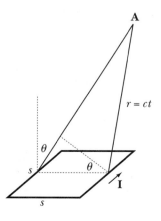

Figure 15.5 Magnetic dipole.

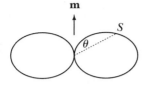

Figure 15.6   Poynting's vector.

and

$$B_\theta = \frac{\mu_0[m]k^2}{4\pi r}\left(\frac{1}{k^2 r^2} - 1 - \frac{i}{kr}\right)\sin\theta$$

These magnetic dipole fields are complementary to those of the electric dipole: E and B are interchanged. Now the magnetic field lines are shown in Figure 15.4, and the electric field lines are perpendicular to the paper. Correspondingly, the angular distribution of the radiation is the same as for the electric dipole, Figure 15.6, but its polarization is perpendicular to that of the electric dipole. For the average radiated Poynting's vector, we find

$$\mathbf{S} = \frac{1}{16\pi^2\epsilon_0 c}\left(\frac{(ks)^2\,I\sin\theta}{r}\right)^2\hat{\mathbf{r}} \tag{15.22}$$

We have set $m = Is^2$, where s^2 is the area of the dipole. As before, use I_{rms} (or m_{rms}) to get average S and P.

We integrate over the sphere as in Section 15.1. The result for the radiated power is

$$P = \frac{2}{3}\frac{1}{4\pi\epsilon_0 c}(k^2 s^2\,I)^2 = \frac{2}{3}\frac{\mu_0}{4\pi c^3}m^2\omega^4$$
$$= 20.0\,(ks)^4\,I^2$$

For the electric dipole, we found $R_{\text{rad}} = 20.0\,(ks)^2$. Now for the magnetic dipole,

$$\boxed{R_{\text{rad}} = 20.0\,(ks)^4} \tag{15.23}$$

s^2 being the area of the dipole.

Electric quadrupole radiation involves a similar formula; see Problem 15-19. Magnetic and electric dipole radiation, and the radiation from a single charge, Section 14.6, all come into the category of dipole radiation, and all have a sine-square angular dependence of Poynting's vector. Quadrupole radiation is significantly different in angular distribution.

▼ EXAMPLE 15-2

A 1-meter-square loop antenna carries rms current of 1 A at 1 MHz; find the average power radiated.

ANSWER The wavelength λ is much longer than the antenna, so the dipole approximation is okay. Then:

$$k = \omega/c = 2\pi f/c$$
$$= 2\pi \times 10^6/3 \times 10^8 = 0.0209 \text{ m}^{-1}$$
$$R_{\text{rad}} = 20.0 \times (0.0209 \times 1)^4 = 3.85 \times 10^{-6}\,\Omega$$
$$P = I^2 R_{\text{rad}}$$
$$= 1^2 \times 3.85 \times 10^{-6} = 3.85\ \mu\text{W}$$

For the electric dipole in the corresponding example of the preceding section, the power was 8.8 mW. These dipoles just don't radiate very much.

Magnetic dipoles typically radiate even less than electric dipoles. The reason is that in the electric dipole all of the current is going the same direction, so the contributions to $\partial\mathbf{A}/\partial t$ all add up; but in the magnetic dipole, currents in opposite sides of the loop are going opposite directions and so their radiated fields tend to cancel. When the dipole sizes approach $\lambda/2$, then they become comparable, but of course at that point the dipole approximation is no longer valid.

15.3.   Half-Wave Antenna

Atoms and nuclei tend to radiate as little dipoles, quadrupoles, and so forth. They are not very effective transmitters by our standards, because they are so small, but they keep at it and they get the job done. The dipole formulas that we have seen suggest that bigger is better, and that is certainly true as long as the dipole approximation holds and even somewhat beyond. It turns out, in fact, that practical antennas are usually about $\lambda/2$ in length.

In Figure 15.7a we see a half-wave antenna and its pattern of radiation, super-

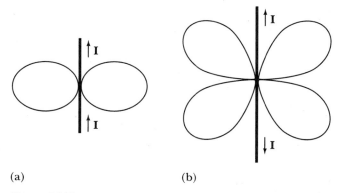

(a)   (b)

Figure 15.7   (a) Half-wave antenna. (b) Full-wave antenna.

posed on the same figure. It resonates at the frequency corresponding to $\lambda/2$, so the current I forms a standing wave, with current going the same direction at the same time throughout. The full-wave antenna also resonates at that frequency, but by contrast, in Figure 15.7b, the currents in opposite halves of the antenna are going in opposite directions. Consequently, the fields cancel perfectly in the median plane. The resulting complications in the radiation pattern are a nuisance for most practical work.

We now examine the radiation from a **half-wave antenna,** Figure 15.8a. (It is sometimes called a "dipole antenna," but this is a misnomer.) We shall simplify things by assuming that we observe the radiation at distance $r \gg \lambda$. For the field of the element dy, we use the θ component of the radiated dipole field, Equation 15.13 with $s = dy$:

$$dE = \frac{-ik\,[I]\,\sin\theta\,dy}{4\pi\epsilon_0 cr} \tag{15.24}$$

We assume that the current in the antenna is a standing wave, Figure 15.8b (which is an approximation because energy loss distorts the half sine wave):

$$\mathrm{Re}\,(I) = I_0\cos(ky)\cos(\omega t)$$

This is easiest to deal with as the sum of two waves traveling in opposite directions:

$$I(t) = \frac{I_0}{2}\,(e^{i(ky-\omega t)} + e^{i(-ky-\omega t)}) \tag{15.25}$$

For the retarded value $[I]$, we replace ωt with $\omega(t - r/c) = \omega t - kr$ here. From Figure 15.8a, $r = r_0 - y\cos\theta$. Thus, gathering it all together,

$$[I] = \frac{I_0}{2}\,(e^{i(ky+kr_0-ky\cos\theta-\omega t)} + e^{i(-ky+kr_0-ky\cos\theta-\omega t)})$$

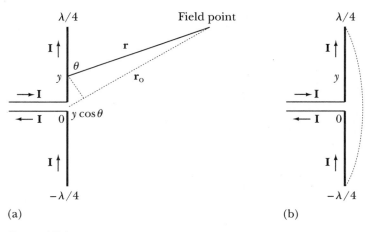

(a)   (b)

Figure 15.8   (a) Half-wave antenna. (b) Standing wave of **I**.

Substituting into Equation 15.24, and integrating, we arrive at

$$E = \int \frac{-ik\, I_0 (e^{i(ky+kr_0-ky\cos\theta-\omega t)} + e^{i(-ky+kr_0-ky\cos\theta-\omega t)}) \sin\theta\, dy}{8\pi\epsilon_0 cr_0}$$

In the denominator we have substituted r_0 for r because at large distances their ratio approaches 1; however, we must keep their difference, $y\cos\theta$, in the imaginary exponent because the phase changes rapidly with r. We can remove various constants from the integral:

$$E = \frac{-ik\, I_0 e^{i(kr_0-\omega t)}\sin\theta}{8\pi\epsilon_0 cr_0} \int_{-\lambda/4}^{\lambda/4} (e^{i(ky-ky\cos\theta)} + e^{i(-ky-ky\cos\theta)})\, dy \qquad \textbf{(15.26)}$$

The integral is trivial at least in principle. Remember that $k\lambda/4 = \pi/2$.

$$
\begin{aligned}
E &= \frac{-ik\, I_0 e^{i(kr_0-\omega t)}\sin\theta}{8\pi\epsilon_0 cr_0} \left(\frac{e^{i(1-\cos\theta)\pi/2} - e^{-i(1-\cos\theta)\pi/2}}{ik(1-\cos\theta)} \right. \\
&\quad \left. + \frac{e^{i(-1-\cos\theta)\pi/2} - e^{-i(-1-\cos\theta)\pi/2}}{ik(-1-\cos\theta)} \right) \\
&= \frac{-I_0 e^{i(kr_0-\omega t)}\sin\theta}{8\pi\epsilon_0 cr_0} \left(\frac{2i\sin\left(\frac{\pi}{2}(1-\cos\theta)\right)}{1-\cos\theta} + \frac{2i\sin\left(\frac{\pi}{2}(1+\cos\theta)\right)}{1+\cos\theta} \right) \\
&= \frac{-i\, I_0 e^{i(kr_0-\omega t)}\sin\theta}{4\pi\epsilon_0 cr_0} \left(\frac{\cos\left(\frac{\pi}{2}\cos\theta\right)}{1-\cos\theta} + \frac{\cos\left(\frac{\pi}{2}\cos\theta\right)}{1+\cos\theta} \right)
\end{aligned}
$$

So the result is an unusual trigonometric function of a trigonometric function,

$$\mathbf{E} = \frac{-i\, I_0 e^{i(kr_0-\omega t)}}{2\pi\epsilon_0 cr} \frac{\cos(\frac{\pi}{2}\cos\theta)}{\sin\theta}\, \hat{\boldsymbol{\theta}} \qquad \textbf{(15.27)}$$

Then, with $E = cB$,

$$\mathbf{B} = \frac{-i\, I_0 e^{i(kr_0-\omega t)}}{2\pi\epsilon_0 c^2 r} \frac{\cos(\frac{\pi}{2}\cos\theta)}{\sin\theta}\, \hat{\boldsymbol{\phi}}$$

And the average Poynting's vector is (with $I_{rms} = I$)

$$\mathbf{S} = \frac{I^2 \cos^2(\frac{\pi}{2}\cos\theta)}{4\pi^2\epsilon_0 cr^2 \sin^2\theta}\, \hat{\mathbf{r}} \qquad \textbf{(15.28)}$$

See Figure 15.7a. The distribution is similar to that of a dipole, but relatively more radiation goes out near the median plane.

Finally, we integrate $\mathbf{S}$ over a sphere to find the average radiated power:

$$
\begin{aligned}
P &= \frac{I^2}{4\pi^2\epsilon_0 cr^2} \int_0^\pi \frac{\cos^2(\frac{\pi}{2}\cos\theta)}{\sin^2\theta} 2\pi r^2 \sin\theta\, d\theta \\
P &= \frac{I^2}{2\pi\epsilon_0 c} \int_0^\pi \frac{\cos^2(\frac{\pi}{2}\cos\theta)}{\sin^2\theta}\, d\theta
\end{aligned}
\qquad \textbf{(15.29)}
$$

The integral is found numerically to be 1.219, so the power is

$$P = 73.1\, I^2$$

and then

$$\boxed{R_{\text{rad}} = 73.1 \ \Omega}$$ (15.30)

If we were to apply the electric dipole formula to an antenna of this length, we would get 197 Ω. The actual value is only a third of this. The difference may be ascribed to the fact that in the true electric dipole there is no destructive interference between various parts of the antenna.

 EXAMPLE 15-3

Find the average power radiated by a half-wave antenna of length 2 m, operating at a frequency of 75 MHz, and carrying current $I_{\text{rms}} = 1$ A.

ANSWER

$$P = I^2 \, R_{\text{rad}}$$
$$= 1^2 \times 73 = 73 \ \text{W}$$

Length and frequency are irrelevant as long as we know that it is a half-wave antenna.

So finally we have achieved a substantial amount of radiated power. The half-wave antenna forms the building block of many practical antenna arrays, as we shall see.

But first, we need to assure ourselves that a given antenna can perform equally well, in some sense, in either transmission (radiation) or reception.

15.4. Antenna Reciprocity

In the next section, Section 15.5, we will be looking at antennas both for transmission (broadcast, radiation) and for reception. So far, we have only been looking at radiating antennas. But it is intuitively reasonable, especially in the light of the usual reversibility of physical laws, that antennas that broadcast well will also receive signals well under corresponding circumstances. We shall now prove a simple **reciprocity** theorem in this connection. (There are many reciprocity theorems, and more general ones, in physics.)

We imagine that we broadcast a signal using current I of a given angular frequency ω in a dipole antenna of length $\mathbf{s}_1$ in the plane of the paper; and we receive this signal as an emf in another dipole of length $\mathbf{s}_2$, arbitrarily located relative to the first, Figure 15.9a. Our purpose is to show that if we exchange functions of the two dipoles, so that $\mathbf{s}_2$ does the broadcasting and $\mathbf{s}_1$ does the receiving, then a current of I in $\mathbf{s}_2$ will produce the same emf in $\mathbf{s}_1$.

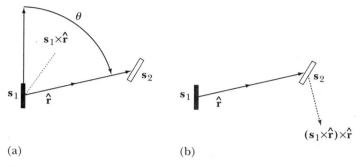

(a)   (b)

Figure 15.9 (a) Two dipoles. (b) Field at s_2.

From Section 15.1, the radiated field from dipole $\mathbf{s}_1$ is

$$E_\theta = -\,i\,\frac{k\,[I]\,s_1\,\sin\theta}{4\pi\epsilon_0 cr}$$

or, in vector form, with $k = \omega/c$,

$$\mathbf{E} = -\,i\,\frac{\mu_0\omega\,[I]\,s_1\,\sin\theta}{4\pi\,r}\,\hat{\boldsymbol{\theta}}$$

Now if we look into the reverse phenomenon, with $\mathbf{s}_2$ radiating, we immediately find that the θ from $\mathbf{s}_2$ to $\mathbf{s}_1$ is just not symmetric in any way, so we need to get rid of it. By contrast, the $\hat{\mathbf{r}}$ is the reverse of the other $\hat{\mathbf{r}}$, so we can keep it. Taking the cross product of the vector $\mathbf{s}_1$ with $\hat{\mathbf{r}}$, Figure 15.9a, we find a vector pointing into the paper, with the magnitude

$$\left|\mathbf{s}_1 \times \hat{\mathbf{r}}\right| = s_1\sin\theta \tag{15.31}$$

which is the sort of thing we need. Then we take the cross product of that with $\hat{\mathbf{r}}$, arriving at a vector in the direction of $\hat{\boldsymbol{\theta}}$, Figure 15.9b, with magnitude

$$\left|(\mathbf{s}_1 \times \hat{\mathbf{r}}) \times \hat{\mathbf{r}}\right| = s_1\sin\theta$$

That's perfect. So in the $\mathbf{E}$ field above we can replace $s_1\sin\theta\,\hat{\boldsymbol{\theta}}$ with $(\mathbf{s}_1 \times \hat{\mathbf{r}}) \times \hat{\mathbf{r}}$, arriving at

$$\mathbf{E} = -\,i\,\frac{\mu_0\omega\,[I]}{4\pi r}\,(\mathbf{s}_1 \times \hat{\mathbf{r}}) \times \hat{\mathbf{r}}$$

Then the emf in $\mathbf{s}_2$ is

$$\text{emf} = \mathbf{E}\cdot\mathbf{s}_2 = -\,i\,\frac{\mu_0\omega\,[I]}{4\pi r}\,((\mathbf{s}_1 \times \hat{\mathbf{r}}) \times \hat{\mathbf{r}})\cdot\mathbf{s}_2$$

We can rearrange this, using the fact that $(\mathbf{A} \times \mathbf{B}) \cdot \mathbf{C} = \mathbf{A} \cdot (\mathbf{B} \times \mathbf{C})$ and then $\mathbf{A} \times \mathbf{B} = -\mathbf{B} \times \mathbf{A}$:

$$\text{emf} = i\,\frac{\mu_0\omega\,[I]}{4\pi r}\,(\hat{\mathbf{r}} \times \mathbf{s}_1)\cdot(\hat{\mathbf{r}} \times \mathbf{s}_2) \tag{15.32}$$

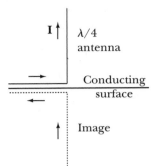

Figure 15.10   Quarter-wave antenna.

At this point the emf is symmetrical in s_1 and s_2, and it doesn't change when you reverse $\hat{r}$. And so that proves the reciprocity theorem for this simple case. If we put a current I in s_1 and receive a signal emf in s_2, then if we put the same current I in s_2, we get the same emf in s_1. Roughly speaking, antennas that broadcast well will receive well.

15.5.   Antenna Arrays

We finally arrived at the half-wave antenna in Section 15.3, and we found that it has a convenient radiation resistance; so why do we bother with anything else? The answer is, most antennas are indeed based directly or indirectly on the half-wave antenna, but we can usually find even better variations, depending on the circumstances.

For example, a **quarter-wave antenna** may be employed next to a conducting surface, Figure 15.10. The resulting antenna is shorter by a factor of two, which is sometimes a considerable advantage. The image generated in the conducting surface combines with the quarter-wave antenna to produce radiation corresponding to a half-wave antenna. For AM broadcast at 1 MHz, for which $\lambda = 300$ m, a half-wave antenna would be 150 meters high; but you can instead build a much less expensive quarter-wave antenna of only 75 meters. Near the sea, it can be placed in salt flats; or in Iowa, where you have chickens but no sea, chicken-wire mesh can be spread on the ground to provide the reflecting surface. Similarly, an antenna protruding from the metal roof of a car is likely to be a quarter-wave antenna. For a given current, only half the universe is filled with radiation, so the effective R_{rad} is $73.1/2 = 37 \, \Omega$.

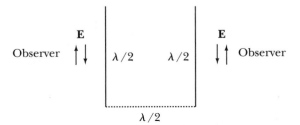

Figure 15.11   Two half-wave antennas.

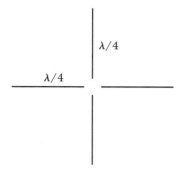

Figure 15.12   Turnstile.

We can also control the direction of the radiation. If we place two half-wave antennas next to each other, Figure 15.11, in the plane of the paper, separated by a distance $\lambda/2$, and we drive them in phase, then for an observer in the plane of the paper, as shown, the fields from the two antennas cancel: There is no radiation in the plane of the paper, and maximum radiation perpendicular to the paper. If on the other hand we drive them 180° out of phase, the situation is reversed: The observers shown find maximum radiation, and there is no radiation perpendicular to the paper.

In the United States, television waves are polarized with their **E** vector horizontal, not by any geographical quirk but by arbitrary decision of the Federal Communications Commission (some European countries have **E** vertical). In order to obtain good radiation through 360° a **turnstile antenna** is employed, Figure 15.12. The two orthogonal half-wave antennas are driven 90° out of phase, so that when one is producing maximum radiation the other is not radiating at all. Such an antenna radiates well in all directions. To suppress unwanted radiation in the vertical direction (Mars buys nothing from us) an identical antenna is mounted at $\lambda/2$ above the first and driven in phase with it, as described in connection with Figure 15.11.

Broadcasters want to spread their signal around. Television receivers, on the other hand, need a directional antenna in order to suppress "ghosts" caused by reflection of the signal from surrounding obstacles. Two half-wave antennas spaced $\lambda/4$ apart are shown in Figure 15.13. If they are driven by oscillators 90° out of phase, with A leading B, as shown, then by the time a wave crest from A reaches B, B will be producing a corresponding wave crest, and the amplitudes will add coher-

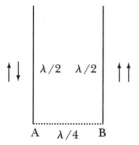

Figure 15.13   Directional antenna.

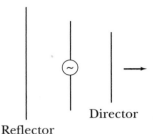

Director

Reflector   **Figure 15.14**   Yagi antenna.

ently, whereas a crest that starts from B toward A will tend to be canceled by A. The pair of antennas will radiate more to the right than the left, and correspondingly as receiving antennas they will be more sensitive to radiation coming from the right.

Building on the above concepts, in the 1920's the Japanese radio engineer Yagi developed the **parasitic array** shown in Figure 15.14. Only the middle element is driven. The reflector is slightly longer than $\lambda/2$, and so it is operating at a frequency above its own resonance; this implies that its signal leads the driving signal (Fig. 13.18). Correspondingly, the director lags. In both cases, the effect is to produce waves preferentially toward the right. Phase relationships are complicated because the fields of the reflector and director are not negligible compared with the driven element. The optimum spacing turns out to be between 0.1λ and 0.2λ. Such an antenna may have several directors, but there is no advantage in having more than one reflector because the field there is too small.

The **log periodic antenna,** Figure 15.15, is a highly directional wide-band antenna commonly used for television reception. As in the Yagi antenna, shorter prongs tend to serve as directors, and longer ones as reflectors, but the actual function changes with frequency. There is no one-to-one correspondence between prong length and TV channel number, but longer arms do correspond generally to lower channel numbers; see Section 14.5 for the frequencies corresponding to various channels.

Antenna **gain** is a measure of directivity: It is the ratio of the maximum intensity S to the average intensity over the entire pattern, $P/4\pi r^2$. For a half-wave antenna, with $\theta = 90°$ at maximum intensity,

$$\text{Gain} = \frac{S}{P/4\pi r^2} = \frac{1}{4\pi^2\epsilon_0 c}\frac{4\pi}{73} = 1.64$$

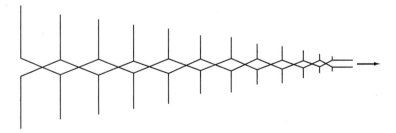

Figure 15.15   Log periodic antenna.

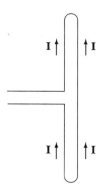

Figure 15.16   Folded dipole.

It is sometimes given in decibels (db):

$$10 \log_{10}(\text{gain}) = 2.1 \text{ db}$$

(a bel is a factor of 10).

Sometimes the 73-ohm impedance (R_{rad}) of the half-wave antenna is simply inconvenient, and one may then find a half-wave antenna variant of different impedance. The **folded dipole,** Figure 15.16, is a half-wave antenna made of "300-Ω twin lead" cable (see Section 17.3) with its ends soldered together. Equal currents I travel in the same direction in the two leads because the voltages at the ends are tied together. Consequently for a given I introduced by the source, $2I$ flows. So the power is greater than that of a simple half-wave antenna by a factor of $2^2 = 4$, and the radiation resistance is $4 \times 73.1 \approx 300 \ \Omega$. This matches standard 300-ohm cable satisfactorily. (It also happens to have a wider bandwidth than a half-wave antenna.)

There are many other kinds and shapes of antenna, some seemingly owing as much to art as to science. There is even a corkscrew shape for producing circularly polarized waves; this can be useful for communication with a spacecraft whose orientation may be arbitrary.

15.6   SUMMARY

We have examined various radiating systems, and we have found that, for most practical purposes, the half-wave antenna is the appropriate building block. An antenna is characterized primarily by its radiation resistance.

The power radiated by an antenna may be expressed as

$$P = I^2 \, R_{\text{rad}} \tag{15.16}$$

For a radiating electric dipole of length s the radiation resistance is

$$R_{\text{rad}} = 20.0 \ (ks)^2 \tag{15.17}$$

For a radiating magnetic dipole it is

$$R_{\text{rad}} = 20.0 \ (ks)^4 \tag{15.23}$$

where now s^2 is the area of the dipole.

For radiating dipoles, the radiation is polarized with the **E** vector parallel to the current, and the angular distribution of the intensity follows a $\sin^2\theta$ law.

For a half-wave antenna the radiation resistance is

$$R_{\text{rad}} = 73.1 \ \Omega \tag{15.30}$$

and the angular distribution follows the form

$$\frac{\cos^2(\frac{\pi}{2} \cos \theta)}{\sin^2\theta} \tag{15.28}$$

Transmission and reception by a given antenna are related by a reciprocity theorem.

▼ PROBLEMS

15-1 An electric dipole antenna of length 25 cm radiates 30 mW at 28 MHz. Find the current.

15-2 A magnetic dipole antenna of diameter 15 cm carries 0.20 A at 57 MHz. Find the power radiated.

15-3 In a half-wave copper antenna of length 1.5 m and diameter 1 mm, with current 1 A, how far do the electrons go (peak-to-peak, not rms)? Assume one conduction electron per atom, and all such electrons are equally involved (we learn about skin effect in the next chapter).

15-4 **(a)** A uniformly charged sphere rotates uniformly; does it radiate? **(b)** A uniformly charged sphere oscillates radially; does it radiate?

15-5 Find the radiation resistance of a 40-cm rod operating at **(a)** 75 MHz **(b)** 750 MHz.

15-6 In the median plane of a radiating electric dipole, find the distance in wavelengths at which the static electric dipole field equals the radiated E field.

15-7 A radiating dipole located at the origin shows maximum intensity at $\theta = 90°$ and polarization (**E** vector) in the ϕ direction. Is the dipole electric or magnetic? What is the orientation of its dipole moment?

15-8 What fraction of the power of a radiating magnetic dipole is emitted within $\pm 45°$ of the median plane?

15-9 For a radiating electric dipole, find the phase velocity for A and for V. Plot them versus r on one graph. (To get all of the time dependence of V into the exponent, set

$$1 + \frac{i}{kr} = \sqrt{1^2 + (1/kr)^2} \ e^{i \arctan \ (1/kr)}$$

then differentiate the exponent.)

15-10 Repeat the preceding problem for E_r and E_θ. (The result for E_θ is rather strange.)

*15-11 The differential equation of the radiating electric dipole field lines is $dr/rd\theta = E_r/E_\theta$. Solve.

15-12 For radiating electric and magnetic dipoles, show that V and $\mathbf{A}$ satisfy the Lorentz condition.

15-13 Find the rms E and B fields in the median plane of a half-wave antenna of length 3 m, radiating 1 kW average, at a distance of 16 km from the antenna.

15-14 A radio station broadcasts total power of 5 kW at 94 MHz using a half-wave antenna. Find the length of the antenna, the current in the antenna, and Poynting's vector in the median plane at a distance of 7 km.

*15-15 For an electric dipole rotating at constant angular velocity ω about an axis perpendicular to its moment $p = Qs$, Figure 15.17, find the average Poynting's vector $\mathbf{S}$ at large distance, and plot it. (It's like an orthogonal pair of dipoles 90° out of phase.)

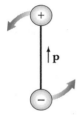

Figure 15.17

15-16 For the preceding problem, find the radiated power P by **(a)** integrating Poynting's vector, and by **(b)** Larmor's formula, Section 14.6, applied to the uniformly centripetally accelerated charges.

15-17 Find the gain or directivity of a radiating **(a)** electric dipole; **(b)** magnetic dipole.

*15-18 A magnetic dipole rotates with its magnetic moment $\mathbf{m}$ at angle θ relative to its angular velocity. Find the power radiated. Evaluate for the earth: $m = 8 \times 10^{22}$ A · m^2, $\theta = 11°$.

15-19 For an oscillating linear quadrupole, Figure 15.18, with $Q = Q_0 e^{i\omega t}$, find Poynting's vector and the radiated power. Sketch the radiation pattern.

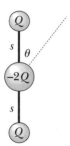

Figure 15.18

RADIATION IN MEDIA

We have looked at general properties of radiation in vacuum in Chapter 14. Then in Chapter 15 we dealt with the production of radiation by various particular types of antenna. Now we shall examine in some detail how existing radiation is modified by the material it passes through, and we will find that the conductivity is the most significant parameter.

16.1. Arbitrary Media; Complex *k*

Actually, there are lots of arbitrary media, so we are going to start cutting back quite soon. We will focus our attention on three special cases that illustrate diverse aspects of radiation in media. The media will be ideal in that they are homogeneous, linear, isotropic, and of course stationary.

We have already derived the wave equations we need for **E** and **B**, straight from Maxwell's equations. From Section 14.3,

$$\nabla^2 \mathbf{E} - \mu\epsilon \frac{\partial^2 \mathbf{E}}{\partial t^2} = \mu \frac{\partial \mathbf{J}}{\partial t} + \nabla \frac{\rho}{\epsilon} \tag{16.1}$$

and

$$\nabla^2 \mathbf{B} - \mu\epsilon \frac{\partial^2 \mathbf{B}}{\partial t^2} = -\mu \nabla \times \mathbf{J}$$

First we shall set $\rho = 0$: We permit free current **J** but no free charge ρ. Next, we assume that *if* there is some **J**, it obeys Ohm's law, $\mathbf{J} = \sigma\mathbf{E}$, with the conductivity σ being a constant. Substituting $\sigma\mathbf{E}$ for **J** in both wave equations, we arrive at

$$\nabla^2 \mathbf{E} - \mu\epsilon \frac{\partial^2 \mathbf{E}}{\partial t^2} - \mu\sigma \frac{\partial \mathbf{E}}{\partial t} = 0 \tag{16.2}$$

and

$$\nabla^2 \mathbf{B} - \mu\epsilon \frac{\partial^2 \mathbf{B}}{\partial t^2} - \mu\sigma \frac{\partial \mathbf{B}}{\partial t} = 0 \qquad \left(\text{We used } \nabla \times \mathbf{E} = -\frac{\partial \mathbf{B}}{\partial t} \text{ too.}\right)$$

It is quite striking that these equations are identical in form. However, if **E** and **B** are to keep up with each other through thick and thin, then it must be so.

Now we introduce the usual-looking plane-wave solutions:

$$\mathbf{E} = \mathbf{E}_0 \, e^{i(kz-\omega t)}$$
$$\mathbf{B} = \mathbf{B}_0 \, e^{i(kz-\omega t)}$$

$$(16.3)$$

These are different from what you saw in Section 14.5, however: *k* is now complex, in order to deal with the characteristics of the medium. Clearly the frequency $f = \omega/2\pi$ doesn't change in different media—if the frequency were lower, where did those missing waves go? No, *f* can't change. But the wavelength $\lambda = 2\pi/k$ *does* change. All of the characteristics of the medium are reflected in the propagation constant (wave number) *k*, which must therefore be very versatile. So we make it complex.

What does an imaginary *k* mean? Well, for a real *k*, $\mathrm{Re}(e^{ikz}) = \cos kz$; this is a sinusoidal wave. But if *k* is imaginary, let's set $k = i\alpha$, so that then $\mathrm{Re}(e^{ikz}) = \mathrm{Re}(e^{-\alpha z}) = e^{-\alpha z}$; it falls off exponentially. So if the wave is being absorbed, we expect to find that *k* has an imaginary part.

To discover how *k* behaves, we will substitute these plane wave solutions into the wave equations. When we differentiate **E** with respect to *z*, an *ik* pops out, leaving **E** otherwise unchanged:

$$\frac{\partial}{\partial z} \mathbf{E}_0 \, e^{i(kz-\omega t)} = ik\,\mathbf{E}_0 \, e^{i(kz-\omega t)}$$

And when we differentiate with respect to *t*, a $-i\omega$ appears in front. So when we substitute the solution, the **E** wave equation becomes:

$$-k^2\mathbf{E} + \mu\epsilon\omega^2\mathbf{E} + i\omega\mu\,\sigma\mathbf{E} = 0 \qquad (16.4)$$

Now we cancel out the **E**'s:

$$k = \sqrt{\epsilon\,\mu\omega^2 + i\omega\mu\,\sigma}$$

and we remind you that $\epsilon = \epsilon_0\epsilon_r$ and $\mu = \mu_0\mu_r$. We employ results from Section 14.5 for free space: $\mu_0\epsilon_0 c^2 = 1$, and $c = \omega/k_0$, using k_0 for the *k* of free space. We finally arrive at

$$\boxed{k = k_0 \sqrt{\epsilon_r\mu_r} \sqrt{1 + \frac{i\sigma}{\omega\epsilon}}} \qquad (16.5)$$

This is the equation that tells you how *k* in the medium is different from k_0 in the air. Obviously the **B** wave equation yields the same result, and as in Section 14.5 we still have

$$B = \frac{k}{\omega}\,E \qquad (16.6)$$

The various *k*'s produce all sorts of interesting phenomena, but it turns out that the conductivity σ is the dominant factor in determining the general nature of the waves. We shall examine three extreme cases:

- For *dielectrics,* σ is negligible.
- For *conductors,* σ is large.
- For *plasma,* σ is imaginary. How can conductivity be imaginary? This will be revealed in Section 16.4.

We begin with dielectrics.

16.2.   Dielectrics; Index of Refraction n

For a dielectric medium, the conductivity σ is essentially zero. Thus, the preceding expression for k becomes simply

$$k = k_0\sqrt{\epsilon_r\mu_r}\,\sqrt{1 + \frac{i\sigma}{\omega\epsilon}} = k_0\sqrt{\epsilon_r\mu_r} \tag{16.7}$$

The phase velocity of waves was found in Section 14.5:

$$v = \frac{\omega}{k}$$

so in the medium

$$v = \frac{\omega}{k_0\sqrt{\epsilon_r\mu_r}} = \frac{c}{\sqrt{\epsilon_r\mu_r}} \tag{16.8}$$

Now you already know that light slows down in glass by a factor called n, the **index of refraction** (refractive index):

$$v = \frac{c}{n}$$

So the theory confidently predicts that

$$\boxed{n = \sqrt{\epsilon_r\mu_r}} \tag{16.9}$$

This is a remarkable triumph of theoretical physics, comparable with the prediction of the speed of light (Section 14.5), using static measurements to predict how rapidly moving, rapidly oscillating waves will behave in dielectric material. **If it works. Let's check it.**

The accompanying table lists predicted and observed n's for various materials. The μ_r's are all essentially 1.00 (Section 10.2). The table shows that the theory works for nonpolar dielectrics but not for polar ones. The reason is, polar molecules have to move so as to align their dipole moments with the **E** field; but at optical frequencies they don't have time (see Fig. 9.10). Whereas in nonpolar molecules, all that is involved is a slight redistribution of electrons, which is very quick. This suggests that in polar molecules n should be smaller than $\sqrt{\epsilon_r}$, and that is generally true. So actually, the theory is well vindicated.

Material	$\sqrt{\epsilon_r}$	n
Air	1.00029	1.00029
Argon	1.00028	1.00028
CO_2 gas	1.00047	1.00045
HCl gas	1.0023	1.00045
Benzene	1.49	1.48
Ethanol	5.3	1.36
O_2 liq	1.21	1.22
Water	9.0	1.33
$(CH_2)_n$	1.60	1.59
NaCl	2.47	1.54
Sulfur	2.0	2.0

An aside: In a microwave oven, molecules in food are subjected to a rapidly varying **E** field (3 GHz). Water is both highly polar and relatively mobile, so it is water molecules that are getting flipped about, and it is the water that absorbs most of the power. Foods that are truly dry do not heat well in microwave ovens.

B increases in magnitude, relative to *E*, inside the dielectric; from Section 14.5,

$$\frac{E}{B} = \frac{\omega}{k} = v = \frac{c}{\sqrt{\epsilon_r \mu_r}} \tag{16.10}$$

using the index of refraction from Eq. 16.9. From this it is easy to show (see Problem 16-3) that, as in air (Section 14.5), the electric and magnetic energy densities are equal:

$$\tfrac{1}{2} \epsilon E^2 = \tfrac{1}{2} B^2 / \mu$$

and Poynting's vector is energy density times velocity:

$$S_{av} = \epsilon E_{rms}^2 \, v = n \epsilon_o c E_{rms}^2 \tag{16.11}$$

One can see *reflections* in glass, like in a mirror but fainter. We can easily calculate the normal reflectance, now that we know the fields inside and out. We just need the **boundary conditions** at the surface for tangential **E** and **B** fields.

Referring to Figure 16.1, when we apply Stokes' theorem to Faraday's law we get

$$\oint \mathbf{E} \cdot d\mathbf{l} = - \int \frac{\partial \mathbf{B}}{\partial t} \cdot d\mathbf{a}$$

so

$$l (E_2 - E_1) = l \Delta z \frac{\partial B}{\partial t} \tag{16.12}$$

where E_1 and E_2 are the tangential components of **E** on the two sides of the boundary. As $\Delta z \to 0$, it must be that $E_2 \to E_1$ unless $\partial B / \partial t \to \infty$ in such a way that the product $\Delta z \, \partial B / \partial t$ remains finite. Which doesn't happen. So $E_1 = E_2$, which means $\Delta E_t = 0$, where E_t means the tangential component of **E**.

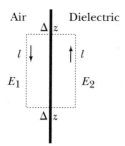

Air | Dielectric

E_1

E_2

Figure 16.1   Stokes' theorem.

The same kind of reasoning doesn't work for **B** in the presence of magnetic material, because there may *be* an infinite current density $\mathbf{J_b}$ in the form of a bound surface current. So we revert to $\nabla \times \mathbf{H} = \mathbf{J} + \partial\mathbf{D}/\partial t$, which ignores $\mathbf{J_b}$. We can still have an infinite free current density $\mathbf{J}$ on the surface of a superconductor, in the form of a surface current **K**, whose contribution doesn't get smaller as $\Delta z \to 0$. But $\partial\mathbf{D}/\partial t$ can't go infinite on us. So $\Delta H_t = K$. We shall here stipulate that the reflectance of a superconductor is 100% and leave that case aside, so $\Delta H_t = 0$.

To summarize, the boundary conditions are

$$\Delta D_n = \sigma_f \quad \text{(from } \nabla \cdot \mathbf{D} = \rho)$$

$$\Delta B_n = 0 \quad \text{(from } \nabla \cdot \mathbf{B} = 0)$$

$$\Delta E_t = 0 \quad \text{(from } \nabla \times \mathbf{E} = -\partial\mathbf{B}/\partial t)$$

$$\Delta H_t = K_f \quad \text{(from } \nabla \times \mathbf{H} = \mathbf{J} + \partial\mathbf{D}/\partial t)$$

(16.13)

where for example $\Delta B_n = 0$ means that the normal component of **B** is continuous across the surface. We won't need the normal components here, D_n and B_n, so we leave them for Problem 16-2.

Since the transverse **E** fields are equal inside and outside the surface, we have (Fig. 16.2a)

$$E_I - E_R = E_T \tag{16.14}$$

and $\Delta H_t = 0$ gives

$$B_I + B_R = B_T/\mu_r$$

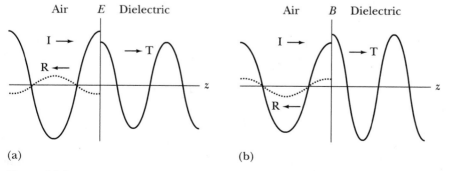

(a)   (b)

Figure 16.2   (a) **E** wave entering dielectric. (b) **B** wave entering dielectric.

where the subscripts refer to Incident, Reflected, and Transmitted waves (these are amplitudes at the surface, without time dependence); and we have used $B = \mu H$ (Fig. 16.2b). Either E or B must change sign on reflection in order that the reflected Poynting's vector $\mathbf{S}_R = \mathbf{E}_R \times \mathbf{B}_R / \mu$ point to the left. We know B_R is plus here because, as we remarked above, B_T is increased relative to E_T, as shown in Figure 16.2. (If the light were going from the dielectric to air, the signs of E_R and B_R would be reversed.) So E changes phase by $180°$, and B does not.

We will set $\mu_r = 1$ because that's the usual case. Substituting $B = k/\omega\, E$ and $k = nk_0$ (and $n = 1$ in air) into what is left of the **B** condition, $B_I + B_R = B_T$, we get

$$\frac{k_0}{\omega} E_I + \frac{k_0}{\omega} E_R = \frac{nk_0}{\omega} E_T$$

$$E_I + E_R = nE_T$$

Now we substitute for E_T in the E condition, $E_I - E_R = E_T$, and solve; the result is

$$\frac{E_R}{E_I} = \frac{n-1}{n+1}$$

So with $S = \epsilon_0 c E^2$ in the air

$$\frac{S_R}{S_I} = \boxed{R = \left(\frac{n-1}{n+1}\right)^2} \tag{16.15}$$

where R is the coefficient of reflection, or the **reflectance.** Transmittance is T = 1 − R.

EXAMPLE 16-1

Find the fraction of normally incident light reflected by a window pane of $n = 1.60$; include only single reflections (and ignore interference).

ANSWER The reflectance is the same at both surfaces.

$$R = \left(\frac{n-1}{n+1}\right)^2$$

$$= \left(\frac{0.6}{2.6}\right)^2 = 0.053 \qquad \text{and} \qquad T = 1 - R = 0.947$$

$$R + TRT = 0.053 + 0.947 \times 0.053 \times 0.947 = 0.101$$

If the transition from air to dielectric is made more gradual, the reflection is reduced in general. One may either enhance or eliminate reflections as desired using various kinds of dielectric coatings. In particular, if you look at the reflection in a camera lens, it is likely to be dark blue or some other faint color rather than the bright white of reflection from an ordinary glass surface; here we wish to reduce unwanted reflections, and this is accomplished with a quarter-wave coating of a dielectric with an index of refraction intermediate between that of glass and air (see Problem 16-4). For enhanced reflection: Silver mirrors have a reflectance of only 93% or so, but one can achieve over 99% for selected wavelengths using multiple dielectric layers.

16.3.   Conductors; Skin Depth δ

The k formula again is (Eq. 16.5)

$$k = k_0\sqrt{\epsilon_r\mu_r}\,\sqrt{1 + \frac{i\sigma}{\omega\epsilon}}$$

Now for copper at 1 MHz,

$$\frac{\sigma}{\omega\epsilon} = \frac{5.8 \times 10^7}{2\pi \times 10^6 \times 8.85 \times 10^{-12}} = 1.0 \times 10^{12} \tag{16.16}$$

This indicates that for good conductors we can neglect the 1 in the k formula. (At 60 Hz the ratio is even bigger.) Then,

$$k = k_0\sqrt{\frac{i\mu_r\sigma}{\omega\epsilon_0}} \tag{16.17}$$

Again for copper at 1 MHz, $k = k_0\sqrt{i\,1.0 \times 10^{12}} \approx 10^6\,k_0$ in magnitude, which means that the wavelength and velocity are small, since $\lambda = 2\pi/k$ and $v = \omega/k$. In this and other ways, a conductor behaves like a dielectric of large ϵ_r (Sections 9.2 and 12.2).
   This also implies a simplification of the wave equation (Section 16.1),

$$\nabla^2\mathbf{E} - \mu\epsilon\frac{\partial^2\mathbf{E}}{\partial t^2} - \mu\sigma\frac{\partial\mathbf{E}}{\partial t} = 0$$

The second term is negligible, and we are left with

$$\nabla^2\mathbf{E} - \mu\sigma\frac{\partial\mathbf{E}}{\partial t} = 0 \tag{16.18}$$

with no second time derivative. (This is reminiscent of Schroedinger's equation in nonrelativistic quantum mechanics:

$$-\frac{\hbar^2}{2m}\nabla^2\psi + V\psi = i\hbar\frac{\partial\psi}{\partial t} \tag{16.19}$$

in that it is a wave equation with no second time derivative, so strong dispersion: Speed depends on frequency.)
   If we set $k_0 = \omega/c$, and use $\mu_0\epsilon_0c^2 = 1$, then Equation 16.17 becomes

$$k = \sqrt{i\omega\sigma\mu} \tag{16.20}$$

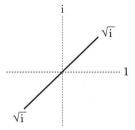

Figure 16.3   Complex plane.

a cute result. But what is $\sqrt{i}$? Well, in view of the Euler formula $e^{i\theta} = \cos\theta + i\sin\theta$,

$$i = e^{i\pi/2} \quad or \quad e^{i(\pi/2 + 2\pi)} \quad (\text{since } e^{2\pi i} = 1)$$

$$\sqrt{i} = i^{1/2} = e^{i\pi/4} \quad or \quad e^{i(\pi/4 + \pi)}$$

so

$$\sqrt{i} = \frac{1+i}{\sqrt{2}} \quad or \quad \frac{-1-i}{\sqrt{2}} \quad (\text{Fig. 16.3})$$

If you square either expression, you get i. The positive one corresponds to wave propagation in the $+z$ direction, which is what we want. So,

$$k = \sqrt{\frac{\omega\sigma\mu}{2}}\,(1+i) \tag{16.21}$$

Now we introduce the **skin depth**

$$\boxed{\delta = \sqrt{\frac{2}{\omega\sigma\mu}}} \tag{16.22}$$

so that

$$k = \frac{1}{\delta} + \frac{i}{\delta}$$

Then the **E** wave is

$$\mathbf{E} = \mathbf{E}_0\,e^{i(kz - \omega t)}$$
$$= \mathbf{E}_0\,e^{i(z/\delta - \omega t)}\,e^{-z/\delta} \tag{16.23}$$

This is a wave that falls to $1/e$ of its original value in a distance δ, and whose phase changes by 1 radian in the same distance δ, Figure 16.4. Energy is dissipated due to the conductivity. The skin depth δ controls both the wiggle and the falloff; but the wave falls off so fast that it hardly gets a chance to wiggle at all. Note that the general shape of the curve is the same regardless of frequency or conductivity, as long as the material is still a good conductor, that is, as long as the 1 in the k formula can be neglected; this is due to the fact that the real and imaginary parts of k are equal.

Re E

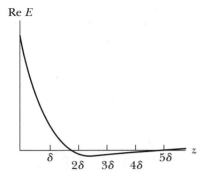

Figure 16.4 E in a conductor.

 EXAMPLE 16-2

Find the skin depth for copper at 60 Hz.

ANSWER $\mu_r \approx 1$ and $\sigma = 5.8 \times 10^7$ S/m. At 60 Hz, $\omega = 2\pi f = 377$ rad/s. So,

$$\delta = \sqrt{\frac{2}{\omega\sigma\mu}}$$

$$= \sqrt{\frac{2}{377 \text{ rad/s} \times 5.8 \times 10^7 \text{ S/m} \times 4\pi \times 10^{-7} \text{ H/m}}}$$

$$= \sqrt{\frac{2 \text{ m}^2}{377/\text{s} \times 5.8 \times 10^7 \text{ A/V} \times 4\pi \times 10^{-7} \text{ V·s/A}}} = 0.85 \text{ cm}$$

So for ordinary house current, there is no point in making a conductor much larger than 1 cm in diameter, because the current does not penetrate to the interior. This is sometimes called the **skin effect.** At high frequency, the skin depth may be microscopic.

 EXAMPLE 16-3

Repeat for seawater.

ANSWER $\mu_r \approx 1$ and $\sigma = 5$ S/m (Section 8.1) and $\omega = 377$ rad/s. So,

$$\delta = \sqrt{\frac{2}{\omega\sigma\mu}} = \sqrt{2/377 \times 5 \times 4\pi \times 10^{-7}} = 29 \text{ m}$$

So for communication with submarines, the lower the frequency, the better.

And now let's look at the magnetic field B. The solution of the wave equation is, as before (Eq. 16.3),

$$\mathbf{B} = \mathbf{B}_0 \, e^{i(kz-\omega t)} = \mathbf{B}_0 \, e^{i(z/\delta - \omega t)} \, e^{-z/\delta}$$

$\mathbf{B}_0$ is a constant vector that is permitted to be complex. We already know $\mathbf{E}$, Equation 16.23; and Section 14.5 assures us that

$$B = \frac{kE}{\omega} \tag{16.24}$$

Also, $k = \sqrt{i\omega\sigma\mu}$, and $\sqrt{i} = e^{i\pi/4}$, so

$$B = \sqrt{\frac{\sigma\mu}{\omega}} \, E_0 \, e^{i(z/\delta - \omega t + \pi/4)} \, e^{-z/\delta} \tag{16.25}$$

and so

$$B_0 = \sqrt{\frac{\sigma\mu}{\omega}} \, E_0 \, e^{i\pi/4}$$

Thus, the magnetic field lags the electric field by $45°$ $(=\pi/4$ radians).

Now in vacuum, $E/B = c = 3 \times 10^8$ m/s. But in a metal, say for copper at 1 MHz, we find

$$|E/B| = \sqrt{\frac{\omega}{\sigma\mu}} = \sqrt{\frac{6.3 \times 10^6}{5.8 \times 10^7 \times 4\pi \times 10^{-7}}} = 300 \text{ m/s}$$

so in a metal B is much larger relative to E, by a factor of a million here. This is due to conduction current in the metal.

Then Poynting's vector is (Eq. 14.44)

$$\mathbf{S}_{\mathrm{av}} = \tfrac{1}{2} (\text{Re } \mathbf{E} \times \mathbf{B}^*/\mu) = \tfrac{1}{2} \sqrt{\frac{\sigma}{2\omega\mu}} \, E_0^2 \, e^{-2z/\delta} \, \hat{\mathbf{z}} \tag{16.26}$$

The power falls off twice as fast as E or B, that is, as $e^{-2z/\delta}$; the energy is rapidly converted to heat inside the metal. That does not mean that the metal is absorbing most of the incident energy, which instead is reflected away; it means that any energy that gets into the metal gets absorbed in a hurry.

The average energy density is almost all magnetic, due to the current; at 1 MHz in copper, as in the above examples,

$$\frac{\tfrac{1}{2} B_0^2/\mu}{\tfrac{1}{2} \epsilon E_0^2} = \frac{\sigma}{\omega\epsilon} = 10^{12} \tag{16.27}$$

Metals *reflect* well; after all, that's how mirrors work. We can calculate the reflectance and transmittance for normal incidence much as we did for dielectrics (Eq. 16.14), starting with (Fig. 16.5)

$$E_{\mathrm{I}} - E_{\mathrm{R}} = E_{\mathrm{T}} \tag{16.28}$$

and

$$B_{\mathrm{I}} + B_{\mathrm{R}} = B_{\mathrm{T}}/\mu_{\mathrm{r}}$$

As in Section 16.2, E changes sign, and B does not, upon reflection.

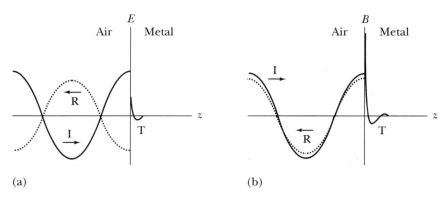

Figure 16.5   **(a) E** at metal surface. **(b) B** at metal surface.

As before, in the **B** condition $B_I + B_R = B_T/\mu_r$ we substitute $B = (k/\omega)\,E$ and $k = nk_0$, and we set $\mu_r = 1$, obtaining

$$\frac{k_0}{\omega} E_I + \frac{k_0}{\omega} E_R = \frac{nk_0}{\omega} E_T$$

However, in this case E_T is small and so things will go better if we get rid of E_R, using the E condition above, $E_I - E_R = E_T$. The result is

$$\frac{E_T}{E_I} = \frac{2}{1 + n} \qquad\qquad\qquad \textbf{(16.29)}$$

Near the beginning of this section we found (Eq. 16.17) $k = k_0 \sqrt{\dfrac{i\,\mu_r\sigma}{\omega\epsilon_0}}$ and so our complex index of refraction is

$$n = \sqrt{\frac{i\,\mu_r\sigma}{\omega\epsilon_0}} \qquad\qquad\qquad \textbf{(16.30)}$$

It is large, on the order of 10^6, so we can neglect the 1 in Equation 16.29:

$$\frac{E_T}{E_I} = \frac{2}{1 + n} \approx \frac{2}{n}$$

We will need B too because we can't just use $S = \epsilon_0 c E^2$ in the medium. So we have

$$B_I + B_R = B_T$$

$$B_I - B_R = B_T/n \qquad \text{(from } E_I - E_R = E_T)$$

which yield

$$\frac{B_T}{B_I} = \frac{2}{1 + 1/n} \approx 2$$

Using $\mathbf{S} = \frac{1}{2}\,\mathrm{Re}\,(\mathbf{E} \times \mathbf{B}^*/\mu)$ we find

$$\frac{S_T}{S_I} = \mathrm{Re}\,\frac{4}{n}$$

So the transmittance is (with $\mu_r = 1$)

$$T = \sqrt{\frac{8\omega\epsilon_0}{\sigma}}$$

(16.31)

(This is sometimes called the Absorptance A, because usually everything that is transmitted by the front surface winds up getting absorbed in the metal.)

 EXAMPLE 16-4

Find the reflection coefficient for mercury ($\sigma = 10^6$ S/m), yellow light ($\lambda = 0.6$ μm), normal incidence.

ANSWER

$$T = \sqrt{8\omega\epsilon_0/\sigma}$$
$$= \sqrt{8 \times 2\pi \times 3 \times 10^8 \times 8.85 \times 10^{-12}/10^6 \times 0.6 \times 10^{-6}} = 0.47$$
$$R = 1 - T = 0.53$$

It is a reasonable number (0.70 is measured), but you should *not* learn to trust DC conductivities at optical frequencies.

The reflection gives nearly identical waves traveling in opposite directions, Figure 16.5, which is the prerequisite for standing waves. Nodes of **E** are located at 0, $\lambda/2$, λ, and so forth from the surface; and nodes of **B** are in between, at $\lambda/4$, $3\lambda/4$, and so on. Wiener's experiment in 1890 involved placing a thin photographic emulsion near the metal surface, Figure 16.6; it was blackened at the antinodes of **E**, indicating that **E** is the effective field even though **E** and **B** contain equal

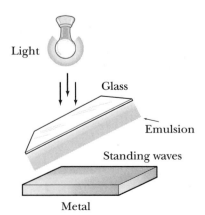

Light

Glass

Emulsion

Standing waves

Metal

Figure 16.6   Wiener's experiment.

energy. (That's because matter consists of electric monopoles, not magnetic ones.) This formed the basis for an 1891 color photography process and for a Nobel Prize for Gabriel Lippmann in 1908: When a mirror is coated with emulsion, exposed to an image, and developed, it tends to reflect the original colors by interference.

16.4. Plasmas; Plasma Frequency ω_p

Plasma is ionized matter, that is, matter in which the electrons and positive charges are not bound together, typically because the temperature is high. Plasma is the fourth state of matter, after solid, liquid, and gas. Most of the matter in the universe is plasma; a star, for example, is almost entirely plasma. A spark is a plasma.

How can the conductivity of a plasma be imaginary? If we combine $\mathbf{J} = \sigma\mathbf{E}$ and $i = e^{i\pi/2}$, we see that an imaginary conductivity implies that E *leads J* by $\pi/2$ radians, or 90°. This is due to the inertia of electrons: When we apply an E field, the electrons begin to accelerate, but it takes time for them to gather speed. (Can E ever *lag J*? In Section 13.2 we found that current may lead or lag voltage. See the following text for more on this.)

We will assume that the electrons are completely free, that is, that there are no collisions and the electrons feel no force other than the $F = eE$ that is due to the electric field. The positive charges are also free, but their mass is at least 1836 times that of the electrons, so they are sluggish and make a negligible contribution to the conductivity.

We begin with Newton's Second Law:

$$F = ma = eE \tag{16.32}$$

(Last time we wrote this, we were doing radiation scattering, Section 14.6. Will scattering be significant here? Answer: The mean distance for scattering will be about $1/N\sigma$, where N is the number of electrons per cubic meter and σ is the Thomson cross section. N can range from around 1 to 10^{28} m^{-3}; let's try 10^{14}. Then $1/N\sigma \approx (10^{14} \times 0.67 \times 10^{-28})^{-1} \approx 10^{14}$ m, which is likely to be much larger than the corresponding plasma; so we shall neglect scattering.)

The **E** field is that of incident radiation, which we will take to be plane waves in the z direction:

$$E = E_0\, e^{i(kz - \omega t)}$$

The velocity of the electrons is found by integrating the acceleration:

$$v = \int a\, dt = \int \frac{eE}{m}\, dt$$

$$= -\frac{eE}{i\omega m} \qquad (\text{average } v = 0)$$

Current density is charge density times speed:

$$J = \rho v = eN\frac{ieE}{\omega m} \tag{16.33}$$

where N is the number of electrons per cubic meter. So the conductivity is

$$\sigma = \frac{J}{E} = \frac{ie^2N}{\omega m} \tag{16.34}$$

We can substitute this σ in our equation for k, Section 16.1:

$$k = k_0 \sqrt{\epsilon_r \mu_r} \sqrt{1 + \frac{i\sigma}{\omega\epsilon}}$$

Plasmas are typically rarefied gases, so we may set $\epsilon_r \mu_r = 1$. Thus,

$$k = k_0 \sqrt{1 + \frac{i\frac{ie^2N}{\omega m}}{\omega\epsilon_0}} = k_0 \sqrt{1 - \frac{e^2N}{\omega^2 m\epsilon_0}} \tag{16.35}$$

We introduce the **plasma frequency** ω_p:

$$\boxed{\omega_p = \sqrt{\frac{e^2N}{m\epsilon_0}}} \tag{16.36}$$

so that

$$\boxed{k = k_0 \sqrt{1 - \left(\frac{\omega_p}{\omega}\right)^2}} \tag{16.37}$$

The behavior of an electromagnetic wave in a plasma depends on whether the frequency of the wave is above or below the plasma frequency. In Figure 16.7a, a wave arrives below the plasma frequency: $\frac{\omega_p}{\omega} > 1$, so k is imaginary, which implies an exponential falloff of the wave inside the plasma; the wave penetrates into the plasma but ultimately all of the energy is reflected. In Figure 16.7b, the wave is just at the plasma frequency, so $k = 0$ and λ is infinite; the field throughout the plasma rises and falls simultaneously. In Figure 16.7c the frequency is above

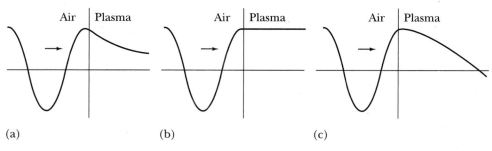

(a)   (b)   (c)

Figure 16.7   (a) Below ω_p. (b) At ω_p. (c) Above ω_p.

the plasma frequency, $\dfrac{\omega_p}{\omega} < 1$, so k is real but $k < k_0$; so λ is larger inside the plasma, and the phase velocity is greater than c (see the following text).

 EXAMPLE 16-5

The ionosphere (the atmosphere at high altitude) transmits frequencies above about 3 MHz; find the electron density.

ANSWER

$$\omega_p = \sqrt{\frac{e^2 N}{m\epsilon_0}} = \sqrt{\frac{(1.6 \times 10^{-19})^2 N}{9.1 \times 10^{-31} \times 8.85 \times 10^{-12}}} = 56.4\sqrt{N}$$

so

$$\omega_p/2\pi = \boxed{f_p \approx 9\sqrt{N}} \tag{16.38}$$

$$N = f_p^2/9^2$$
$$= (3 \times 10^6)^2/9^2 = 1.1 \times 10^{11} \text{ free electrons per cubic meter}$$

Below the plasma frequency of the ionosphere, for example for AM broadcast at 1 MHz, radio waves can travel long distances by bouncing between the earth and the ionosphere; at higher frequencies, such as FM broadcast at 100 MHz, reception must be line-of-sight with the transmission tower. The ionosphere is created by solar radiation, which ionizes air, and so the characteristics of the ionosphere depend on time of day and solar activity.

As in Section 14.5, the phase velocity v_p is

$$v_p = \omega/k$$

$$= \omega/k_0 \sqrt{1 - \left(\frac{\omega_p}{\omega}\right)^2}$$

so

$$v_p = c/\sqrt{1 - \left(\frac{\omega_p}{\omega}\right)^2} \tag{16.39}$$

For a plasma above the plasma frequency, $v_p > c$, Figure 16.8. But don't be alarmed: It is the **group velocity** v_g that determines how fast a signal can be sent. A pure sine wave does not convey information; it is the modulations, the changes in frequency, that carry the information. We shall now examine wave and group velocities.

Figure 16.9a shows a wave group. The waves in the group travel faster than the group itself. Analogously, if you drop a pebble in a pond, and carefully watch the ripples spread out, you can see that little wavelets are coming into being at the

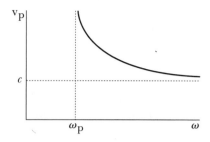

Figure 16.8   Phase velocity.

back of the group, moving forward through the group, and disappearing out the front. Figure 16.9b shows a string of pseudogroups, produced by a pair of sine waves that differ by $2\Delta k$ and $2\Delta\omega$. (For a pure sine wave, as we have emphasized above, ω doesn't change in passing from one medium to another; but here we are starting with two different ω's.) The size of the pseudogroups is proportional to Δk, but the speed of the pseudogroups is nearly independent of it, so this simple model enables us to determine v_g. One wave is given by $e^{i((k-\Delta k)z - (\omega-\Delta\omega)t)}$. The sum of the two waves may be expressed thus:

$$e^{i((k-\Delta k)z - (\omega-\Delta\omega)t)} + e^{i((k+\Delta k)z - (\omega+\Delta\omega)t)} = \tag{16.40}$$

$$e^{i(kz-\omega t)} \left(e^{i(-\Delta kz+\Delta\omega t)} + e^{i(\Delta kz-\Delta\omega t)} \right) = e^{i(kz-\omega t)} \, 2\cos(\Delta kz - \Delta\omega t)$$

The real part of this is

$$2\cos(kz - \omega t)\cos(\Delta kz - \Delta\omega t)$$

This is a sinusoidal wave of wavelength $2\pi/k$, with an envelope of wavelength $2\pi/\Delta k$, Figure 16.9b. The speed of the sinusoidal wave is the phase velocity,

$$v_p = \frac{\omega}{k} \tag{16.41}$$

and the speed of the envelope is the group velocity,

$$v_g = \frac{\Delta\omega}{\Delta k} \implies \frac{1}{dk/d\omega} \tag{16.42}$$

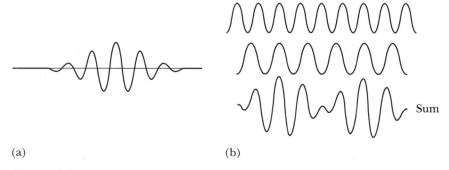

(a) (b)

Figure 16.9   **(a)** Wave group. **(b)** Pseudogroups.

It may easily be shown that

$$\boxed{v_p v_g = c^2}$$ (16.43)

and in fact it is Problem 16-19 at the end of the chapter. So even though $v_p > c$, the group velocity is always less than c, so relativity is still all right.

How well do two waves represent a wave group? The group actually contains a distribution of (an infinite number of infinitesimal) wave amplitudes. The size of the group is approximately $2\pi/\Delta k$ and so it is quite sensitive to the two waves chosen. However, the group velocity $\Delta\omega/\Delta k$ is nearly constant, for small increments $\Delta\omega$ and Δk. So our conclusions regarding group velocity are reliable.

Group velocity and phase velocity differ when the phase velocity depends on frequency. We have

$$k = \frac{\omega}{v_p}$$

If v_p is constant then $v_g = 1/(dk/d\omega) = \omega/k = v_p$. Otherwise the median exhibits **dispersion:** Velocity depends on frequency.

Plasmas are usually rarefied gases. However, . . .

 EXAMPLE 16-6

Find the plasma frequency for sodium.

ANSWER

$$N = 0.97 \frac{g}{cm^3} \times \frac{10^6 \, cm^3}{m^3} \times \frac{1 \, mole}{23 \, g} \times 6 \times 10^{23} \frac{electrons}{mole}$$

$$= 2.5 \times 10^{28} \text{ electrons per cubic meter}$$

We are assuming that sodium is an electron gas of one free electron per atom. Then,

$$f_p = 9\sqrt{N}$$
$$= 9\sqrt{2.5 \times 10^{28}} = 1.4 \times 10^{15} \text{ Hz}$$

so

$$\lambda = 0.21 \, \mu$$

At high frequencies the alkali metals satisfy the plasma requirements adequately; in particular, sodium does become transparent below 0.21 μ in the ultraviolet, as calculated.

Since σ is imaginary for a plasma, we find that $\mathbf{E} \cdot \mathbf{J}$ is imaginary and there is no energy dissipation. This is a consequence of our original assumption that there are no collisions in the plasma, and so there are no energy loss mechanisms. The total current density $\mathbf{J}_T$, including free current $\mathbf{J} = \sigma \mathbf{E}$ and displacement current $\partial \mathbf{D}/\partial t$, is

$$\nabla \times \mathbf{B}/\mu = \sigma \mathbf{E} + \epsilon \frac{\partial \mathbf{E}}{\partial t} = \frac{ie^2 N}{\omega m} \mathbf{E} - \epsilon_0 i \omega \mathbf{E} \tag{16.44}$$

(since $\epsilon_r \approx 1$). This is zero at

$$\omega = \sqrt{\frac{e^2 N}{m \epsilon_0}}$$

So at the plasma frequency there is no current and no magnetic field. Below ω_p the conduction current is dominant, as in a metal but without energy dissipation, and $\mathbf{J}_T$ lags E because of the inertia of the electrons involved. Above ω_p the displacement current is greater, as in a dielectric but with $n < 1$, and $\mathbf{J}_T$ leads E, as in a capacitor.

Above ω_p, where there is a wave and k is real, we have as before (sections 14.5 and 16.2)

$$\frac{E}{B} = \frac{\omega}{k} = v_p = \frac{c}{\sqrt{1 - \left(\dfrac{\omega_p}{\omega}\right)^2}} \tag{16.45}$$

So the ratio of E to B is reduced, compared to free space; and as the frequency approaches ω_p, B approaches zero, as also indicated in the preceding paragraph.

16.5.   Dispersion

In the preceding section we learned quite a bit about plasmas by starting with a simple model, namely, free electrons. In particular, we were able to see that a plasma is a dispersive medium: the speed of a wave depends on its frequency. Now we'll try a more complicated system: we will assume that the electrons in a material are on springs; thus they are attached to atoms and yet are able to oscillate in response to an incident electromagnetic wave. With this crude model we can gain some qualitative understanding of dispersion in non-conductors.

Newton's second law is a good place to begin:

$$F = ma = m \frac{d^2 x}{dt^2} = m \ddot{x} \tag{16.46}$$

The wave is incident in the positive z direction and polarized in the x direction; and an electron of mass m responds by moving in the x direction in "forced damped oscillation."

The force F on the electron consists of the following three parts:

(1) The electric field provides a force $eE_0 e^{i(kz-\omega t)}$; however, although the wave is moving in the z direction, this doesn't really matter because the electron is located at constant z. So we will absorb e^{ikz} into the constant amplitude E_0, and the result is $eE_0 e^{-i\omega t}$.

(2) The spring of spring constant k applies a Hooke's-law restoring force $-kx$. The force holding an electron in an atom is not so simple, of course, but most stable systems, even complicated ones, obey Hooke's law approximately for small x displacements. The equilibrium position of the electron is $x = 0$. The electron of mass m on a spring of constant k will have a resonant frequency given by the expression $\omega_0^2 = k/m$, so this force will be written as $-m\omega_0^2 x$.

(3) The motion is damped by a force proportional to speed, $-m\gamma \dfrac{dx}{dt} = -m\gamma\,\dot{x}$. We can regard this as being primarily due to radiation by the accelerated electron. The form is unsatisfactory in that the radiation actually depends on electron acceleration; however, the term is usually small, and so this treatment provides a useful approximation. For convenience we have included the constant m in this term.

Putting all three parts together, we find the equation of motion of the electron to be

$$F = m\ddot{x} = eE_0e^{-i\omega t} - m\omega_0^2 x - m\gamma\,\dot{x} \tag{16.47}$$

General solutions of this differential equation are fairly complicated; but fortunately all we want is the steady-state result in which, after a long time, the electron oscillates with the frequency of the incident wave, so we set

$$x(t) = x = x_0e^{-i\omega t}$$

When we substitute this into Equation 16.47, we obtain

$$x_0 = \frac{e}{m}\frac{1}{\omega_0^2 - \omega^2 - i\gamma\omega}E_0$$

The dipole moment of the electron is

$$p_0 = x_0\,e = \frac{e^2}{m}\frac{1}{\omega_0^2 - \omega^2 - i\gamma\omega}E_0$$

and if we have N electrons per unit volume participating in this oscillation, we find the dipole moment per unit volume to be

$$P_0 = Np_0 = \frac{Ne^2}{m}\frac{1}{\omega_0^2 - \omega^2 - i\gamma\omega}E_0 \tag{16.48}$$

The N we use here corresponds more or less to the number of atoms per unit volume. Each atom will usually have several electrons, each with its own characteristic ω_0; but here we are interested in only one such electron per atom.

From Section 9.2,

$$P = \epsilon_0\chi_e E = \epsilon_0(\epsilon_r - 1)E$$

so

$$\epsilon_r = 1 + \frac{P_0}{\epsilon_0 E_0}$$

In view of Equation 16.48, this becomes

$$\epsilon_r = 1 + \frac{Ne^2}{\epsilon_0 m} \frac{1}{\omega_0{}^2 - \omega^2 - i\gamma\omega}$$

Then we introduce the complex index of refraction (cf Section 16.2):

$$n = \sqrt{\epsilon_r} = \sqrt{1 + \frac{Ne^2}{\epsilon_0 m} \frac{1}{\omega_0{}^2 - \omega^2 - i\gamma\omega}} \tag{16.49}$$

We can check our work so far by inserting values for the plasma, for which the results are known from the preceding section. For the plasma, there is no dissipation, so $\gamma = 0$; and no resonance is involved, so $\omega_0 = 0$. Thus we find

$$n = \frac{c}{v_p} = \sqrt{1 - \frac{Ne^2}{\epsilon_0 m} \frac{1}{\omega^2}}$$

$$= \sqrt{1 - \frac{\omega_p{}^2}{\omega^2}}$$

which agrees with Equation 16.39, so it checks.

We will need to simplify Equation 16.49. First we will assume that n is nearly equal to 1 in magnitude so that we can apply the usual approximation, $\sqrt{1 + \delta} \approx 1 + \frac{1}{2}\delta$ for small δ. Equation 16.49 then becomes

$$n \approx 1 + \frac{Ne^2}{2\epsilon_0 m} \frac{1}{\omega_0{}^2 - \omega^2 - i\gamma\omega} \tag{16.50}$$

Then, in the vicinity of $\omega \approx \omega_0$,

$$n \approx 1 + \frac{Ne^2}{2\epsilon_0 m} \frac{1}{(\omega_0 + \omega)(\omega_0 - \omega) - i\gamma\omega_0}$$

$$\approx 1 + \frac{Ne^2}{4\epsilon_0 m \omega_0} \frac{1}{(\omega_0 - \omega) - i\gamma/2}$$

Breaking this into real and imaginary parts,

$$n_r = 1 + \frac{Ne^2}{4\epsilon_0 m \omega_0} \frac{\omega_0 - \omega}{(\omega_0 - \omega)^2 + \gamma^2/4} \tag{16.51}$$

and

$$n_i = \frac{Ne^2}{4\epsilon_0 m \omega_0} \frac{\gamma/2}{(\omega_0 - \omega)^2 + \gamma^2/4} \tag{16.52}$$

These are plotted in Figure 16.10. They are for one electron resonance. Other electron states may produce similar resonances at other points in the spectrum, and the corresponding values of $n_r - 1$ and n_i usually just add up. The resonances of most bound electrons lie in the ultraviolet.

Physical significance: As Figure 16.10 indicates, for **normal dispersion** the index of refraction n_r increases with frequency. For example, in the visible spectrum, blue light refracts more than red.

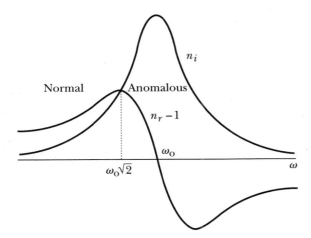

Figure 16.10   Dispersion and attenuation at a resonance.

In the vicinity of a resonance, on the other hand, we encounter **anomalous dispersion:** the slope of the $n_r - 1$ curve goes negative.

Above the resonance, the index of refraction is less than 1, suggesting a phase velocity exceeding c, as we encountered before in the case of plasmas. For X-rays, above the ultraviolet resonances, that is what is found.

The imaginary part of n corresponds to absorption (since $n = (c/\omega)k$, and Section 16.1 mentions that imaginary k corresponds to absorption). Maximum n_i occurs at resonance, as is intuitively reasonable: the electrons on their little springs prefer that particular frequency, and they absorb it and thereby increase the amplitude of their oscillations. In particular, cool gases absorb spectral lines that they would emit if they were hot; the dark Fraunhofer lines in the solar spectrum illustrate this. Note that γ is typically much smaller than ω_0: line widths are narrow.

Far from resonance, $n_r - 1 \gg n_i$ and the index of refraction is almost entirely real.

So the above model gives us interesting information about the behavior of the index of refraction, even though quantum mechanical considerations have been ignored.

16.6 SUMMARY

In order to get an overview of radiation propagation in materials, we have looked in detail at three extreme cases involving qualitatively different conductivities.

For electromagnetic waves in an ideal material,

$$k = k_0 \sqrt{\epsilon_r \mu_r} \sqrt{1 + \frac{i\sigma}{\omega \epsilon}} \qquad (16.5)$$

For dielectrics, σ is negligible.
For conductors, σ is large.
For plasma, σ is imaginary.

In a *dielectric,* the index of refraction is

$$n = \sqrt{\epsilon_r \mu_r} \tag{16.9}$$

Waves are reduced in speed, and in wavelength, by this factor. The reflection coefficient for normal incidence is

$$R = \frac{S_r}{S_i} = \left(\frac{n-1}{n+1}\right)^2 \tag{16.15}$$

In a *conductor,* the skin depth, and the radian length, is

$$\delta = \sqrt{\frac{2}{\omega \sigma \mu}} \tag{16.22}$$

Waves penetrate into a conductor by this distance. Their energy is dissipated by Joule heating. The transmission coefficient for normal incidence is

$$T = \sqrt{\frac{8\omega\epsilon_0}{\sigma}} \tag{16.31}$$

In a *plasma,*

$$k = k_0 \sqrt{1 - \left(\frac{\omega_p}{\omega}\right)^2} \tag{16.37}$$

and the plasma frequency is

$$\omega_p = \sqrt{\frac{e^2 N}{m\epsilon_0}} \quad (\text{so } f_p \approx 9\sqrt{N}) \tag{16.36}$$

Waves are transmitted only above the plasma frequency. The phase velocity v_p may exceed c, but the group velocity v_g is less than c:

$$v_p v_g = c^2 \tag{16.43}$$

There is no energy loss mechanism in an ideal plasma: the ions don't interact.

◤ PROBLEMS

16-1 Find the phase velocity of electromagnetic waves in benzene.

16-2 Show that at a surface $\Delta D_n = \sigma$ and $\Delta B_n = 0$.

16-3 For electromagnetic waves in a dielectric, show that the electric and magnetic energy densities are equal, and show that Poynting's vector equals the energy density times the (phase) velocity.

16-4 Find the thickness and index of refraction of a dielectric layer that eliminates normal reflection of green light of wavelength 550 nm from glass of $n = 1.60$.

16-5 Calculate the reflection from a window pane including all reflections. Evaluate for $n = 1.55$.

16-6 For a dielectric, find n for R = T.

16-7 For a dielectric, derive T from the fields and show that R + T = 1.

16-8 Estimate the frequency above which seawater is no longer a good conductor in the context of the equation for k. (At optical frequencies it is essentially a dielectric.)

16-9 Find the skin depth for electromagnetic waves of 1 kHz in seawater. (For communication with submarines, frequencies as low as 20 Hz may be employed.)

16-10 Find the phase velocity in copper at 1 MHz. (For comparison, sound travels at about 3600 m/s in copper.)

16-11 Find the skin depth in copper at 1 MHz.

16-12 A silver-coated plastic component works almost as well as solid silver at 3 GHz. Find the thickness of silver if Poynting's vector falls by a factor of 20 inside the silver.

16-13 Find the transmission coefficient for normal incidence on copper at 3 GHz.

16-14 In a half-wave copper antenna of length 1.5 m and diameter 1 mm, with current 1 A, how far do the electrons go? Assume one conduction electron per atom, and all such electrons within one skin depth of the surface are equally involved. (See Problem 15-3.)

16-15 Show that the radiation energy lost in a good conductor equals the Joule heating.

*16-16 Show that the radiation pressure on a good conductor, for normal incidence, is $2S/c$, by: **(a)** applying the Maxwell stress tensor (specify your surface); **(b)** integrating the force on the current inside the conductor.

16-17 Find the plasma frequency for interstellar space having 10^5 electrons/m^3.

16-18 Estimate the wavelength below which rubidium becomes transparent.

16-19 Show that $v_p v_g = c^2$ for a plasma.

16-20 Two waves, of frequency 3 MHz and 4 MHz, traverse a plasma of $N = 10^{10}$ electrons/m^3; find the average v_p, v_g, and distance from one wave pseudogroup to the next.

16-21 For Figure 16.10, verify: **(a)** The peak of n_i is twice as high as the maximum value of $n_r - 1$; **(b)** the two curves cross at $\omega = \omega_0 - \gamma/2$.

*16-22 For X-rays of wavelength 0.13 nm, incident on a silver surface, find **(a)** the index of refraction, and **(b)** the critical angle for total external reflection.

16-23 Verify Cauchy's equation for gases in visible light:

$$n_r - 1 = A\left(1 + \frac{B}{\lambda^2}\right),$$

A and B being constants.

CHAPTER

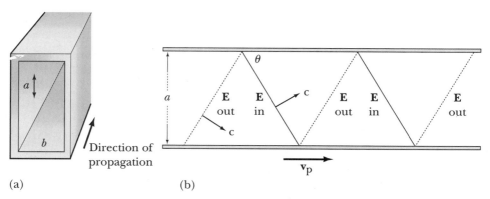

17

GUIDED WAVES

This is the fourth and final chapter devoted to the production and propagation of electromagnetic radiation. Here we will be guiding the waves along conducting cylinders (general ones, not necessarily circular); the ubiquitous coaxial cable will be included.

17.1. Rectangular Waveguide; Cutoff Wavelength 2*a*

Broadcasters want their signal to spread out all over the place; the more people pick it up, the better. But in many practical applications one wishes to have more control over the electromagnetic waves in one's own jurisdiction. A hollow rectangular waveguide provides this: The waves can't get out unless you let them out. Waves are inserted and removed using some sort of little antenna inside the waveguide. (For circular cylinders, things get more complicated; see for example Portis, p. 517.)

Figure 17.1a shows a rectangular waveguide. It is typically made of brass. In the usual mode, the waves bounce back and forth between the sides separated by distance *a*. Figure 17.1b gives a cross-sectional view, with some **E** waves.

Figure 17.1   (a) Hollow rectangular waveguide. (b) Wave fronts.

The electric field **E** points into or out of the page. These are **TE** waves: **T**rans-verse **E**lectric field. The **E** field is transverse to the direction in which the waves are propagating, that is, down the waveguide. The **B** field is generally perpendicular to **E**; and so there is a substantial component of **B** that points in the direction of the waveguide. There can exist also **TM** waves in which the **M**agnetic field **B** is entirely **T**ransverse to the direction of propagation; however, the TE mode shown is lowest in frequency, by a factor of two, and consequently it is usually employed to the ex-clusion of all others. The waves are reflected at the metal sides of the waveguide. As we mentioned in Section 16.3, the **E** wave changes phase by 180° on reflection: If it points into the page before reflection, it points out afterwards.

The whole zig-zag pattern slides unchanging down the waveguide at phase veloc-ity v_p. The "**E** in" and "**E** out" waves always meet at the metal surface. The reason is that the tangential E_t field must be nearly zero at the surface of a good conductor, as discussed in Sections 8.2, 16.2, and 16.3. (E_t must be the same inside the conducting material as outside, and it is kept at low values inside, even for short times, by the large conductivity.) So the "**E** out" continues to cancel the "**E** in" at the surface.

The phase velocity exceeds the speed of light:

$$v_p = \frac{c}{\sin \theta} \tag{17.1}$$

This is an example of the "scissors paradox" of special relativity: When a straight line of speed c approaches another line at angle θ, Figure 17.2, the point of contact moves at speed $c/\sin \theta$, but no information can thereby be transmitted faster than light. The situation is similar to what we encountered in Section 16.4 in connection with plasmas. In this case, the signal velocity (or group velocity) is the speed at which a photon would progress down the guide if it were bouncing back and forth:

$$v_s = c \sin \theta \tag{17.2}$$

As in the case of the plasma, we have

$$v_p v_s = c^2 \tag{17.3}$$

So a waveguide is highly dispersive: Different frequencies travel at different speeds, and wave shapes get mangled in the process. By contrast, the transmission lines to be discussed starting in Section 17.3 do not have this problem.

The distance λ_g may be called the guide wavelength; it is the distance from one peak to the next along the waveguide, Figure 17.3. It is longer than λ_0, the wavelength in air; from the figure,

$$\lambda_g = \frac{\lambda_0}{\sin \theta}$$

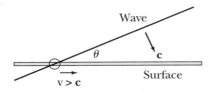

Figure 17.2   Scissors paradox.

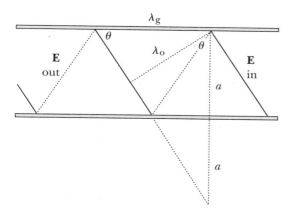

Figure 17.3   Cutoff wavelength.

So

$$c = \lambda_0 f \quad \text{and} \quad v_p = \lambda_g f$$

(The subscript g is common, but it is unfortunate in that one would wish to find v_p associated with λ_p.) The angle θ depends on the wavelength, Figure 17.3:

$$\cos \theta = \frac{\lambda_0}{2a} \tag{17.4}$$

Since $\cos \theta$ cannot exceed 1, this implies a **cutoff wavelength:**

$$\boxed{\lambda_0 \le 2a} \tag{17.5}$$

Waves longer than this cannot proceed down the waveguide. Since $b \approx a/2$, longer waves also cannot proceed in the other polarization mode. At cutoff, $\theta = 0$: The wave fronts are parallel to the sides of the guide, and they just bounce up and down without making any progress down the guide.

Usually, a waveguide is operated with $a < \lambda_0 < 2a$. Above $\lambda_0 = 2a$, there is no transmission. And below $\lambda_0 = a$, other modes can enter, with other speeds and phase relationships, producing unnecessary complications. So, typically, one stays around $\lambda_0 = 1.5a$.

◤ EXAMPLE 17-1

Find a for a rectangular waveguide suitable for the FM band, 100 MHz.

ANSWER For $f = 100$ MHz, we have

$$\lambda_0 = c/f$$
$$= 3 \times 10^8/100 \times 10^6 = 3 \text{ m}$$

Then we would use

$$a = \lambda_0/1.5 = 3/1.5 = 2 \text{ m}$$

It is a walk-in waveguide, too large for practical applications.

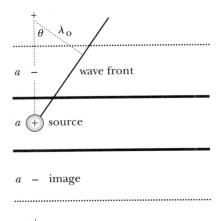

Figure 17.4   Interference construction.

Rectangular waveguides may be used above 2 GHz; below this, one uses coaxial cables, discussed in Sections 17.3 and 17.4. The rectangular waveguides transmit more power with less loss at very high frequency, but with large dispersion; whereas the transmission lines cover a very wide range of frequency with good frequency characteristics (little dispersion).

Here is another view of the waves in the waveguide, Figure 17.4. A line source is perpendicular to the paper and shown as ⊕. It produces an opposite image in each plane of the waveguide (Section 12.2), and they produce more images, as shown. The situation is analogous to that of a diffraction grating. At angle θ at which the extra distance to the wave front shown is one λ_0, constructive interference occurs, and the wave propagates. So we have, as before, $\cos \theta = \lambda_0/2a$. Clearly, if $\theta = 90°$, then the waves from the + and − images cancel perfectly, and there is no wave. This is one way of understanding why there is no wave straight down the waveguide (i.e., no TEM wave; see below). If we have $n\lambda_0$ instead of λ_0 in Figure 17.4, then the whole picture becomes more complicated, with several different speeds for a given frequency; usually we simply avoid these difficulties by keeping $\lambda_0 > a$.

There is also a wave at the same angle going upwards, so the complete figure looks like Figure 17.5. Each wave proceeds in accordance with the equations found previously.

When the waveguide is not quite matched to the load (Section 13.3), reflection occurs; power transfer is less than optimal, and a standing wave is set up. To measure this effect, one observes the "voltage standing wave ratio" (VSWR), the ra-

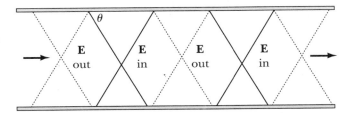

Figure 17.5   Wave fronts.

tio of maximum to minimum voltage signal on a suitable probe inserted into the waveguide at various points. For no reflection, the VSWR should be 1, which is usually what we would like.

In a rectangular waveguide, one may have TE waves, or (possibly) TM, but *not* TEM. If both Electric and Magnetic fields are Transverse, the waves would be going straight down the guide. It happens that the TEM wave cannot then satisfy the various conditions placed upon it inside a hollow conductor; see Section 17.2 for boundary conditions. What actually happens, when a TEM wave enters a hollow waveguide, is in effect a diffraction process, and the disturbance in the formerly plane wave spreads inward. For short wavelength radiation, such as visible light, the wave may penetrate hundreds of meters into the waveguide; longer waves are bent more at the entrance. We return to this point in Section 17.4.

17.2.   Fields in the Rectangular Waveguide

Figure 17.6 shows a more detailed picture of the **E** and **B** lines in the TE wave discussed before (Fig. 17.5).

We now present the **E** and **B** fields for the general TE and TM cases in the rectangular waveguide. Details of the derivations will be left for the problems. We place the x axis along the side labeled a, Figure 17.7, and the z axis in the direction of propagation.

This is a boundary value problem, and the **boundary conditions** (at the metal surfaces) are

$$\mathbf{E} \times \hat{n} = 0, \quad \text{so} \quad E_z = 0$$

$$\mathbf{B} \cdot \hat{n} = 0, \quad \text{so} \quad \partial B_z / \partial x = 0 \text{ at the } x = 0 \text{ surface} \tag{17.6}$$

$$\partial B_z / \partial y = 0 \text{ at the } y = 0 \text{ surface}$$

E_z is zero because tangential E is zero at the surface of a good conductor (Sections 8.2, 16.2, and 16.3). The B_z condition is a little more subtle. The magnetic field does not enter far into a good conductor at high frequency (Section 16.3); and $\nabla \cdot \mathbf{B} = 0$, so the normal component of B is zero next to the surface. (A tangential component is allowed because of surface currents.) Now consider a point just out-

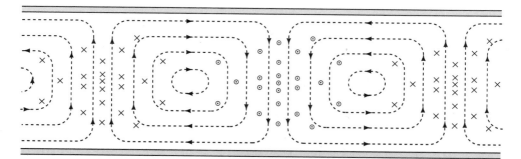

Figure 17.6   Lines of **E** in (x), **E** out (⊙), and **B** (dashed).

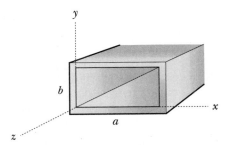

Figure 17.7   Rectangular waveguide.

side of the $x = 0$ surface. For Ampère's law with no J, we have

$$\nabla \times \mathbf{B} = \mu\epsilon \frac{\partial \mathbf{E}}{\partial t}$$

and the y component is

$$\frac{\partial B_x}{\partial z} - \frac{\partial B_z}{\partial x} = \mu\epsilon \frac{\partial E_y}{\partial t} \tag{17.7}$$

Here $E_y = 0$ because it's tangential, and $\partial B_x/\partial z = 0$ because this normal component stays zero as you move along the surface in the z direction. So the tangential component B_z need not be zero, but its x derivative $\partial B_z/\partial x$ is zero.

This takes care of the x and y directions; we expect standing-wave forms corresponding to reflections. In the z direction, we will be interested only in traveling waves of the type $e^{i(kz-\omega t)}$. The wave equations are (Section 14.3), with $\mathbf{J} = 0$ and $\rho = 0$,

$$\nabla^2 \mathbf{E} - \mu\epsilon \frac{\partial^2 \mathbf{E}}{\partial t^2} = 0$$

$$\nabla^2 \mathbf{B} - \mu\epsilon \frac{\partial^2 \mathbf{B}}{\partial t^2} = 0 \tag{17.8}$$

Solutions of the wave equations, consistent with the boundary conditions, are the following:

TE Wave:

$$E_x = -\frac{i\omega}{k_c^2} \frac{n\pi}{b} B_0 \cos \frac{m\pi x}{a} \sin \frac{n\pi y}{b} \exp(i(k_g z - \omega t))$$

$$E_y = \frac{i\omega}{k_c^2} \frac{m\pi}{a} B_0 \sin \frac{m\pi x}{a} \cos \frac{n\pi y}{b} \exp(i(k_g z - \omega t))$$

$$E_z = 0$$

$$B_x = -\frac{ik_g}{k_c^2} \frac{m\pi}{a} B_0 \sin \frac{m\pi x}{a} \cos \frac{n\pi y}{b} \exp(i(k_g z - \omega t)) \tag{17.9}$$

$$B_y = -\frac{ik_g}{k_c^2} \frac{n\pi}{b} B_0 \cos \frac{m\pi x}{a} \sin \frac{n\pi y}{b} \exp(i(k_g z - \omega t))$$

$$B_z = B_0 \cos \frac{m\pi x}{a} \cos \frac{n\pi y}{b} \exp(i(k_g z - \omega t))$$

TM Wave:

$$E_x = \frac{ik_g}{k_c^2} \frac{m\pi}{a} E_0 \cos\frac{m\pi x}{a} \sin\frac{n\pi y}{b} \exp(i(k_g z - \omega t))$$

$$E_y = \frac{ik_g}{k_c^2} \frac{n\pi}{b} E_0 \sin\frac{m\pi x}{a} \cos\frac{n\pi y}{b} \exp(i(k_g z - \omega t))$$

$$E_z = E_0 \sin\frac{m\pi x}{a} \sin\frac{n\pi y}{b} \exp(i(k_g z - \omega t))$$

(17.10)

$$B_x = -\frac{i\omega\mu\epsilon}{k_c^2} \frac{n\pi}{b} E_0 \sin\frac{m\pi x}{a} \cos\frac{n\pi y}{b} \exp(i(k_g z - \omega t))$$

$$B_y = \frac{i\omega\mu\epsilon}{k_c^2} \frac{m\pi}{a} E_0 \cos\frac{m\pi x}{a} \sin\frac{n\pi y}{b} \exp(i(k_g z - \omega t))$$

$$B_z = 0$$

When you substitute these solutions into the wave equations, you find that the guide propagation constant (or wave number) is

$$k_g = \sqrt{k_0^2 - k_c^2} = \frac{2\pi}{\lambda_g}$$

(17.11)

where

$$k_0 = \mu\epsilon\omega^2, \qquad \text{the } k \text{ for a plane wave in the medium}$$

and

$$k_c = \sqrt{\left(\frac{m\pi}{a}\right)^2 + \left(\frac{n\pi}{b}\right)^2}, \qquad \text{the } k \text{ at cutoff}$$

(17.12)

The m and n are integers. Modes are designated in the form TE_{mn}; for example, TE_{10} is the usual lowest-frequency mode. In this case, we see that the fields E_x and B_y are zero. There is no TE_{00} mode (it would correspond to a TEM wave); and for TM_{mn} waves both $m = 0$ and $n = 0$ are forbidden.

Since $k_g < k_0$, and $v_p = \omega/k_g$, we see that the phase velocity of the wave in the guide is greater than that in the infinite undisturbed medium, and so may be greater than the speed of light in vacuum. We also see that k_g goes imaginary when $k_0 = k_c$ at cutoff; this results in an exponential falloff inside the waveguide, rather like the situation in a plasma, Section 16.4.

Relative sizes and phases of the fields are obtained using Maxwell's equations. For example, the i in front corresponds to a phase shift of 90°.

Energy flow is most readily calculated using $S_{av} = \frac{1}{2}\text{Re}(E \times B^*/\mu)$.

17.3.   Transmission Lines; Characteristic Impedance Z_0

Transmission lines have two separate conductors rather than one, and these two conductors need not be at the same potential. Consequently (see Section 17.4), they are able to support TEM waves; in other words, **E** and **B** are both perpendicular to the direction of propagation, and the waves go straight down the transmission line. The category includes coaxial cable, twin lead, strip line, and generally any cylindrical pair of conductors of arbitrary cross section. They are used below about 3 GHz, and they work all the way down to DC (direct current, zero frequency).

Why are there two different kinds of waveguides? Answer: At high power and high frequency, as in radar, the rectangular waveguide is used because losses are small in the air that fills it, and heat due to surface currents is readily dissipated through the metal walls. However, the rectangular waveguide is quite dispersive (phase speed depends on frequency) and its frequency range is limited. If you send a single sharp pulse down a rectangular waveguide you get a "whistler" at the other end: Higher frequencies arrive before lower ones. Rectangular waveguides are inflexible and difficult to work with, and at frequencies below 2 GHz the size of the rectangular waveguide becomes prohibitive.

By contrast, the transmission line has little dispersion and it works over a wide range of frequencies; but energy deposited in central conductors is less easily disposed of.

A transmission line can be regarded as a limiting case of an LC network, Figure 17.8a, with infinitely many infinitely small capacitors and inductors. Its inductance per unit length is L', and its capacitance per unit length is C'. These values are distributed uniformly along the transmission line, such as the coaxial cable of Figure 17.8b.

We suppose that the current I and voltage V are changing. In a given length Δz, the voltage changes by ΔV due to the inductance. In general $V = -L\, dI/dt$, so here

$$\Delta V = -L'\Delta z\frac{\partial I}{\partial t}$$

For small Δz this becomes

$$\frac{\partial V}{\partial z} = -L'\frac{\partial I}{\partial t} \tag{17.13}$$

Similarly, the current changes by ΔI in the length Δz: Less I goes out than goes in if the capacitance is being charged. In general $Q = CV$ and $dQ/dt = -\Delta I$, so

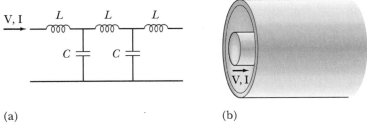

(a)   (b)

Figure 17.8   (a) LC network. (b) Coaxial cable.

here

$$\Delta I = - C' \Delta z \frac{\partial V}{\partial t}$$

For small Δz this becomes

$$\frac{\partial I}{\partial z} = - C' \frac{\partial V}{\partial t} \tag{17.14}$$

We substitute an arbitrary wave into both of the above differential equations:

$$V = V(kz - \omega t)$$

and by Ohm's law

$$I = \frac{V(kz - \omega t)}{Z_0} \tag{17.15}$$

where Z_0 is permitted to be complex (but does not avail itself of this opportunity). The role of the resistance usually played by R is here assigned to the **characteristic impedance** Z_0.

For the first differential equation (Eq. 17.13), $\partial V / \partial z = -L' \, \partial I / \partial t$, we get (with V' being the derivative of V)

$$k \, V' = -L' \frac{-\omega V'}{Z_0}$$

so

$$\frac{\omega}{k} = \boxed{v = \frac{Z_0}{L'}} \tag{17.16}$$

For the second (Eq. 17.14), $\partial I / \partial z = -C' \partial V / \partial t$, we find

$$\frac{kV'}{Z_0} = -C' \, (-\omega \, V')$$

so

$$\frac{\omega}{k} = \boxed{v = \frac{1}{C' Z_0}} \tag{17.17}$$

(Mnemonic: Remember that the time constant is $\tau = RC$, and the time per unit length $1/v$ is the impedance Z_0 times C' per unit length.)

These equations imply that Z_0 is real. It is remarkable that elements that are only capacitive and inductive can produce an impedance Z_0 that is purely resistive over a wide frequency range. They also imply that the speed of the wave is (Section 14.5)

$$v = \frac{1}{\sqrt{\epsilon \mu}} = \frac{1}{\sqrt{L' C'}} \tag{17.18}$$

assuming that the electromagnetic waves are confined to the medium separating the two (perfect) conductors. Usually, $\mu_r = 1$, and $v = c/\sqrt{\epsilon_r} \approx \frac{2}{3}c$.

 EXAMPLE 17-2

Find the characteristic impedance of strip line $w = 1$ cm wide, separated by $s = 1$ mm of plastic of dielectric constant 2.3, Figure 17.9.

Figure 17.9 Strip line.

ANSWER It is a long parallel-plate capacitor (Section 9.4):

$$C = \frac{\epsilon \, w \, \Delta z}{s} \qquad (\Delta z \text{ is length as above})$$

so

$$Z_0 = \frac{1}{vC'} = \frac{s\sqrt{\epsilon_r}}{c\epsilon \, w} = \boxed{\frac{377}{\sqrt{\epsilon_r}} \frac{s}{w}}$$

$$= \frac{377}{\sqrt{2.3}} \frac{0.001}{0.01} = 25 \, \Omega$$

a typical result. For high-frequency waves, this strip line behaves electrically like a 25-ohm resistor. To terminate a wave, connect the end of the line to a real 25-ohm resistor; the wave thinks the resistor is just more strip line, and it plunges in and perishes.

If the end of a transmission line is not terminated by a resistance equal to the characteristic impedance, then waves are reflected in a fashion analogous to the reflection of waves in a string. If the two conductors simply end, so that the resistance is infinite, then the reflection is of the same sign; if they are shorted together, so that the resistance is zero, then the reflection is inverted. Impedance-matching transformers (see Section 10.7) are often used to avoid reflections and effect maximum power transfer. This effect is easily demonstrated with an oscilloscope and a 10-m cable having a potentiometer (variable R) at the far end.

For the coax, Figure 17.10,

$$C' = \frac{2\pi\epsilon}{\ln(r_2/r_1)}$$

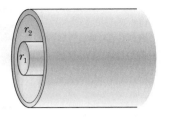

Figure 17.10

and

$$Z_0 = \boxed{\frac{60}{\sqrt{\epsilon_r}} \ln \left(\frac{r_2}{r_1} \right)} \qquad \text{typically 50 to 100 } \Omega \qquad \textbf{(17.19)}$$

For twinlead, Figure 17.11, with radius $a \ll$ separation s,

$$C' = \frac{\pi \epsilon}{\ln (s/a)}$$

and

$$Z_0 = \boxed{\frac{120}{\sqrt{\epsilon_r}} \ln \left(\frac{s}{a} \right)} \qquad \text{typically 300 } \Omega \qquad \textbf{(17.20)}$$

The coaxial cable and twinlead seldom stray far from their typical values because their geometrical parameters are confined within natural logarithms.

17.4.   Fields in the Coaxial Cable

Transmission lines can support TE waves somewhat as hollow conductors do, but the frequencies are around 10 GHz, because the typical dimensions are only a centimeter or two. Usually one operates transmission lines below cutoff for TE waves. These waveguides can support TEM waves from a few GHz down to DC with relatively low loss and little dispersion.

We shall examine the coaxial cable in some detail because it is the commonest and perhaps simplest of the transmission lines. Figure 17.12 shows the fields for this case.

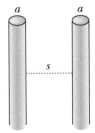

Figure 17.11

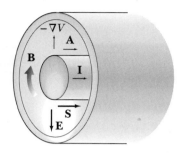

Figure 17.12   Coaxial cable fields.

We solve for the electric and magnetic fields using the following cute trick. For transverse fields, progressing as sinusoidal waves in the z direction, we may write

$$\mathbf{E} = \mathbf{E}_0(x, y) \, e^{i(kz-\omega t)} \qquad \text{and} \qquad \mathbf{B} = \mathbf{B}_0(x, y) \, e^{i(kz-\omega t)}$$

These resemble the corresponding equations of Section 14.5 except in that now we permit $\mathbf{E}_0$ and $\mathbf{B}_0$ to be functions of x and y; they are still transverse, so there is no z component, and all the z and t dependence is still in the exponential factor. So we have a wave that slides along the cable without changing shape.

First we apply Gauss's law with $\rho = 0$, $\nabla \cdot \mathbf{E} = \rho/\epsilon_0 = 0$, obtaining

$$\frac{\partial E_{0x}}{\partial x} + \frac{\partial E_{0y}}{\partial y} = 0 \tag{17.21}$$

after canceling out the common factor of $e^{i(kz-\omega t)}$. Then we do the z component of Faraday's law: $(\nabla \times \mathbf{E})_z = -\dfrac{\partial B_z}{\partial t} = 0$. This yields:

$$\frac{\partial E_{0y}}{\partial x} - \frac{\partial E_{0x}}{\partial y} = 0$$

So the vector $\mathbf{E}_0(x, y)$ has zero divergence and zero curl. Consequently it is the solution of a two-dimensional electrostatic problem in free space. We have converted what might seem to be a difficult three-dimensional problem into a two-dimensional electrostatic one. And that is the above-mentioned trick. Needless to say, we can do the same thing with $\mathbf{B}$. So all that remains is to write out static solutions, with traveling-wave appendages:

$$\mathbf{E} = \frac{\lambda_0}{2\pi\epsilon\rho} \, e^{i(kz-\omega t)} \, \hat{\boldsymbol{\rho}} \qquad \text{Problem 2-23}$$

$$\mathbf{B} = \frac{\mu I_0}{2\pi\rho} \, e^{i(kz-\omega t)} \, \hat{\boldsymbol{\phi}} \qquad \text{Problem 3-20}$$

$$V = \frac{\lambda_0}{2\pi\epsilon} \ln \frac{r_2}{\rho} \, e^{i(kz-\omega t)} \qquad \text{Problem 6-12}$$

$$\mathbf{A} = \frac{\mu I_0}{2\pi} \ln \frac{r_2}{\rho} \, e^{i(kz-\omega t)} \, \hat{\mathbf{z}} \qquad \text{Section 6.3}$$

(17.22)

We can employ the usual relationships to show that

$$E = vB \qquad I = v\lambda \qquad V = vA_z$$

where $v = 1/\sqrt{\epsilon\mu} = c$ in free space.

We find the characteristic impedance Z_0 of the coaxial cable using the above formulas for V, λ, and I:

$$Z_0 = \frac{V}{I} = \frac{1}{2\pi\epsilon_0 c \sqrt{\epsilon_r}} \ln \frac{r_2}{r_1}$$

$$= \frac{60}{\sqrt{\epsilon_r}} \ln \frac{r_2}{r_1}$$

(17.23)

as found in the preceding section using C'.

Poynting's vector is obtained as usual using $\mathbf{S}_{av} = \frac{1}{2} \mathrm{Re}(\mathbf{E} \times \mathbf{B}^*/\mu)$ (Section 14.5). If we integrate the result over the space in the cable, we obtain the power $P = VI$; this is left for Problem 17-17.

Now we can return to the question of why there are no TEM waves in hollow conductors. We have just shown above that TEM waves involve a solution of a two-dimensional static problem. A hollow conductor is an equipotential, so $\mathbf{E} = -\nabla V$ must be zero inside, in the electrostatic case, which means there can be no wave inside the hollow conductor. But if there is a separate conductor inside, at a different potential, then there can be a nonzero $\mathbf{E} = -\nabla V$ inside, and that is the case for transmission lines, such as coaxial cables.

And all that is very well, except for one thing: If you look again, you will discover that we seem to have forbidden the existence of plane waves in free space, including the light waves that you are now using in order to read about their nonexistence. It was in Section 14.5 that we learned that $V = 0$ everywhere in infinite plane waves. But all of the above reasoning still applies, leading to the conclusion that $\mathbf{E} = -\nabla V = 0$, so no wave. Where did we go wrong?

The answer is that we have ignored the role of $\partial \mathbf{A}/\partial t$ in the equation $\mathbf{E} = -\nabla V - \partial \mathbf{A}/\partial t$ (Section 6.2). It is remarkable that for plane waves in space, $V = 0$ everywhere, and $\mathbf{E}$ is entirely due to $-\partial \mathbf{A}/\partial t$, whereas for the waves in a transmission line $\mathbf{E}$ is entirely due to $-\nabla V$. In each case we have $\mathbf{B} = \nabla \times \mathbf{A}$, but for the infinite plane wave $\mathbf{A}$ is transverse (Section 14.5), whereas in the transmission line $\mathbf{A}$ is longitudinal (this section, above). In an enclosed space, the specification that $\mathbf{B}$ must be transverse means that the surface current density $\mathbf{K}$ is longitudinal (Section 3.2) and $\mathbf{A}$ is longitudinal. That means that in this case the above reasoning is okay, because a longitudinal $\mathbf{A}$ means a longitudinal $\partial \mathbf{A}/\partial t$, which doesn't contribute to a transverse $\mathbf{E}$. So there is no TEM wave in a hollow conductor.

◤ EXAMPLE 17-3

For TEM radiation of 1400 W/m^2 (solar power), find $\mathbf{E}$ and ΔV across 1 cm (the size of a waveguide), for a hollow waveguide, and for plane waves in free space.

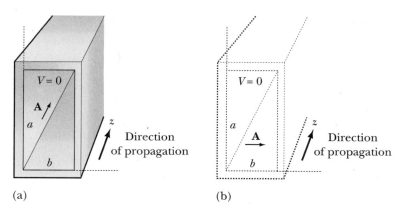

(a)   (b)

Figure 17.13   **(a)** Hollow rectangular waveguide. **(b)** Imaginary guide in free space.

ANSWER   From Section 14.5:

$$E = \sqrt{S/\epsilon_0 c}$$
$$= \sqrt{1400 \times 377} = 726 \text{ V/m in both cases}$$

Waveguide, Figure 17.13a: $A = A_z$, longitudinal so no contribution to transverse E, so

$$\Delta V = Eb$$
$$= 726 \times 0.01 = 7 \text{ V, impossible because the guide is an equipotential.}$$

Free space, Figure 17.13b: $\Delta V = 0$ (Section 14.5), and yet

$$E = 726 \text{ V/m still, due to transverse } A$$

17.5 SUMMARY

Generally speaking, waveguides fall into two categories, depending on whether they consist of one conductor or two separate ones. In hollow waveguides, there are no TEM waves, and the frequency range is relatively limited; typically they are used at high frequencies. Transmission lines carry TEM waves and display a characteristic impedance; their wide frequency range extends from a gigahertz down to DC (direct current).

Rectangular waveguides are normally operated in lowest TE mode, for which the cutoff wavelength is

$$\lambda_0 \leq 2a \tag{17.5}$$

All other modes have a higher cutoff frequency.
    Signal speed is

$$v_s = c \sin \theta \tag{17.2}$$

where θ is the angle between a wave front and the side of the waveguide:

$$\cos \theta = \frac{\lambda_0}{2a} \qquad (17.4)$$

Phase velocity is

$$v_p = \frac{c}{\sin \theta} \qquad (17.1)$$

Thus,

$$v_p v_s = c^2 \qquad (17.3)$$

Transmission lines normally operate in TEM mode, which requires two conductors. Their characteristic impedance is

$$Z_0 = \frac{1}{vC'} \qquad (17.17)$$

Signal speed is

$$v = \frac{1}{\sqrt{\epsilon\mu}} = \frac{1}{\sqrt{L'C'}} \qquad (17.18)$$

Sources and loads should usually match the impedance of associated transmission lines.

In a transmission line, the fields correspond to two-dimensional static fields multiplied by a plane-wave factor of $e^{i(kz-\omega t)}$.

PROBLEMS

17-1 Find the cutoff frequency for the waveguide dimensions: $a = 2.3$ cm, $b = 1$ cm, Figure 17.14.

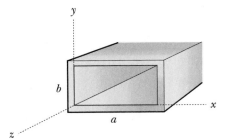

Figure 17.14

17-2 For the waveguide of the preceding problem, find the cutoff frequency when it is filled with oil of $\epsilon_r = 2.2$.

17-3 For the waveguide of the first problem, operating at 10 GHz, find the wavelength in air, the guide wavelength, and the phase and signal speeds.

17-4 For a rectangular waveguide, show that **(a)** $v_p = \omega/k_g$; **(b)** $v_s = \dfrac{1}{dk_g/d\omega}$; and
(c) $v_p v_s = c^2$, as in the case of a plasma, Section 16.4.

17-5 For the TE_{10} mode of a rectangular waveguide, find $\mathbf{S}_{av}$ and show that the energy flows in the direction of the z axis.

*17-6 For the TE_{10} mode of a rectangular waveguide, find the energy flux averaged over the interior; show that it equals the product of the average field energy density times the signal speed.

17-7 For the rectangular waveguide, TM_{11} mode, find the cutoff wavelength in terms of a and b. Then find the TE and TM cutoff frequencies for $a = 2.3$ cm, $b = 1.0$ cm.

*17-8 Verify the TE and TM equations for the rectangular waveguide, Equations 17.9 and 17.10.

17-9 Find the characteristic impedance of twin lead consisting of a pair of wires of diameter $2a = 1.0$ mm, separated by $s = 7$ mm, imbedded in plastic foam of $\epsilon_r \approx 1.0$, Figure 17.15. Find the signal speed. (See Problem 9-17.)

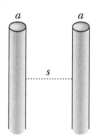

a   *a*

s

Figure 17.15

17-10 Find the characteristic impedance of coaxial cable of inner and outer diameters 1.3 mm and 4.6 mm, filled with dielectric of $\epsilon_r = 2.3$. Find the signal speed.

17-11 A 125-ohm coax has outer diameter 0.4 cm and $\epsilon_r = 3$. Find the inner diameter; is it reasonable? Find the phase and signal speeds.

17-12 Find the characteristic impedance of strip line of width 60 cm, Figure 17.16, separated by 1 cm of helium gas.

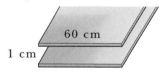

60 cm

1 cm

Figure 17.16

17-13 For the coaxial cable, find the characteristic impedance Z_0 using L' and not C'.

17-14 For a transmission line, show that the current equals the linear charge density times the speed of propagation ($I = \lambda v$).

17-15 Write the E and B fields for a maximum signal of V_0 and I_0 in a coaxial cable.

17-16 For a coaxial cable, derive the inductance per unit length, $L' = \mu \ln(b/a)/2\pi$, from the capacitance per unit length, $C' = 2\pi\epsilon/\ln(b/a)$.

17-17 For a coaxial cable, show that the power $\int S\, da = VI$.

17-18 For a coaxial cable, show that $E = vB$, $I = v\lambda$, and $V = vA_z$.

17-19 For a coaxial cable, assume uniform current in a thickness δ (Section 16.3) in the inner and outer conductors of conductivity σ (Section 8.1) and of radii r_1 and r_2, respectively, Figure 17.17.
(a) Find the resistance per unit length R'.
(b) Evaluate for copper at 1 MHz, with inner and outer radii 0.5 mm and 2 mm, and plastic of $\epsilon_r = 2.2$.
(c) Compare with the characteristic impedance Z_0.

Figure 17.17

CHAPTER

18

RELATIVITY IN
ELECTROMAGNETISM

In preceding chapters we have mentioned relativistic effects but deferred their discussion; now at last we will deal with such phenomena directly. This will bring us to a deeper understanding of the underlying bond between electricity and magnetism.

18.1. Introduction to Relativity; β, γ

We humans live life on a certain scale and at a certain speed. Our intuition develops in this setting. It is not surprising that at other sizes and speeds, phenomena are quite different from our ordinary experience. At very tiny distances and masses, things get fuzzy and objects behave in ways that are not only bizarre but are virtually indescribable in everyday terms; this is the realm of quantum mechanics. At large sizes, space-time curves in ways that can be defined mathematically but hardly visualized; this is general relativity applied to cosmology.

And for objects moving at high speed v relative to the observer, we encounter the following phenomena, dealt with in the field of special relativity (we employ the customary Greek letters "beta," β, and "gamma," γ; and c is the speed of light):

- **Time dilation:** Moving clocks run slow by the factor

$$\gamma = \frac{1}{\sqrt{1 - \beta^2}} \qquad \text{with} \qquad \beta = \frac{v}{c} \qquad (18.1)$$

- **Lorentz contraction:** Moving meter sticks are contracted by γ in the direction of motion.
- **Mass increase:** Moving masses are more massive by the factor γ (but see Section 18.5).

And of course other quantities may be different for things in motion. For example, mass density increases by γ^2, because of the mass increase and the Lorentz contraction. On the other hand, transverse length doesn't change at all; a meter stick held transverse to the direction of motion gets skinnier and heavier, but it stays the same length. We will explore these and other phenomena in the remainder of this chapter. **Notice:** $\gamma \geq 1$, always. Also, these phenomena are real, not apparent; they remain after correction for the finite speed of light, and so forth.

It is worth remarking that the strange predictions of special relativity have been subjected to the most rigorous scrutiny by generations of experimentalists who are well aware that a Nobel prize awaits the one who detects the slightest flaw. We shall examine the experimental record as we proceed through the chapter.

▼ EXAMPLE 18-1

A spaceship clock runs at half its **proper rate** (that is, half of its rate in its own **rest frame**). (**a**) How fast is the ship traveling? (**b**) How fast is your watch running as observed by someone on the spaceship?

ANSWER (**a**) The question means, of course, how fast is the ship traveling *relative to the observer?* The ship clock runs slow by a factor of 2, so $\gamma = 2$. Solving for β,

$$\beta = \sqrt{1 - \frac{1}{\gamma^2}} = \sqrt{1 - \frac{1}{2^2}} = 0.866$$

On your pocket calculator, to get β you can enter 2 and then press, in order: $\frac{1}{x}$, arccos, sin. Then if you like, starting with β, you can reverse the process to retrieve γ. So,

$$v = 0.866 \times 3 \times 10^8 = 2.60 \times 10^8 \text{ m/s}$$

At this speed an object is contracted to half its proper length and has twice its rest mass.

(**b**) Your watch is slow by a factor of two in the ship frame of reference. The relativistic effects are reciprocal: Each observer finds the other's clocks and meter sticks to be affected, but you find nothing wrong with your own equipment.

Who is right? Each observer is correct in his own frame of reference—it is an egocentric kind of theory, because there is no way to tell who is "really" moving. You may be stationary relative to the earth, but the earth itself is whirling through space in a dizzying path. In the 19th century many thought that at least the "ether" could be regarded as fixed for some purposes, but those days are gone (see Section 18.3).

Relativity theory is based on **Einstein's Postulates** (1905; and see Fig. 18.25):

(**1**) The *Principle of Relativity:* The laws of physics are the same in all inertial reference frames. Uniform motion is relative. Colloquially, you can't tell how fast you're going. This idea goes back at least to Newton and Galileo (1642). (An **inertial frame** is one where if you place an object it stays put, like a well-trained dog; see Section 18.4 for more on this.)

(2) Invariant speed of light: The speed of light in vacuum is independent of the (supposed) uniform motion of the observer or source. You can't catch up with light. This one originated with Einstein, and it seems unbelievable, but it has been tested rather directly using rapidly moving particles. Using a particle accelerator you can create π^0 mesons, which promptly decay into gamma ray photons. If you produce the π^0 mesons at various energies and look at the time-of-flight of the emergent photons over a known path in vacuum, you find that the photons are traveling at speed c regardless of the speed of the π^0 mesons (Alvager).

(Experimental verification of the second postulate for visible light is complicated by the fact that in a material medium of index of refraction n, light of wavelength λ is continuously absorbed and reradiated coherently, in the forward direction, with a characteristic "extinction length" $\lambda/2\pi(n-1)$, which is about 0.3 mm in air and a light-year in interstellar space. So the speed of light depends on the speed of the medium it is passing through, Section 18.3.)

To illustrate Einstein's Postulates, we shall now derive the time dilation. We construct a "light-clock," Figure 18.1a, with a photon, or a light pulse, that goes ticktock back and forth between two mirrors. The time between tick and tock is $t = y/c$, where y is the separation of the mirrors. Now if we move to the left at speed v relative to the light-clock, the situation is that of Figure 18.1b, in our frame of reference (where time is primed). Our intuition says that the photon has to go faster in order to take the same time between mirrors. But no: Einstein's **second postulate** says that the photon goes the same speed, so it takes longer between mirrors. From the Pythagorean theorem,

$$(ct)^2 + (vt')^2 = (ct')^2 \tag{18.2}$$

so

$$t' = t\frac{1}{\sqrt{1 - \dfrac{v^2}{c^2}}} = \gamma\, t$$

so the clock takes longer to tick in the primed frame, as advertised.

Now what relevance has the above to other clocks, such as my wristwatch, or my heart, which function on very different principles? The answer is, Einstein's

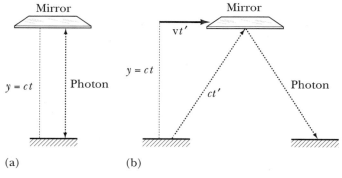

(a)   (b)

Figure 18.1   (a) Light-clock. (b) Moving light-clock.

Calvin and Hobbes   by Bill Watterson

Figure 18.2

first postulate says that all clocks must show the same time dilation; otherwise you could tell how fast you were going by comparing their rates. Note the complementary contributions of Einstein's two postulates; both were needed in order to complete this derivation of the time dilation.

The time dilation in particular has been thoroughly verified experimentally and has even found its way into the popular press, Figure 18.2. Elementary particles such as mesons of known lifetime have been observed at various speeds, and their rates of decay are inversely proportional to γ, as expected, for γ values from one to hundreds. The rates of atomic-beam clocks are measurably altered by transportation in high-speed aircraft; see Section 18.4. No such experimental effect had ever been observed at the time of Einstein's 1905 paper, which makes his contribution all the more amazing: It is good to be able to *explain* an observation, but it is simply astonishing to *predict* a phenomenon never seen before.

18.2.   Lorentz Transformation. Simultaneity

We have just derived the time dilation using Einstein's Postulates, and one could proceed farther down that road, but there's a better way. Relativistic effects are more elegantly dealt with using the **Lorentz transformation** between inertial reference frames.

An **event** is characterized by particular space and time coordinates; it could be like a flashbulb or a firecracker. Its coordinates are related between frames (coordinate systems) by the Lorentz transformation:

Inverse:

$$x' = \gamma x - \beta\gamma ct \qquad x = \gamma x' + \beta\gamma ct' \qquad\qquad \textbf{(18.3)}$$

$$y' = y \qquad\qquad y = y'$$

$$z' = z \qquad\qquad z = z'$$

$$ct' = \gamma ct - \beta\gamma x \qquad ct = \gamma ct' + \beta\gamma x'$$

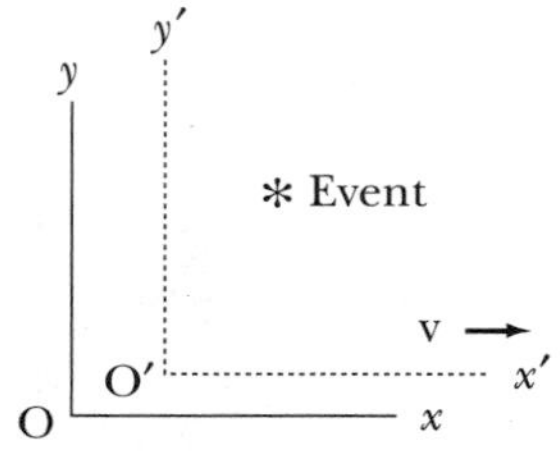

Figure 18.3   Event transformation.

The unprimed and primed coordinate systems are shown in Figure 18.3. Their origins, O and O', coincide at $ct = ct' = 0$. The primed frame moves to the right at velocity $v = c\beta$ relative to the unprimed fame. The inverse transformation is obtained by solving the Lorentz transformation equations for the unprimed coordinates; v still refers to the velocity of the primed frame in the unprimed frame. We'll discuss derivation in a minute.

(It would perhaps be reasonable, but it is not customary, to use v' in the inverse Lorentz transformation, where v' is the speed of the unprimed frame in the primed frame. If we did so, we would use $v' = -v$, and $\beta' = -\beta$, and $\gamma' = \gamma$. The result would be:

(not customary)

$$x = \gamma'x' - \beta'\gamma'ct'$$

$$y = y'$$

$$z = z'$$

$$ct = \gamma'ct' - \beta'\gamma'x'$$

(18.4)

Then the inverse transformation is the same as the direct, except that all the primes are on the other side of the equations.)

◥ EXAMPLE 18-2

A flash of light occurs at $x = 1$ m, $y = 1$ m, $z = 1$ m, and $ct = 1$ m (so $t = 3.33 \times 10^{-9}$ s). Locate this event in the primed frame, which moves at $v/c = 0.60$ to the right.

ANSWER $\beta = 0.60$ and $\gamma = 1.25$, so $\beta\gamma = 0.75$. So,

$$x' = \gamma x - \beta\gamma\, ct = 1.25x - 0.75\, ct = 0.50 \text{ m}$$

$$y' = y = 1 \text{ m}$$

$$z' = z = 1 \text{ m}$$

$$ct' = \gamma\, ct - \beta\gamma\, x = 1.25\, ct - 0.75x = 0.50 \text{ m} \qquad (\text{so } t' = 1.67 \times 10^{-9} \text{ s})$$

The Lorentz transformation may be contrasted with the old, classical, intuitive Galilean one for this event:

<table>
<tr><th>Lorentz</th><th>Galilean</th><th></th></tr>
</table>

$$x' = \gamma x - \beta\gamma\, ct = 0.50 \text{ m} \qquad x' = x - \beta\, ct = 0.40 \text{ m} \qquad (18.5)$$

$$y' = y \qquad\qquad\quad = 1 \text{ m} \qquad\quad y' = y \qquad\quad = 1 \text{ m}$$

$$z' = z \qquad\qquad\quad = 1 \text{ m} \qquad\quad z' = z \qquad\quad = 1 \text{ m}$$

$$ct' = \gamma\, ct - \beta\gamma\, x = 0.50 \text{ m} \qquad ct' = ct \qquad\quad = 1 \text{ m}$$

The x coordinate transforms in somewhat the same way in both transformations, in that x' depends on both x and ct. The striking difference is that, in relativity, time changes as you change reference frames; it depends on x as well as ct. Before relativity, time was regarded as absolute, that is, as being the same in all reference frames.

Now let's talk about derivations. First, there can be no change in the transverse directions, y and z, by the following argument. An observer at y_0, Figure 18.4, is located on the y axis at, say, 1 meter from the origin, in his unprimed reference frame. And similarly, an observer at y'_0 is 1 meter from the origin in his frame. Now suppose that, as in Figure 18.4a, the primed frame moves along the x axis toward the unprimed, as observed by y_0, and suppose that that motion results in a contraction along the y' axis, as shown. As the y and y' axes pass through each other, the y_0 observer marks the y' axis with a piece of chalk. The mark will be *above* the observer at y'_0, on the y' axis. That, at least, is how y_0 sees it.

But the observer at y'_0 will see things differently, Figure 18.4b. By symmetry, the contraction will occur along the y axis, according to the y'_0 observer. (By "symmetry" we imply that, since only relative (uniform) motion has significance, relativistic effects must be reciprocal: Each observer sees the same effect in the other's reference frame.) Thus, he finds that the mark made by y_0 is *below* him. When we stop the relative motion and bring the y and y' axes together for comparison, we encounter a hopeless contradiction: The chalk mark cannot be both above and below y'_0. To avoid this contradiction, it is necessary that there be no contraction transverse to the motion. Thus we establish the equations $y' = y$ and $z' = z$ in the Lorentz transformation.

For the other equations, we will need first to establish the Lorentz contraction, Section 18.1. This is done by turning the light-clock sideways (Fig. 18.1). Details are left for Problem 18-4.

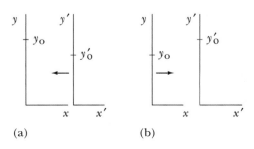

(a)   (b)

Figure 18.4   No transverse change in length.

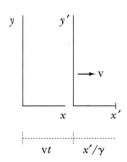

Figure 18.5   Lorentz transformation.

Now, referring to Figure 18.5, suppose the origins coincided at time $t = 0$. Then the point marked as x' is located at distance $vt + x'/\gamma$ on the x axis: vt due to the relative motion of the axes, and x'/γ due to the Lorentz contraction of the x' distance in the unprimed frame. Thus, for this point,

$$x = vt + x'/\gamma$$

Solving for x':

$$x' = \gamma x - \gamma v t = \gamma x - \beta \gamma ct \tag{18.6}$$

which is the first of the Lorentz transformation equations, shown in Equation 18.3. Consequently, symmetry requires that

$$x = \gamma x' + \beta \gamma ct' \tag{18.7}$$

Solving Equations 18.6 and 18.7 for ct', we easily obtain

$$ct' = \gamma ct - \beta \gamma x$$

which is the fourth of the equations; now we've derived them all.

The Lorentz transformation is supported by the following argument. Imagine that a flash occurs at the origin at time $t = t' = 0$. Then a spherical wave of light moves out at speed c in each reference frame, Figure 18.6, according to Einstein's second postulate. Surprisingly, in each frame the sphere is centered on the origin.

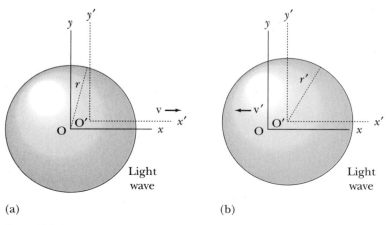

(a)   (b)

Figure 18.6   **(a)** In the unprimed frame. **(b)** In the primed frame.

The radius r is given by

$$r^2 = x^2 + y^2 + z^2 \tag{18.8}$$

and of course

$$r'^2 = x'^2 + y'^2 + z'^2$$

Since the light travels at speed c in both frames, we have $r = ct$ and $r' = ct'$. So we may write, for a point on the wave front,

$$x'^2 + y'^2 + z'^2 - (ct')^2 = x^2 + y^2 + z^2 - (ct)^2$$

The Lorentz transformation satisfies this equation; the Galilean does not.

We can derive the **time dilation** with trivial mathematics using the Lorentz transformation. We want to show that the time between "tick" and "tock" is longer in the frame in which the clock is moving. So we have now two events: the tick and the tock. For each event, the time transforms as

$$ct' = \gamma ct - \beta \gamma x$$

The difference in time $\Delta t = t_2 - t_1$, for the two events at times t_1 and t_2, transforms as

$$\Delta ct' = \gamma \, \Delta ct - \beta \gamma \, \Delta x$$

(with $\Delta x = x_2 - x_1$). To simplify things, we will want $\Delta x = 0$. Then the clock is standing still in the unprimed frame, and the result is

$$\Delta t' = \gamma \, \Delta t \tag{18.9}$$

which is the time dilation again: The time between tick and tock is longer in the primed frame, where the clock is moving.

The argument is circular, of course, in that we used the time dilation to derive the form of the Lorentz transformation. It is the two Einstein postulates that one must revert to as a basis for the theory.

And now let's do the **Lorentz contraction.** We suppose we have a meter stick moving at speed v in the unprimed frame. We will find its length by marking the ends simultaneously in the unprimed frame, in which it is moving. The difference in x position transforms as

$$\Delta x' = \gamma \, \Delta x - \beta \gamma \, \Delta ct$$

Again, we simplify by setting $\Delta t = 0$, because we mark the ends simultaneously, with $\Delta t = 0$, in the unprimed frame. Then in the primed frame it is stationary, and we have

$$\Delta x' = \gamma \, \Delta x \tag{18.10}$$

so the meter stick is longer in the primed frame, where it is stationary.

The mathematics is trivial when you use the Lorentz transformation here, but you have to watch out about what frame you're in. The clock was moving in the *primed* frame, and the meter stick was moving in the *unprimed* frame. That was only for mathematical convenience, of course; the effects could be derived starting in either frame.

Note well: We marked the ends of the meter stick simultaneously in the un-primed frame, but that was *not* simultaneous in the primed frame:

$$\Delta ct' = \gamma \, \Delta ct - \beta\gamma \, \Delta x$$

and $\Delta t = 0$, so the difference in time between marking one end and the other is

$$\Delta ct' = -\beta\gamma \, \Delta x$$

In the primed frame it doesn't matter when we mark the two ends, because the meter stick isn't moving in that frame, so we can mark them at different times. This is an example of the **relativity of simultaneity:** Events that are simultaneous in one frame need not be simultaneous in another.

 **EXAMPLE 18-3**

The *Pole and Barn Paradox,* Figure 18.7. A runner with 4-meter pole held horizontal, moving with $\gamma = 2$, approaches a barn of width 4 meters, with a door on each side. In the farmer's reference frame the pole is only 2 meters long and so when the runner is inside, the farmer can shut both doors simultaneously, and open them again, without hitting the pole. But in the runner's frame the pole is 4 meters and the barn is only 2 meters, so the doors cannot be shut simultaneously. Who is right?

Figure 18.7   Pole and barn.

ANSWER Both are right, but the door shutting that is simultaneous in the farmer's frame is not simultaneous in the runner's frame. In the runner's frame the door on the right shuts and opens first; then, after the pole is through the left door, it shuts and opens.

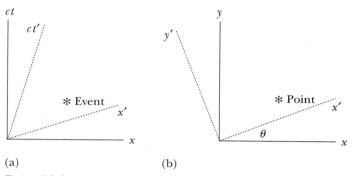

Figure 18.8   (a) Lorentz transformation. (b) Rotation.

The Lorentz transformation is reminiscent of a coordinate rotation, Figures 18.8 and 1.11:

$$x' = \gamma x - \beta \gamma \, ct \qquad x' = x \cos \theta + y \sin \theta$$

$$ct' = \gamma ct - \beta \gamma x \qquad y' = y \cos \theta - x \sin \theta$$

(An event, defined in space and time, is analogous to a point in ordinary space.) The γ is like $\cos \theta$, and $\beta \gamma$ is kind of like $\sin \theta$ except the sign doesn't quite work right. Correspondingly, the x and ct axes go opposite directions. The rotation mixes the x's and y's, and the Lorentz transformation mixes the x's and ct's. What is remarkable about this is that some time transforms into space and vice versa. There is a kind of equivalence between the space coordinate x and the time coordinate ct. That is the basic idea behind the four-dimensional description of the world in relativity theory. The Lorentz transformation represents a kind of rotation in four-dimensional space-time. The radius is the same, so . . .
For the point,

$$x'^2 + y'^2 + z'^2 = x^2 + y^2 + z^2 \tag{18.11}$$

whereas for the event,

$$x'^2 + y'^2 + z'^2 - (ct')^2 = x^2 + y^2 + z^2 - (ct)^2$$

as before, Figure 18.6, in connection with the light flash in two frames. This quantity is **invariant:** It's the same in both frames. Once again, the ct coordinate is behaving in a way analogous to an ordinary space coordinate *except* for that nagging minus sign. (Some treatments get rid of it by using ict instead of ct as the fourth dimension; see Section 18.8.)

In view of this, Minkowski proposed the fusion of space and time into one entity called space-time. Newton, in *Principia* (1687), had clearly stated his own understanding of space and time: "Absolute space, in its own nature, without relation to anything external, remains always similar and immovable. . . . Absolute, true, and mathematical time, of itself, and from its own nature, flows equably without relation to anything external." By contrast, Minkowski, in an address at Cologne (1908; see Lorentz, App. B), opined: "Henceforth space by itself, and time by itself, are doomed to fade away into mere shadows, and only a kind of union of the two will preserve an independent

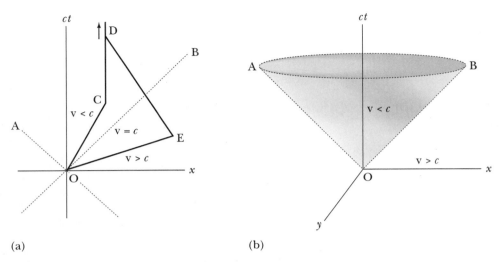

(a)   (b)

Figure 18.9   (a) Space-time diagram. (b) Light cone.

reality." Figure 18.8a is a Minkowski diagram, or a **space-time diagram,** plotting distance versus time; it has the convenient feature that it doesn't change as time progresses.

Figure 18.9 shows another space-time diagram. The line A-O-B is called the "light cone." A photon starting at the origin follows the light cone. The path O-C-D is the path or **world line** of a car that started from the origin at a given velocity $v < c$ and then stopped at point C. The path O-E-D is the path of no material object, because anything traveling from O to E goes faster than light: Its slope is less than that of the light cone. But it could be the path of a shadow, or of the dot on the screen of an oscilloscope.

The **interval** Δs between two events is

$$\Delta s = \sqrt{(\Delta ct)^2 - (\Delta x)^2} \tag{18.12}$$

(The word "interval" is defined in other ways in some texts.) The interval between events at O and C is sometimes called "time-like" because $ct > x$, so one person can attend both events by moving at $v < c$. The interval between O and E is called "space-like" because a person can't get from one event to the other. And the interval from O to B is "light-like."

The concept of the unity of space and time is central to further work in relativity, especially general relativity (which we shall at least mention in Section 18.4), and so we have provided a brief introduction. But we will have little further need for space-time diagrams in our development here.

18.3. Velocity Addition. The Ether

The speed of light represents an absolute upper speed limit for all known objects. ("Tachyons" are possibly permitted particles traveling faster than light, but none has been found.) Since the factor γ goes to infinity as v approaches c, the mass of a material particle (having nonzero rest mass) would increase without limit, which would make it impossible to accelerate it all the way to the speed of light. Photons have no rest mass, so they are condemned to wander the universe always at the speed of light. Neutrinos may have a small rest mass, but if so then their γ must be large whenever they have substantial energy; so it's hard to tell whether they're moving at the speed of light or just very close to it.

But can we find a transformation that will give us a speed greater than light?

For example, suppose a bullet is fired to the right at speed $c\beta_x = 0.8c$ in a spaceship, the unprimed frame, which is itself moving at speed $c\beta'_s = 0.8c$ to the right in the primed frame of the observer, Figure 18.10. Is the bullet going faster than light in the observer's frame?

The relativistic velocity addition formula is again somewhat counterintuitive. We don't just add up the β's and get 1.6. For two events differing in x coordinate and time, the tick and tock of a moving clock, we have

$$\Delta x' = \gamma\,\Delta x - \beta\gamma\,\Delta ct$$
$$\Delta ct' = \gamma\,\Delta ct - \beta\gamma\,\Delta x \tag{18.13}$$

Dividing, we obtain

$$\frac{\Delta x'}{\Delta ct'} = \frac{\gamma\,\Delta x - \beta\gamma\,\Delta ct}{\gamma\,\Delta ct - \beta\gamma\,\Delta x}$$

so

$$\beta'_x = \frac{\beta_x - \beta}{1 - \beta\beta_x} \tag{18.14}$$

Now in Figure 18.10 the primed frame is moving to the left in the unprimed frame, so $\beta = -0.8$. Then,

$$\beta'_x = (0.8 + 0.8)/(1 + 0.8 \times 0.8) = 1.60/1.64 = 0.976$$

We might be inclined to say: "We almost made it that time; let's try 0.9." But it doesn't work. The formula is so constructed that you can't exceed the speed of light unless you start with speed v > c.

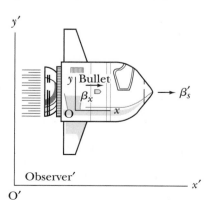

Figure 18.10   Spaceship.

In the y direction, starting with the Lorentz transformation as before, we get

$$\frac{\Delta y'}{\Delta ct'} = \frac{\Delta y}{\gamma \Delta ct - \beta \gamma \Delta x}$$

$$\beta'_y = \frac{\beta_y}{\gamma(1 - \beta\beta_x)}$$

(18.15)

To generalize a little, we will use $\beta_\parallel$ instead of β_x to represent motion parallel to the relative velocity of the coordinate axes, and $\beta_\perp$ for motion perpendicular to that. So we write

$$\boxed{\beta'_\parallel = \frac{\beta_\parallel - \beta}{1 - \beta\beta_\parallel}}$$

(18.16)

and

$$\boxed{\beta'_\perp = \frac{\beta_\perp}{\gamma(1 - \beta\beta_\parallel)}}$$

We can get acceleration $a_\parallel$ immediately by differentiation (at constant β):

$$\beta'_x = \frac{\beta_x - \beta}{1 - \beta\beta_x}$$

$$d\beta'_x = \frac{1 - \beta^2}{(1 - \beta\beta_x)^2} d\beta_x$$

We shall simplify at this point: Let our object be moving slowly in the unprimed frame, so β_x is small, and $\beta \approx -\beta'_x$. We divide by $dt' = \gamma\, dt$ (from the time dilation) arriving at

$$\boxed{a'_\parallel = a_\parallel / \gamma^3}$$

(18.17)

and similarly

$$\boxed{a'_\perp = a_\perp / \gamma^2}$$

for small values of $\beta_\parallel$ and $\beta_\perp$ (but the speed of the primed frame β may be large).

▼ EXAMPLE 18-4

An apple drops from rest with an acceleration of $a_y = 9.8 \text{ m/s}^2$. It leaves marks at 1-second intervals. Use $y' = \frac{1}{2} a'_y t'^2$ and $v'_y = a'_y t'$ to find the positions of the marks, and the vertical speeds at the marks, in a primed frame moving in the $+x$ direction with $\gamma = 2$.

ANSWER

$$a'_y = a_y/\gamma^2$$
$$= 9.8/2^2 = 2.45 \text{ m/s}^2$$
$$t' = \gamma t = 2 \text{ s}, 4 \text{ s, etc.} \quad \text{(due to time dilation)}$$

So

$$y' = \tfrac{1}{2} a'_y t'^2 = 4.9 \text{ m}, 19.6 \text{ m, etc.} \quad \text{just like } y$$
$$v'_y = a'_y t' = 4.9 \text{ m/s}, 9.8 \text{ m/s, etc.}$$

which is half of v_y at corresponding points, as one would also know of course from the velocity transformation formula: for small $\beta_\parallel$, $\beta'_\perp = \beta_\perp/\gamma = \beta_\perp/2$.

And of course

$$v'_x = -0.866c = -2.6 \times 10^8 \text{ m/s}$$

and

$$x' = -v'_x t' = -5.2 \times 10^8 \text{ m}, -10.4 \times 10^8 \text{ m, etc.}$$

The velocity addition formulas played an important role in the early history of relativity and ether theory. In particular, there was a famous experimental result, before Einstein came along, called the Fizeau ether drag experiment. It involves the following calculation:

Suppose we pass a beam of light through moving water, in the direction of flow; does it go faster? And by how much?

The light travels in stationary water at $\beta_x = \dfrac{1}{n}$, where n is the index of refraction, 1.33 for water. The water moves with $\beta_w = -\beta$ in the laboratory frame. Then in the lab the light travels at

$$\beta'_x = \frac{\beta_x + \beta_w}{1 + \beta_x \beta_w} \approx (\beta_x + \beta_w)(1 - \beta_x \beta_w) \approx \beta_x + \left(1 - \frac{1}{n^2}\right)\beta_w \quad (18.18)$$
$$= \beta_x + 0.43\,\beta_w$$

So the speed of the water is *partially* added to that of the light.

Now what is astonishing is that that particular value, 0.43, was predicted by the Fresnel ether theory in the early 19th century, long before relativity. Fresnel's theory postulated that light is carried by a medium called the **ether,** which permeates all space, and it predicted that the ether would be partially carried along by materials. When Fizeau measured the "convection coefficient" in 1851 and obtained the predicted 43%, it constituted a spectacular triumph for the ether theory. It is one of the more extraordinary coincidences in the history of science that relativity theory predicts the same result, and this coincidence probably delayed relativity for years.

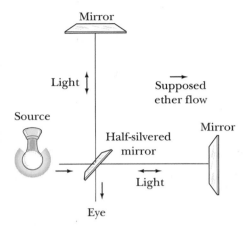

Figure 18.11   Michelson interferometer.

After Fizeau's experiment, which was repeated with great care by Michelson and Morley in 1886, the ether theory was very strong indeed, and it appeared feasible to actually measure the speed of the earth through the ether. This was attempted by Michelson and Morley in 1887 using the Michelson interferometer, Figure 18.11. A *white* light beam was split and sent in two mutually perpendicular directions; each arm was made 11 meters long by multiple reflections. After traversing a total path of 22 meters in each arm, the light was recombined, and colored interference fringes were observed. (We here pay homage parenthetically to the extraordinary skill of Michelson in making the two arms of the interferometer equal to within one wavelength of light.) If the ether flows through the system, then the light beams get out of phase, and that phase shift, observable as a fringe shift, changes as the interferometer is rotated. (See Problem 18-9.)

The experiment was done, and there was no changing fringe shift. In view of the previous successes of the ether theory, it proved difficult to reconcile the various experimental facts until Einstein simply did away with the ether in 1905 (Fig. 18.25).

The accompanying table (from Panofsky and Phillips) summarizes a great deal of experimental information related to various theories of light propagation. It is especially intended for those of us who suspect, "Surely there must be something they haven't thought of." As is immediately apparent, the special theory of relativity is the only theory that agrees (it gets straight A's) with all of the experiments cited.

Here is a brief description of the various types of experiment listed.

Aberration: The apparent position of stars shifts slightly due to the earth's motion about the sun.

Fizeau convection coefficient: We discussed it a few paragraphs back.

Michelson-Morley: Ditto.

Kennedy-Thorndike: Michelson made the arms of his interferometer equal; Kennedy and Thorndike used monochromatic light and made them as unequal as possible.

Moving sources and mirrors: The speed of light (detected by interference fringes) does not depend on motion of sources or mirrors.

Theory \ Experiment	Light Propagation Experiments							Experiments from Other Fields				
	Aberration	Fizeau convection coefficient	Michelson-Morley	Kennedy-Thorndike unequal arms	Moving sources and mirrors	De Sitter spectroscopic binaries	Michelson-Morley, using sunlight	Variation of mass with velocity	Radiation from moving charges	Meson decay at high velocity	Trouton-Noble capacitor	Unipolar induction, using permanent magnet
Ether theories — Stationary ether, no contraction	A	A	D	D	A	A	D	D	A	N	D	D
Stationary ether, Lorentz contraction	A	A	A	D	A	A	A	A	A	N	A	D
Ether attached to ponderable bodies	D	D	A	A	A	A	A	D	N	N	A	N
Emission theories — Original source	A	A	A	A	A	D	D	N	D	N	N	N
Ballistic	A	N	A	A	D	D	D	N	D	N	N	N
New source	A	N	A	A	D	D	A	N	D	N	N	N
Special theory of relativity	A	A	A	A	A	A	A	A	A	A	A	A

Legend: A, the theory agrees with experimental results; D, the theory disagrees with experimental results; N, the theory is not applicable to the experiment.

	Emission Theory	Classical Ether Theory	Special Theory of Relativity
Reference System	No special reference system	Stationary ether is special reference system	No special reference system
Velocity Dependence	The velocity of light depends on the motion of the source	The velocity of light is independent of the motion of the source	The velocity of light is independent of the motion
Transformation Equations	Inertial frames are connected by a Galilean transformation	Inertial frames are connected by a Galilean transformation	Inertial frames are connected by a Lorentz transformation

De Sitter spectroscopic binaries: If the speed of light depended on the speed of the source, then the fluctuating light from distant binary stars would show anomalies.

Michelson-Morley, using sunlight: The source is of course irrelevant in relativity theory.

Variation of mass with velocity: Rapidly moving particles display increased inertia or mass; we'll discuss this in Section 18.5.

Radiation from moving charges: The intensity, frequency, and angular distribution are correctly predicted by relativity theory.

Meson decay at high velocity: The time dilation has been checked for mu mesons stemming from cosmic rays, and for many other kinds of unstable particles.

Trouton-Noble: A charged capacitor moving through the ether should experience a torque.

Unipolar induction, using permanent magnet: We talked about this, in the homopolar generator, in Section 4.3.

Three general types of theory are also presented.

Ether theories: The light travels at speed c relative to the ether. The Lorentz contraction may be grafted onto the ether theory to explain away Michelson-Morley.

Emission theories: The photons jump off at speed c relative to the source. Ballistic: At a mirror, they bounce elastically. New source: They move at c relative to the mirror.

Relativity theory: The light travels at speed c relative to the observer.

18.4.   Inertial versus Accelerated Frame

In Section 18.1 we remarked that in an inertial reference frame an object stays put. However, it is well known that an apple, if released, falls toward the center of the earth, and yet we usually treat the earth as an inertial frame. So we need to devise a better criterion to distinguish inertial from accelerated frames. The **Twin Paradox** provides an interesting and natural path to this criterion.

The Twin Paradox (or Clock Paradox) is the most famous paradox in science; after more than three-quarters of a century, papers continue to be published resolving it in various ways or, occasionally, attacking relativity itself because of it (Dingle). A **paradox** is an apparent contradiction ("apparent paradox" would be a tautology). Paradoxes can be instructive as well as amusing. If you feel that a paradox needs resolving, it's an indication that you are missing a necessary intuitive concept. If you already understand the concepts involved, it may not even be clear to you that there is a paradox at all; one man's paradox is another man's truism.

The Twin Paradox goes like this: One twin goes on a long journey, let's say to the star Altair, 16 light-years away, at a speed of 2.60×10^8 m/s corresponding to $\gamma = 2$, Figure 18.12a. The round trip time is

$$2 \times 16 \text{ light-years} \times \frac{9.46 \times 10^{15} \text{ m}}{1 \text{ light-year}} \times \frac{1 \text{ s}}{2.60 \times 10^8 \text{ m}}$$

$$\times \frac{1 \text{ yr}}{3.15 \times 10^7 \text{ s}} = 37.0 \text{ years} \quad \textbf{(18.19)}$$

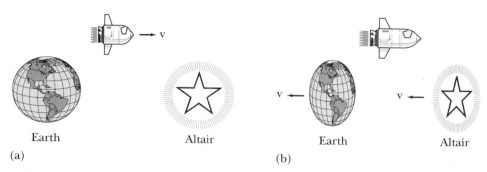

Figure 18.12   **(a)** Earth reference frame. **(b)** Ship reference frame.

as seen by the terrestrial twin. (We colloquially use the phrase "as seen by" to mean "in the reference frame of," implying a complete analysis, including correction for the finite speed of light; what the twin sees with the naked eye is quite a different story.)

On the other hand, the traveling twin's clock has been slowed by the time dilation, $\gamma = 2$, both going and coming back, and so only 18.5 years have elapsed on board the spaceship. The traveler's age must keep pace with the traveling clock; otherwise it would be possible to tell how fast the spaceship was going, and as remarked in Section 18.1, that is a no-no.

In the ship frame ("as seen by" the ship occupants), Figure 18.12b, there is no time dilation of the ship's clock; instead, the trip takes a relatively short time because the distance to the star is only 8 light-years, due to the Lorentz contraction. The time dilation and Lorentz contraction are complementary, in that what is explainable as a time dilation in one frame is due to a Lorentz contraction in the other.

So at the end of the trip, the twins are reunited, and it is an observable *fact* that the traveling twin has aged less. All of the above is correct and in accordance with theory and experiment.

The *paradox* enters when we ask: Is not all motion relative? It is certainly true that earth clocks run slow, in the reference frame of the spaceship, during uniform relative motion (inertial frames). If so, why would it not be equally valid to maintain that the earth took the round trip, and that the spaceship was stationary? In that case, the terrestrial twin should age less. This appears to contradict the former argument.

The simplest bare-bones *resolution* of the paradox is just the observation that the spaceship had to accelerate; otherwise it could never return at all. So for part of the trip it was not in a sufficiently inertial frame, and the time dilation is inadequate to explain observations in a noninertial frame. The earth did not accelerate significantly. This removes the seeming symmetry between the frames, and so there is no contradiction. (How much is "significant"? We return to this question at the end of this section.)

To summarize:

Fact: The traveling twin ages less.

Paradox: If all motion is relative, how can there be any asymmetry in aging?

Resolution: Not all motion is relative. The traveler didn't remain in a given inertial frame, so the twins' experiences are not symmetric (time dilation is inadequate).

Although we have disposed of the Twin Paradox in principle, there are still several intriguing questions. One is: How do the twins know which one is taking the trip? It is *not* true that the one that feels the acceleration is the one that is on the journey: The traveling twin can complete the voyage with little "felt" acceleration, especially if part of the necessary acceleration is provided by gravitational fields. And the terrestrial twin feels the acceleration due to the earth's gravity g all the time and yet does not move in a significant way. The remarkable fact is, no measurements taken inside a closed spaceship (that is, without prior information concerning external masses) will enable either twin to tell who is taking the trip. They must be able to see the "fixed stars," that is, the overall mass distribution of the universe. The kind of acceleration that is significant in this context is acceleration relative to the fixed stars. The concept involved here is called "Mach's principle"; Einstein noted it in his 1916 paper on general relativity, and it has since been elaborated by Sciama and others.

Other questions are: Why doesn't the earth's gravity have any significant effect? And what *does* the traveling twin see in the accelerated reference frame? There is some connection between an accelerated frame and a gravitational field; indeed, a **pseudogravitational field** appears in an accelerated frame: If we accelerate in the $+x$ direction, then loose objects seem to "fall" toward the $-x$ direction, in our accelerated frame. But we can answer the relevant questions without having to stray too far into general relativity.

Special relativity includes inertial frames and the Lorentz transformation; it is what we have dealt with so far. **General relativity** involves gravitational fields. But where does the accelerated reference frame lie? Surprisingly, two different informed opinions coexist:

(1) At the beginning level, authors usually assert: "The special theory of relativity is restricted to transformations between nonaccelerating reference frames" (Fisher) and ". . . the general theory is concerned with accelerated reference frames and gravity" (Tipler).

(2) For workers in the field, including Synge (and Eddington), "The special theory of relativity is the theory of flat space-time, interpreted physically . . . The theory of gravity based on the curvature of space-time is Einstein's general theory of relativity." The accelerated frame is flat, and curvature cannot be transformed in or out, so this places it in special relativity.

Without taking sides on the question of definitions, we shall have a look at the accelerated reference frame here, carefully, and with small speed, so $\gamma \approx 1$. Let us suppose . . .

To begin with, we, as observer, are at the origin of the O frame. Our wristwatch reads "0" at that time. For convenience, we have all clocks read ct rather than just t.

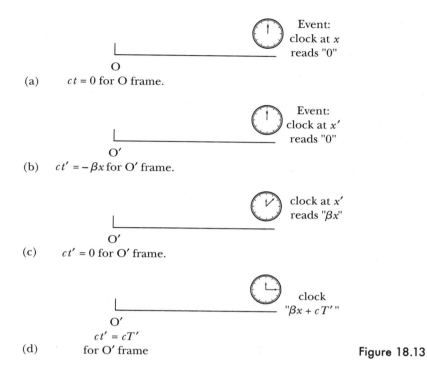

(a) $ct = 0$ for O frame.

(b) $ct' = -\beta x$ for O' frame.

(c) $ct' = 0$ for O' frame.

(d) $ct' = cT'$
for O' frame

Figure 18.13

A clock located at x reads "0" at $ct = 0$ in the O frame; that represents an *event* in the unprimed frame, Figure 18.13a.

The O' frame moves to the right at speed v (in the O frame). In the O' frame, the clock reads "0" at $ct = 0$, so $ct' = 0 - \beta x$, for small v, Figure 18.13b. That means that when the above-mentioned *event* occurs, time in the O' frame is given by $ct' = -\beta x$. The clock is not synchronized with the time in the O' frame.

So when ct' reaches 0, the clock has advanced to "βx" in the primed frame. That's the situation just after we step from O to O' at the origin at $ct = ct' = 0$. We find that the clock has jumped forward by βx, Figure 18.13c, as observed by us. Note that we do not remain in an inertial reference frame during the step, so the usual time dilation is inadequate to explain our own observations of the clock.

At this point, we have stepped from O to O', undergoing a brief acceleration. Our wristwatch still reads "0." However, the clock at x, which is now x', has suddenly jumped forward from "0" to "βx." This is a general rule for a Lorentz transformation; for v > 0, clocks at higher values of x move ahead, that is, read a later time, and those at lower x move backwards. This phenomenon occurs all along the x axis from $-\infty$ to ∞. It implies that when we accelerate from one inertial frame to another, clocks in the direction of our acceleration tick more rapidly, and those behind us run more slowly or even go backwards.

All of the above is nothing but a simple Lorentz transformation. If it seems puzzling, you should have another look at it.

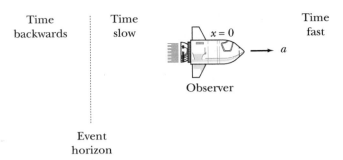

Figure 18.14   Accelerated reference frame.

Now we need to establish an average rate of acceleration. To do this, we coast for an arbitrary time cT'. Then the clock reads "$\beta x + cT'$." Our wristwatch says $ct' = cT'$, Figure 18.13d.

The ratio of the clock rate f_c to our wristwatch time rate f is

$$\frac{f_c}{f} = \frac{\beta x + cT'}{cT'} = \frac{ax}{c^2} + 1$$

using $a = v/T' = c\beta/T'$ for our average acceleration. Rearranging, with $\Delta f = f_c - f$,

$$\boxed{\frac{f_c}{f} - 1 = \frac{\Delta f}{f} = \frac{ax}{c^2}} \tag{18.20}$$

That was for one step. Now we take another step in the same way, and so on. In the limit of small step size and many steps, we arrive at continuous acceleration, with the same formula. This is sometimes called the gravitational frequency shift. **Higher clocks run faster** (in the pseudogravitational field mentioned before). Again, this development does not require that x be small or even positive; if x is negative, then the clock "advance" βx is negative, and the clock runs backwards if it's far enough away.

The above simple procedure carries us unambiguously into the (uniformly) **accelerated reference frame**, Figure 18.14. Where $x = -c^2/a$, so $\Delta f = -f$, we find zero frequency; this implies the presence of an **event horizon** where all motion ceases.

▼ EXAMPLE 18-5

For acceleration $g = 9.8 \text{ m/s}^2$, find the distance to the event horizon.

ANSWER

$$x = -c^2/a$$
$$= -(3 \times 10^8)^2/9.8 = -9.18 \times 10^{15} \text{ m} = -0.97 \text{ light-years}$$

So if an atom bomb explodes at distance x, we have only to accelerate away with $a = c^2/x$; then the gamma rays are at our event horizon and they can never catch up with us.

Note well: We are accelerating the observer, not the clock. An ideal clock is unaffected by its own acceleration.

The frequency shift explains the observations of the traveling twin in the Twin Paradox. While the ship moves away from the earth, the terrestrial clocks run slow by the relativistic factor γ, in the ship's frame; and the same is true while the ship returns to the earth. But during the period of acceleration at Altair, the earth clocks are effectively above the spaceship clocks in the pseudogravitational field, and so the earth clocks run fast by a factor ax/c^2, where x is 16 light-years. The terrestrial twin grows old rapidly, in the frame of the traveling twin, so much so that after the round trip is completed, the traveler is younger than the other. In this sense the time dilation is complementary to the gravitational frequency shift: The difference in aging that is due to time dilation in the earth frame is largely due to gravitational frequency shift in the ship frame.

Two frequently asked questions: (1) Why aren't we concerned about the acceleration of the ship when it is near the earth? Answer: Because in that case x is small, so the effect is negligible. (2) Why can't we make the acceleration a negligible by making it of very short duration? Answer: If it takes less time, then a itself must be proportionately larger, in order to get the ship turned around.

The gravitational shift also may be invoked in the case of the Pole and Barn Paradox, Figure 18.7. In order to get into a frame moving to the right, the runner had to accelerate toward the right. During the acceleration, clocks to the right ran faster, and so the barn door that is farther to the right shuts and opens sooner, in the runner's frame. So we can explain the fact that clocks to the right are ahead in this case *either* with the Lorentz transformation *or* with the gravitational frequency shift.

The **Principle of Equivalence** states that the effects of gravitation should be indistinguishable from those of acceleration of the reference frame over a sufficiently small region. For a *large* region a uniform gravitational field *cannot* be replaced by an accelerated frame: For a stationary object in the uniform field, the proper acceleration is the same everywhere, but for the accelerated frame, the proper acceleration is inversely proportional to distance from the event horizon. To put it another way, the uniform gravitational field involves curved space-time, but the accelerated frame is flat (Desloge). But for small regions the results obtained above should be applicable in the case of a true gravitational field.

 **EXAMPLE 18-6**

For an airplane traveling at 300 m/s at an altitude of 10,000 m, Figure 18.15, compare the time dilation and the gravitational frequency shift. Ignore terrestrial rotation, which has a significant effect on the results.

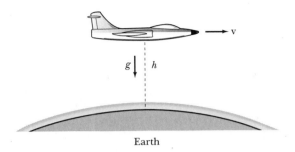

Figure 18.15   Hafele and Keating experiment.

ANSWER For the time dilation,

$$\gamma - 1 = \frac{1}{\sqrt{1 - \beta^2}} - 1 \approx \frac{1}{1 - \beta^2/2} - 1 \approx 1 + \beta^2/2 - 1 \approx \beta^2/2$$

$$= (300/3 \times 10^8)^2/2 = 0.5 \times 10^{-12} \qquad \text{slower}$$

For the gravitational effect,

$$ax/c^2 = gh/c^2$$

$$= 9.8 \times 10^4/(3 \times 10^8)^2 = 10^{-12} \qquad \text{faster}$$

So the gravitational effect is around twice as great in this case. This experiment was conducted by Hafele and Keating in 1972: They carried Hewlett-Packard cesium beam atomic clocks around the world in commercial airliners, once east and once west, confirming both aspects of these time effects.

   And now. The earth is not an inertial frame: If we release an apple, it falls. So why does the earth's gravitational field matter in the Hafele-Keating experiment but not matter in the Twin Paradox? The answer is, it falls off inversely as the square of the distance, so it is effective for only a short distance, on an astronomical scale. In other words, the pseudogravitational field that springs up in the ship frame during acceleration extends undiminished for many light-years, but the earth's gravitational field loses most of its effect within a few thousand miles. The ax that appears in the gravitational frequency shift formula is the gravitational potential, which is gh for the constant field near the earth, but more generally may be expressed as $- Gm/r$. (It is analogous to electrostatic potential $Q/4\pi\epsilon_0 r$, and it is zero at infinity.) Putting in numbers (see back cover),

$$\frac{\Delta f}{f} = \frac{ax}{c^2} \Longrightarrow \frac{Gm}{r c^2}$$

$$= \frac{6.7 \times 10^{-11} \times 6 \times 10^{24}}{6.4 \times 10^6 \times (3 \times 10^8)^2} = 7 \times 10^{-10}$$

which may or may not be negligible; it implies that the earth is practically an inertial reference frame for the Twin Paradox, which involves much larger effects, but not for the Hafele-Keating experiment, which can detect smaller ones. Thus we now have a quantitative criterion for deciding whether a given frame qualifies as an **inertial reference frame.**

18.5.   **Momentum and Energy**

At high speed, mass may be regarded as increasing by a factor of γ. One argument goes as follows. Consider a glancing collision between two identical particles of **rest mass** m_0 (their mass in a reference frame in which they are at rest). In Figure 18.16a the lower particle has no x velocity, and v of the upper particle is much greater than v_y of the lower one. In Figure 18.16b we view the same collision in a frame in which the upper particle has no x velocity. Now the point is, if we wish to preserve the law of conservation of momentum, then by symmetry both particles in both frames must have the same y component of momentum p_y. For the rapidly moving particle in either case, v_y is reduced by a factor of γ according to Section 18.3; so its mass must be increased by the same factor, in order that the product $p_y = mv_y$ be conserved.

So it is convenient to define the relativistic mass as $m = \gamma m_0$, where m_0 is the rest mass, the mass in the rest frame of the object. Then momentum is simply mv. The relativistic increase in mass is just the mass of the kinetic energy. (An alternative approach uses four-velocity, below.)

It turns out that the components of momentum transform in the same way as corresponding displacement components (x, y, z), that is, via the Lorentz transformation. Consider for example the x component:

$$p_x = m \frac{\Delta x}{\Delta t} = \gamma m_0 \frac{\Delta x}{\Delta t} = \gamma m_0 \frac{\Delta x}{\gamma \Delta \tau} = m_0 \frac{\Delta x}{\Delta \tau} \qquad (18.21)$$

Here we have introduced the **proper time** $\tau = t/\gamma$, the time in the rest frame of a clock. The rest mass m_0 and proper time τ are invariant: No matter how fast you go, measurements made in the rest frame stay the same. So p_x is just Δx times a

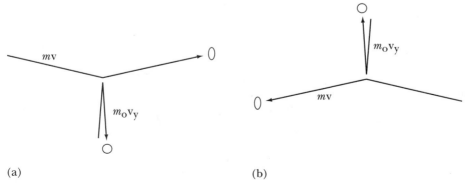

(a)   (b)

Figure 18.16   Collision between identical particles, two views.

couple of invariants, and so p_x and Δx will transform in the same way, that is, according to the Lorentz transformation.

The quantity $\Delta x/\Delta\tau$ is the x component of the **four-velocity,** which is useful in relativity theory. Ordinary velocity involves quantities like dx/dt, where x and t are both measured in the observer's reference frame. But for the four-velocity, the distance x is measured in the frame of the observer, while the time τ is measured in the frame of the moving object. Consequently, at the speed of light, the four-velocity is infinite. The momentum may be defined either as relativistic-mass times velocity, or as rest-mass times four-velocity:

$$p = (\gamma m_0)\frac{dx}{dt} = m_0\left(\gamma\frac{dx}{dt}\right)$$

It's the same momentum either way; there is no physically significant distinction. Note that there is **no relativistic mass increase** if we use four-velocity. We shall stay with ordinary velocity here because it is more appropriate for practical work in electromagnetism (for example, the Lorentz force is $q\mathbf{v}\times\mathbf{B}$ and it does not go infinite at the speed of light).

Now, we saw above that the x, y, and z components of momentum transform according to the Lorentz transformation. It is convenient to couple them with a fourth quantity analogous to ct. Instead of p_t, we'll call it E/c. Proceeding as above,

$$\text{``}p_t\text{''} = E/c = m\frac{\Delta ct}{\Delta t} = \gamma m_0\frac{\Delta ct}{\gamma\Delta\tau} = m_0\frac{\Delta ct}{\Delta\tau}$$

and so this fourth component of momentum transforms like time. It may be written as $\gamma m_0 c = mc = E/c$, and so $E = mc^2$. E is called **total energy.** We shall use the same symbol E for energy and for electric field, because each is customary, and you should have no difficulty telling which is meant by the context.

For a mass at rest, with $\gamma = 1$, E becomes $m_0 c^2$, which is consequently called **rest energy.** Relativistically the **kinetic energy** is $T = E - m_0 c^2$, that is, total energy minus rest energy. If we expand $E = mc^2$ for low speed, we find

$$E = \gamma m_0 c^2 = m_0 c^2\left(1 - \frac{v^2}{c^2}\right)^{-1/2}$$

$$= m_0 c^2 + \frac{1}{2}m_0 v^2 + \ldots$$

(18.22)

The first term is the rest energy, and the second is the usual nonrelativistic kinetic energy. Clearly, the expression $\frac{1}{2}m_0 v^2$ is inadequate at relativistic speeds.

Ordinary three-dimensional vectors are properly defined not just by the usual "magnitude and direction" but by their transformation properties under coordinate rotation (Section 1.6). Correspondingly, there exists a class of **four-vectors** whose components are transformed from one reference frame to another by the Lorentz transformation. We have already encountered three four-vectors:

Displacement:	x, y, z, ct	
Four-velocity:	$dx/d\tau, dy/d\tau, dz/d\tau, dct/d\tau$	(18.23)
Momentum-energy:	$p_x, p_y, p_z, E/c$	

Later we are going to find others:

$$\text{Current:} \qquad J_x, J_y, J_z, c\rho$$
$$\text{Potential:} \qquad A_x, A_y, A_z, V/c$$

(18.24)

Thus, for example, the momentum-energy four-vector transformation is

$$p'_x = \gamma p_x - \beta\gamma E/c$$
$$p'_y = p_y$$
$$p'_z = p_z$$
$$E'/c = \gamma E/c - \beta\gamma p_x$$

(18.25)

just like the Lorentz transformation for x, y, z, ct.

◤ EXAMPLE 18-7

A proton whose kinetic energy T equals its rest energy m_0c^2 moves at speed v_y in the $+y$ direction, Figure 18.17a. Find its speed v', momentum p', and total energy E' in a primed frame moving at $v = 2.4 \times 10^8$ m/s in the $+x$ direction, Figure 18.17b.

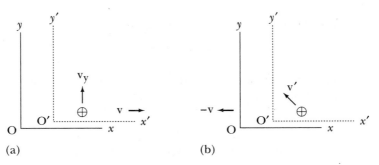

(a)   (b)

Figure 18.17   (a) Proton in unprimed frame. (b) Proton in primed frame.

ANSWER   This is a lengthy but instructive exercise. We will use a y subscript for motion of the proton *in* the unprimed frame to distinguish that from the motion *of* the primed frame.

In the *unprimed frame*, we have $T = m_0c^2$, so

$$E = m_0c^2 + T = 2\ m_0c^2$$

Also,

$$E = \gamma m_0 c^2$$

so it must be that $\gamma_y = 2$ for the proton here.

So its β_y is 0.866. Its momentum is

$$p_y = m v_y = \gamma_y m_0 \beta_y c$$
$$= 2 \times 1.67 \times 10^{-27}\ \text{kg} \times 0.866 \times 3 \times 10^8\ \text{m/s} = 8.68 \times 10^{-19}\ \text{kg} \cdot \text{m/s}$$

and E is

$$E = mc^2 = \gamma_y m_0 c^2$$
$$= 2 \times 1.67 \times 10^{-27}\, kg \times (3 \times 10^8\, m/s)^2 = 3.01 \times 10^{-10}\, J$$

The speed of the primed frame in the unprimed frame is $v = 2.4 \times 10^8\, m/s$, so $\beta = +0.800$ and $\gamma = 1.667$. Then the Lorentz transformation reads (with $\beta_x = 0$)

$$p'_x = -\beta\gamma E/c = -0.8 \times 1.667 \times 3.01 \times 10^{-10}/3 \times 10^8$$
$$= -1.34 \times 10^{-18}\, kg \cdot m/s$$
$$p'_y = 8.68 \times 10^{-19}\, kg \cdot m/s$$
$$p'_z = 0$$
$$E'/c = \gamma E/c = 1.667 \times 3.01 \times 10^{-10}/3 \times 10^8 = 1.67 \times 10^{-18}$$

E': $$E' = 3 \times 10^8 \times 1.67 \times 10^{-18} = 5.02 \times 10^{-10}\, J$$

$$p' = \sqrt{p'^2_x + p'^2_y}$$

p': $$= \sqrt{(-1.34 \times 10^{-18})^2 + (8.68 \times 10^{-19})^2} = 1.60 \times 10^{-18}\, kg\, m/sec$$

$$\gamma' = E'/m_0 c^2 \text{ (for proton in primed frame)}$$
$$= 5.02 \times 10^{-10}/1.67 \times 10^{-27} \times (3 \times 10^8)^2 = 3.34$$

$$\beta' = 0.954$$

v': $$v' = 3 \times 10^8 \times 0.954 = 2.86 \times 10^8\, m/s$$

When we differentiate momentum with respect to time, we get **force,** by definition, just as in the classical case:

$$\mathbf{F} = \frac{d\mathbf{p}}{dt} \tag{18.26}$$

but *not* $\mathbf{F} = m\mathbf{a}$, because now m is the relativistic mass γm_0:

$$\mathbf{F} = \frac{d\mathbf{p}}{dt} = \frac{d}{dt}(\gamma m_0 \mathbf{v}) \qquad \text{but } not \qquad \mathbf{F} = m\mathbf{a} = \gamma m_0 \frac{d\mathbf{v}}{dt}$$

(It is interesting that Newton himself wrote his second law as $F = \dot{p}$, not $F = ma$.)

Force is not part of a four-vector. You can tell this by the transverse component: For a four-vector it's invariant, but for the force it's not. We have

$$F'_\perp = \frac{dp'_\perp}{dt'}$$

Now $\mathbf{p}$ is a four-vector, so p_y is invariant, and $dp'_\perp = dp_\perp$; but $dt' = \gamma\, dt$ (time dilation), so

$$F'_\perp = \frac{dp_\perp}{\gamma dt}$$

$$\boxed{F'_\perp = F_\perp/\gamma} \tag{18.27}$$

regardless of the nature of the force, for small speed in the unprimed frame. We will find this particular equation useful in connection with Figure 18.34, in Section 18.9. (There exists a "four-force" that *is* a four-vector, $d\mathbf{p}/d\tau$, but we won't need it.)

 **EXAMPLE 18-8**

An apple drops from rest with an acceleration of $a_y = 9.8$ m/s^2. Find $F'_\perp$ using an acceleration transformation (Section 18.3).

ANSWER $F_\perp = m_0 a_y$, just as classically, because the object is at rest, or nearly so, in the unprimed frame. Then,

$$F'_\perp = \gamma m_0 \frac{a_y}{\gamma^2} = \frac{m_0 a_y}{\gamma}$$

$$= \frac{F_\perp}{\gamma}$$

as found above using the time derivative of the momentum.

We have already mentioned that $x^2 + y^2 + z^2 - (ct)^2$ is invariant (Section 18.2). For every four-vector there is a corresponding invariant, and the one for the momentum-energy is one that we will find particularly useful in the rest of this section:

$$p_x^2 + p_y^2 + p_z^2 - E^2/c^2 = p^2 - E^2/c^2$$
$$= (\gamma m_0 \beta c)^2 - (\gamma m_0 c^2/c)^2 \qquad \textbf{(18.28)}$$
$$= (m_0 c)^2(\beta^2 \gamma^2 - \gamma^2) = -(m_0 c)^2$$

so

$$p^2 - E^2/c^2 = -(m_0 c)^2 \qquad \text{which is invariant.}$$

Rearranging,

$$E^2 = (m_0 c^2)^2 + (pc)^2$$

When $p = 0$, this turns into the familiar $E = m_0 c^2$ for the rest energy of a mass m_0.

It is customary, and usually easier, to deal with elementary-particle kinematics using **electron-volt units** (Section 6.4) and at the same time units in which

$$\boxed{c = 1}$$

In the above invariant, we set $c = 1$, and while we're at it we introduce new symbols for mass and momentum to remind us that the units are different: $m_0 \Rightarrow M$, and $p \Rightarrow P$. Note that M is used for rest mass even though it has no zero subscript. Now the above equation satisfies the Pythagorean relationship, Figure 18.18:

$$\boxed{E^2 = P^2 + M^2} \qquad \textbf{(18.29)}$$

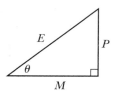

Figure 18.18 $E^2 = M^2 + P^2$.

We shall also be defining kinetic energy T using the equation

$$E = T + M \qquad (18.30)$$

(so $T = \frac{1}{2} mv^2$ is only an approximation).

For calculations involving velocity we can use

$$E = \gamma M \qquad \text{and} \qquad P = \beta \gamma M = \beta E \qquad (18.31)$$

For example, starting with $p = mv$, we proceed to $p = \gamma m_0 v \, c/c = \beta \gamma m_0 c \Rightarrow P = \beta \gamma M$, since $c = 1$ in these units.

Also, this facilitates relativistic calculations involving speeds and eV units:

$$\cos \theta = \frac{M}{E} = \frac{1}{\gamma} \quad \text{and} \quad \sin \theta = \frac{P}{E} = \beta \quad \text{and} \quad \tan \theta = \frac{P}{M} = \beta \gamma \quad (18.32)$$

Now we must work on the new units. First, we remember that an electron volt (eV) is the energy gained by one electron charge in falling through one volt:

$$W = QV = e \times 1 \text{ volt} = 1.602 \times 10^{-19} \text{ joule}$$

So, one eV is 1.602×10^{-19} joules; the E *conversion factor* is numerically equal to e, the elementary charge.

We can relate E to the other quantities as follows. We start with the above invariant,

$$E^2 = (m_0 c^2)^2 + (pc)^2 \qquad (18.33)$$

If $p = 0$, then $m_0 = E/c^2$, and the M *conversion factor* is e/c^2: One eV/c^2 is 1.78×10^{-36} kg. If $m_0 = 0$ (photon), then $p = E/c$, and the P *factor* is e/c: One eV/c is 5.34×10^{-28} kg $\cdot$ m/s.

In the particle-physics laboratory, relativistic particles are often guided using magnetic fields. For a particle of mass m $(= \gamma m_0)$, speed v, momentum $\mathbf{p} = m\mathbf{v}$, and charge $+e$ describing a circle of radius r in a magnetic field $\mathbf{B}$, Figure 18.19, the derivation is the same as in the classical case. We employ similar triangles, along with the fact that the centripetal force is the Lorentz force.

$$\Delta \theta = \frac{\Delta p}{p} = \frac{v \Delta t}{r}$$

Rearranging,

$$F = \frac{\Delta p}{\Delta t} = \frac{pv}{r} = \frac{mv^2}{r} \qquad \text{the centripetal force}$$

$$= evB \qquad \text{the Lorentz force}$$

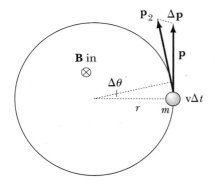

Figure 18.19   Path in magnetic field.

So,

$$p = eBr$$

which is correct relativistically. Now if we change the momentum to electron-volt units (above), we obtain

$$\boxed{P = cBr} \tag{18.34}$$

where P is in eV/c, and c, B, and r are in standard SI units; the equation comprises a peculiar but useful coupling of disparate units.

 EXAMPLE 18-9

A proton describes a circle of radius 1.83 m in a 1.2-T magnetic field. Find its speed, momentum, kinetic energy, and total energy. (Its rest mass, 938 MeV, is given in the back-cover table.)

ANSWER   We begin by finding the momentum directly in eV units.

$$P = cBr$$
$$= 3 \times 10^8 \text{ m/s} \times 1.2 \text{ T} \times 1.83 \text{ m} = 6.59 \times 10^8 \text{ eV/c} = 659 \text{ MeV/c}$$

Then,

$$E = \sqrt{M^2 + P^2}$$
$$= \sqrt{938^2 + 659^2} = 1146 \text{ MeV}$$
$$T = E - M$$
$$= 1146 - 938 = 208 \text{ MeV}$$
$$\beta = P/E$$
$$= 659/1146 = 0.575$$

so

$$v = c\beta$$
$$= 3 \times 10^8 \times 0.575 = 1.73 \times 10^8 \text{ m/s}$$

Experimentally, protons of this speed are in fact found to describe this radius, which implies that their mass really is greater than their rest mass (assuming that the Lorentz force $Q\mathbf{v} \times \mathbf{B}$ doesn't start to lose its grip at high speed).

In order to calculate simple particle reactions or decays we need something like the following:

$$M_c = \sqrt{(E_1 + E_2)^2 - (\mathbf{P}_1 + \mathbf{P}_2)^2} \qquad (18.35)$$

It is the same as $M^2 = E^2 - P^2$, but for two particles. The total energies are added as scalars, and the momenta are added as vectors. M_c is the rest mass of the system of two particles, which is the same as the total energy in the center-of-mass frame, in which they have no net momentum.

▼ EXAMPLE 18-10

An 800-MeV/c proton and a 200-MeV/c pion include an angle of 35°, Figure 18.20. If they result from baryon decay, what is the mass of the baryon? Rest mass of pion is 140 MeV.

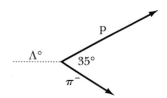

Figure 18.20   Particle decay.

ANSWER For the proton, $E_1 = \sqrt{800^2 + 938^2} = 1233$ MeV
        For the pion, $E_2 = \sqrt{200^2 + 140^2} = 244$ MeV
We find the vector sum of the momenta using the law of cosines:

$$P = \sqrt{P_1^2 + P_2^2 + 2\,P_1 P_2 \cos \theta}$$
$$= \sqrt{800^2 + 200^2 + 2 \times 800 \times 200 \cos 35°} = 970 \text{ MeV/c}$$

So

$$M_c = \sqrt{(E_1 + E_2)^2 - (\mathbf{P}_1 + \mathbf{P}_2)^2}$$
$$= \sqrt{(1233 + 244)^2 - 970^2} = 1115 \text{ MeV}$$

which happens to be the mass of the Λ^0 particle.

▼ EXAMPLE 18-11

The Berkeley Bevatron was a proton accelerator designed to make antiprotons using stationary target protons: $p + p \rightarrow 3p + \bar{p}$. The antiproton $\bar{p}$ is in effect a negative proton with rest mass 938 MeV. Find the threshold total energy E_1 for the reaction.

ANSWER At threshold the $3p + \bar{p}$ are stationary in the center-of-mass frame, so the mass M_c is that of four stationary protons, $4\, M_p$. The incident proton has energy E_1 and momentum P_1. The stationary target proton has energy $E_2 = M_p = 938$ MeV, and it has momentum $P_2 = 0$. So,

$$M_c = 4\, M_p = \sqrt{(E_1 + M_p)^2 - P_1^2} \quad \text{and} \quad P_1^2 = E_1^2 - M_p^2$$

Solving,

$$E_1 = 7\, M_p$$
$$= 7 \times 938 = 6570 \text{ MeV} = 6.57 \text{ GeV}$$

The Bevatron was designed for 7 GeV, and in 1955 it duly yielded antiprotons and Nobel prizes for Chamberlain and Segrè.

18.6. Moving Sources; Doppler Effect

(We revert to ordinary SI units: $c = 3 \times 10^8$ m/s now.)
     When a radiating source moves, the motion may change the radiation observed in frequency and in total power. In preceding chapters we studied radiation from slowly moving sources; now we can extend the development to rapidly moving charges, such as those in particle accelerators.
     In the case of sound, you presumably know already that the frequency f' heard from a moving source is related to the frequency f for the same source at rest by the formula

$$f' = \frac{f}{1 - \beta_s \cos \theta'} \tag{18.36}$$

θ' being the angle between the path of the source and the line from source to observer, Figure 18.21. Here we use

$$\beta_s = \frac{v}{v_s}$$

with v_s being the speed of sound. This is the **Doppler effect.** Approaching sources sound higher pitched.
     If the observer is moving through the air, and the source is stationary, the formula is $f' = f(1 + \beta_s \cos \theta)$. However, this has no relevance here for light for the following reason. For sound there is a medium called the air, and sound moves at about

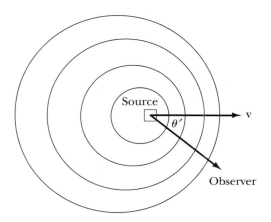

Figure 18.21   Doppler effect.

340 m/s relative to the air; its speed relative to the observer depends on the relative speed of the observer, and correspondingly the form of the Doppler formula depends on relative motion of source and observer. But for light there is no such medium (i.e., no ether) and the speed of light (in vacuum) is always 3×10^8 m/s relative to the observer. Consequently, for light there is only one Doppler formula.

The only formal difference between the Doppler formula for sound for a moving source, and the formula for light, is the time dilation in the latter case. The frequency is reduced by the relativistic time-dilation factor γ. So the formula becomes

$$f' = \frac{f}{\gamma \, (1 - \beta \cos \theta')} \qquad (18.37)$$

where now β has returned to its usual meaning of v/c. The distinction between θ and θ' is insignificant for sound, but it matters here: We want the angle in the observer's reference frame. When $\theta' = 0$, this becomes

$$f' = f\sqrt{\frac{1 + \beta}{1 - \beta}}$$

Despite the time dilation, the frequency is greater for an approaching source.

◢ EXAMPLE 18-12

For the earth in its orbit, $\beta \approx 10^{-4}$ (relative to the stationary sun). Find the corresponding maximum annual Doppler shift in the yellow sodium D line, 589 nm, for a distant star (that is, the difference between receding and approaching).

ANSWER For all Doppler problems, when v and θ are small it may be helpful to employ the approximation

$$v/c \approx \Delta f / f \quad \approx - \Delta \lambda / \lambda \qquad (18.38)$$

where v and Δf are positive for an approaching source. In this case,

$$\Delta\lambda/\lambda = 2\,\text{v}/c$$

$$\Delta\lambda = 589 \text{ nm} \times 2 \times 10^{-4} = 0.118 \text{ nm}$$

which is readily detectable.

The electromagnetic Doppler effect is well established experimentally, and in fact it is in common use in finding the speeds of such diverse objects as stars and cars (radar).

What if the source is traveling faster than light? This can happen in a dielectric medium, where the speed of light is c/n, n being the index of refraction (Section 16.2). As a charged particle passes through the medium, the nearby atoms absorb and radiate energy. For example, in water, $n = 1.33$, so light only goes at $0.75\ c$, and particles traveling at v $> c/n$ can emit **Cerenkov radiation,** Figure 18.22, like a shock wave, or the bow wave of a boat. The angle of the radiation is given by

$$\cos\theta = \frac{c}{n\text{v}} \qquad\qquad (18.39)$$

This phenomenon is employed in "Cerenkov counters," which discriminate among charged particles on the basis of their speeds.

The total power radiated by an accelerated charge in rapid motion is easily found using the Larmor formula (Section 14.6) and relativity. The energy E emitted in a given process increases with γ; this may be regarded as a result of the relativistic mass increase or of the Lorentz transformation. However, the time t required also increases with γ, due to the time dilation, so the net result is, surprisingly, that the total radiated power E/t is independent of γ, for a given

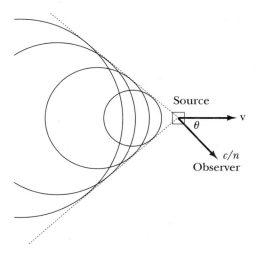

Figure 18.22   Cerenkov radiation.

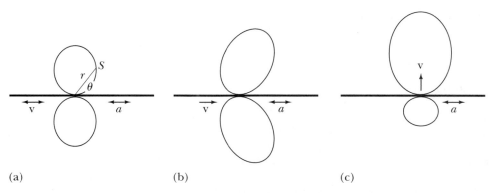

Figure 18.23   Radiation pattern (Poynting's vector) in three different reference frames.

proper acceleration. To find how the power depends on *lab* acceleration (acceleration of the source in the laboratory, that is, with respect to the observer), we need only transform the acceleration, Section 18.3. For a slow-moving source, the Larmor formula is (here we return to W for energy):

$$\frac{dW}{dt} = \frac{2}{3} \frac{Q^2 a^2}{4\pi\epsilon_0 c^3}$$

The corresponding radiation pattern is displayed in Figure 18.23a. As mentioned in Section 14.6, this is a polar plot of Poynting's vector **S**: the distance r is proportional to S at the angle θ. Three dimensionally it is a donut, isotropic in angle ϕ. The pattern doesn't depend on whether v is to right or left, nor on a in the same sense. Figures 18.23b and 18.23c show the same pattern in other reference frames. The radiation is polarized with **E** in the plane of the paper.

For rapid motion in the direction of the velocity we have $a'_{\parallel} = a_{\parallel}/\gamma^3$, so, substituting for a in the above formula, and then replacing $a'_{\parallel}$ with a, we find

$$\frac{dW}{dt} = \frac{2}{3} \gamma^6 \frac{Q^2 a^2}{4\pi\epsilon_0 c^3} \qquad (18.40)$$

Here a is the acceleration in the observer's frame, usually called the lab frame. Figure 18.23b is the corresponding plot for a charge moving to the right at around v = $c/3$. It is a distorted donut because of the Doppler effect. Once again, it doesn't matter whether a is to the right or left. This kind of pattern carries the name **bremsstrahlung** (German: "braking radiation"). In an x-ray machine, electrons are accelerated to 100 keV and smashed into a tungsten target; they produce x-rays of energy up to 100 keV largely through bremsstrahlung.

And for motion perpendicular to acceleration, $a'_{\perp} = a_{\perp}/\gamma^2$, so

$$\frac{dW}{dt} = \frac{2}{3} \gamma^4 \frac{Q^2 a^2}{4\pi\epsilon_0 c^3} \qquad (18.41)$$

for $a = a'_{\perp}$ in the lab frame. Figure 18.23c shows the pattern for a charge moving upward at around v = $c/3$. The pattern corresponds to **synchrotron radiation,** and the increased intensity in the forward direction is sometimes called the "headlight effect." Energy loss by synchrotron radiation introduces a serious limitation on circular electron accelerators; in fact, electron accelerators are often made linear rather than circular because of the heavy losses of energy during centripetal acceleration. For protons the problem is much smaller.

 EXAMPLE 18-13

(a) Find the radiated power, in eV per revolution, for a 600 MeV/c proton circulating in a cyclotron with a 1.2 T magnetic field. (b) Repeat for an electron.

ANSWER This lengthy example provides a review of practical relativistic calculations. In writing out the answer, we shall aim for clarity rather than efficiency.

(a) $P = cBr$, so

$$r = P/cB$$
$$= 600 \times 10^6/3 \times 10^8 \times 1.2 = 1.67 \text{ m}$$

$$E = \sqrt{P^2 + M^2}$$
$$= \sqrt{600^2 + 938^2} = 1113 \text{ MeV}$$

$$\beta = P/E$$
$$= 600/1113 = 0.539$$

$$v = \beta c$$
$$= 0.539 \times 3 \times 10^8 = 1.62 \times 10^8 \text{ m/s}$$

$$a = v^2/r$$
$$= (1.62 \times 10^8)^2/1.67 = 1.57 \times 10^{16} \text{ m/s}^2$$

$$\gamma = E/M$$
$$= 1113/938 = 1.187$$

Now we can calculate the power; it's synchrotron radiation:

$$dW/dt = \frac{2}{3}\gamma^4 \frac{Q^2 a^2}{4\pi\epsilon_0 c^3}$$

$$= 9 \times 10^9 \times \tfrac{2}{3} \times 1.187^4 \frac{(1.6 \times 10^{-19})^2 \times (1.57 \times 10^{16})^2}{(3 \times 10^8)^3}$$

$$= 2.79 \times 10^{-21} \text{ watts}$$

The energy per revolution is power times time:

$$2.79 \times 10^{-21} \times 2\pi r/v = 2.79 \times 10^{-21} \times 6.28 \times 1.67/1.62 \times 10^8 \text{ m/s}$$
$$= 1.80 \times 10^{-28} \text{ joules/rev}$$
$$= 1.12 \times 10^{-9} \text{ eV per revolution, which is negligible}$$

(b) For the electron of the same momentum, r is the same, 1.67m.

$$E = \sqrt{P^2 + M^2}$$
$$= \sqrt{600^2 + 0.511^2} \approx 600 \text{ MeV}$$

$$\beta = P/E \approx 1$$

$$a = v^2/r$$
$$\approx (3 \times 10^8)^2/1.67 = 5.39 \times 10^{16} \text{ m/s}^2$$

$$\gamma = E/M$$
$$= 600/0.511 = 1174$$

$$dW/dt = \frac{2}{3} \gamma^4 \frac{Q^2 a^2}{4\pi\epsilon_0 c^3}$$

$$= 9 \times 10^9 \times \tfrac{2}{3} \times 1174^4 \frac{(1.6 \times 10^{-19})^2 \times (5.39 \times 10^{16})^2}{(3 \times 10^8)^3}$$

$$= 3.14 \times 10^{-8} \text{ watts}$$

The energy per revolution is power times time:

$$3.14 \times 10^{-8} \times 2\pi r/v = 3.14 \times 10^{-8} \times 6.28 \times 1.67/3 \times 10^8$$
$$= 1.10 \times 10^{-15} \text{ joules/rev}$$
$$= 6900 \text{ eV per revolution, which is important}$$

In this example, the electron radiates more than the proton by a factor of 10^{12}, due mostly to its larger γ. If the electrodes of the cyclotron can operate at around 10 kilovolts with a frequency of $v/2\pi r = 29$ MHz, then in each revolution, electrons can pick up 10 keV, but they squander most of it in radiation.

On the other hand, in some applications the radiation of electrons can be an advantage: It provides a powerful and well-defined source of electromagnetic radiation. Note that most of the radiation comes out ahead of the electrons and tangential to their circular orbit: the "headlight effect," see Figure 18.23c.

18.7.   E and B Transformations

Mechanics was relativistic in spirit since at least the time of Galileo, and Newton intended that his natural philosophy should be based on the principle of relativity. By contrast, when Maxwell was developing his theory in the 1860s, the conceptual foundation of light and electromagnetism included the preferred frame of the ether. So it is ironic that the formalism of Maxwell's equations survived the advent of relativity unscathed, while Newton's mechanics had to be recast.

Electromagnetism involves charge, of course, so the first experimental fact we need is that, unlike mass,

Charge is **conserved** and **invariant**.

"Conserved" means you can't create or destroy it; in a given frame, it never changes. "Invariant" means that it is the same in different reference frames; it doesn't change with speed. The usual change-by-γ would easily be detected: For example, in a metal, the negative and positive charges move in different ways and at different speeds, so if charge changed with speed then a piece of metal would change voltage when heated.

By contrast, rest mass is *invariant* but not *conserved*. To illustrate, consider the decay of a neutral K meson:

$$K^0 \longrightarrow \pi^+ + \pi^-$$

Charge is conserved: total of zero charge on each side. But rest mass is not conserved: for K^0, $M = 498$ MeV, but for each pion, $M = 140$ MeV.

And relativistic mass, or total energy, is *conserved* but not *invariant*. This is true even nonrelativistically: if I stand by a stationary car in the O frame, it has no kinetic energy, and in that frame the energy stays the same: it is conserved. But if I walk past it, in the O' frame, I find that it has kinetic energy. I did not give the car the energy; I can hardly get out of bed in the morning. The car has the energy in the O' frame, but not in the O frame: It is not invariant.

Before going into general results, we shall illustrate field transformation with a simple example. When a charge moves parallel to an uncharged wire carrying a current, Figure 18.24a, there is a Lorentz force $Q\mathbf{v} \times \mathbf{B}$, which is directed toward the wire in this case: "Parallel currents attract." To simplify the picture, we show the positive charges as all moving at the same speed v, and the negative charges are stationary.

But if we move along with the positive charge, Figure 18.24b, then v = 0 and there is no Lorentz force. The magnetic field is still there, but it has no effect. The force that was entirely magnetic is now entirely electrostatic. Examining the distribution of charges, we find that the negative charges are moving and thus they show the Lorentz contraction, whereas the plus charges have stopped and so have "Lorentz-relaxed" and are farther apart. The net effect is that an electric field has arisen to supply the force that was due to $\mathbf{v} \times \mathbf{B}$ in the other frame.

This is at first a surprising result and it bears reiteration. **Whenever** we have a Lorentz force $Q\mathbf{v} \times \mathbf{B}$, we can eliminate it by moving to a reference frame in which v = 0. But in that new frame, the electric charge distribution is changed in such a way as to supply the necessary force. The rearrangement of charge is fundamentally relativistic; in the present case it is a matter of Lorentz contractions. Similar considerations contributed to Einstein's relativity, Figure 18.25.

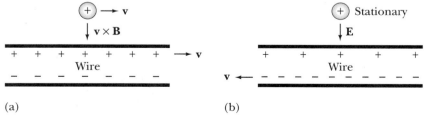

(a)   (b)

Figure 18.24   **(a)** Magnetic force. **(b)** Electrostatic force.

ON THE ELECTRODYNAMICS OF MOVING
BODIES

By A. EINSTEIN

It is known that Maxwell's electrodynamics—as usually understood at the present time—when applied to moving bodies, leads to asymmetries which do not appear to be inherent in the phenomena. Take, for example, the reciprocal electrodynamic action of a magnet and a conductor. The observable phenomenon here depends only on the relative motion of the conductor and the magnet, whereas the customary view draws a sharp distinction between the two cases in which either the one or the other of these bodies is in motion. For if the magnet is in motion and the conductor at rest, there arises in the neighbourhood of the magnet an electric field with a certain definite energy, producing a current at the places where parts of the conductor are situated. But if the magnet is stationary and the conductor in motion, no electric field arises in the neighbourhood of the magnet. In the conductor, however, we find an electromotive force, to which in itself there is no corresponding energy, but which gives rise—assuming equality of relative motion in the two cases discussed—to electric currents of the same path and intensity as those produced by the electric forces in the former case.

Examples of this sort, together with the unsuccessful attempts to discover any motion of the earth relatively to the "light medium," suggest that the phenomena of electrodynamics as well as of mechanics possess no properties corresponding to the idea of absolute rest. They suggest rather that, as has already been shown to the first order of small quantities, the same laws of electrodynamics and optics will be valid for all frames of reference for which the equations of mechanics hold good. We will raise this conjecture (the purport of which will hereafter be called the "Principle of Relativity") to the status of a postulate, and also introduce another postulate, which is only apparently irreconcilable with the former, namely, that light is always propagated in empty space with a definite velocity c which is independent of the state of motion of the emitting body. These two postulates suffice for the attainment of a simple and consistent theory of the electrodynamics of moving bodies based on Maxwell's theory for stationary bodies. The introduction of a "luminiferous ether" will prove to be superfluous inasmuch as the view here to be developed will not require an "absolutely stationary space" provided with special properties, nor assign a velocity-vector to a point of the empty space in which electromagnetic processes take place.

The theory to be developed is based—like all electrodynamics—on the kinematics of the rigid body, since the assertions of any such theory have to do with the relationships between rigid bodies (systems of co-ordinates), clocks, and electromagnetic processes. Insufficient consideration of this circumstance lies at the root of the difficulties which the electrodynamics of moving bodies at present encounters.

Figure 18.25   The first page of Einstein's first relativity paper is displayed here as Figure 18.25. Note that the phenomenon he refers to in the first paragraph is similar to that of Figure 18.24: There is a force $Q\mathbf{v} \times \mathbf{B}$ in one frame and $Q\mathbf{E}$ in another. (*Ann. Der Physik, 17*, 891 (1905); reprinted, Dover Press.)

Note, too, that in Figure 18.24b there is an **E** field, so there must be electric field lines sticking into the wire which aren't present in the other frame. Under other circumstances, it can be the magnetic field lines that appear or disappear when we change frames. So **field lines** can be useful conceptually, but they are lacking in physical reality.

Now, how can there be a significant relativistic effect here, when we learned in Chapter 3 that the drift speed of the electrons is only around $v = 10^{-4}\,\mathrm{m/s}$? The difference due to relativity is only

$$\gamma - 1 = \frac{1}{\sqrt{1 - \beta^2}} - 1 \approx \frac{1}{1 - \beta^2/2} - 1 \approx 1 + \beta^2/2 - 1 \approx \beta^2/2 \approx 10^{-25}$$

(18.42)

which is an exceedingly tiny effect; but we got lots of electrons, so let's pursue it. If 1 ampere flows in the wire, then the linear charge density is 1 C/s $\div$ 10^{-4} m/s = 10^4 C/m of current-carrying electrons. When we take 10^{-25} of this, we get a relativistic remnant of only $\lambda = 10^{-21}$ C/m. Then at a distance of 1 cm, Figure 18.24b,

$$E = \frac{\lambda}{2\pi\epsilon_0 r}$$

$$= \frac{10^{-21}}{2\pi \times 8.85 \times 10^{-12} \times 0.01} \approx 2 \times 10^{-9} \text{ V/m}$$

That was using relativity without magnetism. Now: The magnetic field due to the 1 ampere is

$$B = \frac{\mu_0 I}{2\pi r} = \frac{4\pi \times 10^{-7} \times 1}{2\pi \times 0.01} = 2 \times 10^{-5} \text{ T}$$

Figure 18.24a, so the magnitude of the Lorentz force is Q times

$$vB = 2 \times 10^{-9} \text{ T m/s} = 2 \times 10^{-9} \text{ V/m}$$

which is the same as the E field above. The magnetic field just provides another way of calculating what is really a relativistic effect. The exact equality is fortuitous; this was an order-of-magnitude calculation. We'll do a better job later. But it does show how there may be noticeable relativistic effects even at quite low speeds.

We could actually calculate many magnetic forces using Coulomb's law and special relativity, on the basis of pictures like Figure 18.24. But the old magnetic-field equations and concepts, developed before relativity, are easier to deal with than a relativistic formalism would be.

Now that we have seen an example showing how and why fields must undergo relativistic transformations, we need the general expressions:

$$\boxed{\begin{array}{ll} \mathbf{E}'_{\parallel} = \mathbf{E}_{\parallel} & \mathbf{E}'_{\perp} = \gamma\,\mathbf{E}_{\perp} + \gamma\,\mathbf{v} \times \mathbf{B} \\ \mathbf{B}'_{\parallel} = \mathbf{B}_{\parallel} & \mathbf{B}'_{\perp} = \gamma\,\mathbf{B}_{\perp} - \gamma\,\mathbf{v} \times \mathbf{E}/c^2 \end{array}}$$

(18.43)

We write them as vectors parallel ($\parallel$) or perpendicular ($\perp$) to the relative frame velocity $\mathbf{v}$ because it's easier to grasp the relationships in this form than as x, y, z components:

$$E'_x = E_x \qquad\qquad B'_x = B_x$$
$$E'_y = \gamma E_y - \gamma v B_z \qquad B'_y = \gamma B_y + \gamma v E_z/c^2$$
$$E'_z = \gamma E_z + \gamma v B_y \qquad B'_z = \gamma B_z - \gamma v E_y/c^2$$

We will discuss their derivation later in this section, and more formally in Section 18.8.

Maxwell's equations also involve the sources, **J** and ρ. Their transformation properties are similar to those of momentum and energy, which we learned in Section 18.5. We may write

$$J_x = \rho\,\frac{dx}{dt} = \gamma\rho_0\,\frac{dx}{\gamma d\tau} = \rho_0\,\frac{dx}{d\tau} \tag{18.44}$$

where J_x is the component of current density in the x direction, ρ is charge density (of current-carrying charge), ρ_0 is charge density in the rest frame, and τ is proper time. Since ρ_0 and τ are invariant, the transformation of J_x is the same as that of x. Similarly,

$$\rho = \gamma\rho_0\,\frac{dt}{dt} = \rho_0\,\frac{dt}{d\tau}$$

and the charge density ρ is found to be analogous to time. Thus **J** and $c\rho$ constitute another four-vector and they transform via the Lorentz transformation:

$$\begin{aligned}
J'_x &= \gamma J_x - \beta\,\gamma\,c\rho \\
J'_y &= J_y \\
J'_z &= J_z \\
c\rho' &= \gamma c\rho - \beta\,\gamma\,J_x
\end{aligned} \tag{18.45}$$

Given the similarity between the equation of charge continuity,

$$\nabla\cdot\mathbf{J} + \frac{\partial\rho}{\partial t} = 0$$

and the Lorentz condition,

$$\nabla\cdot\mathbf{A} + \frac{\partial V}{c^2\partial t} = 0$$

it is to be expected that **A** and V/c also constitute a four-vector, and it is in fact true.

◢ EXAMPLE 18-14

Find $E'_\perp$ (called E) in Figure 18.24b, starting with Figure 18.24a, by **(a)** field transformation, and by **(b)** source transformation. (Assume current I and distance r.)

ANSWER **(a)** In Figure 18.24a the magnetic field is $B_\perp = \mu_0 I/2\pi r$. Applying the E transformation equation (Eq. 18.43), with $E_\perp = 0$, we find

$$E'_\perp = 0 + \gamma v B_\perp = \gamma v\,\mu_0 I/2\pi r$$

(b) Applying the source transformation (Eq. 18.45), with $\rho = 0$, we obtain

$c\rho' = 0 - \beta\,\gamma\,J_x,$ so, multiplying by the area of the wire,

$c\lambda' = -\beta\,\gamma\,I$ (the wire is negatively charged in the primed frame)

so

$$E' = \lambda'/2\pi\epsilon_0 r = \beta\,\gamma\,I/2\pi\epsilon_0 cr = \gamma\,\text{v}\,\mu_0 I/2\pi r \qquad \text{as above}$$

using $\epsilon_0\mu_0 c^2 = 1$. So the two different procedures give the same answer.

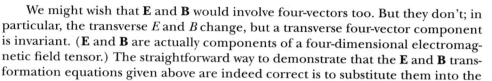

We might wish that **E** and **B** would involve four-vectors too. But they don't; in particular, the transverse E and B change, but a transverse four-vector component is invariant. (**E** and **B** are actually components of a four-dimensional electromagnetic field tensor.) The straightforward way to demonstrate that the **E** and **B** transformation equations given above are indeed correct is to substitute them into the Maxwell equations. When you do so, you wind up with Maxwell's equations again, with primes on everything. Maxwell's equations don't change form under this transformation: They are invariant. But this is not so easy to carry through.

Instead, we shall introduce examples that demonstrate the **E**, **B** transformations with particular clarity; these may have almost the force of proofs, once we accept that the transformation properties of given **E** and **B** fields do not depend on where those fields came from. We shall deal with parallel and perpendicular components separately.

First, consider a pair of parallel plates of uniform surface charge density $\pm\sigma$; in Figure 18.26a the pair is at rest, and in Figure 18.26b we view it in a frame that is moving to the right at $+\mathbf{v}$, relative to Figure 18.26a, so the plates are moving to the left at $\mathbf{v}'$ in the primed frame. The charge density doesn't change, and the spacing doesn't matter, so we find immediately that

$$\mathbf{E}'_{\parallel} = \mathbf{E}_{\parallel} \tag{18.46}$$

which is the first equation in the group called Equation 18.43.

For parallel plates carrying surface current **K**, Figure 18.27, the surface current doesn't change when we move parallel to **B**. This follows from the transformation of **J** and $c\rho$ given above: $J'_y = J_y$ and $J'_z = J_z$. Alternatively, we could argue that the speed v_y of the charges is reduced by a factor of γ (Section 18.3; it goes back to the time dilation), and the density of charge carriers goes up by the same factor (due to Lorentz contraction), so the net effect is that the current density $\mathbf{K} = \mathbf{K}'$ doesn't change. In any case, we arrive at

$$\mathbf{B}'_{\parallel} = \mathbf{B}_{\parallel} \tag{18.47}$$

which is another part of Equation 18.43.

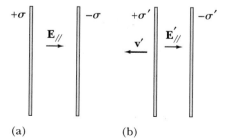

(a)   (b)

Figure 18.26   **(a)** Parallel plates, stationary. **(b)** Same, moving.

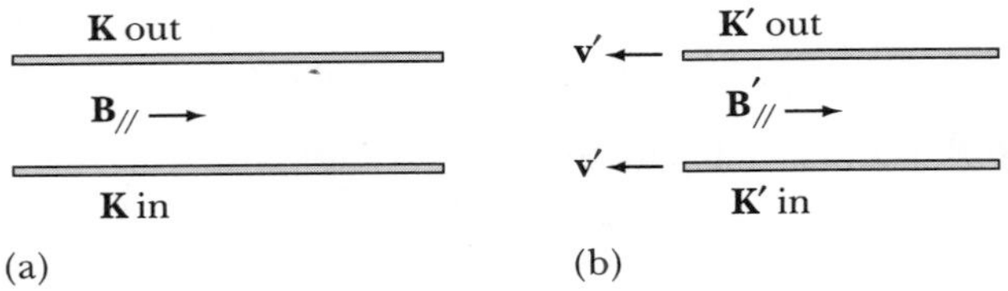

Figure 18.27   (a) Parallel plates, stationary. (b) Same, moving.

And now consider charged plates parallel to the direction of motion, Figure 18.28a. In the primed frame, Figure 18.28b, we can find the charge and current using the J, $c\rho$ transformation again. With $J_x = 0$, we find

$$J'_x = 0 - \beta \gamma c\rho$$

Each plate has a thickness, but we needn't specify it explicitly because we're already about to get rid of it, thus: when we multiply both sides by the plate thickness, we get

$$K'_x = -\gamma \sigma v$$

This result can also be explained as follows: In the primed frame, σ' is larger than σ due to the Lorentz contraction, and when a charge density σ' moves, it introduces a current density $K'_x = K' = \sigma'v'$. This K' results in a field

$$B'_\perp = \mu_0 K' = -\mu_0 \gamma \sigma v = -\mu_0 \gamma \epsilon_0 E_\perp v$$

so

$$\mathbf{B}'_\perp = -\gamma \mathbf{v} \times \mathbf{E}/c^2 \qquad (18.48b)$$

due to movement across a transverse $\mathbf{E}$ field. (Equation 18.48b provides part of one equation in Equation 18.43; the other part will be supplied by Equation 18.48a, which follows.) Remember that $\mathbf{v} = -\mathbf{v}'$: The primed frame is moving at $\mathbf{v}$ to the right in the unprimed frame. Note also that the $\mathbf{E}$ in the cross product no longer needs the $\perp$ symbol, because $\mathbf{v} \times \mathbf{E}_\parallel = 0$ always.

To find out about $E_\perp$ we invoke another J, ρ transformation, still for $J_x = 0$:

$$c\rho' = \gamma c\rho - 0$$

The charge density increases (this is due to Lorentz contraction), so we have

$$\mathbf{E}'_\perp = \gamma \mathbf{E}_\perp \qquad (18.49a)$$

due to transverse $\mathbf{E}$ field.

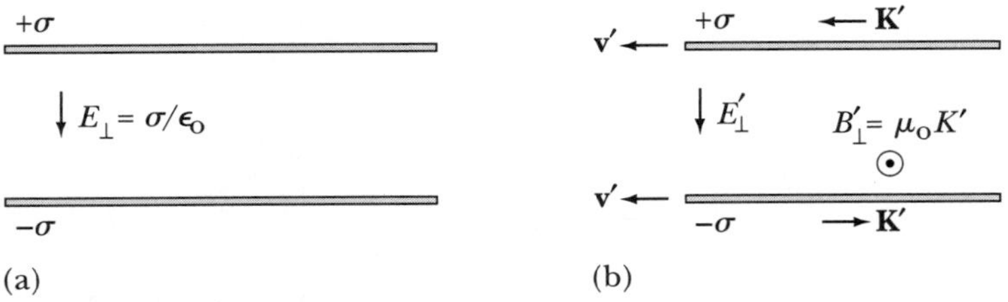

Figure 18.28   (a) Parallel plates, stationary. (b) Same, in primed frame.

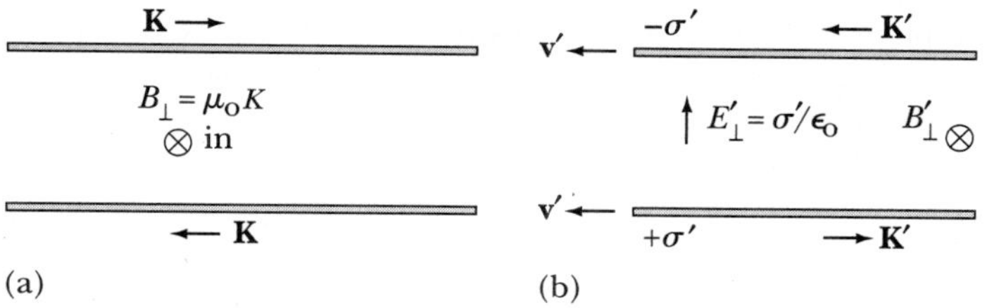

$K \longrightarrow$

$B_\perp = \mu_0 K$
$\otimes$ in

$\mathbf{v} \longleftarrow$ $-\sigma'$ $\longleftarrow K'$

$\uparrow E'_\perp = \sigma'/\epsilon_0$ $B'_\perp \otimes$

$\longleftarrow K$

$\mathbf{v} \longleftarrow$
$+\sigma'$ $\longrightarrow K'$

(a)   (b)

Figure 18.29   (a) Parallel plates, stationary. (b) Same, in primed frame.

If instead the stationary plates are uncharged and carry current, Figure 18.29a, an E' field arises in the primed frame, Figure 18.29b. The charge transformation we need is, with $\rho = 0$

$$c\rho' = 0 - \beta\gamma J_x$$

Multiplying both sides by the plate thickness as before, we find

$$c\sigma' = -\beta\gamma K$$

so

$$c\epsilon_0 E_\perp' = -\gamma v B_\perp/c\mu_0$$

$$E'_\perp = -\gamma \mathbf{v} \times \mathbf{B}_\perp \tag{18.49b}$$

due to motion across a transverse $\mathbf{B}$ field. As for $B'_\perp$, we have (with $\rho = 0$)

$$J'_x = \gamma J_x - 0$$

so

$$\mathbf{B}'_\perp = \gamma\mathbf{B}_\perp \tag{18.48a}$$

Putting together the preceding results, we find

$$\mathbf{B}'_\perp = \gamma\mathbf{B}_\perp - \gamma\mathbf{v} \times \mathbf{E}/c^2 \text{ and } \mathbf{E}'_\perp = \gamma\mathbf{E}_\perp + \gamma\mathbf{v} \times \mathbf{B} \tag{18.50}$$

And this completes the derivation of Equation 18.43 using simple models. If you still want something more formal, see Section 18.8.

Particle trajectories in **crossed E and B fields** provide a particularly pretty illustration of the field transformations. In Figure 18.30a a positive charge is released from rest in a region where $\mathbf{E}$ is perpendicular to $\mathbf{B}$, and it proceeds to describe a

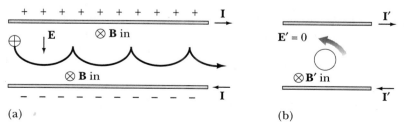

(a)   (b)

Figure 18.30   (a) Crossed **E** and **B**. (b) No **E′**.

cycloid. However, let us move into a reference frame in which $E' = 0$, Figure 18.30b. Then we have merely a charge moving in a circle in a uniform magnetic field. The equation is

$$\mathbf{E}'_\perp = \gamma \mathbf{E}_\perp + \gamma \mathbf{v} \times \mathbf{B} = 0 \qquad (18.51)$$

so

$$\mathbf{E}_\perp = -\mathbf{v} \times \mathbf{B}$$

Therefore, in order to accomplish this obliteration of $\mathbf{E}'$, we must move in such a way that $v = E/B$. Also, $\mathbf{v}$ has to be perpendicular to both fields; otherwise there is an $E_\parallel$ or $B_\parallel$ that doesn't go away. Is this always possible? Alas no, for two reasons:

First, if $\mathbf{E}$ is not perpendicular to $\mathbf{B}$, then we cannot make $\mathbf{E} = -\mathbf{v} \times \mathbf{B}$ no matter how hard we try. This condition is related to one of the invariants of the electromagnetic field:

$$\mathbf{E}' \cdot \mathbf{B}' = \mathbf{E} \cdot \mathbf{B} \qquad (18.52)$$

which is to say, $\mathbf{E} \cdot \mathbf{B}$ is invariant. This is easily demonstrated by substituting the field transformation conditions into the equation. Now if it is our desire to transform away either $\mathbf{E}$ or $\mathbf{B}$, we can do so only from a frame where $\mathbf{E} \cdot \mathbf{B} = 0$ already, that is, where $\mathbf{E}$ is perpendicular to $\mathbf{B}$. (The invariants of Eq. 18.52, here, and 18.53, below, are derived in Problems 18-29 and 18-30.)

Secondly, if E is large or B small, we could face the requirement that $v > c$, which cannot be met. The related invariant is

$$E^2 - c^2 B^2 \qquad (18.53)$$

which can be multiplied by $\frac{1}{2}\epsilon_0$ and written as $\frac{1}{2}\epsilon_0 E^2 - \frac{1}{2}B^2/\mu_0$, making it more readily identifiable as the difference (not the sum) of the electric and magnetic field energies. Now if there is more electric than magnetic energy, then we just aren't going to be able to transform $\mathbf{E}$ away, because then the magnetic energy would have to go negative. (But we might be able to transform the $\mathbf{B}$ field away, if that helps.)

◢ EXAMPLE 18-15

This instructive exercise will review several aspects of electromagnetism. Uncharged parallel plates carrying current $\pm K$ move at velocity $\mathbf{v}$ perpendicular to their surfaces, Figure 18.31. Find the $\mathbf{E}$ and $\mathbf{B}$ fields **(a)** directly from Maxwell's equations, and **(b)** using the field transformations.

ANSWER All fields are $\perp$, so this symbol will be suppressed.

(a) At each plate, the magnetic field $\mathbf{B}$ is changing (for a stationary observer). Faraday's law is $\nabla \times \mathbf{E} = -\partial \mathbf{B}/\partial t$. For the stationary loop shown, of length l,

$$\oint \mathbf{E} \cdot d\mathbf{l} = Bl\, dx/dt$$

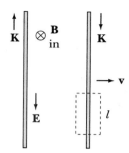

Figure 18.31   Parallel plates, moving.

so

$$E = Bv$$

in the direction shown (according to Lenz's law).

At each plate, the **E** field is changing. Ampère's law is

$$\nabla \times \mathbf{B} = \mu_0 \mathbf{J} + \mu_0 \epsilon_0 \partial \mathbf{E} / \partial t$$

For a loop perpendicular to the page,

$$\oint \mathbf{B} \cdot d\mathbf{l} = \mu_0 K l + \mu_0 \epsilon_0 E l \, dx / dt$$

$$B = \mu_0 K + \frac{v^2}{c^2} B$$

$$B(1 - \beta^2) = \mu_0 K$$

So

$$\boxed{B = \gamma^2 \mu_0 K \qquad \text{and} \qquad E = \gamma^2 \mu_0 K v} \qquad (18.54)$$

(b) We transform to the primed rest frame of the plates. The plates are still uncharged, since the Lorentz transformation for current and charge density says

$$c\rho' = \gamma c\rho - \beta \gamma J_x = 0$$

And there is no changing B' field. Consequently there is no E' field, and the transformation equations become (with $\mathbf{v}' = -\mathbf{v}$)

$$E = \gamma E' + \gamma \mathbf{v}' \times \mathbf{B}' \qquad \longrightarrow \qquad E = \gamma \mathbf{v}' \times \mathbf{B}'$$

$$B = \gamma B' - \gamma \mathbf{v}' \times \mathbf{E}' / c^2 \qquad \longrightarrow \qquad B = \gamma B'$$

The current density is unchanged: $J'_y = J_y$ and so forth. But the plates are thicker by a factor of γ, because they are no longer moving, so they carry more current. So,

$$K' = \gamma K$$

(This may also be regarded as an effect of the transverse velocity transformation, Section 18.3.) So the magnetic field is

$$B' = \mu_0 K' = \gamma \mu_0 K$$

Substituting in the above transformation equations, we immediately find

$$B = \gamma^2 \mu_0 K \quad \text{and} \quad E = \gamma^2 \mu_0 Kv$$

which is in agreement with part (a).

Note that Poynting's vector, $\mathbf{S} = \mathbf{E} \times \mathbf{B}/\mu_0$, points in the direction of $\mathbf{v}$, which is the direction of transportation of the field energy (Section 14.1).

18.8. Electromagnetic Field Tensor

We shall now undertake the expression of electromagnetism in four-dimensional tensor form. There is little new physics to be gleaned thereby, but it will increase our understanding of the interrelationship between electricity and magnetism, and it will underscore the relativistic covariance of Maxwell's formulation: It's consistent with relativity.

We have been writing the Lorentz transformation in the following form (Eq. 18.3):

$$x' = \gamma x - \beta \gamma\, ct$$

$$y' = y$$

$$z' = z$$

$$ct' = \gamma ct - \beta \gamma\, x$$

But now the notation can be made more consistent if we make the time coordinate imaginary: We will number the four coordinates:

$$x \rightarrow x_1, \quad y \rightarrow x_2, \quad z \rightarrow x_3, \quad \text{and} \quad ict \rightarrow x_4$$

Then we will rewrite the Lorentz transformation, using Greek letter subscripts for four-dimensional coordinates:

$$x'_{\mu} = a_{\mu\nu} x_{\nu} \tag{18.55}$$

This represents four equations, one for each value of μ. Summation is implied over the repeated index, ν (Section 1.6). The Lorentz-transformation coefficients can be written in matrix form as:

$$a_{\mu\nu} = \begin{array}{c} \nu = 1 \quad 2 \quad 3 \quad 4 \\ \begin{vmatrix} \gamma & 0 & 0 & i\beta\gamma \\ 0 & 1 & 0 & 0 \\ 0 & 0 & 1 & 0 \\ -i\beta\gamma & 0 & 0 & \gamma \end{vmatrix} \end{array} \tag{18.56}$$

As remarked in Section 18.2, a Lorentz transformation is like a rotation in four-dimensional space-time.

EXAMPLE 18-16

Let's transform the time coordinate, x_4. Setting $\mu = 4$:

$$x'_4 = a_{4\nu}x_\nu$$
$$= a_{41}x_1 \quad + a_{42}x_2 + a_{43}x_3 + a_{44}x_4$$
$$= -i\beta\gamma x_1 + \quad 0 \quad + \quad 0 \quad + \gamma x_4$$

So

$$ict' = -i\beta\gamma x + \gamma ict$$

$$ct' = \gamma ct - \beta\gamma x$$

So this procedure gives the same result as previously, Equation 18.3.

The potential four-vector becomes A_1, A_2, A_3, and $A_4 = iV/c$. The transformation is the same as for x_μ. The Lorentz condition may be written as

$$0 = \nabla \cdot \mathbf{A} + \frac{1}{c^2}\frac{\partial V}{\partial t} = \sum_{\mu=1}^{4}\frac{\partial A_\mu}{\partial x_\mu} \rightarrow \frac{\partial A_\mu}{\partial x_\mu} = \Box \cdot \mathbf{A} \tag{18.57}$$

We have introduced a natural generalization of the ∇ operator, a kind of four-dimensional divergence; with unit vectors omitted,

$$\Box = \frac{\partial}{\partial x} + \frac{\partial}{\partial y} + \frac{\partial}{\partial z} + \frac{\partial}{\partial ict}$$

The square of this operator is the "d'Alembertian":

$$\Box^2 = \nabla^2 - \frac{1}{c^2}\frac{\partial^2}{\partial t^2} = \frac{\partial^2}{\partial x_\mu \partial x_\mu}$$

analogous to the three-dimensional Laplacian. However, the square notation is not much used in practice.

Correspondingly, the current four-vector is J_1, J_2, J_3, and $J_4 = ic\rho$, so the equation of charge continuity, Equation 5.6, is

$$0 = \nabla \cdot \mathbf{J} + \frac{\partial\rho}{\partial t} = \frac{\partial J_\mu}{\partial x_\mu} \tag{18.58}$$

The wave equations for the potentials, Equations 14.21 and 14.23,

$$\nabla^2\mathbf{A} - \frac{1}{c^2}\frac{\partial^2\mathbf{A}}{\partial t^2} = -\mu_0\mathbf{J} \quad \text{and} \quad \nabla^2 V - \frac{1}{c^2}\frac{\partial^2 V}{\partial t^2} = -\frac{\rho}{\epsilon_0}$$

may be condensed into the following:

$$\frac{\partial^2 A_\mu}{\partial x_\nu \partial x_\nu} = -\mu_0 J_\mu \tag{18.59}$$

Now we have just found (Eq. 18.57) that the four-dimensional "divergence" of the potential is zero, a scalar. It is reasonable to ask, what is its "curl"? We will use the symbol $\mathbf{F}$:

$$\text{"}\square \times \mathbf{A}\text{"} = \mathbf{F}$$

In three dimensions, the curl is actually a second-rank tensor masquerading as a vector, Section 1.6. It has only three independent components, and we can ascribe these to the three coordinates by means of an arbitrary right-hand rule. Unfortunately, the four-dimensional "curl" has six independent components, and there are only four coordinate axes, so we are not going to be able to pretend that it is a vector any more; we must deal with it as a second-rank tensor. As in three dimensions, the $\mathbf{F}$ components are of the form:

$$F_{\mu\nu} = \frac{\partial A_\nu}{\partial x_\mu} - \frac{\partial A_\mu}{\partial x_\nu} \tag{18.60}$$

We must evaluate the $F_{\mu\nu}$'s one by one.

For $\mu = \nu = 1$ it's easy: $F_{11} = \dfrac{\partial A_1}{\partial x_1} - \dfrac{\partial A_1}{\partial x_1} = 0.$

For $\mu = 2, \nu = 1$: $F_{21} = \dfrac{\partial A_1}{\partial x_2} - \dfrac{\partial A_2}{\partial x_1} = \dfrac{\partial A_x}{\partial y} - \dfrac{\partial A_y}{\partial x} = -B_z$, since $\mathbf{B} = \nabla \times \mathbf{A}.$

For $\mu = 1, \nu = 2$:
$F_{12} = -F_{21} = B_z$. This tensor is antisymmetric (if you exchange indices, you change the sign).

For $\mu = 4, \nu = 1$:
$$F_{41} = \frac{\partial A_1}{\partial x_4} - \frac{\partial A_4}{\partial x_1} = \frac{\partial A_x}{\partial ict} - \frac{\partial iV/c}{\partial x} = \frac{iE_x}{c}, \text{ since } \mathbf{E} = -\nabla V - \frac{\partial \mathbf{A}}{\partial t}.$$

And so forth. The final result is the **electromagnetic field tensor:**

$$F_{\mu\nu} = \begin{matrix} \nu = & 1 & 2 & 3 & 4 \\ & \begin{vmatrix} 0 & B_z & -B_y & -iE_x/c \\ -B_z & 0 & B_x & -iE_y/c \\ B_y & -B_x & 0 & -iE_z/c \\ iE_x/c & iE_y/c & iE_z/c & 0 \end{vmatrix} \end{matrix} \tag{18.61}$$

It turns out to be easy to write Maxwell's equations in terms of this tensor:

$$\frac{\partial F_{\mu\nu}}{\partial x_\nu} = \mu_0 J_\mu \tag{18.62}$$

and

$$\frac{\partial F_{\mu\nu}}{\partial x_\lambda} + \frac{\partial F_{\nu\lambda}}{\partial x_\mu} + \frac{\partial F_{\lambda\mu}}{\partial x_\nu} = 0 \tag{18.63}$$

Summation over ν is implied in the first of these, Equation 18.62; but there is no summation in Equation 18.63, because there is no index repeated in any given term. To see how these equations work, let us set $\mu = 1$. Then Equation 18.62 becomes:

$$\frac{\partial F_{11}}{\partial x_1} + \frac{\partial F_{12}}{\partial x_2} + \frac{\partial F_{13}}{\partial x_3} + \frac{\partial F_{14}}{\partial x_4} = \mu_0 J_1$$

$$\frac{\partial 0}{\partial x} + \frac{\partial B_z}{\partial y} - \frac{\partial B_y}{\partial z} - \frac{\partial i E_x / c}{\partial ict} = \mu_0 J_x$$

which is the x component of Ampère's law, $\nabla \times \mathbf{B} = \mu_0 \mathbf{J} + \dfrac{1}{c^2} \dfrac{\partial \mathbf{E}}{\partial t}$.

In Equation 18.63, with $\mu = 1$, we will also set $\nu = 2$ and $\lambda = 3$. Then

$$\frac{\partial F_{12}}{\partial x_3} + \frac{\partial F_{23}}{\partial x_1} + \frac{\partial F_{31}}{\partial x_2} = 0$$

$$\frac{\partial B_z}{\partial z} + \frac{\partial B_x}{\partial x} + \frac{\partial B_y}{\partial y} = 0$$

which affirms that $\nabla \cdot \mathbf{B} = 0$. And in fact all of Maxwell's equations are there in Equations 18.62 and 18.63.

The $F_{\mu\nu}$ tensor may be transformed according to the usual rules, Section 1.6, extended to a fourth dimension:

$$F'_{\mu\nu} = a_{\mu\kappa} a_{\nu\lambda} F_{\kappa\lambda} \tag{18.64}$$

which involves a double summation, of course, over κ and λ. So, let us begin.

$\underline{F'_{11}}$: We set $\mu = 1$ and $\nu = 1$, and we refer to Equations 18.56 and 18.61; then

$$F'_{11} = a_{1\kappa} a_{1\lambda} F_{\kappa\lambda} = a_{1\kappa} (a_{11} F_{\kappa 1} + a_{12} F_{\kappa 2} + a_{13} F_{\kappa 3} + a_{14} F_{\kappa 4})$$

$$= a_{1\kappa} (\gamma F_{\kappa 1} + 0\ F_{\kappa 2} + 0\ F_{\kappa 3} + i\beta\gamma F_{\kappa 4})$$

$$= a_{11} (\gamma F_{11} + i\beta\gamma F_{14}) \quad + a_{12} (\gamma F_{21} + i\beta\gamma F_{24}) + a_{13} (\gamma F_{31} + i\beta\gamma F_{34})$$

$$+ a_{14} (\gamma F_{41} + i\beta\gamma F_{44})$$

$$= \gamma (\gamma 0 + i\beta\gamma (-iE_x / c)) + \quad\quad 0\ (\ ) \quad\quad + \quad\quad 0\ (\ )$$

$$+ i\beta\gamma (\gamma iE_x / c + i\beta\gamma 0)$$

so

$$F'_{11} = \gamma i\beta\gamma (-iE_x / c) + i\beta\gamma \gamma iE_x / c = \boxed{0}$$

$\underline{F'_{12}}$: We set $\mu = 1$ and $\nu = 2$; then

$$F'_{12} = a_{1\kappa} a_{2\lambda} F_{\kappa\lambda} = a_{1\kappa} (a_{21} F_{\kappa 1} + a_{22} F_{\kappa 2} + a_{23} F_{\kappa 3} + a_{24} F_{\kappa 4})$$

$$= a_{1\kappa} (0\ F_{\kappa 1} + 1\ F_{\kappa 2} + 0\ F_{\kappa 3} + 0\ F_{\kappa 4})$$

$$= a_{11} F_{12} + a_{12} F_{22} + a_{13} F_{32} + a_{14} F_{42}$$

$$= \gamma B_z \quad + 0\ (\ ) + 0\ (\ ) + i\beta\gamma iE_y / c$$

so

$$F'_{12} = \boxed{\gamma\, B_z - \beta\gamma\, E_y/c}$$

Only fourteen to go. When we finish, the result is:

$$
F'_{\mu\nu} = \begin{vmatrix}
\overset{\nu=1}{0} & \overset{2}{\gamma B_z - \beta\gamma E_y/c} & \overset{3}{-\gamma B_y - \beta\gamma E_z/c} & \overset{4}{-iE_x/c} \\
-\gamma B_z + \beta\gamma E_y/c & 0 & B_x & -i\gamma E_y/c + i\beta\gamma B_z \\
\gamma B_y + \beta\gamma E_z/c & -B_x & 0 & -i\gamma E_z/c - i\beta\gamma B_y \\
iE_x/c & i\gamma E_y/c - i\beta\gamma B_z & i\gamma E_z/c + i\beta\gamma B_y & 0
\end{vmatrix}
$$

$$(18.65)$$

Comparing this with Equation 18.61, we find that the transformed $\mathbf{E}$ and $\mathbf{B}$ fields are indeed given by the electromagnetic field transformation presented in Section 18.7, Equation 18.43:

$$\mathbf{E}'_{\parallel} = \mathbf{E}_{\parallel} \qquad \mathbf{E}'_{\perp} = \gamma\,\mathbf{E}_{\perp} + \gamma\mathbf{v}\times\mathbf{B}$$

$$\mathbf{B}'_{\parallel} = \mathbf{B}_{\parallel} \qquad \mathbf{B}'_{\perp} = \gamma\,\mathbf{B}_{\perp} - \gamma\mathbf{v}\times\mathbf{E}/c^2$$

18.9. Fields of a Uniformly Moving Charge

This book began near the subject of the fields of simple charges, and it ends at more or less the same topic, but with a deeper understanding of what we were missing in terms of the relativistic interrelationship of the electric and magnetic fields.

   Of course, we have already seen the field of an arbitrarily moving charge in Section 14.4. But here we look at a uniformly moving charge in more detail.

   The E field of a stationary charge is, to an extraordinary degree of accuracy,

$$E = \frac{Q}{4\pi\epsilon_0 r^2}$$

entirely in the radial direction, Figure 18.32a. And with corresponding accuracy,

$$B = 0$$

(for a simple charge having no spin). Using that information, we can find the fields of uniformly moving charges by applying the $\mathbf{E}$ and $\mathbf{B}$ transformation equations (Eq. 18.43):

$$\mathbf{E}'_{\parallel} = \mathbf{E}_{\parallel} \qquad \mathbf{E}'_{\perp} = \gamma\,\mathbf{E}_{\perp} + \gamma\mathbf{v}\times\mathbf{B}$$

$$\mathbf{B}'_{\parallel} = \mathbf{B}_{\parallel} \qquad \mathbf{B}'_{\perp} = \gamma\,\mathbf{B}_{\perp} - \gamma\mathbf{v}\times\mathbf{E}/c^2$$

Remember that $\mathbf{v}$ is the speed of the primed frame in the unprimed frame, and so $v' = -v$.

   First we'll look at a couple of easy points. In the direction of motion, we find immediately

$$\mathbf{E}'_{\parallel} = \mathbf{E}_{\parallel} \qquad\qquad (18.66)$$

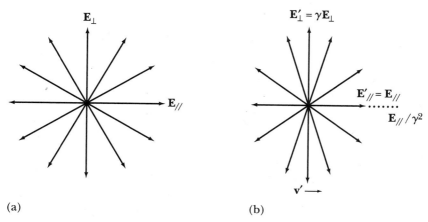

Figure 18.32 (a) Stationary charge. (b) Moving charge.

Yes but where? The answer is, this gives the field $\mathbf{E'}_{\parallel}$ at the Lorentz-contracted position, as indicated at the right in Figure 18.32b. For instance, if we place the charge at one end of a meter stick in the unprimed frame, and the point of observation at the other end, then in the primed frame the meter stick is Lorentz-contracted, and the observation point is closer to the charge. If on the other hand we want the field at the same distance $r' = r$ in the primed frame, then $\mathbf{E'}_{\parallel}$ will be down by a factor of γ^2 because of the inverse-square law, as suggested in Figure 18.32b.

At right angles to the motion we find

$$\mathbf{E'}_{\perp} = \gamma\,\mathbf{E}_{\perp} \qquad (18.67)$$

since $B = 0$. So the $\mathbf{E'}$ field lines are squeezed in the direction of $\mathbf{v}$, as shown in Figure 18.32b. It is as if we drew the field lines on a piece of paper and then moved the paper: The pattern is Lorentz-contracted.

We shall now obtain expressions for the $\mathbf{E}$ and $\mathbf{B}$ fields of a simple charge Q moving at uniform speed $\mathbf{v}$ in the x direction. Because of symmetry about the x axis, we can place the point of observation, (x, y) or (r, θ), in the xy plane.

The charge is stationary in Figure 18.33a, and in Figure 18.33b it is moving to the right along the x' axis at speed v'. (We use v' here to distinguish this velocity from the one in the transformation equations. In order for the charge to move at v', to the right, in the primed frame, the primed frame must move at speed $v = -v'$, to the left, in the unprimed frame.)

First we find the direction of $\mathbf{E'}$. Since $x' = x/\gamma$ (Lorentz contraction), and $y' = y$, we find that

$$\frac{y'}{x'} = \tan\theta' = \gamma \tan\theta$$

And from the transformation equations in Equation 18.43, with $\mathbf{B} = 0$, we find that $E'_x = E_x$, and $E'_y = \gamma E_y$, and this too yields

$$\frac{E'_y}{E'_x} = \tan\theta' = \gamma \tan\theta$$

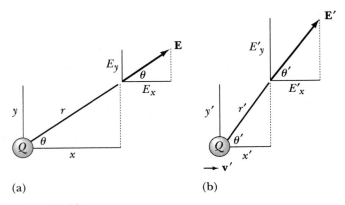

Figure 18.33

So E' is parallel to r'. Since $\mathbf{E}$ is radial in the unprimed frame, $\mathbf{E}'$ is radial in the primed frame: it points directly away from Q.

Now we will find the magnitude of $\mathbf{E}'$. We know what the field is in the un-primed frame, so we will express E' in terms of unprimed variables, and then transform back to the primed frame.

$$E' = \sqrt{E'^2_x + E'^2_y} = \sqrt{E^2_x + (\gamma E_y)^2} \tag{18.68}$$

In the unprimed frame we have

$$E = \frac{Q}{4\pi\epsilon_0 r^2}$$

and

$$E_x = E \cos\theta = E\frac{x}{r} \qquad \text{and} \qquad E_y = E \sin\theta = E\frac{y}{r}$$

So,

$$E' = \frac{Q}{4\pi\epsilon_0 r^3}\sqrt{x^2 + (\gamma y)^2}$$

$$= \frac{Q}{4\pi\epsilon_0} \frac{1}{(x^2 + y^2)^{3/2}}\sqrt{x^2 + (\gamma y)^2}$$

That's the magnitude of E' in terms of quantities in the unprimed frame. Now we go back to the primed coordinates: $x \rightarrow \gamma x'$ and $y \rightarrow y'$.

$$E' = \frac{Q}{4\pi\epsilon_0} \frac{1}{((\gamma x')^2 + y'^2)^{3/2}}\sqrt{(\gamma x')^2 + (\gamma y')^2}$$

$$= \frac{Q}{4\pi\epsilon_0\gamma^3} \frac{\gamma r'}{\left(x'^2 + \dfrac{y'^2}{\gamma^2}\right)^{3/2}} = \frac{Q}{4\pi\epsilon_0\gamma^2} \frac{r'}{(x'^2 + y'^2(1 - \beta^2))^{3/2}} \tag{18.69}$$

We can set $y' = r' \sin \theta'$, and collect factors of r', arriving finally at

$$E' = \frac{Q}{4\pi\epsilon_0 r'^2 \gamma^2 (1 - \beta^2 \sin^2\theta')^{3/2}}$$

(18.70)

   Let's check it with what we said earlier in connection with Figure 18.32b. At $\theta = 0°$, Equation 18.70 becomes

$$E' = \frac{Q}{4\pi\epsilon_0 x'^2 \gamma^2}$$

so as observed before, at a given distance the $E'_{\parallel}$ field is reduced by a factor of γ^2. And at $\theta = 90°$, we get

$$E' = \frac{\gamma Q}{4\pi\epsilon_0 y'^2}$$

(18.71)

so $E'_{\perp}$ is greater due to the motion, as before.
   Once we have E', B' is easy. From Equation 18.43,

$$\mathbf{E'}_{\parallel} = \mathbf{E}_{\parallel} \qquad \mathbf{E'}_{\perp} = \gamma \, \mathbf{E}_{\perp} + \gamma \mathbf{v} \times \mathbf{B}$$

$$\mathbf{B'}_{\parallel} = \mathbf{B}_{\parallel} \qquad \mathbf{B'}_{\perp} = \gamma \, \mathbf{B}_{\perp} - \gamma \mathbf{v} \times \mathbf{E}/c^2$$

With $\mathbf{B} = 0$, we find $B'_x = B'_{\parallel} = 0$, and so

$$\mathbf{B'} = -\gamma \mathbf{v} \times \mathbf{E}/c^2 = +\mathbf{v'} \times \mathbf{E'}/c^2$$

so

$$B' = \frac{\mu_0 Q v' \sin \theta'}{4\pi r'^2 \gamma^2 (1 - \beta^2 \sin^2\theta')^{3/2}}$$

(18.72)

This $\mathbf{B'}$ field is perpendicular to the paper in Figures 18.32b and 18.33b, with direction given by the usual right-hand rule. When v is small, we find (omitting the primes now)

$$B = \frac{\mu_0 Q v \sin \theta}{4\pi r^2}$$

(18.73)

as suggested by the Biot-Savart law, Section 3.2.

 **EXAMPLE 18-17**

Find the electric force, the magnetic force, and the total force for two identical charges Q moving side-by-side at separation r and speed v, Figure 18.34.

Figure 18.34 Two charges.

ANSWER Since r is transverse, the calculations are trivial:
For stationary charges, by Coulomb's law

$$F = \frac{Q^2}{4\pi\epsilon_0 r^2}$$

When the charges move, the force goes *down:* the *total force* is, relativistically (Section 18.5)

$$F' = F'_\perp = F_\perp/\gamma = \frac{1}{\gamma}\frac{Q^2}{4\pi\epsilon_0 r^2}$$

But the electric field goes *up:*

$$E'_\perp = \gamma E_\perp = \gamma \frac{Q}{4\pi\epsilon_0 r^2}$$

So the *electrical force* goes up:

$$F'_e = QE'_\perp = \gamma \frac{Q^2}{4\pi\epsilon_0 r^2}$$

The *magnetic force* is the difference between the above total force and electrical force:

$$F'_m = \left(\gamma - \frac{1}{\gamma}\right)\frac{Q^2}{4\pi\epsilon_0 r^2} = \gamma\beta^2 \frac{Q^2}{4\pi\epsilon_0 r^2}$$

$$= Qv\,\gamma\frac{\mu_0 Qv}{4\pi r^2} = QvB'$$

which corresponds to Equation 18.72 and to the Biot-Savart law (Section 3.2), with γ, at $\theta = 90°$. In this context, the magnetic force could be regarded as a kind of relativistic correction to the electrical force.

Similarly, we can find the scalar and vector potentials of a moving charge by applying the Lorentz transformation to the V of a stationary charge. The results are

$$V = \frac{Q}{4\pi\epsilon_0 r \sqrt{(1 - \beta^2 \sin^2\theta)}} \tag{18.74}$$

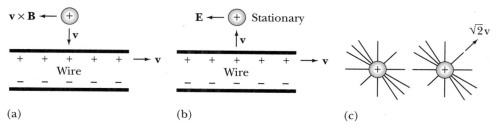

Figure 18.35 (a) Magnetic force. (b) Electrostatic force. (c) The + charges in the wire.

and

$$A = \frac{v}{c^2} V \qquad (18.75)$$

For small β these formulas revert to the usual ones: $V = Q/4\pi\epsilon_0 r$ and $A = \mu_0 Q\mathbf{v}/4\pi r$. Their derivatives yield **E** and **B** in the usual way: $\mathbf{E} = -\nabla V - \partial\mathbf{A}/\partial t$ and $\mathbf{B} = \nabla \times \mathbf{A}$. (They are the "Liénard-Wiechert" potentials for a uniformly moving charge. There exist more general forms expressed in terms of retarded variables.)

Using the E field of Figure 18.32b we can tie up a couple of loose ends. First, returning to Figure 18.24, we have dealt with a charge moving parallel to the wire; but what if the charge moves *towards* the wire, as in Figure 18.35a?

Again, we can transform away the Lorentz force $Q\mathbf{v} \times \mathbf{B}$: We move downwards along with the charge. Now the charge is stationary, Figure 18.35b, and the wire is moving towards it. An **E** field must spring up, as shown, to replace the lost Lorentz force. But where is the left-right asymmetry that can produce an **E** field pointing to the left?

The answer lies in the contracted **E** field shown back in Figure 18.32b. The positive charges in the wire are moving diagonally, and so their **E** fields are contracted as shown in Figure 18.35c. These fields point predominantly to the left. And that "explains" the Lorentz force in this case.

But see how much easier it is simply to apply $\mathbf{F} = Q\mathbf{v} \times \mathbf{B}$ than to try to follow the various relativistic phenomena involved.

The other loose end involves Figure 4.2b, way back in the chapter on Faraday's law. The wire that carries a current and that simultaneously moves perpendicular to its length has an **E** field component parallel to its length, as in Figure 18.35c. The E field is stronger closer to the wire, and so the emf points as shown in Figure 4.2.

18.10   SUMMARY

We have introduced relativity theory, starting with Einstein's postulates and proceeding through the Lorentz transformation to the relativistic transformation of the **E** and **B** fields.

Special relativistic effects include time dilation, Lorentz contraction, and mass increase; in each case the change is by a factor of

$$\gamma = \frac{1}{\sqrt{1 - \beta^2}} \quad \text{with} \quad \beta = \frac{v}{c} \tag{18.1}$$

The Lorentz transformation connects events in inertial reference frames:

$$x' = \gamma x - \beta \gamma \, ct$$

$$y' = y$$

$$z' = z \tag{18.3}$$

$$ct' = \gamma \, ct - \beta \gamma \, x$$

Relativistic velocity addition formulae are

$$\beta'_{\parallel} = \frac{\beta_{\parallel} - \beta}{1 - \beta\beta_{\parallel}} \quad \text{and} \quad \beta'_{\perp} = \frac{\beta_{\perp}}{\gamma(1 - \beta\beta_{\parallel})} \tag{18.16}$$

Higher clocks run faster; the gravitational frequency shift is

$$\frac{\Delta f}{f} = \frac{ax}{c^2} \tag{18.20}$$

Relativistic momentum p and total energy E are

$$p = mv = \gamma \, m_0 v \tag{18.21}$$

and

$$E = mc^2 = \gamma \, m_0 c^2 \tag{18.22}$$

Four-vectors include displacement and time (x, y, z, ct), four-velocity, energy-momentum $(\mathbf{p}, E/c)$, current and charge density $(\mathbf{J}, c\rho)$, and potential $(\mathbf{A}, V/c)$. They are transformed by the Lorentz transformation.

Force is not part of a four-vector, and it is defined in the usual nonrelativistic way:

$$\mathbf{F} = \frac{d\mathbf{p}}{dt} \tag{18.26}$$

For small speed in the unprimed frame,

$$F'_{\perp} = F_{\perp}/\gamma \tag{18.27}$$

For relativistic calculations involving elementary particles, it is convenient to express quantities in electron-volt units with $c = 1$. We have

$$E^2 = M^2 + P^2 \tag{18.29}$$

with $E = \gamma M$ and $P = \beta\gamma M$. A useful practical formula, with P in eV/c and cBr in MKS units, is

$$P = cBr \tag{18.34}$$

For the relativistic Doppler effect,

$$f' = \frac{f}{\gamma(1 - \beta \cos \theta')} \tag{18.37}$$

The electromagnetic field transformation equations are

$$\mathbf{E}'_{\parallel} = \mathbf{E}_{\parallel} \qquad \mathbf{E}'_{\perp} = \gamma \mathbf{E}_{\perp} + \gamma \mathbf{v} \times \mathbf{B}$$

$$\mathbf{B}'_{\parallel} = \mathbf{B}_{\parallel} \qquad \mathbf{B}'_{\perp} = \gamma \mathbf{B}_{\perp} - \gamma \mathbf{v} \times \mathbf{E}/c^2 \tag{18.43}$$

In the case of crossed $\mathbf{E}$ and $\mathbf{B}$ fields, it is possible to "transform away" either $\mathbf{E}$ or $\mathbf{B}$.

The fields of a moving simple charge are

$$\mathbf{E} = \frac{Q\,\hat{\mathbf{r}}}{4\pi\epsilon_0 r^2 \gamma^2 (1 - \beta^2 \sin^2\theta)^{3/2}} \tag{18.70}$$

and

$$\mathbf{B} = \frac{\mu_0 Q v \sin\theta\,\hat{\boldsymbol{\phi}}}{4\pi r^2 \gamma^2 (1 - \beta^2 \sin^2\theta)^{3/2}} \tag{18.72}$$

⬥ PROBLEMS

18-1 Find the frequency of a clock whose proper frequency is 60 Hz and that travels at 2.40×10^8 m/s.

18-2 At what speed is the length of a meter stick reduced by 1 mm?

18-3 A distant cube moves with $\gamma = 2$. Describe and sketch its appearance to the naked eye when it appears to be moving perpendicular to the line of sight.

18-4 Derive the Lorentz contraction by turning a light-clock on its side (see Figure 18.1).

18-5 The positions of the ends of a meter stick are marked simultaneously ($\Delta t = 0$), in its unprimed rest frame, on an x' axis that is moving at speed β relative to the stick; how far apart ($\Delta x'$) are the marks on the x' axis? What is the difference in time $\Delta t'$ between the marks in that frame?

18-6 For Example 18.2, show that the inverse Lorentz transformation returns the original coordinates.

18-7 A line makes an angle θ with the x axis. What is the angle in the primed frame?

18-8 Using the velocity addition formulas, show that $\beta'_{\parallel}{}^2 + \beta'_{\perp}{}^2 \le 1$ if $\beta_{\parallel}{}^2 + \beta_{\perp}{}^2 \le 1$ (and $\beta^2 < 1$).

18-9 In the Michelson-Morley experiment, Figure 18.11, calculate the fringe shift for terrestrial speed $\beta = 10^{-4}$ relative to the supposed ether. Assume average $\lambda = 550$ nm.

18-10 In Figure 18.10, suppose that the bullet is instead fired upward, in the y direction. Find its β' in the observer's frame.

18-11 Clocks at rest at points A and B are synchronized. A bullet is fired from A toward B, a distance of 40 m, at 600 m/s. In its reference frame, which clock leads, and by how much?
 (a) By Lorentz transformation;
 (b) By gravitational frequency shift during the bullet's acceleration in the gun barrel (ignore earth's gravity).

18-12 At what altitude does a satellite clock keep the same time as a terrestrial one, Figure 18.36? (Ignore terrestrial rotation.)

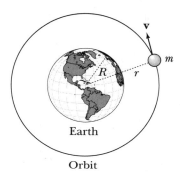

Figure 18.36

18-13 For a K^+ meson of momentum 600 MeV/c and rest mass 494 MeV, what is its speed? If its proper lifetime is 1.24×10^{-8} s, what is its lifetime?

18-14 Colliding-beam particle accelerators provide more center-of-mass energy (but they require careful focussing). **(a)** When two protons each of 30 GeV total energy collide head-on, what energy is available in the center-of-mass system? **(b)** For the same center-of-mass energy, what energy must an incident proton have in a collision with a stationary proton?

18-15 An electron linear accelerator (SLAC, Stanford Linear Accelerator) is 3000 m long. Find the time for an electron to traverse the length of the accelerator at 40 GeV **(a)** in the lab frame and **(b)** in the electron frame. **(c)** Find the length of the accelerator in the electron frame.

***18-16** In the Compton effect, a gamma-ray photon of energy hf collides elastically with a free electron, Figure 18.37. Show that the change in photon wavelength is

$$\Delta\lambda = \frac{h}{m_e c}(1 - \cos\theta).$$ Evaluate for 100 keV gamma rays at 90°.

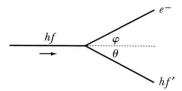

Figure 18.37

18-17 Recalculate the decay of Figure 18.20 on the assumption that the Λ^0 may be a K^0 and the p may be a π^+ meson. Is the result consistent with the known rest mass of the K^0, 498 MeV?

18-18 The Doppler effect partially masks the gravitational red shift in solar spectral lines. Compare the red shift with the Doppler shift at the edge of the sun's disk, on its equator. The sun rotates once every 25 days (at the equator; 35 days at the poles). Ignore terrestrial gravity and motion.

18-19 A source of proper wavelength 500 nm is receding at speed 2×10^8 m/s; what wavelength is observed?

18-20 A motorist is arrested for running a red light. The motorist explains that because of the Doppler effect, the red (630 nm) appeared green (550 nm). The judge changes the charge to speeding and fines the motorist $5 for each km/hr in excess of the speed limit. What is the fine (to two significant figures)?

*18-21 A 600-MeV electron decelerates uniformly in such a way that all of its kinetic energy is radiated as bremsstrahlung. What is the distance over which it decelerates? (Set $\gamma = (m_e c^2 + W)/m_e c^2$.)

18-22 Large parallel conducting plates are separated by 3 cm and charged to a potential difference of 20 kV, Figure 18.38. Find $\mathbf{E}'$, $\mathbf{B}'$, V and $\mathbf{A}'$ between the plates in a frame moving to the right at 340 m/s.

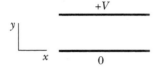

Figure 18.38

18-23 A long solenoid of N' turns per meter carries current I. Find $\mathbf{E}'$ and $\mathbf{B}'$ in a primed reference frame moving at speed v perpendicular to the axis, Figure 18.39. Sketch the solenoid and the fields in the primed frame.

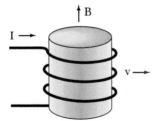

Figure 18.39

18-24 Repeat the preceding problem for speed parallel to the axis.

18-25 For Figure 18.39, set $N' = 10^4$, $I = 2$A, v = 15 m/s, radius 5 cm. Find the magnitude and direction of $\mathbf{E}'$ and $\mathbf{B}'$.

18-26 Large parallel plates carry surface current $K = \pm 10^3$ A/m, Figure 18.31. Their speed is v = 340 m/s perpendicular to the plates, and their separation is 5 cm. Find $\mathbf{E}$ and $\mathbf{B}$ (including direction) between, and outside of, the plates.

18-27 A proton starts from rest in crossed $\mathbf{E}$ and $\mathbf{B}$ fields with $E = 10^6$ V/m and $B = 10^{-3}$ T. Show how to transform to a primed frame in which either E' or B' is zero, and find the magnitude of the nonzero field there. Sketch the proton's trajectory in both frames.

18-28 A velocity selector permits particles to pass if they experience no deflection in its crossed **E** and **B** fields. It is set to select K^+ mesons of momentum 600 MeV/c and rest mass 494 MeV. Its E field is 10^7 V/m; what is its B?

18-29 Show that $\mathbf{E} \cdot \mathbf{B}$ is invariant.

18-30 Show that $E^2 - c^2 B^2$ is invariant.

18-31 Show that $S^2 - c^2 (dW/dv)^2$ is invariant, where **S** is Poynting's vector and dW/dv is electromagnetic energy density. (You can do this simply by showing that it can be constructed using the invariants of the preceding two problems.)

*__18-32__ Show that the velocity **v** of the frame in which **E**′ is parallel to **B**′ and perpendicular to **v** is given by:

$$\frac{\mathbf{v}}{1 + v^2/c^2} = \frac{\mathbf{E} \times \mathbf{B}}{E^2/c^2 + B^2}$$

*__18-33__ Show that when a magnetic dipole **m** moves at speed v $\ll$ c, it is accompanied by an electric dipole moment of $\mathbf{p}' = \mathbf{v} \times \mathbf{m}/c^2$. Sketch the magnetic dipole as a square current loop traveling perpendicular to its axis, and show the accumulations of positive and negative charge.

18-34 Referring to the preceding problem: Show that there is no corresponding magnetic dipole for a moving electric dipole.

18-35 Two parallel line charges of linear charge density λ, separated by r, extend to $\pm\infty$ in the x direction. Find the electric, magnetic, and total force per unit length between them in the primed frame of an observer moving at uniform speed v in the x direction.

*__18-36__ Actually, this is a project rather than a problem. Derive Maxwell's equations from Coulomb's law plus relativity. You might have a look at such texts as Purcell and Lorrain, and the article of Neuenschwander with its references (Appendix B: Bibliography). In this book relevant concepts are found in Sections 2.4, 3.1, 3.3, 4.1, 5.1, 14.4, 18.7, and 18.9. State your simplifying assumptions, to the extent that you yourself recognize them.

FROM *HERBLOCK'S HERE AND NOW* (SIMON & SCHUSTER, 1955)

ANSWERS

Appendix A:
Answers To Odd-Numbered Problems

Chapter 1

1-1 -5; $78\hat{\mathbf{x}} + 13\hat{\mathbf{z}}$; $93.6°$.

1-3 Apply "bac-cab" to each term.

1-5 $(\mathbf{A} - \mathbf{B}) \cdot (\mathbf{A} - \mathbf{B}) = \mathbf{C} \cdot \mathbf{C}$

1-7 Intuitively: If the parallelepiped $\mathbf{A} \cdot \mathbf{B} \times \mathbf{C}$ includes no volume then its vectors are coplanar.

1-9 $(\mathbf{r} - \mathbf{A})$ is perpendicular to $\mathbf{A}$. $A_x (x - A_x) + A_y (y - A_y) + A_z (z - A_z) = 0$

1-11 $\mathbf{A} \times \mathbf{B} + \mathbf{B} \times \mathbf{C} + \mathbf{C} \times \mathbf{A} + (\mathbf{C} - \mathbf{A}) \times (\mathbf{B} - \mathbf{A}) = 0$.

1-13 $\nabla \cdot (\mathbf{A} \times \mathbf{B}) = \mathbf{B} \cdot (\nabla \times \mathbf{A}) - \mathbf{A} \cdot (\nabla \times \mathbf{B}) = 0 + 0 = 0$.

1-15 $\nabla f = (1 - y)\,\hat{\mathbf{x}} - x\hat{\mathbf{y}} + 6z\hat{\mathbf{z}}$.

1-17 $\nabla \times \mathbf{A} = -y\,\hat{\mathbf{z}}$; $\nabla \cdot \mathbf{A} = 1 - x + 6z$.

1-19 $\nabla \cdot r^n \,\hat{\mathbf{r}} = \dfrac{1}{r^2} \dfrac{\partial r^2 r^n}{\partial r} = 0$ for $n = -2$.

1-21 $r\hat{\boldsymbol{\phi}}$ and $r^2\hat{\boldsymbol{\phi}}$ on $r = 1$ sphere.

1-23 $\nabla \cdot y\hat{\mathbf{y}} = 1$, a constant. Thus, the volume integral is $\int \nabla \cdot \mathbf{A}\, dv = s^3$. For the surface integral, the only contribution is on the top surface, so $\int \mathbf{A} \cdot d\mathbf{a} = s^3$.

1-25 Inside front cover: $\int \nabla f\, dv = \oint f\, d\mathbf{a}$, where v is volume. In this case, $f = 1$, so $\nabla f = 0$.

1-27 $1, 2, 3$.

1-29 (a) $\frac{1}{2} = \frac{1}{2}$; (b) $1 \neq 0$; (c) In part (a), $\mathbf{A}$ has zero curl, while in part (b) $\nabla \times \mathbf{A} = \hat{\mathbf{z}}$.

1-31 Write its length as $(x'_i)^2 = (x_i)^2$. Insert $x'_i = c_{ij}x_j = c_{ik}x_k$. Then, $(x'_i)^2 = c_{ij}c_{ik}x_jx_k = \delta_{jk}x_jx_k = (x_j)^2 = (x_i)^2$.

Chapter 2

2-1 $F = 4.7 \times 10^{-4}$ N, attractive.

2-3 $E = \dfrac{Q}{4\pi\epsilon_0 r^2}$, so $Q = -4.5 \times 10^5$ C $\to \sigma = -8.85 \times 10^{-10}$ C/m^2.

2-5 $\dfrac{3 \times 10^{-9}}{4\pi\epsilon_0 x^2} = \dfrac{5 \times 10^{-9}}{4\pi\epsilon_0(0.3 - x)^2}$; $x = 0.131$ m.

2-7 $E = 764$ N/C, at $67.9°$ up from the $+x$ axis.

2-9 $E = 0$ N/C by symmetry.

2-11 $E = \dfrac{Q}{4\pi\epsilon_0 r^2} \cos\theta$ for the z component, so $E = \dfrac{az\lambda}{2\epsilon_0\,(z^2 + a^2)^{3/2}}$ along the z

axis. Check: $E \to 0$ for small z; $E \to \dfrac{2\pi a\lambda}{4\pi\epsilon_0 z^2} = \dfrac{Q}{4\pi\epsilon_0 z^2}$ like a point

charge for $z \gg a$.

2-13 $E = \dfrac{\lambda}{2\pi\epsilon_0 r} = 2400 \text{ N/C.}$

2-15 $\nabla \cdot \mathbf{E} = \rho/\epsilon_0$.

$\int \nabla \cdot \mathbf{E}\, dv = \oint \mathbf{E}\cdot d\mathbf{a} = \int \rho/\epsilon_0\, dv.$

<u>Inside:</u> $4\pi r^2 E_i = \tfrac{4}{3}\pi r^3 \rho/\epsilon_0$

$$E_i = \frac{\rho r}{3\epsilon_0} = \frac{Qr}{4\pi\epsilon_0 R^3}, \text{ since } Q = \tfrac{4}{3}\pi R^3 \rho.$$

<u>Outside:</u> $4\pi r^2 E_o = \tfrac{4}{3}\pi R^3 \rho/\epsilon_0$

$$E_o = \frac{Q}{4\pi\epsilon_0 r^2} = \frac{\rho R^3}{3\epsilon_0 r^2}.$$

2-17   For a Gaussian cylinder of length l:

<u>Inside:</u> $E = \dfrac{1}{2\pi r l\epsilon_0} \displaystyle\int_0^r \dfrac{\sigma}{r}\, 2\pi r l\, dr = \dfrac{\sigma}{\epsilon_0} = \dfrac{\lambda}{2\pi\epsilon_0 R}$, a constant.

<u>Outside:</u> $E = \dfrac{1}{2\pi r l\epsilon_0} \displaystyle\int_0^R \dfrac{\sigma}{r}\, 2\pi r l\, dr = \dfrac{\sigma R}{\epsilon_0 r} = \dfrac{\lambda}{2\pi\epsilon_0 r}.$

2-19 $E = \displaystyle\int \dfrac{\cos\theta\, dQ}{4\pi\epsilon_0 R^2} = \sigma/4\epsilon_0$ (which is half the field of an infinite plane).

2-21   For integration of Coulomb's law we will need the surface charge density $\sigma = Q/4\pi R^2$. Then,

$$E = \int \frac{\cos\varphi\, dQ}{4\pi\epsilon_0 r^2} = \int \frac{(z - R\cos\theta)\sigma 2\pi R\sin\theta\, R d\theta}{4\pi\epsilon_0(z^2 + R^2 - 2zR\cos\theta)^{3/2}}$$

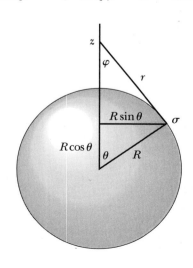

so

$$E_o = \frac{Q}{4\pi\epsilon_o z^2} \quad \text{and} \quad E_i = 0$$

Gauss's law: For E_o we use a Gaussian sphere of radius z containing charge Q; whereas for E_i, $Q = 0$.

2-23 From Section 2.3, $E_i = \dfrac{\lambda}{2\pi\epsilon_o r}$; $E_o = 0$; $\lambda = 5.6 \times 10^{-7} \, \text{C/m}$.

2-25 Superpose a uniform cylinder of radius R, of the same ρ, without a cavity, on another of radius $R/2$, of opposite sign.

$$E = \frac{5\rho R}{12\epsilon_0} \qquad \text{(which is } \tfrac{5}{6} \text{ of } E \text{ without the cavity)}$$

2-27 This figure is an enlargement of Figure 2.34. $\mathbf{A} + \mathbf{B}$ is constant and horizontal, just like the $R/2$ line. And the field we want is $\mathbf{E} = \mathbf{A} + \mathbf{B}$. We can simply evaluate it at point c, the center of the large sphere.

$$E = \frac{\rho R}{6\epsilon_0}, \qquad \text{constant and horizontal}$$

(Or express it in Cartesian components.)

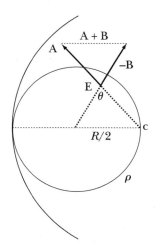

2-29 $\nabla \cdot (x\,\hat{\mathbf{x}}) = 1$; see Figure 2.24c.
$\rho = \epsilon_0 \nabla \cdot \mathbf{E} = 8.85 \times 10^{-12} \, \text{C/m}^3$.

2-31 Surround q with eight one-unit cubes. The total flux of q/ϵ_0 is divided evenly among the 24 outside faces. So, $\Phi_e = q/24\epsilon_0$ or zero.

2-33 $F = \displaystyle\int E \cos\theta \, dQ = \int_0^R \int_0^{\pi/2} \frac{\rho r}{3\epsilon_0} \cos\theta \, \rho \, 2\pi r \sin\theta \, r \, d\theta \, dr = \frac{3}{16} \frac{Q^2}{4\pi\epsilon_0 R^2}$

2-35 By symmetry, along the line shown as "a" there is an E component only in the direction of "a," and we need only consider the effect of the two charges shown here. When we omit Q and $4\pi\epsilon_0$, and set $a = 1$ and $b = \sqrt{2}$, the total for both charges becomes

$$E = \frac{1 + x}{((1 + x)^2 + 2)^{3/2}} - \frac{1 - x}{((1 - x)^2 + 2)^{3/2}}$$

This is most easily dealt with numerically. The result is shown in graphical form. E is always positive to the right of the center, so there is no rest for a positive charge inside the cube.

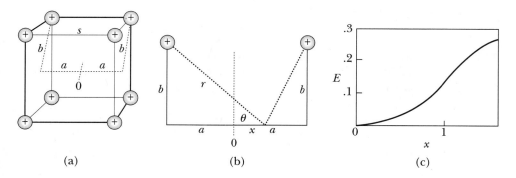

(a)   (b)   (c)

2-37 $\dfrac{d\theta(x)}{dx} = \delta(x).$ Then, $\displaystyle\int d\theta(x) = \int \delta(x)\,dx.$ So, $\theta(+\infty) - \theta(-\infty) = 1.$

Chapter 3

3-1 $3A \times 1\,\dfrac{C}{s \cdot A} \times \dfrac{1\,e^-}{1.6 \times 10^{-19}C} \times \dfrac{1\text{ mole}}{6.02 \times 10^{23}e^-} \times \dfrac{63.5g}{\text{mole}} \times \dfrac{1\text{ cm}^3}{8.9g}$

$\times \dfrac{1}{\pi(0.05\text{ cm})^2} \times \dfrac{0.01\text{ m}}{\text{cm}} = 2.8 \times 10^{-4}\text{ m/s}.$

3-3 $F = \mu_0 I^2/4\pi = 10^{-7}\text{ N}$ (independent of s), repulsive.

3-5 $B = \dfrac{\mu_0 I}{4\pi} \displaystyle\int \dfrac{dy\cos\varphi}{r^2} = \dfrac{\mu_0 I}{4\pi a}\dfrac{L}{\sqrt{L^2 + a^2}} = 1.1 \times 10^{-6}\text{ T}\left(\rightarrow \dfrac{\mu_0 I}{2\pi a}\text{ for }\infty\text{ wire}\right).$

By the right-hand rule, $\mathbf{B}$ points into the paper, along the negative z axis.

3-7 $B = \dfrac{\mu_0 I}{4\pi} \displaystyle\int \dfrac{dl \times \hat{\mathbf{r}}}{r^2} = \dfrac{\mu_0 I}{4r} = 3.1 \times 10^{-5}\text{ T}$, about half the earth's magnetic field.

3-9 $F = QvB = 2.0 \times 10^{-10}\text{ N}$ downward.

3-11 $v = \dfrac{E}{B} = 2.0 \times 10^6\text{ m/s}$ Eastward.

3-13 $B = \mu_0 K/2 = 2.51 \times 10^{-6}\text{ T}.$

3-15 $\nabla \times \mathbf{B} = \mu_0 \mathbf{J} + \mu_0 \epsilon_0 \dfrac{\partial \mathbf{E}}{\partial t}$.

So, $\oint \mathbf{B} \cdot d\mathbf{l} = \int \mu_0 \mathbf{J} \cdot d\mathbf{a}$.

<u>Inside:</u> $2\pi r B_i = \mu_0 NI$, so $B_i = \dfrac{\mu_0 NI}{2\pi r} = 0.012$ T.

<u>Outside:</u>   no net current passes through the Amperian loop, so $B_o = 0$.

3-17 For $d\mathbf{l}$ use $K\,dz$. Then, $B = \dfrac{\mu_0 K a^2}{2} \displaystyle\int \dfrac{dz}{(z^2 + a^2)^{3/2}} = \mu_0 K$

inside the solenoid; so in view of Ampère's law, $B = 0$ outside.

3-19 $B = \mu_0 K = 1.26 \times 10^{-3}$ T.

3-21 See Problem 2-25. $B = \dfrac{5\mu_0 JR}{12}$.

3-23 $\nabla \times (x\hat{\mathbf{x}}) = 0$, so $J = 0$. See Figure 2.24c. It is not physically possible for B because it has a divergence of one unit.

3-25 From Section 3.2, $B = \dfrac{\mu_0 I a^2}{2(z^2 + a^2)^{3/2}}$ for a circular loop; so,

$$B = \frac{8\mu_0 NI}{5^{3/2}\, R} = 1.8 \times 10^{-3} \text{ T}$$

for the conditions given. (It's only 36 times the earth's magnetic field.)

3-27 Start with a single loop, Section 3.2, and set $a = x$ and $dI = w\sigma x\,dx$.

$$B = \frac{\mu_0 w\sigma}{2}\left[\sqrt{z^2 + R^2} + \frac{z^2}{\sqrt{z^2 + R^2}} - 2z \right] \qquad (\to \mu_0 w\sigma R/2 \text{ at } z = 0)$$

$$= 5.7 \times 10^{-12} \text{ T, exceedingly small.}$$

3-29 $B = \mu_0 ev/4\pi r^2 = 12$ T, which is enormous.

3-31 Appose an identical solenoid, pointing in the same direction, so that horizontal B components add and vertical ones cancel. The field at point b will be doubled. The flux through the end (perpendicular to a surface across the end) will be doubled. Vertical field components will disappear.

3-33 $\mathbf{F} = I\oint d\mathbf{l} \times \mathbf{B} = I\int (d\mathbf{a} \times \nabla) \times \mathbf{B}$ from inside front cover,

$= 0$ because $\mathbf{B}$ is constant.

3-35 $B = \dfrac{\mu_0 K}{2\pi} \displaystyle\int_{-w/2}^{w/2} \dfrac{dx}{x + a} = \dfrac{\mu_0 K}{2\pi} \ln \dfrac{a + w/2}{a - w/2} \to \dfrac{\mu_0 K w}{2\pi a}$ as for a wire.

Chapter 4

4-1 $|\text{emf}| = N\dfrac{\Delta\Phi}{\Delta t} = 17\,\dfrac{0.60^2 \times 0.06 \cos 25°}{4}$

$= 0.083$ V, clockwise by Lenz's law.

4-3 emf $= Blv \sin\theta = 0.100$ V.

4-5 $|\text{emf}| = N\dfrac{\Delta\Phi}{\Delta t} = 57\,\dfrac{\pi\, 0.02^2\, 2 \times 1.3}{0.1} = IR$, so $I = 0.12$ A.

4-7 $|\text{emf}| = \Delta\Phi/\Delta t = 8.1 \times 10^{-6}$ V.

4-9 $|\text{emf}| = lvB_1 - lvB_r = lv\,\dfrac{\mu_0 I}{2\pi}\left(\dfrac{1}{0.04} - \dfrac{1}{0.065}\right)$

       $= 5.77 \times 10^{-8}$ V, a very small value.

4-11 $\text{emf} = 5.77 \times 10^{-8} = \dfrac{\mu_0(dI/dt)}{2\pi}\, l\ln r\,\Big|_{0.04}^{0.065}$;   so, $\dfrac{dI}{dt} = 11.9 \text{ A/s}.$

4-13 $\nabla \times (17\hat{\boldsymbol{\phi}}/\rho) = 0$ so **B** is constant in time, except at $\rho = 0$. See Figure 3.24a.

4-15 $\text{emf} = NaB\omega \cos \omega t$; maximum emf 2.6 V.

4-17 $\tau = BIr^2/2 = 7.5 \times 10^{-5}\ N \cdot m.$

4-19   For low-lying orbits, $\dfrac{mv^2}{r} \approx mg$, so v $= \sqrt{rg} = 7900\ m/s$. Then emf $=$

       $B/v = 7.9$ kV.

Chapter 5

5-1 $E_0 = cB_0$ and $\mu_0\epsilon_0 c^2 = 1.$

5-3 $\rho = 0$; see Figure 3.24c. $\nabla \times \mathbf{E} = -\partial\mathbf{B}/\partial t = -\hat{\mathbf{z}}$ is okay now but wasn't okay in electrostatics.

5-5   (a) $\nabla \cdot \mathbf{E} = 2$; constant positive charge density inside a negative cylindrical shell; (b) $\nabla \cdot \mathbf{E} = 2\hat{\mathbf{z}}$; B field increasing linearly with time, in the $-z$ direction, inside a circular solenoid; (c) $\nabla \times \mathbf{E} = \hat{\boldsymbol{\phi}}$; interior of a cylinder having charge density inversely proportional to distance from the axis and accelerating uniformly in the $+z$ direction.

5-7   The equation of charge continuity gives us

$$\partial\rho/\partial t = -\nabla \cdot \mathbf{J} = -2J_0/r;$$

     and

$$I = -\dot{Q} = 4\pi R^2 J_0.$$

5-9   Biot-Savart gives $B = \dfrac{\mu_0 Qv \sin\theta}{4\pi r^2}$ (Section 3.2).

     The fraction of the electric flux "Φ_e" subtended by the loop is $(1 - \cos\theta)/2$. Then,

$$\frac{d\Phi_e}{dt} = \frac{\partial\Phi_e}{\partial z}\frac{dz}{dt} = \frac{Q}{2\epsilon_0}\frac{\rho^2}{r^3}\text{v}$$

     So Ampère's law, with $J = 0$, becomes

$$2\pi\rho B = \mu_0\,\frac{Q}{2}\,\frac{\rho^2}{r^3}\,\text{v}, \qquad \text{so}$$

$$B = \frac{\mu_0 Qv \sin\theta}{4\pi r^2} \qquad \text{as above}$$

5-11 $J = \rho R/3\tau$; $E = \rho R/3\epsilon_0$;

     $\epsilon_0 \partial E/\partial t = -\rho R/3\tau = -J$;

     $B = 0.$

5-13 $\nabla \cdot (\nabla \times \mathbf{E}) = -\nabla \cdot \mathbf{J}_m - \nabla \cdot \dfrac{\partial \mathbf{B}}{\partial t}$

$0 = \nabla \cdot \mathbf{J}_m + \dfrac{\partial \rho_m}{\partial t}$

5-15 $\dfrac{(a)}{(b)} = \dfrac{\dfrac{Q_m}{4\pi r^2}\,ev}{\dfrac{e^2}{4\pi\epsilon_0 r^2}} \approx \dfrac{\epsilon_0 c Q_m}{e} = 68.$

Chapter 6

6-1 $B = \mu_0 N' I = 0.0251$ T, constant;
and so $W = \pi r^2 l\, B^2/2\mu_0 = 7.9 \times 10^{-3}$ J; same.

6-3 $A = Br/2 = 1.5 \times 10^{-3}$ T $\cdot$ m.

6-5 $\mathbf{E}$ outside $= \dfrac{Q}{4\pi\epsilon_0 r^2}\,\hat{\mathbf{r}};$ $\mathbf{E}$ inside $= \dfrac{Qr}{4\pi\epsilon_0 R^3}\,\hat{\mathbf{r}}.$

V outside $= \dfrac{Q}{4\pi\epsilon_0 r};$ V inside $= \dfrac{Q}{4\pi\epsilon_0 R}\left(\dfrac{3}{2} - \dfrac{r^2}{2R^2}\right).$ See Figure 8.5a.

6-7   From Problem 6-5, $V_i = \dfrac{Q}{4\pi\epsilon_0 R}\left(\dfrac{3}{2} - \dfrac{r^2}{2R^2}\right).$ Then,

$$W = \int_0^R \tfrac{1}{2}\, V_i\, \rho\, 4\pi r^2 dr = \dfrac{3}{5}\,\dfrac{Q^2}{4\pi\epsilon_0 R} = 8.3 \times 10^{-3}\, \text{J}.$$

6-9 $V_o = \displaystyle\int_r^R E\, dr = \int_r^R \dfrac{\lambda\, dr}{2\pi\epsilon_0 r} \to \infty$ logarithmically as $R \to \infty.$
(So $V_i \to \infty$, too.)
Or: $V_o = \displaystyle\int_{-L}^L \dfrac{\lambda\, dy}{4\pi\epsilon_0 \sqrt{r^2 + y^2}} = \dfrac{\lambda}{4\pi\epsilon_0}\ln \dfrac{L + \sqrt{L^2 + r^2}}{-L + \sqrt{L^2 + r^2}} \to \infty$
logarithmically as $L \to \infty.$

6-11 $A = \dfrac{\mu_0 I}{2\pi}\ln(r_2/r_1) = 3.3 \times 10^{-7}$ T $\cdot$ m.

6-13   (a)   emf $= -\dot{\Phi};$
(b) $2\pi r \dot{A} = \dot{\Phi}$ and $E = -\dot{A}$ and emf $= 2\pi r E.$

Either way, emf $= \pi r^2 \dfrac{\Delta B}{\Delta t} = 0.022$ V.

6-15 $B = \dfrac{\mu_0 I}{2a} = 6.3 \times 10^{-5}$T; $A = 0$ by symmetry.

6-17   Outside, $\mathbf{A} = R^2 B/2r\,\hat{\boldsymbol{\theta}}$ and $\nabla \times (\partial \mathbf{A}/\partial t) = 0.$
Inside, $\mathbf{A} = rB/2\,\hat{\boldsymbol{\theta}}$ and $\nabla \times (\partial \mathbf{A}/\partial t) = \partial \mathbf{B}/\partial t.$

6-19 $V = \dfrac{\lambda}{2\pi\epsilon_0}\ln \dfrac{4 + 2\sqrt{5}}{1 + \sqrt{5}}.$

6-21 $\frac{1}{2}mv^2 = \dfrac{qQ}{4\pi\epsilon_0 r}$ yields $r = 30$ fm $= 30 \times 10^{-15}$m.

6-23 See Problem 6-7. $W = \dfrac{3}{5}\dfrac{Q^2}{4\pi\epsilon_0 R}\left(1 - 2\,\dfrac{1/2^2}{1/2^{1/3}}\right) = 360$ MeV.

6-25 $V = \displaystyle\int_0^R \dfrac{\sigma 2\pi\rho\, d\rho}{4\pi\epsilon_0 r} = \dfrac{\sigma}{2\epsilon_0}\left(\sqrt{R^2 + z^2} - z\right)$ $(\Rightarrow \pi R^2\sigma/4\pi\epsilon_0 z$ for

small R). At $z = 0$, $V = R\sigma/2\epsilon_0$; for $V = 10^4$V, $\sigma = 7.1 \times 10^{-6}$C/m^2.

6-27 $A = \dfrac{\mu_0 I}{2\pi} \ln \dfrac{r_-}{r_+}$ in the direction of the closer current.

6-29 For a field line, $\dfrac{dy}{dx} = \dfrac{E_y}{E_x}$.

The solution here is a circle centered at y_c:

$$x^2 + (y - y_c)^2 = R^2 = y_c^2 + a^2$$

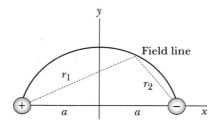

6-31 $F = \dfrac{Q^2}{8(4\pi\epsilon_0 R^2)}$ (which is the force between two point charges of $Q/2$
separated by $\sqrt{2}R$).

6-33 $\nabla \times \mathbf{E} = 0$.

6-35 $V = 2\displaystyle\int_0^{\pi/2}\int_0^r \dfrac{\sigma\rho d\rho\, d\theta}{4\pi\epsilon_0\rho} = \dfrac{\sigma}{2\pi\epsilon_0}\int_0^{\pi/2} r\, d\theta = \dfrac{\sigma R}{\pi\epsilon_0}$.

6-37 The final energy is $\frac{1}{2}mv^2 = \dfrac{(eRB)^2}{2m} = 9.9 \times 10^{-13}$J $= 6.2$ MeV.

The cyclotron frequency is $f = \dfrac{eB}{2\pi m} = 1.83 \times 10^7$ revolutions per second.

So it arrives at 6.2 MeV in $\dfrac{6.2 \text{ MeV}}{4 \text{ keV} \times 1.83 \times 10^7 \text{ s}^{-1}} = 85 \ \mu$s.

Chapter 7

7-1 $p = Qs = 6.0 \times 10^{-11}$ C·m; $V = 3.4$ V;
$E_r = 18.5$ V/m; $E_\theta = 5.3$ V/m; $E = 19.2$ V/m.

7-3 $m = Ia = 5.65 \times 10^{-6}$ A·m^2; $\mathbf{A} = 2.1 \times 10^{-12}$ T·m $\hat{\boldsymbol{\phi}}$;
$\mathbf{B} = 19.3 \times 10^{-12}$ T $\hat{\mathbf{r}} + 5.6 \times 10^{-12}T\hat{\boldsymbol{\theta}}$; $B = 20.1 \times 10^{-12}$ T.

7-5 Monopole: $Q = \int \rho \, dv = 10^{-9}$ C;

Dipole: $\mathbf{p} = \int \mathbf{r}\rho \, dv = 0.04 \times 3 \times 10^{-9}\hat{\mathbf{z}} + 0 \times (-2 \times 10^{-9})\hat{\mathbf{z}}$
$= 0.12 \times 10^{-9}\hat{\mathbf{z}}$ C·m;

Quadrupole: $Q_z = \int (3z^2 - r^2)\rho \, dv = (3 \times 0.04^2 - 0.04^2) \times 3 \times 10^{-9}$
$= 9.6 \times 10^{-12}$ C·m².

7-7 $V = \dfrac{p\cos\theta}{4\pi\epsilon_0 r^2} = -1.40$ V.

7-9 $m = \displaystyle\int_0^a \pi r^2 \sigma\omega r \, dr, \quad \mathbf{m} = \frac{1}{4}\pi a^4 \sigma\omega\hat{\mathbf{z}}.$

7-11 $\mathbf{p} \times (\nabla \times \mathbf{E}) = 0$, so either $\mathbf{B}$ is constant in time, or $\partial\mathbf{B}/\partial t$ is parallel to $\mathbf{p}$.

7-13 The equipotential surfaces are $r^2 = C_1 \cos\theta$, by inspection of Equation 7.4. The field lines point in the direction of the field:

$$\frac{r\,d\theta}{dr} = \frac{E_\theta}{E_r} = \frac{\sin\theta}{2\cos\theta} \qquad \text{so} \qquad r = C_2 \sin^2\theta$$

7-15 (a) For $\theta = 0$, $m = \dfrac{4\pi R^3 B_r}{2\mu_0} = 7.8 \times 10^{22}$ A·m²;

(b) $I = \dfrac{m}{\pi R^2} = 6.1 \times 10^8$ A.

7-17 (a) $\mathbf{F}_2 = \nabla(\mathbf{p}_2 \cdot \mathbf{E}_1) = -\dfrac{6p_1 p_2}{4\pi\epsilon_0 r^4}\,\hat{\mathbf{r}}$ at $\theta = 0$, attractive; and $\tau = \mathbf{p} \times \mathbf{E} = 0$.

(b) $\mathbf{F}_2 = \dfrac{3p_1 p_2}{4\pi\epsilon_0 r^4}\,\hat{\mathbf{r}}$ at $\theta = 90°$, repulsive; and $\tau = \mathbf{p} \times \mathbf{E} = 0$.

(c) $\mathbf{F}_2 = \nabla(\mathbf{p}_2 \cdot \mathbf{E}_1) = \dfrac{3p_1 p_2}{4\pi\epsilon_0 r^4}\,\hat{\boldsymbol{\theta}}$, to the right, at $\theta = 0$;

$\mathbf{F}_1 = -\dfrac{3p_1 p_2}{4\pi\epsilon_0 r^4}\,\hat{\boldsymbol{\theta}}$, to the left.

$\tau_1 = p_1 p_2 / 4\pi\epsilon_0 r^3$ ccw (vector out of page); $\tau_2 = 2p_1 p_2 / 4\pi\epsilon_0 r^3$ ccw;
couple $\tau = |\mathbf{r} \times \mathbf{F}| = 3p_1 p_2 / 4\pi\epsilon_0 r^3$ cw (into page).
For all cases, $\Sigma\mathbf{F} = 0$ and $\Sigma\tau = 0$.

7-19 $\Delta z = \dfrac{F_z l^2}{4kT} = 0.61$ mm.

If μ_B is in the x direction (perpendicular to z and v), then

$$F_x = \mu_B \frac{\partial B_x}{\partial x} = -\mu_B \frac{\partial B_z}{\partial z}$$

(since $\dfrac{\partial B_y}{\partial y} = 0$), so the maximum transverse (x) force may be equal in magnitude to the vertical (z) force; however, the average transverse force is zero because the dipole precesses.

7-21 (a) $B \approx \mu_0 N' I = 0.015$ T;
(b) 0.0075 T;

Now use the dipole approximation, with $m = NI\pi r^2 = 2.3$ A $\cdot$ m^2:

(c) $B_r = \dfrac{2\mu_0 m \cos \theta}{4\pi r^3} = 4.6 \times 10^{-10}$ T at $\theta = 0$;

(d) $B_\theta = \dfrac{\mu_0 m \sin \theta}{4\pi r^3} = 2.3 \times 10^{-10}$ T at $\theta = 90°$.

7-23 $-4\pi\epsilon_0 \nabla V = -(\mathbf{p} \cdot \mathbf{r}) \nabla \dfrac{1}{r^3} - \dfrac{1}{r^3} \nabla (\mathbf{p} \cdot \mathbf{r})$

$4\pi\epsilon_0 \mathbf{E} = 3(\mathbf{p} \cdot \mathbf{r}) \dfrac{\hat{\mathbf{r}}}{r^4} - \dfrac{\mathbf{p}}{r^3}$

7-25   **(a)** $Q_z = \displaystyle\int (3z'^2 - r'^2)\rho \, dv = -qa^2$.

   (b) $V = 0.438 \dfrac{q}{4\pi\epsilon_0 a}$.   **(c)** $V = 0.447 \dfrac{q}{4\pi\epsilon_0 a}$.

Chapter 8

8-1 $\sigma = \dfrac{l}{R\pi r^2} = 3.2 \times 10^{-3}$ S/m.

8-3 $\sigma = \epsilon_0 E = 8.85$ μC/m^2.

8-5 $R = \dfrac{l}{\sigma a} = 2 \, \Omega$; $V = 14$ V.

8-7 $\dfrac{E_H}{E_\sigma} = \sigma R_H B = 3.2 \times 10^{-3}$.

8-9 $E = \dfrac{V}{a(1 - a/R)}$; $\dfrac{dE}{da} = 0$ at $a = R/2$.

8-11 $m = \displaystyle\int_0^\pi \pi(R \sin \theta)^2 \, dI = \tfrac{4}{3}\pi R^3 \omega \, \epsilon_0 V = 4.8 \times 10^{-8}$ A$\cdot$m^2.

8-13 $R = \dfrac{120^2}{700} = 20.6 \, \Omega = \dfrac{l}{\sigma \pi r^2}$;   so diameter $= 2r = 0.30$ mm.

8-15 $E = \dfrac{V_0}{R}$ and $J = \dfrac{\sigma V_0}{R}$

8-17 $V = \dfrac{\lambda}{2\pi\epsilon_0} \ln \dfrac{r_2}{r_1}$ and $I = 2\pi r J l$,   so $\dfrac{V}{I} = R = \dfrac{\ln(r_2/r_1)}{2\pi\sigma l}$.

8-19 $W = \dfrac{1}{8\pi\epsilon_0}\left(\dfrac{q^2}{r} + \dfrac{(Q-q)^2}{R}\right)$.   Then,

$\dfrac{dW}{dq} = 0$ at $\dfrac{q}{r} = \dfrac{Q-q}{R}$,   which implies equal V's

8-21   For acceleration from rest, $v^2 = 2ax = 2\dfrac{eE}{m}x$, which implies that $J = \sigma \sqrt{E}$; this does not conform to experiment.

8-23 $\hat{\mathbf{r}} \times \left(E_0 \hat{\mathbf{z}} + \dfrac{p \sin \theta}{4\pi\epsilon_0 r^3}\hat{\boldsymbol{\theta}}\right) = 0$ so $r = \left(\dfrac{p}{4\pi\epsilon_0 E_0}\right)^{1/3}$. It's a conducting sphere.

8-25 Project the spherical shell onto the disk; the disk is still an equipotential.

$$\sigma(\rho) = \frac{2\sigma_0 a}{\sqrt{a^2 - \rho^2}} = \frac{Q}{2\pi a \sqrt{a^2 - \rho^2}}$$

(which increases without limit at the edge of the disk).

Chapter 9

9-1 $E_i = E_o/\epsilon_r = 4.3 \times 10^5 \text{ V/m}$;
$D_i = D_o = \epsilon_0 E_o = \epsilon E_i = 8.85 \times 10^{-6} \text{ C/m}^2$;
$P = \chi_e \epsilon_0 E_i = D - \epsilon_0 E_i = 5.0 \times 10^{-6} \text{ C/m}^2$;
$\rho_b = 0$; $\sigma_b = P = \div 5.0 \times 10^{-6} \text{ C/m}^2$ on the upper surface ($-$on lower).

9-3 $N = 3.94 \times 10^{28} \text{ atoms/m}^3$; $P = \chi_e \epsilon_0 E_i = 2.66 \times 10^{-4} \text{ C/m}^3$;

$$Qs = p = \frac{P}{N}, \quad \text{so} \quad s = 2.6 \times 10^{-15} \text{ m (only } 10^{-5} \text{ of an atomic diameter).}$$

9-5 Replace the dielectric with disks of bound surface charge $\sigma = \pm P$ (Problem 2-20).

$$E = E_1(z) - E_1(z + s) \rightarrow -\frac{\partial E_1}{\partial z} s = \frac{\sigma s a^2}{2\epsilon_0 z^3 (1 + (a/z)^2)^{3/2}}.$$

For large z, $E \rightarrow \dfrac{\pi a^2 \sigma s}{2\pi\epsilon_0 z^3} \rightarrow \dfrac{p}{2\pi\epsilon_0 r^3}$, as for a dipole at $\theta = 0$, Section 7.1.

Similarly (Problem 6-25),

$$V = -\frac{\partial V_1}{\partial z} s = \frac{\sigma s}{2\epsilon_0}\left(1 - \frac{1}{\sqrt{1 + (a/z)^2}}\right)(= E_1 s \text{ as expected}).$$

For large z, $V \rightarrow \dfrac{\pi a^2 \sigma s}{4\pi\epsilon_0 z^2} \rightarrow \dfrac{p}{4\pi\epsilon_0 r^2}$, as for a dipole at $\theta = 0$.

9-7 $\tau = RC = 4 \times 10^{-4} \text{ s}$.

9-9 $V = \displaystyle\int \frac{\lambda}{2\pi\epsilon r} dr = \frac{\lambda}{2\pi\epsilon} \ln(r_2/r_1)$;

$$C' = \frac{\lambda}{V} = \frac{2\pi\epsilon}{\ln(r_2/r_1)} = 185 \text{ pF/m}.$$

9-11 $V = \displaystyle\int_\infty^R \frac{Q}{4\pi\epsilon_0 r^2} dr + \int_R^0 \frac{\rho r}{3\epsilon} dr$

with $Q = \frac{4}{3}\pi R^3 \rho$;   so,

$$V = \left(1 + \frac{1}{2\epsilon_r}\right)\rho R^2/3\epsilon_0$$

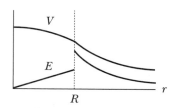

9-13 D is unaffected by ϵ_r, so $D = \dfrac{Q}{4\pi(2R)^2};$ $E = \dfrac{Q}{4\pi\epsilon_0\epsilon_r(2R)^2};$

$P = \dfrac{(1 - 1/\epsilon_r)Q}{4\pi(2R)^2}.$

9-15 $C = \dfrac{\epsilon_0\epsilon_r a}{s} = 72$ pF.

9-17   The voltage difference between the two wires is twice that found for either one relative to the central plane. So, $C' = \dfrac{\lambda}{V} = \dfrac{\pi\epsilon}{\ln(s/a)} = 40$ pF/m.

9-19 $a = -\dfrac{\epsilon_0 A}{2m}\dfrac{V^2}{s^2}$, and it is constant (since E is constant). So,

$$t = \sqrt{-\dfrac{s}{a}} = \dfrac{s^{3/2}}{V}\sqrt{\dfrac{2m}{\epsilon_0 A}}$$

9-21 $C = \dfrac{\epsilon_0 a}{2s}(\epsilon_r + 1) \rightarrow \dfrac{\epsilon_0 a}{s}$ at $\epsilon_r = 1, \rightarrow \infty$ as $\epsilon_r \rightarrow \infty.$

9-23 $F = -\dfrac{dW}{dx} = \dfrac{V_0^2 w\epsilon_0}{2s}\dfrac{\epsilon_r - 1}{\left[(\epsilon_r - 1)\dfrac{x}{w} + 1\right]^2} = 2.0 \times 10^{-3}$ N.

9-25 E is smaller in the dielectric. $V = \displaystyle\int_\infty^{2R} \dfrac{Q\,dr}{4\pi\epsilon_0 r^2} + \int_{2R}^R \dfrac{Q\,dr}{4\pi\epsilon r^2}.$

$C = \dfrac{8\pi\epsilon_0\epsilon_r R}{\epsilon_r + 1} \rightarrow 4\pi\epsilon_0 R$ at $\epsilon_r = 1, \rightarrow 4\pi\epsilon_0(2R)$ as $\epsilon_r \rightarrow \infty$

9-27   The Lorentz force $q\mathbf{v} \times \mathbf{B}$ provides an effective E field directed radially outward.

$$E = vB = \omega rB; \quad P = (\epsilon_r - 1)\epsilon_0\omega rB;$$
$$\rho_b = -\nabla\cdot\mathbf{P} = -2(\epsilon_r - 1)\epsilon_0\omega B;$$
$$\sigma_b = (\epsilon_r - 1)\epsilon_0\omega aB = P \text{ at edge of disk.}$$

Total bound charge: $\pi a^2 s\,\rho_b + 2\pi as\,\sigma_b = 0.$

9-29   From Section 6.3, with $L = l/2,$

$$V = \dfrac{\lambda}{2\pi\epsilon_0}\ln\left(\dfrac{L + \sqrt{L^2 + a^2}}{a}\right) \longrightarrow \dfrac{\lambda}{2\pi\epsilon_0}\ln\dfrac{l}{a} \text{ for small } a;$$

So, in agreement with Section 9.4, $C = \dfrac{2\pi\epsilon_0 l}{\ln(l/a)} = 56$ pF.

9-31 $V_1 + V_2 = \dfrac{Q}{C_1} + \dfrac{Q}{C_2} = \dfrac{Q}{C}.$

9-33   The free current is $J_f = -d\sigma/dt$. Also, $\sigma = D$. So the displacement current dD/dt is equal and opposite to the free current J_f. Then:

$$\nabla \times \mathbf{B} = \mu_0\mathbf{J}_f + \mu_0\partial\mathbf{P}/\partial t + \mu_0\epsilon_0\partial\mathbf{E}/\partial t = \mu_0\mathbf{J}_f + \mu_0\partial\mathbf{D}/\partial t = 0$$

9-35   At the center $V = \int_0^a \dfrac{\sigma 2\pi\rho \, d\rho}{4\pi\epsilon_0 \rho} = \dfrac{Q}{8\epsilon_0 a}$,   so $C = 8\epsilon_0 a$.

Chapter 10

10-1 $B_i = B_o = 10^{-3}$ T;
$H_i = H_o/\mu_r = B/\mu = 4.0$ A/m;
$M = (\mu_r - 1)H = B/\mu_0 - H = 790$ A/m;
$J_b = 0.$

10-3 $H_i = H_o = B/\mu_0 = 8.0 \times 10^5$ A/m;
$B_i = \mu_0(H + M) = 2.3$T;
$J_b = 0; \quad K_b = M = 10^6$ A/m.

10-5   Think loop, $I = Ms$.

$A = 0$ by symmetry; $B = \dfrac{\mu_0 M s a^2}{2\,(z^2 + a^2)^{3/2}}$ (see Section 3.2, Equation 3.9).

At large z, $B = \dfrac{\mu_0 m}{2\pi r^3}$, as for a dipole at $\theta = 0$, Section 7.2.

10-7 $\Phi = 2 \int_a^s Bl \, dr = \dfrac{\mu_0 Il}{\pi} \ln(s/a)$

$L' = \dfrac{\Phi}{lI} = \dfrac{\mu_0}{\pi} \ln\,(s/a) = 6.4 \times 10^{-7}$ H/m.

10-9 $\tau = \dfrac{L}{R} = 5$ ms.

$t = 3\tau = 0.015$ second.

10-11 $\Phi = \int_a^{a+s} \dfrac{\mu NI}{2\pi r} s \, dr = \dfrac{\mu NI}{2\pi} s \ln \dfrac{a+s}{a}$

$L = \dfrac{N\Phi}{I} = \dfrac{\mu N^2}{2\pi} s \ln \dfrac{a+s}{a} = 100$ mH.

10-13 $\Phi = \dfrac{\mu I}{2\pi} \int_0^r 2 \int_0^\pi \dfrac{\rho \, d\theta}{R + \rho \cos\theta} \, d\rho = \mu I \,(R - \sqrt{R^2 - r^2})$.

$M = \dfrac{N\Phi}{I} = \mu\,(R - \sqrt{R^2 - r^2})$.

10-15 $I_2 = \dfrac{450}{15000} = 30$ ma; $I_1 = \dfrac{15000}{120}\,30$ ma $= 3.75$ A; $N_2/N_1 = 125:1$.

10-17   With gap: $L = \dfrac{100 \times 0.097 \times 10^{-4}}{2} = 0.50$ mH;   no gap: $L = 0.80$ mH.

10-19 $\oint \mathbf{H} \cdot d\mathbf{l} = 0$

$B = \mu_0 H_m + \mu_0 M = -\dfrac{s}{2\pi R} B + \mu_0 M \approx \mu_0 M = 0.13$ T.

It is nearly independent of gap size, for small gap.

10-21 $M = \dfrac{N_2\Phi}{I} = \mu N_1 N_2 \dfrac{r^2}{2R} = 5.9 \times 10^{-4}$ H.

10-23 At the center, $B = \mu_0 \mu_r N' I = 0.64$ T.
At the end, $B/2 = 0.32$ T (superposition; compare with Problem 3-31).

10-25 From Section 3.5, $B = \mu J r/2$; it is increased by a factor of μ_r.
$H = J r/2$; it is the same as for a copper wire.
From Equation 10.15, $J_b = \chi_m J$ (same direction as J).
$K_b = M = -\chi_m J a/2$ (direction opposite to J).

The total bound volume current $\pi a^2 J_b$ is 200 times the free current, and the total bound surface current $2\pi a K_b$ is equal and opposite, so the total bound current is zero.

10-27 From Equation 10.28, $L \approx \dfrac{\mu_0 l}{2\pi} \ln \dfrac{l}{a} = 18 \times 10^{-6}$ H.

10-29 Loop a has inductance $L_a = \dfrac{N_a \Phi_a}{I_a}$. It passes flux $k_a \Phi_a$ through loop b. So the mutual inductance is

$$M_{ba} = \frac{N_b k_a \Phi_a}{I_a} = \frac{N_b k_a L_a}{N_a}$$

Similarly,

$$M_{ab} = \frac{N_a k_b L_b}{N_b}$$

Since $M_{ba} = M_{ab}$, the product is $M^2 = k_a k_b L_a L_b$.

10-31 (a) Capacitor: $\frac{1}{2} \epsilon_0 \epsilon_r E^2/\rho = 100$ J/kg.
(b) Iron inductor: $\frac{1}{2} B^2/\mu_0 \mu_r \rho = 1.0$ J/kg.
(c) Cryogenic inductor: $\frac{1}{2} B^2/\mu_0 \rho = 5100$ J/kg.

(d) Lead-acid battery: 12 V $\times$ 50 A·hr $\times$ $3600 \, \dfrac{\text{s}}{\text{hr}} \div 20$ kg

$$= 1.1 \times 10^5 \text{ J/kg}.$$

(e) Flywheel: $\frac{1}{2} I \omega^2/m = 1400$ J/kg.
(f) Gasoline: heat of combustion 5×10^7 J/kg $\times$ 10% efficiency $= 5 \times 10^6$ J/kg.

10-33 Pressure $P = \dfrac{1}{2\mu_0} B^2$, which yields $B \approx 0.90$ T. When the magnet is isolated, the field at the end is only half of this (Problem 3-31), so $B_0 \approx 0.45$ T.

10-35 Meissner effect: no flux escapes. So

$$I = Q_m/L = h/eL.$$

Chapter 11

11-1 A dot represents a partial time derivative here.

$$\nabla \times \mathbf{B} = \mu_0 \mathbf{J}_f + \mu_0 \mathbf{J}_b + \mu_0 \dot{\mathbf{P}} + \mu_0 \epsilon_0 \dot{\mathbf{E}}$$

$$\nabla \times \mathbf{B} - \mu_0 \mathbf{J}_b = \mu_0 \mathbf{J}_f + \mu_0 (\dot{\mathbf{P}} + \epsilon_0 \dot{\mathbf{E}})$$

$$\nabla \times \mathbf{H} = \mathbf{J}_f + \dot{\mathbf{D}}$$

$$\nabla \times \mathbf{B} = \mu \mathbf{J}_f + \mu\epsilon \frac{\partial \mathbf{E}}{\partial t}$$

Chapter 12

12-1 $0 = \oint V_3 \nabla V_3 \cdot d\mathbf{a}$ because V_3 is zero on the boundary. Apply the divergence theorem:

$$0 = \int \nabla \cdot (V_3 \nabla V_3) \, dv$$

$$= \int ((\nabla V_3)^2 + V_3 \nabla^2 V_3) \, dv$$

$$= \int (\nabla V_3)^2 \, dv + 0 \qquad (\text{because } \nabla^2 V_3 = 0)$$

So $\nabla V_3 = 0$ everywhere. $V_3 = 0$ at boundary, so $V_3 = 0$ throughout.

12-3 Proceding as in Section 12.1, the energy to bring Q from infinity is $Qq/4\pi\epsilon_0 R^2$, and the energy to bring the sphere is qV_{avg}. So,

$$V_{\text{avg}} = Q/4\pi\epsilon_0 R^2$$

just as if the charge Q were located at the center.

12-5 From Equation 12.10, $\displaystyle\int_0^\infty \frac{-QD2\pi\rho d\rho}{2\pi(D^2 + \rho^2)^{3/2}} = -Q.$

12-7 $F = \dfrac{QQ'}{4\pi\epsilon_0} \left(\dfrac{1}{(D-b)^2} - \dfrac{1}{D^2} \right)$, attractive;

$$V = -\frac{Q'}{4\pi\epsilon_0 a} = \frac{Q}{4\pi\epsilon_0 D}.$$

12-9 (Problem 12-7): $F = \dfrac{QQ'}{4\pi\epsilon_0} \left(\dfrac{1}{(D-b)^2} \right)$, attractive;

$$V = \frac{Q}{4\pi\epsilon_0 a} \quad (\text{remember Fig. 8.2}).$$

(Problem 12-8): $F = \dfrac{QQ'}{4\pi\epsilon_0} \left(\dfrac{1}{(D-b)^2} \right) = 4.1 \times 10^{-7} \, \text{N}$, attractive;

$V = 0$, since it's still grounded.

12-11 The forces due to positive images all cancel out.

$$F = \frac{Q^2}{4\pi\epsilon_0} \left(\frac{1}{(2s-2D)^2} + \frac{1}{(4s-2D)^2} + \frac{1}{(6s-2D)^2} + \cdots \right.$$

$$\left. - \frac{1}{(2D)^2} - \frac{1}{(2s+2D)^2} - \frac{1}{(4s+2D)^2} - \cdots \right)$$

12-13 D is twice the distance from center of sphere to plane, and $r = a/D$. Then,

$$Q = 4\pi\epsilon_0 aV \quad = 1.11 \times 10^{-9} \text{ C}$$

$$Q' = -rQ \quad = -0.250\, Q$$

$$Q'' = \frac{r^2}{1 - r^2}\, Q = 0.067\, Q$$

$$Q''' = \frac{r^3}{1 - 2r^2}\, Q = -0.018\, Q$$

$$Q'''' \approx (\text{approx})\ 0.005\, Q$$

$$Q - Q' + Q'' - Q''' = 1.35\, Q = 1.49 \text{ nC total charge on sphere.}$$

12-15 $F = \dfrac{3p^2}{4\pi\epsilon_0 (2D)^4} = \dfrac{3p^2}{64\pi\epsilon_0 D^4}$, attractive.

12-17   At large D, the field E due to the charges $+Q$ and $-Q$ becomes

$$E \to 2\,\frac{Q}{4\pi\epsilon_0 D^2}, \qquad \text{which is held constant}$$

and the dipole moment of the sphere becomes

$$p = 2\,b\,Q' = 2\,\frac{a^2}{D}\,\frac{aQ}{D} \to 4\pi\epsilon_0 a^3\, E, \qquad \text{which is constant}$$

12-19   Cylindrical:

$$V = P(\rho)\,\Phi(\phi) \text{ with } \frac{\rho}{P}\,\frac{d}{d\rho}\,(\rho\, P') = C \text{ and } \frac{1}{\Phi}\,\Phi'' = -C;$$

Spherical:

$$V = R(r)\,\Theta(\theta) \text{ with } \frac{1}{R}\,\frac{d}{dr}\,(r^2\, R') = C \text{ and } \frac{1}{\Theta \sin\theta}\,\frac{d}{d\theta}\,(\sin\theta\,\Theta') = -C.$$

Solutions are presented in Section 12.3.

12-21   Now we need both positive and negative exponentials.

$$V(x, y) = \sum_{\text{odd } n}^{\infty} (C_n e^{-n\pi x/b} + D_n e^{n\pi x/b}) \sin\frac{n\pi y}{b}$$

with $C_n = \dfrac{4V_0}{n\pi}\left(\dfrac{1}{1 - e^{-2n\pi a/b}}\right)$ and $D_n = -\dfrac{4V_0}{n\pi}\left(\dfrac{e^{-2n\pi a/b}}{1 - e^{-2n\pi a/b}}\right).$

12-23   Inside: $V(r, \theta) = \sum_0^{\infty} A_n r^n P_n(\cos\theta)$

with $A_m = \dfrac{2m + 1}{2R^m} \displaystyle\int_0^{\pi} V(R, \theta)\, P_m(\cos\theta)\sin\theta\, d\theta.$

Outside: $V(r, \theta) = \sum_0^{\infty} \dfrac{B_n}{r^{n+1}}\, P_n(\cos\theta)$

with $B_m = \dfrac{2m + 1}{2}\, R^{m+1} \displaystyle\int_0^{\pi} V(R, \theta)\, P_m(\cos\theta)\sin\theta\, d\theta.$

12-25 $V_i = \dfrac{\sigma_0 r \cos\theta}{3\epsilon_0} = \dfrac{\sigma_0 z}{3\epsilon_0}$, implying uniform E field inside.

$V_o = \dfrac{\sigma_o a^3 \cos\theta}{3\epsilon_0 r^2}$, which is a perfect dipole outside.

12-27 $V_i = -\dfrac{2E_o}{\epsilon_r + 1}\rho\cos\phi = -\dfrac{2E_o}{\epsilon_r + 1}x;$

and $V_o = -E_o\rho\cos\phi + \dfrac{\epsilon_r - 1}{\epsilon_r + 1}\dfrac{a^2 E_o}{\rho}\cos\phi.$

12-29 The field **H** is analogous to **E**: $\nabla\cdot\mathbf{H} = \rho_m/\mu_o$, and $\nabla\times\mathbf{H} = 0$, so we set $\mathbf{H} = -\nabla V_m$. Correspondingly, $H = \dfrac{Q_m}{4\pi\mu_o r^2}$ and $V_m = \dfrac{Q_m}{4\pi\mu_o r}$. (Don't worry, these magnetic monopoles come out in the wash.) The magnetic dipole moment is $m = Q_m s/\mu_o$, s being the monopole separation. Proceding as in Section 9.1, we find:

Inside: $\mathbf{H}_i = -\dfrac{M}{3}\,\hat{\mathbf{z}}$, so, with $\mathbf{B} = \mu_o\mathbf{H} + \mu_o\mathbf{M}$, we get $\mathbf{B}_i = \dfrac{2\mu_o M}{3}\,\hat{\mathbf{z}}$, a uniform field.

Outside: $\mu_o\mathbf{H}_o = \mathbf{B}_o = \dfrac{2\mu_o M\cos\theta}{3}\dfrac{R^3}{r^3}\,\hat{\mathbf{r}} + \dfrac{\mu_o M\sin\theta}{3}\dfrac{R^3}{r^3}\,\hat{\boldsymbol{\theta}}$, a dipole field.

At the surface at $\theta = 0$, $B_i/B_o = 1$ because $\nabla\cdot\mathbf{B} = 0$. At $\theta = 90°$, $B_i/B_o = 2$, which is not obvious.

Chapter 13

13-1 $t = \tau\ln(10) = 46$ s to $1/10$; $t = \infty$ to zero volts (it's asymptotic).

13-3 $f = \dfrac{\omega}{2\pi} = \dfrac{1}{2\pi\sqrt{LC}}$; so $L = 100\ \mu H$.

13-5 $I(t) = \dfrac{V_o}{\omega' L}\,e^{-Rt/2L}\sinh\omega' t.$

To find ω', substitute this $I(t)$ in the differential equation for I; a difference between this case and the underdamped one is: $d\cos x = -\sin x\,dx$, but $d\cosh x = +\sinh x\,dx$.

13-7 Superposition: (**a**) Introduce 1 mA at point a; by symmetry, 0.25 mA will flow from a to b. (**b**) Instead, remove 1 mA from point b; again, 0.25 mA flows from a to b. (**c**) Superpose procedures (a) and (b); now 0.5 mA flows through the 1-K resistor from a to b. The voltage is $IR = 0.5$ V, the current is 1 mA (in at a, out at b), so the resistance is

$$R = 0.5 \text{ V}/1 \text{ mA} = 0.5 \text{ K}$$

13-9 $f_n = \dfrac{\omega_n}{2\pi} = 5400 \text{ Hz} \left(\approx \dfrac{1}{2\pi\sqrt{LC}} \right).$

Energy $\sim I^2 \sim e^{-Rt/L}$ at large t, so $\dfrac{L}{R} \ln(10) = 1.24 \times 10^{-4}$ s.

13-11 $R = \sqrt{\dfrac{4L}{C}} = 1900 \ \Omega.$

13-13 $X_C = -1/\omega C = -995 \ \Omega.$

13-15 $f_0 = \dfrac{1}{2\pi\sqrt{LC}} = 145 \text{ kHz}.$

13-17 $I = \dfrac{V}{Z} = \dfrac{VR}{R^2 + \omega^2 L^2} + i\left(V\omega C - \dfrac{V\omega L}{R^2 + \omega^2 L^2} \right).$

$I = (0.0263 - 0.0086i)$ A; $I_{\text{rms}} = 0.0277$ A.

13-19   At minimum $|I|$ of 0.0233 A, $\omega = 407.90$ rad/s (find this one numerically);

for real I, $\omega = \sqrt{\dfrac{1}{LC} - \left(\dfrac{R}{L}\right)^2} = 396$ rad/s;

and $\omega_0 = 1/\sqrt{LC} = 408.25$ rad/s.

At large Q, or small R, all three ω's become ω_0.

13-21 $Z = \dfrac{1}{\dfrac{1}{R + i\omega L} + i\omega C} = \dfrac{R}{(1 - \omega^2 LC)^2 + (\omega CR)^2}$

$\qquad + i\dfrac{\omega L(1 - \omega^2 LC) - \omega CR^2}{(1 - \omega^2 LC)^2 + (\omega CR)^2}.$

13-23   Series: $Z = R + i\left(\omega L - \dfrac{1}{\omega C} \right)$ becomes $Z = R.$

Parallel: From Problem 19, with Z, and $\omega^2 LC = 1$:

$$Z = \dfrac{R}{(\omega CR)^2} = \dfrac{L}{RC} = Q\omega L$$

In the parallel case, large R reduces the impedance because it reduces Q.

13-25   Low frequency, $f \ll \dfrac{1}{RC}.$

13-27 $\dfrac{1}{Z} = \dfrac{1}{R + i\omega L} + \dfrac{1}{R + 1/i\omega C} = \dfrac{1}{R}\left(\dfrac{1}{1 + i\omega L/R} + \dfrac{1}{1 + R/i\omega L} \right) = \dfrac{1}{R}.$

Chapter 14

14-1 $S = EB \sin \theta/\mu_0 = 7.0 \times 10^{11}$ W/m² downward.

14-3 $B = r\mu_0\epsilon_0 \dot{E}/2$

$S = EB/\mu_0 = r\epsilon_0 E\dot{E}/2$

area $\times S = 2\pi r s S = \pi r^2 s\epsilon_0 E\dot{E} = $ volume $\times \dfrac{d}{dt}\left(\dfrac{1}{2}\epsilon_0 E^2\right)$.

14-5 On the inner conductor: $\lambda = \dfrac{2\pi\epsilon_0 V}{\ln(r_2/r_1)}$;

Power $= \displaystyle\int EB\, 2\pi r\, dr/\mu_0 = 7.5$ W $= VI$.

14-7 Pressure $= \dfrac{S}{c}$;

$F = \dfrac{\text{Power}}{c} = \dfrac{10^7}{3 \times 10^8} = 0.033$ N. (The area is irrelevant.)

14-9 $F = \dfrac{S}{c}\pi r^2 = G\dfrac{M}{R^2}\dfrac{4}{3}\pi r^3\rho$

$r = 0.12$ μm.

14-11 $dF_x = \Sigma T_{ij}da_j = T_{xx}da_x = \dfrac{\epsilon}{2} E_x^2 \cos\theta\, da$

$dF_y = T_{yy}da_y = -\dfrac{\epsilon}{2} E_x^2 \sin\theta\, da$

so $dF_y/dF_x = -\tan\theta.$

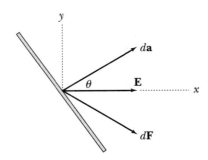

14-13 $\nabla \times (\nabla \times \mathbf{B}) = \mu_0\nabla \times \mathbf{J} + \mu_0\epsilon_0 \dfrac{\partial \nabla \times \mathbf{E}}{\partial t}$

$\nabla(\nabla\cdot\mathbf{B}) - \nabla^2\mathbf{B} = -\nabla^2\mathbf{B} = \mu_0\nabla \times \mathbf{J} - \mu_0\epsilon_0 \dfrac{\partial^2 \mathbf{B}}{\partial t^2}$

14-15 (a) $L = \displaystyle\int_0^\pi\int_a^\infty \dfrac{Q}{4\pi\epsilon_0 r^2}\dfrac{\mu_0 m}{4\pi r^3}\sin\theta\dfrac{1}{\mu_0 c^2} r\sin\theta\, 2\pi r\sin\theta\, r\, dr\, d\theta$

$= \dfrac{Qm}{6\pi\epsilon_0 c^2 a} = \dfrac{2}{9}\mu_0 QMa^2.$

(b) Mechanical angular momentum is $L = I\omega$, with moment of inertia $I = \frac{2}{5} ma^2$, m being mass now. Density is ρ. So,

$$\omega = \frac{L}{I} = \left(\frac{\frac{2}{9}\mu_0 Q M a^2}{\frac{2}{5}(\frac{4}{3}\pi a^3 \rho) a^2} \right) = \frac{5}{3}\frac{VM}{c^2 a^2 \rho} = 1.9 \times 10^{-7} \text{ rad/s } (\approx 1.0 \text{ rev/year})$$

14-17 $\displaystyle\int \nabla \cdot \mathbf{A} \, dv = -\int \frac{\dot{V}}{c^2} \, dv$

$$A_{rms} = \frac{r}{3c^2}\omega V_{rms} = 3.4 \times 10^{-15} \text{ T} \cdot \text{m}$$

14-19 $B = \sqrt{\mu_0 S/c} = 5.41 \times 10^{-9}$ T.

$$|\text{emf}| = \pi r^2 \omega B = 0.107 \text{ V}$$

14-21 $\mathbf{E} = E_0 e^{i(kz-\omega t)} (\hat{\mathbf{x}} - \hat{\mathbf{y}}) + E_0 e^{i(kz-\omega t - \theta)} (\hat{\mathbf{x}} + \hat{\mathbf{y}})$
is a wave linearly polarized at θ relative to the x axis.

14-23 $r = c^2/a \sin\theta = 9.2 \times 10^{15}$ m ≈ 1 light-year.

14-25 $\displaystyle -\frac{dW}{dt} = \frac{2}{3}\frac{e^2 a^2}{4\pi\epsilon_0 c^3} = \frac{2}{3}\frac{e^4 B^2}{4\pi\epsilon_0 c^3 m^2} \frac{2W}{m}$

Integrating, $\displaystyle -\ln\frac{W}{W_0} = \frac{4}{3}\frac{e^4 B^2}{4\pi\epsilon_0 c^3 m^3} t = 0.87\, t$;

When $W/W_0 = 90\%$, $t = 0.12$ s.

14-27 $\displaystyle -\frac{\Delta W}{W} = \frac{\frac{2}{3}\frac{e^2 a^2}{4\pi\epsilon_0 c^3} t}{\frac{1}{2} mv^2} = \frac{2}{3}\frac{e^2 v}{4\pi\epsilon_0 c^3 ms} = 3.7 \times 10^{-6}$.

Chapter 15

15-1 At 28 MHz, $\lambda = 11$ m, so the dipole approximation is appropriate.
$R_{rad} = 20\,(ks)^2 = 0.43\ \Omega$;
$I = \sqrt{P/R_{rad}} = 0.26$ A.

15-3 $v_{rms} = 9.4 \times 10^{-5}$ m/s.
$2x_0 = 4.25 \times 10^{-13}$ m, less than one atomic diameter.

15-5 **(a)** Dipole: $k = 1.57$ m^{-1}; $R_{rad} = 7.9\ \Omega$;
(b) Half-wave: $R_{rad} = 73\ \Omega$.

15-7 Magnetic dipole, $\theta = 0°$.

15-9 *For A:* $A = \dfrac{\mu_0}{4\pi r} I_0 e^{i(kr - \omega t)} s$.

For phase velocity v_p, the exponent is constant, so set its derivative equal to zero:

$$0 = k\, dr - \omega\, dt$$

$$v_p = \frac{dr}{dt} = \frac{\omega}{k} = c$$

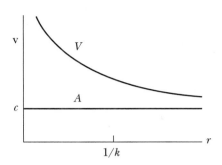

For V: $V = \dfrac{s \cos \theta}{4\pi\epsilon_0 cr} I_0 e^{i(kr-\omega t)} \left(1 + \dfrac{i}{kr}\right)$

$= \dfrac{s \cos \theta}{4\pi\epsilon_0 cr} \sqrt{1 + \dfrac{1}{(kr)^2}}\, I_0 e^{i(kr-\omega t+\arctan(1/kr))}$

Differentiating the exponent, $v_p = \dfrac{dr}{dt} = c\left(1 + \dfrac{1}{(kr)^2}\right)$, which is greater

than c; but no material can thus move faster than light.

15-11 See Appendix C: $\sin^2\theta = \dfrac{L}{\dfrac{1}{kr}\cos(kr-\omega t) + \sin(kr-\omega t)}$.

L is an integration constant: different L's for different lines.

15-13 $P = I^2 R_{\text{rad}}$, so $I = 3.70$ A;

$E = \dfrac{I\cos(\frac{\pi}{2}\cos\theta)}{2\pi\epsilon_0 cr} = 0.0139$ V/m;   and $B = E/c = 4.6 \times 10^{-11}$ T(same

as for a wire).

15-15 We place the point of observation S in the y-z plane at r,θ. The $\mathbf{E}$ fields
there are mutually perpendicular. Consequently, the Poynting vectors
$\epsilon_0 cE^2\, \hat{\mathbf{r}}$ just add up. The total is

$$\mathbf{S}_{\text{av}} = \dfrac{\omega^4 p^2}{32\pi^2\epsilon_0 c^3 r^2}(1 + \cos^2\theta)\hat{\mathbf{r}}$$

It's a peanut.

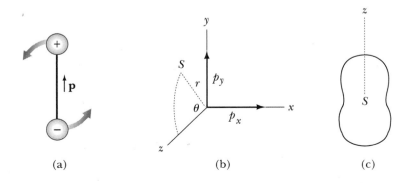

(a)   (b)   (c)

15-17 Their angular distribution is the same. For the electric dipole at $\theta = 90°$,

$$S = \frac{1}{16\pi^2\epsilon_0 c}\left(\frac{ksI}{r}\right)^2 \qquad \text{and} \qquad P = \frac{2}{3}\frac{1}{4\pi\epsilon_0 c}(ks\,I)^2$$

so, $\dfrac{S}{P/4\pi r^2} = 1.5.$

15-19 Consider only the phase difference.

For each dipole: $E_d = -\dfrac{k\omega[Q]s\sin\theta}{4\pi\epsilon_0 cr}.$

For the quadrupole: $E_q = \dfrac{\partial E_d}{\partial r}\,\Delta r = -i\,\dfrac{\omega^3[Q]s^2\sin\theta\cos\theta}{4\pi\epsilon_0 c^3 r}.$

So,

$$\mathbf{S}_{av} = \tfrac{1}{2}\,\mathrm{Re}\,(\mathbf{E}\times\mathbf{B}^*/\mu_0) = \frac{Q_0^2 s^4\omega^6 \sin^2\theta\cos^2\theta}{32\pi^2\epsilon_0 c^5 r^2}\,\hat{\mathbf{r}}$$

and

$$P_{av} = \int S_{av}\,d(\text{area}) = \frac{Q_0^2 s^4\omega^6}{60\pi\epsilon_0 c^5} = \frac{Q_{zo}^2\omega^6}{960\pi\epsilon_0 c^5}$$

where the quadrupole moment (Section 7.3) is $Q_{zo} = 4\,Q_o s^2$.

Chapter 16

16-1 $v_p = c/n = 2.03\times10^8\text{ m/s}.$

16-3 $E = vB = \dfrac{cB}{\sqrt{\epsilon_r\mu_r}} = \dfrac{B}{\sqrt{\epsilon_0\mu_0}\sqrt{\epsilon_r\mu_r}}$

so $\tfrac{1}{2}\epsilon E^2 = \tfrac{1}{2}B^2/\mu$

and $S = EB/\mu = E\dfrac{E\sqrt{\epsilon_r\mu_r}}{c\mu} = \epsilon E^2 v.$

16-5 $R = \left(\dfrac{n-1}{n+1}\right)^2 = 0.0465;$

$R + T^2\,(R + R^3 + R^5 + \dots) = \dfrac{2R}{R+1} = 0.089.$

16-7 $E_I + E_R = nE_T$ and $E_I - E_R = E_T$.

So, $\dfrac{E_T}{E_I} = \dfrac{2}{n+1} \to T = \dfrac{4n}{(n+1)^2}.$

$R = \left(\dfrac{n-1}{n+1}\right)^2$ so $R + T = 1.$

16-9 $\delta = \sqrt{\dfrac{2}{\omega \sigma \mu}} = 7.1$ m.

16-11 $\delta = 6.6 \times 10^{-5}$ m ($\approx$ human hair).

16-13 $T = \sqrt{\dfrac{8\omega \epsilon_0}{\sigma}} = 1.5 \times 10^{-4}.$

16-15   Power per unit area $= S_{av} = \frac{1}{2}\sqrt{\dfrac{\sigma}{2\omega \mu}}\, E_0{}^2\, e^{-2z/\delta};$

Power loss per unit volume $= -\dfrac{\partial S}{\partial z} = \sigma E_{rms}{}^2 = JE =$ Joule heating.

16-17 $f = 9\sqrt{N} = 2800$ Hz.

16-19 $k = \dfrac{\omega}{c}\sqrt{1 - \left(\dfrac{\omega_p}{\omega}\right)^2} = \dfrac{1}{c}\sqrt{\omega^2 - \omega_p{}^2}$

$v_p = \dfrac{\omega}{k}$ and $v_g = \dfrac{1}{dk/d\omega} = \dfrac{c}{\omega}\sqrt{\omega^2 - \omega_p{}^2} = \dfrac{kc^2}{\omega}$

So, $v_p v_g = c^2.$

16-21   (a) Starting with Eq. 16.51, $\dfrac{d(n_r - 1)}{d(\omega_0 - \omega)} = 0$ at $\omega_0 - \omega = \gamma/2.$ Then by

inspection, $\dfrac{n_i \text{ at } \omega_0 - \omega = 0}{n_r - 1 \text{ at } \omega_0 - \omega = \gamma/2} = 2.$

(b) By inspection, $n_r - 1 = n_i$ at $\omega_0 - \omega = \gamma/2.$

16-23   We are below the ultraviolet resonances, so

$n_r - 1 \approx \dfrac{\omega_p{}^2}{2}\,\dfrac{1}{\omega_0{}^2 - \omega^2} \approx \dfrac{\omega_p{}^2}{2\omega_0{}^2}\left(1 + \dfrac{\omega^2}{\omega_0{}^2}\right) = A\left(1 + \dfrac{B}{\lambda^2}\right).$

Chapter 17

17-1 $f_c = c/2a = 6.5$ GHz.

17-3 $\lambda_0 = \dfrac{c}{f} = 3.0$ cm;

$\cos \theta = \dfrac{\lambda_0}{2a} \to \theta = 49.3°;$ $\lambda_g = \dfrac{\lambda_0}{\sin \theta} = 4.0$ cm;

$v_s = c \sin \theta = 2.3 \times 10^8$ m/s; $v_p = \dfrac{c}{\sin \theta} = 4.0 \times 10^8$ m/s.

17-5   For TE_{10}: $m = 1$, $n = 0$, and $k_c = \dfrac{\pi}{a}$. The fields are in Equation 17.9.

So,

$$\mathbf{S}_{av} = \tfrac{1}{2}\operatorname{Re}(\mathbf{E} \times \mathbf{B}^*/\mu) = \frac{1}{2\mu}\left(\frac{a}{\pi}\right)^2 \omega k_g B_0{}^2 \sin^2\left(\frac{\pi x}{a}\right)\hat{\mathbf{z}}$$

In this case, $\mathbf{E} \times \mathbf{B}$ has no y component, and its x component is imaginary.

17-7 From Equation 17.12, with $m = n = 1$, $\lambda_c = 2a/\sqrt{1 + \dfrac{a^2}{b^2}}$, whereas for TE_{10}, $\lambda_c = 2a$. Substituting $a = 2.3$ cm and $b = 1.0$ cm, we find $f_c = 6.5$ GHz for the TE_{10} wave and $f_c = 16$ GHz for the TM_{11}.

17-9 $Z_0 = \dfrac{120}{\sqrt{\epsilon_r}}\ln\dfrac{s}{a} = 317\ \Omega;\quad v_s = 3 \times 10^8$ m/s.

17-11 $Z_0 = \dfrac{60}{\sqrt{\epsilon_r}}\ln\dfrac{r_2}{r_1}$; diameter $= 2r_1 = 110\ \mu$m, the size of a human hair, barely practical. $v_s = v_p = 1.7 \times 10^8$ m/s.

17-13 $L' = \dfrac{\mu\ln(b/a)}{2\pi}$; $\quad Z_0 = L'v = \dfrac{60}{\sqrt{\epsilon_r}}\ln(b/a)$.

17-15 $\mathbf{E} = \dfrac{V_0 e^{i(kz-\omega t)}}{\rho\ln(r_2/r_1)}\hat{\boldsymbol{\rho}}$ and $\mathbf{B} = \dfrac{\mu I_0 e^{i(kz-\omega t)}}{2\pi\rho}\hat{\boldsymbol{\phi}}$

17-17 $\displaystyle\int EB\,da/\mu = \int_{r_1}^{r_2}\dfrac{\lambda}{2\pi\epsilon r}\dfrac{\mu I}{2\pi r\mu}\,2\pi r\,dr = I\dfrac{\lambda}{2\pi\epsilon}\ln\dfrac{r_2}{r_1} = VI.$

17-19 (a) $R' = \dfrac{1}{2\pi\sigma\delta}\left(\dfrac{1}{r_1} + \dfrac{1}{r_2}\right)$;

(b) $\delta = \sqrt{\dfrac{2}{\sigma\omega\mu}} = 6.6 \times 10^{-5}$ m, so $R' = 0.104\ \Omega/$m;

(c) $Z_0 = \dfrac{60}{\sqrt{\epsilon_r}}\ln\dfrac{r_2}{r_1} = 56\ \Omega.$

For usual lengths (less than 100 meters), the resistance is unimportant.

Chapter 18

18-1 $\gamma = 1/\sqrt{1 - v^2/c^2} = 1.67.$ $f' = f/\gamma = 36$ Hz

18-3 It appears to be a cube of the same size, rotated by arc cos $(1/\gamma) = 60°$:

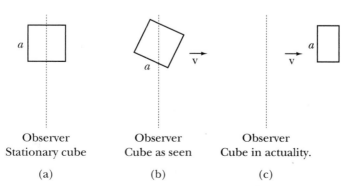

Observer   Observer   Observer
Stationary cube   Cube as seen   Cube in actuality.
(a)   (b)   (c)

18-5 $\Delta x' = \gamma \Delta x - \beta \gamma \Delta x = \gamma$ meters;

and $\Delta ct' = \gamma \Delta ct - \beta \gamma \Delta x = -\beta \gamma$ meters.

18-7 $\tan \theta' = \gamma \tan \theta$.

18-9 Time for round trip $\perp$ ether: $\dfrac{2l}{\sqrt{c^2 - v^2}}$.

Time for round trip $\parallel$ ether: $\dfrac{l}{c - v} + \dfrac{l}{c + v}$;

Difference, in terms of distance traveled by light: $2l\gamma(\gamma - 1) \approx l\beta^2 \Rightarrow 0.20$ fringes.

When the device is rotated through $90°$ the sign changes, so the total shift would be 0.40 fringes.

18-11 (a) $\Delta ct' = \gamma \Delta ct - \beta \gamma \Delta x$, and $\Delta t = 0$, so $\Delta t' = 0.26$ ps;

(b) $\dfrac{\Delta f}{f} \approx \dfrac{\Delta t}{t} = \dfrac{ah}{c^2}$ and $v = at$, so $\Delta t = \dfrac{v \Delta x}{c^2} = 0.26$ ps.

The clock at B leads.

18-13 $\dfrac{P}{M} = \beta \gamma = 1.45$; $v = 2.32 \times 10^8$ m/s;

$\gamma = 1.57$; $\tau = \gamma \times 1.24 \times 10^{-8}$ s $= 1.95 \times 10^{-8}$ s.

18-15 $\gamma = \dfrac{40 \times 10^3 \text{ MeV}}{0.51 \text{ MeV}} = 7.8 \times 10^4$ and $v \approx c$.

(a) $t = \dfrac{x}{v} = 10^{-5}$ s;

(b) $t' = t/\gamma = 1.3 \times 10^{-10}$ s;

(c) $l = 3000/\gamma = 3.8$ cm.

18-17 $P = 970$ MeV/c as before;

$E_1 + E_2 = 1056$ now;

so $M_c = 417$ MeV, too small to be K° at 498 MeV.

18-19 $\lambda' = \lambda \sqrt{\dfrac{1 - \beta}{1 + \beta}} = 1118$ nm.

18-21 $-\dfrac{dW}{dt} = \frac{2}{3} \gamma^6 \dfrac{e^2 a^2}{4\pi\epsilon_0 c^3}$.

Integrating, $\dfrac{m_e c^2}{5} \approx \frac{2}{3} \dfrac{e^2 a^2 t}{4\pi\epsilon_0 c^3}$.

Now $a^2 t = \dfrac{c^3}{2x}$ (same as classical). Consequently,

$$x = \frac{5}{3} \dfrac{e^2}{4\pi\epsilon_0 m_e c^2} = \frac{5}{3} r_e = 4.7 \times 10^{-15} \text{ m}$$

The assumption of uniform acceleration is artificial, but the order-of-magnitude distance is reasonable.

18-23 $E' = \gamma v B = \gamma v \mu_0 N' I$ and $B' = \gamma B = \gamma \mu_0 N' I.$
18-25 $E' = 0.38$ V/m out of page; $B' = 0.025$ T up.

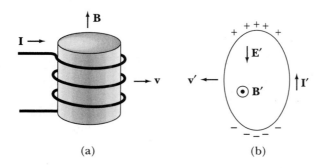

(a)   (b)

18-27 For $B' = 0$, v $= c^2 B/E = 9 \times 10^7$ m/s, and $E' = E/\gamma = 0.954 \times 10^6$ V/m. $E' = 0$ would be impossible here because B is too weak; in other words, $E/B > c.$

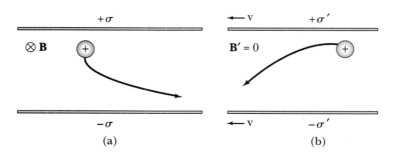

(a)   (b)

18-29 $\mathbf{E' \cdot B'} = \mathbf{E'}_{\parallel} \cdot \mathbf{B'}_{\parallel} + \mathbf{E'}_{\perp} \cdot \mathbf{B'}_{\perp}$
$\qquad = \mathbf{E}_{\parallel} \cdot \mathbf{B}_{\parallel} + \gamma^2 (\mathbf{E}_{\perp} + \mathbf{v} \times \mathbf{B}) \cdot (\mathbf{B}_{\perp} - \mathbf{v} \times \mathbf{E}/c^2)$
$\qquad = \mathbf{E}_{\parallel} \cdot \mathbf{B}_{\parallel} + \gamma^2 (\mathbf{E}_{\perp} \cdot \mathbf{B}_{\perp} + 0 + 0 - \beta^2 \mathbf{E}_{\perp} \cdot \mathbf{B}_{\perp})$
$\qquad = \mathbf{E} \cdot \mathbf{B}.$

18-31 $S^2 - c^2 (dW/dv)^2 = (\mathbf{E} \times \mathbf{B}/\mu)^2 - c^2 (\epsilon E^2/2 + B^2/2\mu)^2$
$\qquad = E^2 B^2/\mu^2 - (\mathbf{E} \cdot \mathbf{B}/\mu)^2 - c^2 (\epsilon E^2/2)^2 - c^2 (B^2/2\mu)^2$
$\qquad\quad - 2c^2 \epsilon E^2 B^2/4\mu$
$\qquad = (\mathbf{E} \cdot \mathbf{B}/\mu)^2 - c^2 (\epsilon E^2/2 - B^2/2\mu)^2$
$\qquad = \dfrac{1}{\mu^2} (\mathbf{E} \cdot \mathbf{B})^2 - \dfrac{c^2 \epsilon^2}{4} (E^2 - c^2 B^2)^2.$

18-33 From the Lorentz transformation,

$$V'/c = \gamma V/c - \beta \gamma A_x$$

For $\gamma \approx 1$,

$$V' = \mathbf{v} \cdot \frac{\mu_0 \mathbf{m} \times \hat{\mathbf{r}}}{4\pi r^2} = \frac{\mathbf{v} \times \mathbf{m}}{4\pi \epsilon_0 c^2 r^2} \cdot \hat{\mathbf{r}}$$

which is the electric dipole potential corresponding to a dipole moment of $\mathbf{p}' = \mathbf{v} \times \mathbf{m}/c^2$.

Again from the Lorentz transformation,

$$c\rho' = \gamma c\rho - \beta\gamma J_x = -\beta\gamma J_x \text{ here.}$$

Integrating over the cross section of a current-carrying wire,

$$c\lambda' = -\beta\gamma I \approx -vI/c$$

The magnitude of the corresponding dipole moment p' is

$$p' = \lambda' a^2 = \frac{v}{c^2} Ia^2 = \frac{v}{c^2} m$$

Examination of the directions involved reveals the vector relationship: $\mathbf{p}' = \mathbf{v} \times \mathbf{m}/c^2$.

18-35 The transverse force between corresponding sections of the lines is reduced by a factor of γ, as all such forces are; but the charge density λ increases due to the Lorentz contraction, and so the force per unit length is independent of the velocity:

$$F/l = F'/l' = \frac{\lambda^2}{2\pi\epsilon_0 r} = \frac{1}{\gamma^2} \frac{\lambda'^2}{2\pi\epsilon_0 r'}$$

(with $r' = r$, of course). The electric force is

$$F'_e/l' = \frac{\lambda'^2}{2\pi\epsilon_0 r'}$$

repulsive, and the magnetic force is attractive,

$$F'_m/l' = -\frac{\mu_0 I'^2}{2\pi r'} = -\frac{\mu_0 \lambda'^2 v'^2}{2\pi r'} = -\beta^2 \frac{\lambda'^2}{2\pi\epsilon_0 r'}$$

so we do find that $F'_e + F'_m = F'$, as we must.

Appendix B:
Bibliography

Books, Electromagnetism

Griffiths, David J. *Introduction to Electrodynamics,* 2nd ed. Prentice-Hall, 1989.

Heald, Mark A., and Terry B. Marion. *Classical Electromagnetic Radiation,* 3rd ed. Saunders, 1995. Gaussian units.

Lorrain, Paul, Dale P. Corson, and François Lorrain. *Electromagnetic Fields and Waves,* 3rd ed. Freeman, 1988.

Jackson, J. D. *Classical Electrodynamics,* 2nd ed. Wiley, 1975. Gaussian units.

Maxwell, James Clerk. *A Treatise on Electricity and Magnetism,* 3rd ed. Clarendon, 1891; Dover, 1954.

Ohanian, Hans C. *Classical Electrodynamics.* Allyn and Bacon, 1988. Gaussian units.

Page, Leigh, and N. Adams. *Electrodynamics.* Van Nostrand, 1940, Chapter 3.

Panofsky, Wolfgang K. H., and Melba Phillips. *Classical Electricity and Magnetism,* 2nd ed. Addison-Wesley, 1962.

Portis, Alan M. *Electromagnetic Fields.* Wiley, 1978.

Purcell, Edward M. *Electricity and Magnetism.* McGraw-Hill, 1965; 2nd ed., 1985. Gaussian units.

Reitz, John R., Frederick J. Milford, and Robert W. Christy. *Foundations of Electromagnetic Theory,* 4th ed. Addison-Wesley, 1992.

Sommerfeld, A. *Electrodynamics.* Academic, 1952, p. 45.

Wangsness, Roald K. *Electromagnetic Fields,* 2nd ed. Wiley, 1986.

Books, Other

Albert Einstein, Philosopher-Scientist, edited by P. A. Schilpp. Harper, 1951, with Einstein's autobiographical notes.

Arfken, G. *Mathematical Methods for Physicists,* 3rd ed. Academic, 1985.

Eddington, A. S. *The Mathematical Theory of Relativity,* Cambridge, 1924, p. 16.

Feynman, Richard P., Robert B. Leighton, and Matthew Sands. *The Feynman Lectures on Physics.* Addison-Wesley, 1964.

Lorentz, H. A., A. Einstein, H. Minkowski, and H. Weyl. *The Principle of Relativity,* 1952. Dover, a collection of the original papers.

Sciama, D. W. *The Unity of the Universe.* Doubleday, 1961.

Synge, J. L. *Relativity: The Special Theory.* North-Holland, 1966, pp. 34–35.

Taylor, E. F., and J. A. Wheeler. *Spacetime Physics,* 2nd ed. Freeman, 1992.

Tipler, P. A. *Physics,* 2nd ed. Worth, 1982, p. 930.

Whittaker, E. T. *A History of the Theories of Aether and Electricity.* Tomash/American Institute of Physics, 1987.

Articles

Alvager, T., et al. *Phys. Letters* 12 (1964):260. Speed of photons from π^0 decay.

Alvarez, Luis, et al. *Phys. Rev. D* 4 (1971):3260. Search for monopoles in moon rocks.

Adawi, I. *Am. J. Phys.* 44 (1976):762. Thomson's monopoles.

Alstrøm, P., et al. *Am. J. Phys.* 50 (1982):697. Stern-Gerlach.

Bartlett, D. F., and Y. Su. *Am. J. Phys.* 62 (1994):683. Uniqueness.

Boebinger, G., et al. *Sci. Am.* 272 (June 1995):58. World-record magnets.

Cabrera, B. *Phys. Rev. Lett.* 48 (1982):1378. Magnetic monopole.

Cohen, E. Richard, and B. N. Taylor. *Phys. Today* August 1997: BG7. Physical constants; see also R. A. Nelson. *Phys. Today* August 1997: BG13. Metric practice.

Desloge, E. *Am. J. Phys.* 57 (1989):1121. Principle of equivalence.

Dick, B. G. *Am. J. Phys.* 41 (1973):1289. Point charge between conducting plates.

Dingle, Herbert. *Nature* 217 (1968):19. The case against the special theory of relativity.

Fisher, G. P. *Am. J. Phys.* 40 (1972):1772. On deriving Maxwell from relativity.

Furry, W. H. *Am. J. Phys.* 37 (1969):621. Momentum in electromagnetic field.

Good, R. H. *Am. J. Phys.* 50 (1982):232. Uniformly accelerated reference frame.

Good, R. H. *Am. J. Phys.* 65 (1997):155. Conducting disk and conducting needle.

Griffiths, D. J. *Am. J. Phys.* 60 (1992):187. Uniformly polarized objects.

Griffiths, D. J. *Am. J. Phys.* 60 (1992):979. Dipoles at rest.

Gutmann, F. *Rev. Mod. Phys.* 20 (1948):457. Electrets.

Hafele, J. C., and R. E. Keating. *Science* 177 (1972):166. Flying-clock experiment.

Heering, P. *Am. J. Phys.* 60 (1992):988. Coulomb's experiments.

Herrmann, F. *Am. J. Phys.* 61 (1993):119. Energy path in capacitor.

Janah, A., et al. *Am. J. Phys.* 56 (1988):1036. Feynman's formula.

Marshall, E. *Science* 212 (1981):644. Elf.

Michelson, A. A., and E. W. Morley. *Am. J. Sci.* 34 (1887):333.

Nahin, P. J. *Sci. Am.* 262 June 1990:122. Heaviside.

Neuenschwander, D., et al. *Am. J. Phys.* 60 (1992):35. Biot-Savart to Maxwell using relativity.

Price, P. B., et al. *Phys. Rev. Lett.* 35 (1975):487. Magnetic monopole.

Robinson, A. *Science* 235 (1987):633. Fundamental constants.

Rubin, L. G., and P. A. Wolf. *Phys. Today* 37 (August 1984): 24. High magnetic field

Slepian, J. *Am. J. Phys.* 19 (1951):87. Non-closure of B field lines.

Wise, M. N. *Science* 203 (1979):1310. Maxwell's model.

Appendix C:
Computations and Plots

This appendix will introduce computational procedures that are useful in various kinds of electromagnetic problems. Sample programs are provided. They can be run from DOS; for example, simply type `lines` to see the result in Figure C2b, below.

Equipotentials and Field Lines

We shall begin by finding differential equations, or actually, finite-difference equations, which we can use to plot equipotentials and electric field lines. We limit ourselves here to the two-dimensional representations that we can print on paper; so there is no z dependence.

Electric field lines (Section 2.3) point in the direction of the **E** field, Figure C1. This means that for a position increment $(\Delta x, \Delta y)$ along the **E** field line,

$$\frac{\Delta y}{\Delta x} = \frac{E_y}{E_x} \tag{C1}$$

The procedure is as follows: You plot a point (x, y) somewhere in space where you want a field line to be; this choice is somewhat arbitrary. You calculate E_x and E_y at that point. You choose an appropriate Δx and use Equation C1 to find the corresponding Δy. You plot the new point $(x + \Delta x, y + \Delta y)$. You calculate E_x and E_y at that point, and determine the next increments. And so on. Experience will inform you as to whether you want to choose Δx and calculate Δy or vice versa, as well as how big to make the increment.

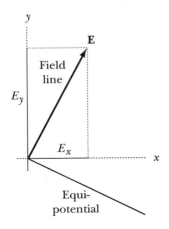

Figure C1.

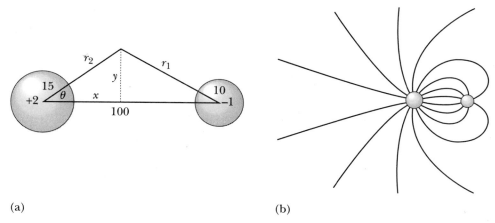

(a)   (b)

Figure C2.

Similarly, we can follow an equipotential (Section 6.3 and Fig. C1) across the paper. For an equipotential, the change in $V(x, y)$ is zero:

$$\Delta V = 0 = \frac{\partial V}{\partial x} \Delta x + \frac{\partial V}{\partial y} \Delta y$$

so

$$\frac{\Delta y}{\Delta x} = -\frac{\dfrac{\partial V}{\partial x}}{\dfrac{\partial V}{\partial y}} = -\frac{E_x}{E_y} \tag{C2}$$

for an equipotential. The result is the negative reciprocal of that for the field line, Equation C1, implying that the two are mutually perpendicular, as is reasonable intuitively.

Now we apply Equation C1 to find some electric field lines in the example illustrated in Figure C2: a positive line charge of two arbitrary units located at distance 100 units from a negative line charge of one unit. The principles and techniques illustrated can be carried over to other kinds of charge distributions. Note that electric field lines typically begin on charges, whereas equipotentials typically occur in free space.

In this C++ listing, the numbers on the left refer to explanatory notes that follow (they are *not* line numbers).

```
(1)     //    Lines                        // title
        #include <graphics.h>              // header files
(2)     #include <conio.h>
        #include <math.h>

        main()
        { int   gdriver=DETECT,gmode;
```

```
(3)     initgraph(&gdriver,&gmode, "");    // graphics mode

        float xc=getmaxx()/3, yc=getmaxy()/2,
(4)           th, r2, r22, r12, x, y, dx, dy, Ex, Ey;

        circle(xc,yc,15); circle(xc+100,yc,10);

(5)     for (th=M_PI/16; th<2*M_PI; th+=M_PI/8)
        { r2 = 15;                          // begin on +2
          x = r2*cos(th); y = r2*sin(th);

        do
(6)       { r22 = r2*r2;
(7)         r12 = (x-100)*(x-100) + y*y;
            Ex = 2*x/r22 - (x-100)/r12;     // calculate E
(8)         Ey = 2*y/r22 - y/r12;
(9)         if (fabs(Ex)>fabs(Ey))
            { dx = Ex/fabs(Ex);
                dy = dx*Ey/Ex;   }          // diff. eq.
            else
            { dy = Ey/fabs(Ey);
                dx = dy*Ex/Ey; }
            x+=dx; y+=dy;                    // increment
            putpixel(xc+x,yc+y,15);         // plot
(10)        r2 = hypot(x,y); }
          while (r12>144 && r2<xc); }

        getch();                            // wait for keystroke
        closegraph(); }
```

Notes

These sample programs are written in Borland C++ on an IBM-type personal computer with vga graphics, but they should be readily adapted to other C++ environments. They have been somewhat compacted in order to facilitate their presentation here; as you enter them into your computer, you can adjust them to suit your own style. (If the compiler warns you about returning a value, you can ignore that.) After a program has successfully run, the machine automatically stores an executable (.exe) file that can be run directly from DOS simply by typing in the name of the program.

If you are unsure of the significance of a particular step, modify it and see what happens.

(1) The "//" double-slash precedes titles or comments having no effect on operation.
(2) Three header files are required for the various operations involved in this program.

(3) For the Borland C++ system used here, graphics mode is usually invoked with the two lines involving "gmode." The file EGAVGA.BGI must be on the disk so that the system can find it; and every time you invoke C++, you must go to Options, Linker, Libraries, and turn on the Graphics library (Standard Run Time should also be on).

(4) getmaxx( ) and getmaxy( ) find out how large your screen is.

(5) M_PI is a built-in value of π. "th" means θ. So this "for" loop begins at $\theta = \pi/16$; it continues while $\theta < 2\pi$; and θ is incremented by $\pi/8$ with each iteration. ("+=" means "increment by.") The result will be 16 evenly spaced field lines starting out at the surface of the +2 line charge.

(6) The "inside" loop is a "do . . . while" loop that continues while the line is not too close to the −1 charge and not too far from +2. ("&&" means "and.")

(7) We use r22 for r_2^2, and r12 for r_1^2.

(8) The fields are $1/r$-type for line charges. So for example, just for the +2, omitting $2\pi\epsilon_0$,

$$E_x = E\cos\theta = \frac{-2}{r_2}\ \frac{x}{r_2} = \frac{2*x}{r22}$$

(9) if . . . else: one of the increments dx (actually means Δx) or dy will be ± 1 unit, and the other will be smaller. "fabs" is "floating-point absolute-value." Equation C1 is seen here in two alternative forms:

$$dy = dx\ \frac{E_y}{E_x} \quad \text{or} \quad dx = dy\ \frac{E_x}{E_y}$$

depending on whether it is dx or dy that we have set equal to ± 1 unit.

(10) hypot(x,y) means $\sqrt{x^2 + y^2}$.

Radiating Dipole

This plot appears in Section 15.1 and is reproduced here, Figure C3. The differential equation for electric field lines in spherical coordinates is

$$\frac{r\ d\theta}{dr} = \frac{E_\theta}{E_r}$$

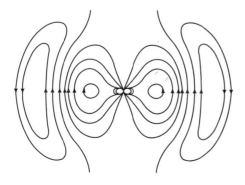

Figure C3.

The real parts of the E field are found in Section 15.1 (Eq. 15.12):

$$\text{Re } E_r = \frac{I_0 ks \cos \theta}{4\pi\epsilon_0 cr} \left(\frac{2}{kr} \cos(kr - \omega t) - \frac{2}{(kr)^2} \sin(kr - \omega t) \right)$$

$$\text{Re } E_\theta = \frac{I_0 ks \sin \theta}{4\pi\epsilon_0 cr} \left(\frac{1}{kr} \cos(kr - \omega t) + \left(1 - \frac{1}{(kr)^2}\right) \sin(kr - \omega t) \right)$$

(15.12)

So,

$$\frac{r \, d\theta}{dr} = \frac{\sin \theta \left(\dfrac{1}{kr} \cos(kr - \omega t) + \left(1 - \dfrac{1}{(kr)^2}\right) \sin(kr - \omega t) \right)}{\cos \theta \left(\dfrac{2}{kr} \cos(kr - \omega t) - \dfrac{2}{(kr)^2} \sin(kr - \omega t) \right)}$$

We could plot this as a differential equation. However, it turns out that we can do better: We can find a solution in closed form. Separating variables,

$$\frac{2 \cos \theta}{\sin \theta} d\theta = \frac{\dfrac{1}{kr} \cos(kr - \omega t) + \left(1 - \dfrac{1}{(kr)^2}\right) \sin(kr - \omega t)}{\cos(kr - \omega t) - \dfrac{1}{kr} \sin(kr - \omega t)} d(kr)$$

There are logarithmic forms on both sides; thus,

$$\sin^2\theta = \frac{C}{\dfrac{1}{kr} \sin(kr - \omega t) - \cos(kr - \omega t)}$$

(C3)

where C is a constant of integration. Different values of C correspond to different field lines, and different times t give different pictures. We can now easily introduce values of r and calculate corresponding values of θ (but for some values of r there is no line, so no θ). In Figure 15.4, C $= \pm 0.15$ for the largest field line, and C $= \pm 1.35$ for the smallest. It happens that for $1 < C < 1.15$, the loops break off but later disappear, and for C > 1.15 loops don't break off at all.

The following program plots this equation point by point, providing an animated view of the development of electric field lines from a radiating dipole. The inside loop increments kr from 0.05 to 10 along one field line. After that, C is incremented and another line is drawn. When C has gone through its five values, one frame is complete. A delay of 20 milliseconds permits the user to see the result, after which the frame is erased, the time ωt is incremented, and the process begins anew. For single-step, press s; to escape, press the escape key.

The equation C3 is visible in the middle of this listing; we employ fabs (floating-point absolute-value) so as to include both positive and negative values of C. We pick up keystrokes on the fly using kbhit( ), which is true if a key has been pressed. "!" means "not." "fr" is a scale factor.

```
// Dipole

#include <graphics.h>
#include <conio.h>
#include <math.h>
#include <dos.h>

main()
{ int gdriver=DETECT,gmode;
  initgraph(&gdriver,&gmode,"");

  char ch='d';
  float xc=getmaxx()/2,  yc=getmaxy()/2,  fr=yc/11,
      C, s2, kr, wt, x, y;

  while  (ch!=(char)27)                       // Escape key
    for (wt=0;  wt<2*M_PI;  wt+=M_PI/64)
    { for (C=.15; C<=1.35; C+=.3)
        for (kr=.05; kr<=10; kr+=.05)

        { s2 = fabs( C / (sin(kr-wt)/kr - cos(kr-wt)) );
          if (s2<1)
          { x = fr*kr*sqrt(s2);
            y = fr*kr*sqrt(1-s2);
            putpixel(xc+x,yc+y,15);           // plot
            putpixel(xc-x,yc-y,15);
            putpixel(xc+x,yc-y,15);
            putpixel(xc-x,yc+y,15); } }

      if (kbhit()) ch=getch();
      if (ch=='s') ch=getch();                // single step
      delay(20);
      cleardevice();                          // erase
      if (ch==(char)27) break; }              // Escape key

  closegraph(); }
```

Moving Charge

In Section 14.6 we drew the electric field lines produced by an accelerated charge, Figure C4. Such pictures are easily produced in real-time animation using the "emission theory" algorithm (Page, bibliography; this however has nothing to do with the emission theories of Section 18.3).

We can explain this using a gedanken experiment. Begin with a stationary charge having eight straight electric field lines, spaced at 45°, in the plane of the pa-

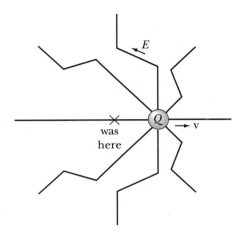

Figure C4.

per. Replace the field lines with tubes carrying photons; the photons do not touch the sides of the tubes, and the photons could be used, without the tubes, to trace out the field lines. The same is true in an inertial reference frame moving relative to the first frame: Because of the Lorentz contraction, the tubes are no longer spaced at 45°, and the same must be true of the photon trajectories, and the same is also true of the field lines, as we have seen in Section 18.9. During acceleration of the charge, the lines of photons must stay continuous, and by Gauss's law so must the field lines. So we can trace out the electric field lines of an arbitrarily moving point charge using these "quasi-photons," which start at the charge and move in straight lines at the speed of light. When the charge moves, velocity components of the quasi-photons are found using the relativistic velocity addition formulas of Section 18.3:

$$\beta'_x = \frac{\beta_x - \beta}{1 - \beta\beta_x} \quad \text{and} \quad \beta'_y = \frac{\beta_y \sqrt{1 - \beta^2}}{1 - \beta\beta_x}$$

The accompanying program provides for 200 quasi-photons along each of three field lines; the other five lines are obtained using symmetry. The x position array is declared as

$$\text{float x}[3][200] = \{0\}$$

which also zeros all array elements. Each quasi-photon begins life at the charge (xc, yc) at j = 0. With each iteration it is erased, its position is incremented, its position and speed are moved from one array element to the next, and it is replotted, until finally it falls off the end of the array at j = 200.

Charge speed vc is controlled with the left and right arrows. i++ increments i by 1. For single-step, press s; to escape, press the escape key.

```
// Charge

#include <iostream.h>
#include <graphics.h>
#include <conio.h>
```

```
#include <math.h>
#include <dos.h>

main()
{ int gdriver=DETECT,gmode;
  initgraph(&gdriver,&gmode,"");

  int  i, j;
  char ch='d';
  float maxx=getmaxx(), xc=maxx/3, yc=getmaxy()/2,
        vc=0, vxo, vyo,
        x[3][200]={0}, y[3][200]={0}, vx[3][200]={0},
        vy[3][200]={0};

  cout.precision(2); cout.setf(ios::showpoint);
                                              // output format
  cout.setf(ios::fixed);  cout.setf(ios::showpos);

  while (ch!=(char)27)                        // escape key
  { for (i=0; i<=2;  i++)
    { for  (j=199;  j>=0;  j--)

        { putpixel(x[i][j],  yc+y[i][j], 0);       // erase
          putpixel(x[i][j],  yc-y[i][j], 0);
          if (j<199)
          { x[i][j+1] = x[i][j] + vx[i][j];        //increment
            y[i][j+1] = y[i][j] + vy[i][j];
            vx[i][j+1] = vx[i][j];
            vy[i][j+1] = vy[i][j];
            putpixel(x[i][j+1], yc+y[i][j+1], 15); // plot
            putpixel(x[i][j+1], yc-y[i][j+1], 15); } }

      x[i][0] = xc;
      vxo = (i-1)*.707; vyo = 1-abs(i-1)*.293;
      vx[i][0] = (vxo+vc) / (1+vc*vxo);          // vel.add.
      vy[i][0] = vyo*sqrt(1-vc*vc) / (1+vc*vxo); }
    line(0,yc,maxx,yc);
    xc = xc + vc;                              // move charge
    if  (xc>=maxx) xc = 1;                      // wrap-around
    if  (xc<=0) xc = maxx-1;
    if  (kbhit())
    { ch=getch();
        if (ch=='K' &&  vc>=-.95) vc-=.05;      // arrows
        if (ch=='M' && vc<=+.95) vc+=.05;  }
    gotoxy(5,24); cout<< "v/c = " << vc          // output
          << "  Use arrows, s, esc.";
```

```
   if (ch=='s') ch=getch();                              // single step
   delay(10); }

closegraph(); }
```

Relaxation

Boundary-value problems can often be solved numerically using the method of **relaxation.** It works as follows. The solutions of Laplace's equation $\nabla^2 V = 0$ tend to be smooth in the sense that the potential V at every point is the average of that at neighboring points. So if we know the potential on the boundary of a region, we can fill in the interior values just by repeatedly smoothing out the potential.

We will demonstrate the principle mathematically in two dimensions; extension to three dimensions is easy. We need to show that the potential $V(x, y)$ at the center of the group of square cells shown in Figure C5 is the average of the four neighboring potentials:

$$V(x, y) = \frac{V(x + h, y) + V(x - h, y) + V(x, y + h) + V(x, y - h)}{4}$$

where h is the distance from one cell to another. To do this, we expand V in a Taylor series; for the cell on the right:

$$V(x + h, y) \approx V(x, y) + \frac{\partial V}{\partial x} h + \frac{1}{2} \frac{\partial^2 V}{\partial x^2} h^2 + \frac{1}{6} \frac{\partial^3 V}{\partial x^3} h^3 + \dots$$

On the left,

$$V(x - h, y) \approx V(x, y) - \frac{\partial V}{\partial x} h + \frac{1}{2} \frac{\partial^2 V}{\partial x^2} h^2 - \frac{1}{6} \frac{\partial^3 V}{\partial x^3} h^3 + \dots$$

The terms in h already cancel, as do those for h^3. As for the other two cells:

$$V(x, y + h) \approx V(x, y) + \frac{\partial V}{\partial y} h + \frac{1}{2} \frac{\partial^2 V}{\partial y^2} h^2 + \frac{1}{6} \frac{\partial^3 V}{\partial y^3} h^3 + \dots$$

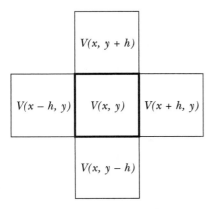

Figure C5.

and

$$V(x, y - h) \approx V(x, y) - \frac{\partial V}{\partial y} h + \frac{1}{2} \frac{\partial^2 V}{\partial y^2} h^2 - \frac{1}{6} \frac{\partial^3 V}{\partial y^3} h^3 + \ldots$$

Adding up the four equations, we get

$$V(x, y) \approx \frac{4V(x, y) + \frac{\partial^2 V}{\partial x^2} h^2 + \frac{\partial^2 V}{\partial y^2} h^2 + \ldots}{4}$$

The two terms in h^2 add up to zero, because in two dimensions Laplace's equation is

$$\nabla^2 V = \frac{\partial^2 V}{\partial x^2} + \frac{\partial^2 V}{\partial y^2} = 0$$

So we do indeed get a total of $V(x, y)$ for the average of the four neighbors, through terms in h^3.

(This reminds us of the fact that the voltage at the center of a sphere is the average voltage on its surface, Section 12.1; but the result here is only an approximation, growing more accurate as h is reduced.)

We can now try relaxation on an example from Section 12.3:

▼ EXAMPLE

There are two semi-infinite planes of $V = 0$ located at $y = 0$ and $y = b$, parallel to the xz plane, extending in the $+x$ direction, Figure C6a. On the yz plane the potential is V_0, a constant. Find the potential V in the region between the planes, for $x > 0$.

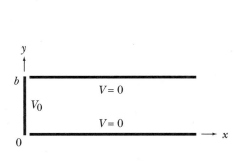

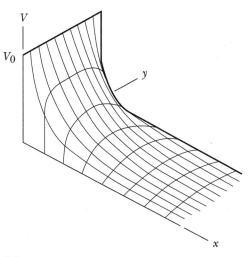

(a)   (b)

Figure C6.

We fill the space between the plates with a grid of square cells all originally at potential $V = 0$. We go to each cell in turn and set its potential equal to the average of its neighbors' V's. The result of the first pass isn't correct yet, and isn't even symmetrical, because the cells changed value as we passed through. So we do it again, hoping the values will converge to a stable and correct state, as indeed they do after around $n = 100$ iterations.

In the following program, $V_0 = 100$ volts. We have arbitrarily set the horizontal size of the grid to 20 cells, and the vertical size Y is selected by the user. For $Y = 10$ the result is shown here. Does it seem to be consistent with Figure C6b?

	0	0	0	0	0	0	0	0	0	0	0	0	0	0	0	0	0	0	0	0
100	49	28	18	12	9	6	4	3	2	2	1	1	0	0	0	0	0	0	0	0
100	67	46	32	23	17	12	9	7	5	4	3	2	1	1	0	0	0	0	0	0
100	76	56	42	31	23	17	13 9	7	5	4	3	2	1	1	0	0	0	0	0
100	80	62	47	36	27	20	15 11 8	6	5	3	2	2	1	1	0	0	0	0
100	81	64	50	38	29	22	17 12 9	7	5	4	3	2	1	1	0	0	0	0
100	81	64	50	38	29	22	17 12 9	7	5	4	3	2	1	1	0	0	0	0
100	80	62	47	36	27	20	15 11 8	6	5	3	2	2	1	1	0	0	0	0
100	76	56	42	31	23	17	13 9	7	5	4	3	2	1	1	0	0	0	0	0
100	67	46	32	23	17	12	9	7	5	4	3	2	1	1	0	0	0	0	0	0
100	49	28	18	12	9	6	4	3	2	2	1	1	0	0	0	0	0	0	0	0
	0	0	0	0	0	0	0	0	0	0	0	0	0	0	0	0	0	0	0	0

Y = 10
N = 126

The Fourier-series solution, from Section 12.3, is

$$V(x, y) = \frac{4}{\pi} V_0 \sum_{\text{odd } n}^{\infty} \frac{1}{n} e^{-n\pi x/b} \sin \frac{n\pi y}{b} \tag{12.35}$$

The Relax program is easily modified to employ this equation. The result is almost the same. Remember to include Math.h, set $b = Y + 1$, and use only n odd.

For next iteration, press Enter; to escape, press the escape key.

```
//  Relax

#include <iostream.h>
#include <graphics.h>
#include <conio.h>
#include <stdlib.h>

main()
{ int   gdriver=DETECT,gmode;
  initgraph(&gdriver,&gmode,"");

  int x, y, n=0, vxy, Y;
  char ch='d', Ych[21];
```

```
float V[22][22]={0};

gotoxy(10,10); cout<< "Enter a number <21 : ";
cin.get(Ych,21);                        // input
Y = atoi(Ych);
cleardevice();
if (Y<=0 || Y>20 Y = 10;
gotoxy(10,Y+4); cout<< "Y = " << Y
      << ". Press Enter or Esc.";

for  (y=1; y<=Y; y++)
{ V[0][y] = 100;                        // initialize
  gotoxy(4,y+1); cout<< "100";   }
for (x=1; x<=20; x++)
{ gotoxy(3*x+5,1);   cout<< "0";
  gotoxy(3*x+5,Y+2);   cout<< "0"; }

while (ch!=(char)27)                     // Escape key
{ for (y=1; y<=Y; y++)
    for (x=1; x<=20; x++)

    { vxy = V[x][y];                     // relax
      gotoxy(3*x+5,y+1); cout<< vxy;
      V[x][y] = (V[x-1][y]+V[x][y-1]+V[x+1][y]+V[x]
      [y+1])/4; }

  moveto(520,15);
  lineto(51,15); lineto(51,16*Y+15); lineto(520,16*Y+15);
  gotoxy(10,Y+5); cout<< "n = " << n;
  ch=getch();                            // wait for keystroke
  n++; }

closegraph(); }
```

Appendix D:
Dimensions and Units

A **dimension** is the kind of quantity involved in a measurement: John's height is a "length." Dimensions are measured in **units:** John's height is 70 "inches" or 1.78 "meters." This book employs the **SI** system of units (French: Système International), the international system of units established in 1960 (Cohen, 1997). It is also called the rationalized MKSA system (or just MKS) for the initials of Meter-Kilogram-Second-Ampere. The word "rationalized" in this context implies a redistribution of factors of 4π in the electromagnetic equations. The four fundamental units ("SI base units") of interest in electromagnetism are:

Meter: (since 1983) The distance traveled by light in a time of $(299{,}792{,}458)^{-1}$ second.

Kilogram: The mass of the international prototype, made of platinum-iridium, maintained at Sèvres, France.

Second: (since 1967) The time for 9,192,631,770 periods of a cesium-133 atomic-beam clock.

Ampere: The current flowing in each of two infinite parallel wires, of negligible cross section, separated by one meter, if the force between them is 2×10^{-7} N/m; the equation is (Section 3.2)

$$\frac{F}{l} = \frac{\mu_0 II'}{2\pi r}$$

The coulomb (Section 2.1) is defined in terms of the ampere: A coulomb is the quantity of electric charge transported in one second by a current of one ampere.

There are three *defined* physical constants:

The speed of light is exactly $c = 299{,}792{,}458$ m/s.
The permeability of vacuum is $\mu_0 = 4\pi \times 10^{-7}$ N/A^2.
The permittivity of vacuum is $\epsilon_0 = 1/\mu_0 c^2$.

One other system of units is still commonly employed: the **Gaussian system.** It involves centimeter-gram-second units (cgs) along with an unrationalized mixture of two old systems: electrostatic units (esu) and electromagnetic units (emu). Usually, esu is used for charge, current, and electric field, and emu is used for magnetic field. In esu, the unit of charge Q is the "statcoulomb" ("stat" from "electrostatic"), which is so defined that Coulomb's law is

$$F = \frac{QQ'}{r^2}$$

with force in dynes and separation r in centimeters. In emu, the unit of current I is the "abampere" ("ab" from "absolute"); and the force per centimeter between two parallel wires, of current I and I' and separation r, is

$$\frac{F}{l} = \frac{2II'}{r}$$

The abcoulomb is an abampere-second, and it equals 3×10^{10} statcoulombs, where 3×10^{10} cm/s is the speed of light in cgs units. In esu, the formula becomes

$$\frac{F}{l} = \frac{2II'}{rc^2}$$

There is not complete agreement on the Gaussian units: some authors (Panofsky & Phillips, p. 467) use emu in the current density J, while others use esu. The esu usage predominates in recent texts, and that is what we shall present here. You can tell which one is being used by checking dimensions in any equation involving J; to return to esu, substitute J/c.

The following table displays some of the more important equations of electromagnetism in SI and Gaussian units.

	SI	**Gaussian**
Coulomb's law	$F = QQ'/4\pi\epsilon_0 r^2$	$F = QQ'/r^2$
Biot-Savart law	$\mathbf{B} = (\mu_0/4\pi)\int \dfrac{\mathbf{J} \times \hat{\mathbf{r}}}{r^2}\, dv$	$\mathbf{B} = (1/c)\int \dfrac{\mathbf{J} \times \hat{\mathbf{r}}}{r^2}\, dv$
Parallel currents	$F/l = \mu_0 II'/2\pi r$	$F/l = 2II'/rc^2$
Maxwell's equations	$\nabla \cdot \mathbf{E} = \rho_t/\epsilon_0$	$\nabla \cdot \mathbf{E} = 4\pi\rho_t$
	$\nabla \cdot \mathbf{B} = 0$	$\nabla \cdot \mathbf{B} = 0$
	$\nabla \times \mathbf{E} = -\partial\mathbf{B}/\partial t$	$\nabla \times \mathbf{E} = -(1/c)\,\partial\mathbf{B}/\partial t$
	$\nabla \times \mathbf{B} = \mu_0 \mathbf{J}_t + \mu_0\epsilon_0\partial\mathbf{E}/\partial t$	$\nabla \times \mathbf{B} = (1/c)(4\pi\mathbf{J}_t + \partial\mathbf{E}/\partial t)$
Maxwell's equations in material	$\nabla \cdot \mathbf{D} = \rho_f$	$\nabla \cdot \mathbf{D} = 4\pi\rho_f$
	$\nabla \cdot \mathbf{B} = 0$	$\nabla \cdot \mathbf{B} = 0$
	$\nabla \times \mathbf{E} = -\partial\mathbf{B}/\partial t$	$\nabla \times \mathbf{E} = -(1/c)\,\partial\mathbf{B}/\partial t$
	$\nabla \times \mathbf{H} = \mathbf{J}_f + \partial\mathbf{D}/\partial t$	$\nabla \times \mathbf{H} = (1/c)(4\pi\mathbf{J}_f + \partial\mathbf{D}/\partial t)$
Constitutive relations	$\mathbf{D} = \epsilon_0\mathbf{E} + \mathbf{P}$	$\mathbf{D} = \mathbf{E} + 4\pi\mathbf{P}$
	$\mathbf{B} = \mu_0\mathbf{H} + \mu_0\mathbf{M}$	$\mathbf{B} = \mathbf{H} + 4\pi\mathbf{M}$
	$\mathbf{D} = \epsilon_r\epsilon_0\mathbf{E}$ and $\mathbf{P} = \chi_e\epsilon_0\mathbf{E}$	$\mathbf{D} = \epsilon\mathbf{E}$ and $\mathbf{P} = \chi_e\mathbf{E}$
	$\epsilon_r = 1 + \chi_e$	$\epsilon = 1 + 4\pi\chi_e$
	$\mathbf{B} = \mu_r\mu_0\mathbf{H}$ and $\mathbf{M} = \chi_m\mathbf{H}$	$\mathbf{B} = \mu\mathbf{H}$ and $\mathbf{M} = \chi_m\mathbf{H}$
	$\mu_r = 1 + \chi_m$	$\mu = 1 + 4\pi\chi_m$
Lorentz force	$\mathbf{F} = Q\mathbf{E} + Q\mathbf{v} \times \mathbf{B}$	$\mathbf{F} = Q\mathbf{E} + (1/c)\,Q\mathbf{v} \times \mathbf{B}$
Energy density	$dW/dv = \mathbf{E} \cdot \mathbf{D}/2 + \mathbf{B} \cdot \mathbf{H}/2$	$dW/dv = \mathbf{E} \cdot \mathbf{D}/8\pi + \mathbf{B} \cdot \mathbf{H}/8\pi$
Poynting's vector	$\mathbf{S} = \mathbf{E} \times \mathbf{H}$	$\mathbf{S} = (c/4\pi)\,\mathbf{E} \times \mathbf{H}$
Larmor formula	$P = (2/3)\,q^2 a^2/4\pi\epsilon_0 c^3$	$P = (2/3)\,q^2 a^2/c^3$
Potentials	$\mathbf{E} = -\nabla V - \partial\mathbf{A}/\partial t$	$\mathbf{E} = -\nabla V - (1/c)\,\partial\mathbf{A}/\partial t$
	$\mathbf{B} = \nabla \times \mathbf{A}$	$\mathbf{B} = \nabla \times \mathbf{A}$
	$V = \displaystyle\int \rho\, dv/4\pi\epsilon_0 r$	$V = \displaystyle\int \rho\, dv/r$
	$\mathbf{A} = \mu_0\displaystyle\int \mathbf{J}\, dv/4\pi r$	$\mathbf{A} = (1/c)\displaystyle\int \mathbf{J}\, dv/r$
Lorentz condition	$\nabla \cdot \mathbf{A} = -(1/c^2)\,\partial V/\partial t$	$\nabla \cdot \mathbf{A} = -(1/c)\,\partial V/\partial t$

Here are some conversions between SI and Gaussian units.

		SI	Gaussian
Length	l	meter (m)	100 centimeters
Mass	m	kilogram (kg)	10^3 grams
Time	s	second (s)	second
Force	F	newton (N)	10^5 dyne
Energy	W	joule (J)	10^7 erg
Power	P	watt (W)	10^7 erg/s
Charge	Q	coulomb (C)	3×10^9 statcoulomb (or esu)
Current	I	ampere (A)	3×10^9 statampere
Resistance	R	ohm (Ω)	$(9 \times 10^{11})^{-1}$ s/cm or statohm
Scalar potential (electrostatic)	V	volt (V)	1/300 statvolt
Vector potential	A	tesla meter	10^6 gauss $\cdot$ cm
Electric field	E	volt per meter	$(3 \times 10^4)^{-1}$ statvolt/cm or dyne/statcoulomb
H field	H	ampere per meter	$4\pi \times 10^{-3}$ oersted (oersted = gauss)
Magnetic field	B	tesla (T)	10^4 gauss
Magnetic flux	Φ	weber (Wb)	10^8 maxwell or gauss $\cdot$ cm^2
Capacitance	C	farad (F)	9×10^{11} centimeter or statfarad
Inductance	L	henry (H)	$(9 \times 10^{11})^{-1}$ s^2/cm or stathenry
Electric dipole moment	p	coulomb meter	3×10^{11} statcoulomb $\cdot$ cm
Magnetic dipole moment	m	ampere meter^2	10^3 gauss $\cdot$ cm^3 or erg/gauss

Index

Permittivity of free space	$\epsilon_0 = 1/\mu_0 c^2 = 8.854 \times 10^{-12}$ F/m or C^2/N $\cdot$ m^2
	$1/4\pi\epsilon_0 = 8.99 \times 10^9$ N $\cdot$ m^2/C^2
Permeability of free space	$\mu_0 = 4\pi \times 10^{-7}$ H/m or N/A^2 or T $\cdot$ m/A
Speed of light	$c = 2.9979 \times 10^8$ m/s
Electronic charge	$e = 1.602 \times 10^{-19}$ C
Mass of electron	$m_e = 9.11 \times 10^{-31}$ kg or 0.511 MeV
Mass of proton	$m_p = 1.673 \times 10^{-27}$ kg or 938 MeV
Classical electron radius	$r_e = e^2/4\pi\,\epsilon_0 mc^2 = 2.82 \times 10^{-15}$ m
Planck's constant	$h = 6.63 \times 10^{-34}$ J $\cdot$ s
	$\hbar = h/2\pi = 1.055 \times 10^{-34}$ J$\cdot$s
Bohr magneton	$\mu_B = \hbar e/2m_e = 9.27 \times 10^{-24}$ A $\cdot$ m^2 or J/T
Nuclear magneton	$\mu_N = \hbar e/2m_p = 5.05 \times 10^{-27}$ A$\cdot$m^2 or J/T
Fine structure constant	$\alpha = e^2/4\pi\,\epsilon_0\hbar c = 1/137$
Boltzmann constant	$k = 1.38 \times 10^{-23}$ J/K
Avogadro's number	$N_A = 6.02 \times 10^{23}$ mol^{-1}
Atmospheric pressure	atm $= 1.01 \times 10^5$ Pa
Gas law constant	$R = 8.31$ J/mol $\cdot$ K
Gravitational constant	$G = 6.67 \times 10^{-11}$ N $\cdot$ m^2/kg^2
Gravitational acceleration	$g = 9.81$ m/s^2
Mass of the earth	5.97×10^{24} kg
Radius of the earth	6.38×10^6 m
Earth–sun distance	1.50×10^{11} m
Mass of the sun	1.99×10^{30} kg
Radius of the sun	6.96×10^8 m
Light-year	9.46×10^{15} m
Parsec	3.26 light-year